1st International Conference on Bioenergy and Environmentally Sustainable Agriculture Technology (ICoN BEAT 2019)

E3S Web of Conferences Volume 226 (2021)

Malang, Indonesia
7 – 8 November 2019

Editors:

Roy Hendroko Setyobudi
Aris Winaya
Juris Burlakovs
Maizirwan Mel
Olga Anne

ISBN: 978-1-7138-2440-4

Printed from e-media with permission by:

Curran Associates, Inc.
57 Morehouse Lane
Red Hook, NY 12571

Some format issues inherent in the e-media version may also appear in this print version.

For additional information, please contact EDP Sciences – Web of Conferences at the address below.

EDP Sciences – Web of Conferences
17, Avenue du Hoggar
Parc d'Activité de Courtabœuf
BP 112
F-91944 Les Ulis Cedex A
France

Phone: +33 (0) 1 69 18 75 75
Fax: +33 (0) 1 69 28 84 91

contact-edps@webofconferences.org

Additional copies of this publication are available from:

Curran Associates, Inc.
57 Morehouse Lane
Red Hook, NY 12571 USA
Phone: 845-758-0400
Fax: 845-758-2633
Email: curran@proceedings.com
Web: www.proceedings.com

TABLE OF CONTENTS

ANALYSIS OF METHODS FOR DETERMINING THE CHARACTERISTICS OF A SINGLE SPATIAL ELECTROMAGNETIC FIELD 1
Khalikov Abdulhak

THE STUDY OF THE ELECTRICAL CONDUCTIVITY OF APPLES AND GRAPES AS AN OBJECT OF ELECTRICAL PROCESSING 10
Radjabov Abdurahman, Ibragimov Matkarim, Eshpulatov Nodir

IMPROVEMENT GENERATIVE GROWTH OF COFFEA ARABICA L. USING PLANT GROWTH REGULATORS AND PRUNING 18
Astri Muliasari Ade, Kemala Dewi Ratih, Fatchur Rochmah Hidayati, Rakoto Malala Andoniana, Gamawati Adinurani Praptiningsih

IMPROVING MARGINS OF THE INDONESIAN SEAWEED SUPPLY CHAIN UPSTREAM PLAYERS: THE APPLICATION OF THE KAIZEN APPROACH 27
Heri Purnomo Agus, Kusumawati Rinta, Pratitis Asri, Alimin Ilham, Wibowo Singgih, Rimmer Mike, Paul Nick

DETERMINANTS OF TECHNICAL IN-EFFICIENCIES IN SWAMP RICE FARMING - CIAMIS DISTRICT, INDONESIA 38
Yuniawan Isyanto Agus, Sudrajat Sudrajat, Nurdin Yusuf Muhamad

CHEMICAL CHARACTERISTICS AND VIABILITY OF STARTER CULTURES OF FREEZE–DRIED SWEET POTATO EXTRACT–SUPPLEMENTED SYNBIOTIC YOGURT 43
Intan Niken Tari Agustina, Budi Handayani Catur, Hartati Sri, Damat Damat, Stankevièa Karina

BIOMASS ENHANCEMENT OF STEVIA REBAUDIANA BERTONI SHOOT CULTURE IN TEMPORARY IMMERSION SYSTEM (TIS) RITA® BIOREACTOR OPTIMIZED IN TWO DIFFERENT IMMERSION PERIODS 53
Christela Melviana Agustine, Rachmi Esyanti Rizkita, Mel Maizirwan, Hendroko Setyobudi Roy

PHYSICAL TREATMENT OF OIL PALM SHELL FOR BRIQUETTE PRODUCTION AS BIOENERGY AT REMOTE AREA 62
Rohsari Indah Utami Amaliyah, Suwandi Suwandi, Alun Mustafa Yoga, Mel Maizirwan

THE POTENTIAL OF CASHEW APPLE JUICE AS ANTI HYPERCHOLESTEROL AGENT ON WHISTAR RATS (RATTUS NORVEGICUS BERKENHOUT, 1769) 69
Asmawati Asmawati, Marianah Marianah, Yaro Abubakar, Hendroko Setyobudi Roy

THE ABILITY OF WATER HYACINTH (EICHHORNIA CRASIPES MART.) AND WATER LETTUCE (PISTIA STRATIOTES LINN.) FOR REDUCING POLLUTANTS IN BATIK WASTEWATER 77
Muchtasjar Bunyamin, Hadiyanto Hadiyanto, Izzati Munifatul, Vincevièa–gaile Zane, Hendroko Setyobudi Roy

POTENTIAL ANALYSIS OF LOW ECONOMIC VALUE FISH IN LAMONGAN REGENCY, EAST JAVA, INDONESIA 83
Anam Choirul, Harini Noor, Damat Damat, Wahyudi Ahmad, Witono Yuli, Kuswardhani Nita, Azus Shony Azar Moh, Anne Olga, Rachmawati Diana

TECHNO-ECONOMIC ANALYSIS OF PHOTOVOLTAIC UTILIZATION FOR LIGHTING AND COOLING SYSTEM OF FERRY RO/RO SHIP 500 GT 95
Faturachman Danny, Yandri Erkata, Tri Pujiastuti Endang, Anne Olga, Hendroko Setyobudi Roy, Yani Yahya, Susanto Herry, Purba Washington, Krido Wahono Satriyo

MORPHOLOGICAL AND PHYSIOLOGICAL RESPONSES OF INDIGOFERA TINCTORIA L. TO LIGHT INTENSITY 105
Setyaningrum Desy, Theresia Sri Budiastuti Maria, Pujiasmanto Bambang, Purnomo Djoko, Supriyono Supriyono

UTILIZATION OF TOFU INDUSTRY WASTE AND BANANA PLANT WASTE FOR GROWING MEDIUM OF BROWN OYSTER MUSHROOMS (PLEUROTUS CYSTIDIOSUS [JACQ. FR.] P.KUMM.) 113
Indratmi Dian, Dian Kurniasari Yossy, Hartawati Hartawati, Ikhwan Ali

THE TECHNICAL DESIGN CONCEPT OF HI-TECH COOK STOVE FOR URBAN COMMUNITIES USING NON-WOOD AGRICULTURAL WASTE AS FUEL SOURCES 121
Yandri Erkata, Novianto Bangun, Fridolini Fridolini, Hendroko Setyabudi Roy, Wibowo Haryo, Krido Wahono Satriyo, Abdullah Kamaruddin, Purba Washington, Adhi Nugroho Yogo

DETERGENTS EFFECT ON EGG HATCHABILITY, MORPHOMETRY AND LARVAL BONE STRUCTURE OF NATIVE INDONESIAN FISH: WADER PARI (RASBORA LATERISTRIATA BLEEKER, 1854) 130
Umi Abida Farahsani, Alam Parvez, Retnoaji Bambang

MODELLING BIODIESEL SUPPLY CHAIN: CURRENT STATE AND OPPORTUNITIES FOR FUTURE RESEARCH 138
Tupa Ronauli Silalahi Fitriani, Mangihut Simatupang Togar, Parlindungan Siallagan Manahan, Horas Manahan Sinaga Rizal

VINASSE AS CULTIVATION MEDIUM OF CHLORELLA SP. TO PRODUCE POLY–HYDROXY BUTYRATE IN VARIOUS LIMITED LOW–COST PRIMARY NUTRIENT 151
Prima Indra Budianto Gregorius, Mukti Wibowo Yari, Hadiyanto Hadiyanto, Widayat Widayat, Arfian Anditya Sudjarwo Wisnu

RELATIONSHIP OF GROUP DYNAMICS AND FISHERMEN INDEPENDENCE IN FISHERIES AGRIBUSINESS ATTAINING MAXIMUM SUSTAINABLE YIELD 158
Rosyidi Hamdi, Tarik Ibrahim Jabal, Sutawi Sutawi, Anne Olga, Rachmawati Diana

DEGRADATION OF PHORBOL ESTERS ON THE JATROPHA CURCAS LINN. SEED BY BIOLOGICAL DETOXIFICATION 164
Handajani Hany, Rahman Hakim Riza, Adhywirawan Sutaro Ganjar, Ronald Mavuso Boy, Chang Zhong–wen, Andriawan Soni

DURIAN RIND MICRO COMPOSTER MODEL: A CASE OF KAMPUNG DURIAN, NGROGUNG, PONOROGO, INDONESIA 170
Setyaningrum Haris, Rukminastiti Masrifah Atika, Susilo Adib, Haryadi Imam

EXPLORATION AND EFFECTIVENESS OF TRICHODERMA SP. FROM JEMBER AND TRENGGALEK, EAST JAVA, INDONESIA CACAO PLANTATION AS A BIOLOGICAL CONTROL OF PHYTOPHTHORA PALMIVORA 174
Sukorini Henik, Wirasdenty Aigahayunindy Feby, Dani Septia Erfan, Khewkhom Netnapis

VARIABILITY OF FUSARIUM OXYSPORUM F. SP. LYCOPERSICI FROM DIFFERENT ALTITUDES IN EAST JAVA, INDONESIA 181
Sukorini Henik, Dani Septia Erfan, Khewkhom Netnapis

DESIGN OF ROTARY DRYER FOR SAND DRYING USING BIOMASS ENERGY SOURCES 190
Susanto Herry, Hendroko Setyobudi Roy, Sugiyanto Didik, Chan Yefri, Yandri Erkata, Krido Wahono Satriyo, Abdullah Kamaruddin, Burlakovs Juris, Widodo Wahyu, Adhi Nugroho Yogo, Yaro Abubakar

THE USE OF PROBIOTIC AND ANTIOXIDANTS TO IMPROVE WELFARE AND PRODUCTION OF LAYER DUCK AT COMMERCIAL FARMS FOR GLOBAL WARMING MITIGATION 202
Suswoyo Imam, Ismoyowati Ismoyowati, Widodo Wahyu, Vincevièa–gaile Zane

ENTRAPMENT FORMULATION FOR ENV-TM GENE BASED ON CHITOSAN LOW MOLECULAR WEIGHT AS A JEMBRANA DISEASE VIRUS VACCINE CANDIDATE 209
Lesmana Rahayu Indra, Kusumawati Asmarani

ENVIRONMENT IDENTIFICATION OF PLANTATION PLANT DEVELOPMENT FOR COMPETITIVE IN SUKOHARJO DISTRICTS, INDONESIA 213
Wardani Irma, Rosana Dewi Tria

CONCEPTUAL DESIGN OF INVENTORY ANALYSIS SOFTWARE TO SUPPORT THE LIFE CYCLE ASSESSMENT IN PALM OIL PRODUCTION 221
Siregar Kiman, Supriyanto Supriyanto, Saka Rani Devitra, Nurdiansyah Yanuar, Wijayanto Feri

FASCIOLOSIS INFECTION LEVEL OF VARIOUS BREED CATTLE IN BATU AND PUJON DISTRICT, EAST JAVA, INDONESIA 226
Zalizar Lili, Rahmawati Khusnul, Yaro Abubakar

FACTORS THAT INFLUENCE FARMER'S BEHAVIOR TOWARDS RISK 231
Nurdin Yusuf Muhamad, Yuniawan Isyanto Agus, Sudradjat Sudradjat

EFFECTIVENESS OF MYCORRHIZA, PLANT GROWTH PROMOTING RHIZOBACTERIA AND INORGANIC FERTILIZER ON CHLOROPHYLL CONTENT IN GLYCINE MAX (L.) CV. DETAM-4 PRIDA 237
Muhammad Muhammad, Isnatin Umi, Soni Peeyush, Gamawati Adinurani Praptiningsih

MICROALGAE MICROBIAL FUEL CELL (MMFC) USING CHLORELLA VULGARIS AND "BATIK" WASTEWATER AS BIOELECTRICITY 245
Faizi Polontalo Nadiyah, Alifsmara Joelyna Falvocha, Malik Islam Filardli Abdullah, Hadiyanto Hadiyanto, Akmar Zakaria Zainul

ADDITION OF MORINGA OLEIFERA LAM. LEAVES FLOUR FOR INCREASING THE NUTRITIONAL VALUE OF MODIFIED CASSAVA FLOUR–BASED BREAKFAST CEREAL 253
Wely Asmoro Novian, Intan Niken Tari Agustina, Widyastuti Retno, Kurnia Setiawan Chandra

ADOPTION LEVEL OF INTEGRATED FARMING SYSTEM BASED ON RICE–CATTLE AND ITS DETERMINANTS RELATED TO SUSTAINABLE AGRICULTURE 258
Kurniati Novitri, Sukiyono Ketut, Purmini Purmini, Oryza Sativa Mutyarsih

ECO-AGRICULTURE AND FARMING IN THE ANTHROPOCENE EPOCH: A PHILOSOPHICAL REVIEW 267
Kala Mahaswa Rangga, Widhianto Agung, Hasanah Nurul

THE STRATEGY OF SALT BUSINESS DEVELOPMENT: A CASE STUDY IN SUMENEP, INDONESIA 278
Diananing Rika, Destryana Amilia, Santosa Ribut, Illi Mohamad Puad Noor, Christela Melviana Agustine

ANALYSIS OF TAKING DECISION OF FARMERS IN CHOOSING RICE CULTIVARS: CASE OF PAKEL DISTRICT, TULUNGAGUNG, INDONESIA 287
Dewi Oryza Sativa Rima, Tarik Ibrahim Jabal, Sutawi Sutawi

PYROLYSIS OF WATER HYACINTH [EICHHORNIA CRASSIPES (MART.) SOLMS] FOR LIQUID SMOKE PRODUCTION 296
Dwi Ratnani Rita, Hadiyanto Hadiyanto, Widiyanto Widiyanto, Mel Maizirwan

EFFECT AND EFFECTIVITY OF GRANULAR ORGANIC FERTILIZER ON GROWTH AND YIELD OF LOWLAND RICE 301
Budiono Rohmad, Nur Aziz Fuad, Dwi Purbajanti Endang, Turkadze Tsitsino, Gamawati Adinurani Praptiningsih

TECHNO ECONOMIC ANALYSIS OF BIOMASS TO METHANOL PLANT BASED ON GASIFICATION OF PALM EMPTY FRUIT BUNCH 308
Heryadi Rudy, Muhammad Nur Syukri

RISK ANALYSIS OF PARTNERSHIP BROILER FARM IN BLITAR DISTRICT, EAST JAVA, INDONESIA 319
Nur Wulan Mulatmi Septi, Devi Anggraini Apriliana, Shaywont Pratami Febrianti, Motalova Martina

HYDROLYSIS OF CELLULOSE FROM OIL PALM EMPTY FRUIT BUNCH USING ASPERGILLUS NIGER 326
Sugiwati Sri, Suaidah Suaidah, Triwahyuni Eka, Muryanto Muryanto, Andriani Yosie, Abimanyu Haznan

LIGNIN AND CELLULOSE CONTENT OF FERMENTED RICE STRAW WITH ASPERGILLUS NIGER (VAN TIEGHEM) AND TRICHODERMA MUTAN AA1 333
Sukaryani Sri, Ainul Yakin Engkus, Wahyu Harinta Yos, Vincevièa–gaile Zane, Dwi Purbajanti Endang

MULTIPLICATION ARBUSCULAR MYCORRHIZAL FUNGI IN CORN (ZEA MAYS L.) WITH POTS CULTURE AT GREENHOUSE 338
Sukmawati Sukmawati, Adnyana Adnyana, Nengah Suprapta Dewa, Proborini Meitini, Soni Peeyush, Gamawati Adinurani Praptiningsih

MODIFIED OFF–SEASON TECHNOLOGY TO THE FLOWERING TIME AND FRUIT YIELD OF ARUMANIS MANGO (MANGIFERA INDICA L.) 348
Husen Syarif, Ishartati Erny, Muhidin Muhidin, Dwi Siskawardani Devi, Rizky Anjar, Syaifudin Akhmad, Onthong Jumpen

ENTREPRENEURSHIP ORIENTATION, ECO-INNOVATION, INFORMATION AND COMMUNICATION TECHNOLOGY (ICT) LEARNING ADOPTION CAPABILITY: A CASE STUDY OF FOOD SME'S IN CENTRAL JAVA, INDONESIA 354
Didiek Wiet Aryanto Vincent, Kondo Kunio, Wismantoro Yohan, Nurtantyo Andono Pulung

POTENTIALS OF GAS EMISSION REDUCTION (GHG) BY THE GLASS SHEET INDUSTRY THROUGH ENERGY CONSERVATION 361
Purba Washington, Yandri Erkata, Hendroko Setyobudi Roy, Susanto Hery, Krido Wahono Satriyo, Siregar Kiman, Adhi Nugroho Yogo, Yaro Abubakar, Abdullah Kamaruddin, Jani Yahya, Faturahman Danny

THE USE OF RUMEN CONTENTS AS BIO–ACTIVATORS FOR FERMENTATION IN GOAT MANURE FERTILIZER PRODUCTION 373
Pancapalaga Wehandaka, Suyatno Suyatno, Sedlacek David

EXPLORATION AND CHARACTERIZATION OF "UWI" PLANT (DIOSCOREA SP.) IN EAST JAVA UPLANDS, INDONESIA 378
Wuryantoro Wuryantoro, Mustika Wardhani Ratna, Rekyani Puspitawati Indah, Gamawati Adinurani Praptiningsih, Mohammad Yamin Bohari

SWEET POTATOES: DEVELOPMENT AND POTENTIAL AS ALTERNATIVE FOOD INGREDIENTS IN KARANGANYAR REGENCY, INDONESIA 386
Silvana Arianti Yoesti, Wahyu Harinta Yos

STUDY OF AGRONOMIC CHARACTERISTICS OF ROBUSTA COFFEE AT COFFEE PLANTATIONS IN TEMANGGUNG, INDONESIA 392
Theresia Maria Astuti Yohana, Rahayu Enny, Nugraha Budi Santosa Tri, Pratama Putra Dian, Solifudin Agus, Wijayanti Yureana, Fittkow Marcus

Author Index

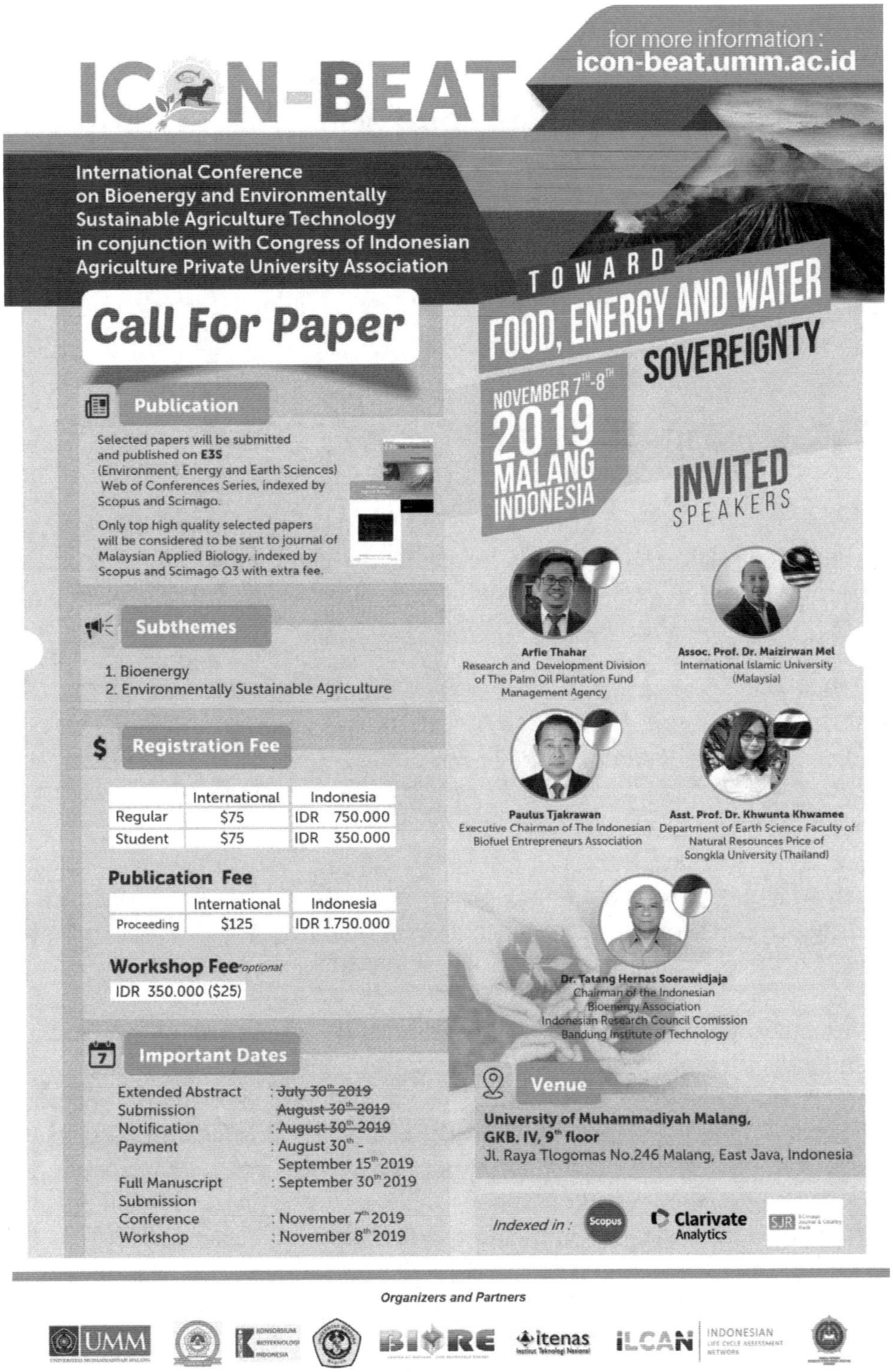

Website : http://icon-beat.umm.ac.id/

UNIVERSITAS MUHAMMADIYAH MALANG

PREFACE: the 1st ICoN-BEAT 2019

On behalf of the organizing committee of the 1st International Conference on Bioenergy and Enviromentally Sustainable Agriculture Technology (ICoN-BEAT 2019) and conjunction with the 2nd Congress Indonesian Agriculture Private University Association (APTSIPI - *Asosiasi Perguruan Tinggi Swasta Ilmu Pertanian Indonesia*). It is an honor and delight to welcome all participants to this conference which be held at the Campus III, University of Muhammadiyah Malang (UMM), Indonesia on November 7th to 8th, 2019. The conference theme is "Bioenergy and Environmentally Sustainable Agriculture Technologies: Toward Food, Energy and Water Sovereignty".

This conference has been organized by Faculty of Agriculture and Animal Science, University of Muhammadiyah Malang. Two main topics have been discussed, i.e. bioenergy and other renewable energy, and environmentally sustainable agriculture. We are perceived confidently that this conference will provide positive influence and contribute to develop the academic field.

Supporting two main topics, the 1st ICoN - BEAT invited five experts in the fields of energy, environment and agriculture from Indonesia, Malaysia, and Thailand. The speakers are Mr. Arfie Thahar (Research and Development Division of the Indonesian Palm Oil Plantation Fund Management Agency, Jakarta), Assoc. Prof. Dr. Maizirwan Mel (Department of Biotecnology Engineering, International Islam University, Malaysia), Asst. Prof. Dr. Khwunta Khwamee (Department of Earth Science, Faculty of Natural Resources, Price of Songkla University, Thailand), Mr. Paulus Tjakrawan (Executive Chairman of the Indonesian Biofuel Entrepreneurs Association, Jakarta), and Assoc. Prof. Dr. Tatang Hernas Soerawidjaja (Chairman of the Indonesian Bioenergy Association, Indonesian Research Council Commission, and Department of Chemical Engineering - Bandung Institute of Technology).

A pride because the number of participants who already send the paper about 116 presenters. After a rigorous selection process, the Scientific & Editorial Board of the 1st ICoN - BEAT 2019 decided to publish 51 papers in E3S Web of Conferences, an open-access proceedings in environment, energy and earth sciences, managed by EDP Sciences, Paris, France and indexed on Scopus, Scimago, Conference Proceedings Citation Index-Science (CPCI-S) of Clarivate Analytics's Web of Science, and DOAJ (Directory of Open Access Journals).

The Proceeding of the 1st ICoN - BEAT 2019, consists of 51 selected papers, amount 38 papers were the results of joint research by Indonesian and overseas scholars. In the collaboration research, 71 institutions were involved 20 of which were from abroad Indonesia. The overseas institutions are from: Australia, Austria, Czech, Estonia, Eswani, Georgia, Germany, India, Japan, Latvia, Lithuania, Madagascar, Malaysia, the Netherlands, P.R. China, Sweden, Taiwan - ROC, Thailand, the United Kingdom, and Uzbekistan. Each of the papers submitted in E3S Web of Conferences was reviewed by at least two experts using the double-blind system. The published papers have passed all necessary improvement requirements in accordance to the Web of Conferences standard, reviewer's comments, SI (*Système International d'Unités*), similarity tests by Turnitin program (with the highest threshold of 20 %), 90 % of references must be at least dated from 15 years, and reflected on Google, as well as editing procedure by professional editors from four countries (Estonia, Indonesia, Lithuania, and Malaysia).

Last but not least, I personally would like to thank you the official committees, organizing partners, and scientific & editorial board. Special thanks as well to our co-host partners: i) APTSIPI, ii) ILUNET (*Ikatan Alumni Energi Terbarukan*) University of Darma Persada, Jakarta, iii) Konsorsium Bioteknologi Indonesia, iv) Merdeka University of Madiun, v) University of Veteran Bangun Nusantara, Sukoharjo, vi) C- BIORE (Center of Biomass and Renewable Energy), vii) ITENAS (*Institut Teknologi Nasional*), viii) ILCAN (Indonesian Life Cycle Assessment Network), ix) *Rumah Paper Kita* as editing and proofreading services for supporting the 1st ICoN - BEAT 2019.

Finally, I would like to express my gratitude feeling for your participations, and please prepare yourself to gain the treasure of knowledge from the passionate experts. Then share the valuable enlightenment for a better future. It is my pleasure to see many of you in the 1st ICoN - BEAT 2019, and see you again in the 2nd ICoN - BEAT 2021.
Stay safe and stay healthy in COVID-19 pandemic.

With warmest regards
Malang - Indonesia, Dec 12, 2020
in the COVID-19 outbreak

Dr. Ir. Listiari Hendraningsih, M.P.
Conference Chair

ICoN-BEAT
TOWARD
FOOD, ENERGY AND WATER
SOVEREIGNTY

SCIENTIFIC & EDITORIAL BOARD

PARTNERS the 1st ICoN-BEAT 2019

PUBLICATION PARTNERS

Malaysian Applied Biology
Biologi Gunaan Malaysia

IJRED International Journal of Renewable Energy Development

E3S Web of Conferences

THE UNIVERSITY OF AGRICULTURE, PESHAWAR - PAKISTAN

THE HASHEMITE UNIVERSITY

EVENT PARTNERS

BRIZZI
uang isi ulang BRI

sawit

ACADEMIC PARTNERS

UMM
UNIVERSITAS MUHAMMADIYAH MALANG

APTSIPI

KONSORSIUM BIOTEKNOLOGI INDONESIA

ILUNET

UNIVERSITAS MERDEKA MADIUN

itenas
Institut Teknologi Nasional

iLCAN
INDONESIAN LIFE CYCLE ASSESSMENT NETWORK

FAKULTAS PERTANIAN
UNIVERSITAS BANGUN NUSANTARA
SUKOHARJO

BIORE
CENTER OF BIOMASS AND RENEWABLE ENERGY

1st ICoN-BEAT 2019: CONFERENCE PHOTOS

UMM
UNIVERSITAS MUHAMMADIYAH MALANG

VENUE:
The 3rd UMM Campus

The 4th GKB-UMM
Venue
ICoN BEAT 2019

Roy Hendroko Setyobudi
(The Principal Editor)
In front of 1st ICoN BEAT's billboard

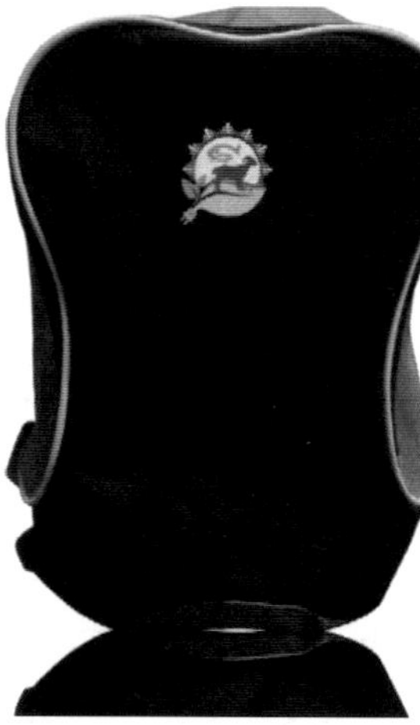

Schedule of IcoN-BEAT 2019

Venue : Main Hall GKB 4, University of Muhammadiyah Malang, Indonesia

Date	Time	Event	Location
Thursday, 7th November 2019	07.00-08.00	**Registration** (Registration for conference and workshop-coaching clinic)	Main Hall GKB 4 (9th Floor)
	08.00-08.05	**Opening Ceremony** Opening by Master of Ceremony	Main Hall GKB 4 (9th Floor)
	08.05-08.10	National Anthem & Sang Surya	
	08.10-08.20	Welcome Dance	
	08.20-08.35	**Speech** Dean of Agriculture and Animal Science Faculty UMM (Assoc. Prof. Dr. Ir. David Hermawan, MP., IPM.)	

Right :	**Aris Winaya** (The Conference Advisor)
Center :	**Listiari Hendraningsih** (The Conference Chair)
Left :	**Roy Hendroko Setyobudi** (The Principal Editor)

Welcome Dance

Welcome Speech

The honorable President of University of Muhammadiyah Malang, Dr. Fauzan, M.Pd

THE PLENARY

SESSION

Arfie Thahar, Research and Development Division of the Palm Oil Plantation Fund Management Agency

SESSION

Assoc. Prof. Dr. Maizirwan Mel, Dept. Biotecnology Engineering, International Islam University, Malaysia

SESSION

Paulus Tjakrawan, Excecutive Chairman of the Indonesian Biofuel Entrepreneurs Association

SESSION

Asst. Prof. Dr. Khwunta Khwamee,
Dept. of Earth Science, Faculty of Natural Resources, Price of Songkla University, Thailand

SESSION

Assoc. Prof. Dr. Tatang Hernas Soerawidjaja,
Chairman of the Indonesian Bioenergy Association, Indonesian Research Council Commission, Bandung Institute of Technology

SESSION

Awarding

for invited speakers

The 1st ICoN BEAT

The Dean of Faculty Agriculture and Animal Science – University of Muhammadiyah Malang, Invited Speakers, Principal Editor, and Participans

PARALLEL
SESSION

PARALLEL SESSION

POSTER SESSION

Awarding

for best participants

Workshop and Coaching Clinic
1st ICoN BEAT's Manuscripts

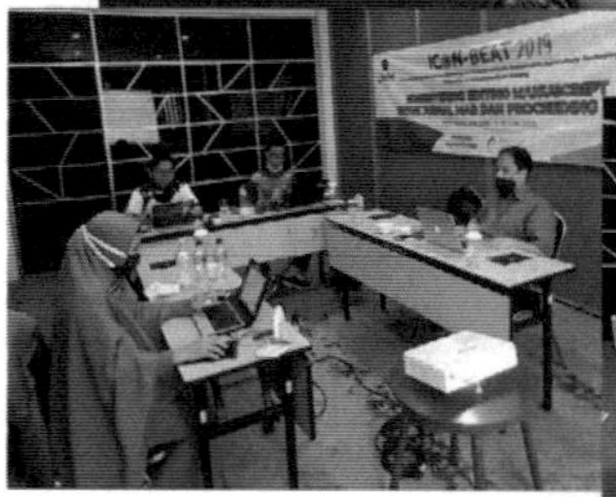

Consignment Refereeing & Editing of 1st ICoN BEAT's Manuscripts

Refereeing and Editing in Era COVID-19 Pandemic

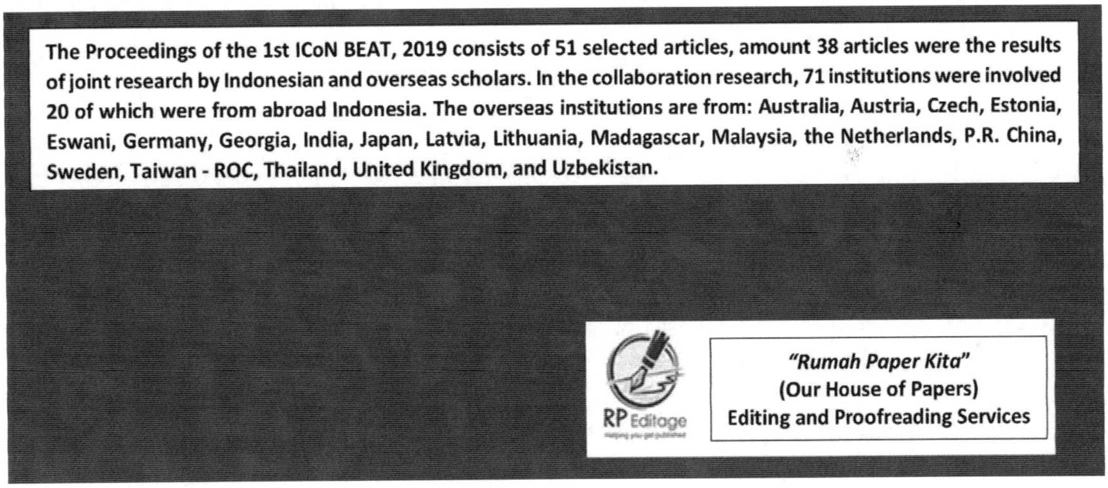

Please email to the Contact Person below
if you need a proceedings cover the 1st ICoN BEAT - 2019 above:

Roy Hendroko Setyobudi
Email: roy_hendroko@hotmail.com
Phone/ WA: +62 815 9555 028

Statement of Peer review

In submitting conference proceedings to *Web of Conferences*, the editors of the proceedings certify to the Publisher that

1. They adhere to its **Policy on Publishing Integrity** in order to safeguard good scientific practice in publishing.
2. All articles have been subjected to peer review administered by the proceedings editors.
3. Reviews have been conducted by expert referees, who have been requested to provide unbiased and constructive comments aimed, whenever possible, at improving the work.
4. Proceedings editors have taken all reasonable steps to ensure the quality of the materials they publish and their decision to accept or reject a paper for publication has been based only on the merits of the work and the relevance to the journal.

Title, date and place of the conference

Title: The 1st International Conference on Bioenergy and Environmentally Sustainable Agriculture Technology (ICoN BEAT 2019).
Date: 7-8 November 2019
Place: GKB IV - University Muhammadiyah Malang, Campus III, Malang, East Java, Indonesia.
Website: http://icon-beat.umm.ac.id/

Proceedings editor(s):

Principal Editor: Roy Hendroko Setyobudi (Malang, IDN)
Board of Editor: Aris Winaya (Malang, IDN), Juris Burlakovs (Tartu, EST), Maizirwan Mel (Kualalumpur, MYS), and Olga Anne (Klaipėda, LTU)

Date and editor's signature

Malang, Indonesia, Oct 28, 2020
Mawlid Nabawi in Era COVID-19 Pandemic

(Roy Hendroko Setyobudi)

17, avenue du Hoggar - PA de Courtabœuf – BP 112 - 91944
Les Ulis Cedex A (France) Tél. : 33 (0)1 69 18 75 75 -
Fax : 33(0)1 69 07 45 17 – www.edpsciences.org

Analysis of Methods for Determining the Characteristics of a Single Spatial Electromagnetic Field

Abdulhak Khalikov*

Tashkent Institute of Railway Engineering 1001676, Tashkent, Uzbekistan

Abstract. This paper analyzes the methods for determining the characteristics of a single spatial electromagnetic field; considerations combine all–electric, magnetic, and electromagnetic fields into a single electromagnetic theory of a spatial field. On the basis of the concept of a single field of force spatial interaction of material bodies, it is analytically established that in a physical vacuum only waves of its polarization actually exist, transporting in the space of a vacuum sphere its excitation energy, which when force interacting with a certain physical characteristic (electrical, magnetic or gravitational) material body, creates a dynamic response of the parameters of this body, which is recorded in the experiment as a real flow of energy with responsible of the physical nature.

Keywords: Electric, gravitational, magnetic, unified spatial electromagnetic field wave equations.

1 Introduction

Contradictions and paradoxes in the theory began to be considered as only strange and harmless exceptions in the harmonious theory [1]. Mathematics of the theory played a significant role. The consequence of excessive mathematics of this theory turned the electromagnetic theory from the physical to the mathematical. The theory of motion in time, the energy of rotation is in many ways similar to Einstein's theory of relativity. In fact, this theory of relativity has a dual character. The first is the peak of human knowledge; the second is the theory of movement in time. In this case, it is not necessary to know this theory, since the theory of motion in time allows you to look at the theory of relativity from the new side and makes it simple and clear [2].

The theory of movement in time is an effective tool for the knowledge of nature. The theory of motion is inextricably linked with the gravitational field. The non–explosive part of electrodynamics and gravity leads to the creation of a single spatial field, which Einstein dreamed of.

The aim of the study is to combine all the fields: Electric, magnetic, electromagnetic into a single electromagnetic theory of the spatial field. With the single spatial electromagnetic field support, such phenomena as the production of fine materials that cannot be obtained by any separators, and if possible, only with a large amount of electricity, are easily explained.

* Corresponding author: Xalikov_abdulxak@mail.ru

Recently, researcher pay more attentionin nanomaterials and nanotechnologies field, working with objects those dimensions are commensurate with the correlation radius of a physical phenomenon (the mean free path of electrons, photons, the coherence length in a superconductor, the dimensions of a magnetic domain of a conductor or nucleus solid phase) [3].

Based on the concept of a single field of force spatial interaction of material bodies, only waves of its polarization really exist in a physical vacuum, transporting its excitation energy in the vacuum medium space analytically and unambiguously. When it interacting with a force, has a certain physical characteristic (electrical, magnetic, gravitational) of the material body, creates a dynamic response of the parameters of this body, which is recorded in the experiment as a real flow energy of the corresponding physical nature [4].

The paradoxes of electrodynamics and the ways of the solution, as well as attempts to bring some electrodynamics positions to a single electromagnetic spatial field, are indicated [4, 5].

2 Material and methods

2.1 Parameter description

Analytical expressions describing different in physical nature electric, magnetic and gravitational forces are identical [6].
The formula of gravitational interaction:

$$\bar{F}_{grav} = \gamma \frac{m_1 \cdot m_2}{r^2} \cdot \frac{\bar{r}}{r} \quad (1)$$

Where is the $\gamma = \frac{1}{4\pi\gamma_0}$ gravitational constant.

Gravitational field strength: $$G_{\text{grav}} = \frac{F}{m} = \gamma \frac{m}{r^2} = \frac{m}{4\pi\gamma_0 r^2}, \quad (2)$$

The dimension of the gravitational field strength is the same as the acceleration m s^{-1}.

From (2):, $$\gamma_0 = \frac{m}{4\pi G_{\text{grav}} r^2} \quad (3)$$

Dimension $$[\gamma_0] = \frac{\kappa g}{m/_{s^2} \cdot m^2} = \frac{\kappa g \cdot s^2}{m^2},$$

γ_0 is a gravitational capacity

Electric capacity analog: $$\mathrm{G} = \frac{\mathrm{m}}{\varphi_{\text{grav}}}, \quad (4)$$

φ_{grav} –is the gravitational potential.

In this regard, the dimensioncan write: $$[\gamma_0]=\left[\frac{\kappa g\cdot s^2}{м^2}\right]=\left[\frac{\kappa g}{B^2}\right].$$

Mass Planck: $$m_{pl}=\frac{\sqrt{\hbar c}}{G_{\text{grav}}}=\sqrt{4\pi\lambda_0\hbar c}=2{,}176\cdot10^{-8}\kappa g,$$

$\hbar=\dfrac{h}{2\pi}$ is the modified Planck constant.

Formula (2) is presented in the form:

$$F_{\text{grav}}=\frac{m_1m_2}{m_{pl}^2}\cdot\frac{m_{Пл}^2}{4\pi r^2\gamma_0}=\frac{m_1m_2}{m_{pl}^2}\cdot\frac{4\pi\gamma_0\hbar c}{4\pi r^2\gamma_0}=\frac{m_1m_2\hbar c}{m_{pl}^2r^2} \quad (5)$$

Denote: $$A_{grav}=\frac{m_1m_2}{m_{pl}^2}, \quad (6)$$

Then: $$F_{grav}=A_{grav}\frac{\hbar c}{r^2}. \quad (7)$$

For electric charges, as well as for gravitational:

$$F_{ЭЛ}=\frac{q_1q_2}{q_{plel}}\cdot\frac{q_{ПлЭЛ}}{4\pi r^2\varepsilon_0}=\frac{q_1q_2}{q_{plel}^2}\cdot\frac{4\pi\varepsilon_0\hbar c}{4\pi r^2\varepsilon_0}=A_{el}\frac{\hbar c}{r^2}, \quad (8)$$

Planck's electric charge: $q_{plel}=\sqrt{4\pi\varepsilon_0\hbar c}=1{,}875\cdot10^{-18}Kl$ so:

$$A_{el}=\frac{q_1q_2}{q_{pl}^2}. \quad (9)$$

In the interaction of magnetic charges for a magnetic field:

$$q_{pl\,маg}=\sqrt{4\pi\mu_0\hbar c}=7{,}066\cdot10^{-16}vb,$$

$$F_{mag}=\frac{q_1q_2}{q_{plmag}}\cdot\frac{q_{plmag}}{4\pi r^2\mu_0}=\frac{q_{1м}q_{2м}}{q_{plmag}^2}\cdot\frac{4\pi\mu_0\hbar c}{4\pi r^2\mu_0}=A_{mag}\frac{\hbar c}{r^2}. \quad (10)$$

2.2 The parameter theory

Based on Formulas (7, 8, 10) represent the equations of a single spatial electromagnetic interaction. These equations indicate the force interaction of material bodies.

The electromagnetic field is characterized by four–vectors $\bar{E}, \bar{D}, \bar{B}, \bar{H}$.
For environments with constant electrical and magnetic permeability:

$$\bar{D}=\varepsilon\varepsilon_0\bar{E}, \quad (11)$$

$$\bar{B}=\mu\mu_0\bar{H}, \quad (12)$$

Therefore, when calculating, it suffices to determine only two vectors.

Usually determined from $\bar{E}$ and $\bar{H}$ the Maxwell equation.

$$rot\,\bar{H} = \gamma\,\bar{E} + \varepsilon\varepsilon_0 \frac{\partial \bar{E}}{\partial t}, \tag{13}$$

$$rot\,\bar{E} = -\mu\mu_0 \frac{\partial \bar{H}}{\partial t}. \tag{14}$$

For the uniqueness of definition $\bar{E}$ and $\bar{H}$, these equations are not enough.

Therefore, the divergence must also be known $\bar{E}$ and $\bar{H}$. By the Gauss theorem in the differential Formula(7–10):

$$div\,\bar{D} = \rho. \tag{15}$$

With a constant value ε: $div\,\bar{E} = \dfrac{\rho}{\varepsilon\varepsilon_0}$. (16)

The main property of the magnetic field is its solenoidality:

$$div\,\bar{B} = 0. \tag{17}$$

By equation (17) can be determined:

$$div\,\bar{H} = 0. \tag{18}$$

At the interface of two media values if ε, μ, γ vary in jumps; therefore, a discontinuity of the field vector continuity will take place on the interface, and the boundary conditions obtained for constant fields of time remain valid for the instantaneous values of the projections of the electromagnetic field vectors.

It is possible to show the basis of the primary basic laws of electromagnetism–Coulomb's law of the interaction of fixed electric point charges:

$$\bar{F} = \frac{q_1 q_2}{4\pi\varepsilon\varepsilon_0 r^2}\bar{1}_r \tag{19}$$

and the law of conservation of electric charge:

$$div\,\bar{\sigma} + \frac{\partial \rho}{\partial t} = 0, \tag{20}$$

a chain of consecutive physical and mathematical arguments can be used to construct a system of Maxwell's electrodynamics' equations.

The functionality of the Coulomb's law consists in the fact that it describes the power interaction of stationary charged material bodies spaced apart in space. The consequence of this interaction is tension:

$\bar{E} = \frac{\bar{F}}{q_0}$, (21) q_0 is the trial point charge.

Because $\bar{E}$ inversely proportional r^2, then the integral of the function over a sphere of any radius is constant: $r^2\left(\frac{1}{r^2}\right)\cdot 4\pi r^2 = 4\pi$ and when using the solid angle, the flux of the vector of electric induction field $\bar{D} = \varepsilon\varepsilon_0 \bar{E}$ through an arbitrary closed surface S, which is identically equal to the total external charge q in volume V_S, covered by this surface

$$\Phi_e = \phi_S, \quad DdS = \int_{US} \rho dV = q \qquad (22)$$

This is the Gauss theorem. If there is an external charge, it does not mean that there is no electric field, since the charges are positive and negative, and the indicated field can be generated by neutral electric sources (electric dipoles). The property of the electric field distinguishes it from the field of Newton, in which the source of the field is gravitational masses, having only one sign – positive [11].

In the system of electrodynamics differential equations, the Gauss theorem in differential form describes the result of the electric polarization of the medium, where with electrical neutrality $(\rho = 0)$ has the appearance $div\,\bar{D} = 0$. Based on equation (21), one can derive the continuity equation.

The law of conservation of charge: The change in charge at a given point in space $\frac{\partial \rho}{\partial t}$ only possible due to the movement of charge from the outside $div\,\bar{\sigma}$.
Divergence is the volume density of the flux of a vector field at a given point

Then: $$div\left(\bar{\sigma} + \frac{\partial \bar{D}}{\partial t}\right) = 0 \qquad (23)$$

Given that $divrot\,\bar{A} = 0$ we will receive:

$$rot\,\bar{H} = \bar{\sigma} + \frac{\partial \bar{D}}{\partial t}.$$

This is the law of total current: Electrical conduction and displacement currents generate an eddy magnetic field, the strength lines of which intensity vectors span the lines of these currents.

In the field of implementation of moving charges and time–varying electric fields $rot\,\bar{H}(r) \neq 0$ function $\bar{H}(r)$ is purely vortex [12].

According to the definition of divergence, one can derive an equation,

$div\,\bar{B} = 0$.

Herefore, equation (17) describes not only the vortex properties $\bar{H}(r)$ but its potential version with no field sources $rot\,\bar{H}(r)=0$. These equations are the result of medium polarization.

The electromagnetic field is not a field in the form adequate for this concept. There is a dynamic interaction between electric and magnetic fields. Such interaction can only be dynamic, a change in one is accompanied by a change in the other. Rotating, for example, two circles in two mutually perpendicular directions and a sinusoid itself, it turns out, hence two types of tension – statistical, according to the expanding sphere and dynamic, according to harmonic oscillation.

The common gravitational, electric, magnetic fields and electromagnetic waves is the environment in which they form and propagate. The Spatio–temporal environment, manifesting physical properties, behaves like any other elastic medium; it does not manifest itself as it does until it is affected. When it is affected, it responds adequately – it forms the fields of the corresponding intensity, there by fulfilling the role of the working medium for transmitting the impact [13].

Based on the concept of a unified field of force spatial interaction of material bodies [14], it was analytically established that in a physical vacuum only waves of its polarization actually exist and transfer its excitation energy in the space of a vacuum sphere, which, when interacting with a force, has a certain physical characteristic (electrical, magnetic) or gravitational) material body creates a dynamic response of the parameters of this body, which is recorded in the experiment as a real flow of energy corresponding physical nature. Based on the definition, the force field represents a region of space, at each point of which the corresponding force interactions are determined.

3 Finite element analysis

There are four types of fundamental interactions, and, consequently, four types of force fields: Gravitational, weak interactions, electromagnetic fields of strong (nuclear) interactions.

Magnetic force: $\bar{F}_{mag}=q\left[\bar{v}\,\bar{B}\right]$, electric field vectors $\bar{E}$ and magnetic induction $\bar{B}$ are components of a single electromagnetic field, that is, the electromagnetic field is a two-component $\left\{\bar{E}\,\bar{B}\right\}$. The relationship between the components of the electromagnetic field:

$$\bar{B}=\frac{1}{c^2}\left[\bar{v}\,\bar{B}\right], \tag{24}$$

where $c=3\cdot10^8\ м/s$ - the speed of light; $\bar{v}$ - The speed of the charge.

At the electromagnetic field on moving with speed $\bar{v}$ charge acting electromagnetic force:

$$\bar{F}_{em}=\bar{F}_{el}+\bar{F}_{mag}=q\bar{E}+q\left[\bar{v}\,\bar{B}\right]. \tag{25}$$

The combination of gravitational, electric, magnetic, electromagnetic fields is a single spatial field. From the side of a single spatial field, a force acts [13-14]:

$$\bar{F} = \bar{F}_{grav} + \bar{F}_{el} + \bar{F}_{mag} + \bar{F}_{elmag} \quad (26)$$

The general formula of a single field of force interaction of bodies in space can be written in the form:

$$\bar{F}(r) = A \cdot \frac{\hbar c}{r^2} \bar{r} = -grad \quad , \quad \bar{F}(r) = (A \cdot \frac{\hbar c}{r^2} r) = -grad U(r) \quad . \quad (27)$$

Here $\hbar = \frac{\hbar c}{2\pi}$ - modified Planck's constant, $c = \frac{1}{\sqrt{\varepsilon_0 \mu_0}}$ - the speed of light in a vacuum, $U(r)$ - potential energy, A is a dimensionless factor.

For different types of fields: $A^{el} = \frac{q_1^e q_2^e}{q_{pl}^{e^2}}$; $A^{mag} = \frac{q_1^m q_2^m}{q_{pl}^{m^2}}$; $A^{grav} = \frac{m_1 m_2}{q_{pl}^2}$,
is determined by the products of the local physical parameters of two fixed interacting point bodies (electrically q^e , magnetic q^m charges, m- masses), normalized to the square of the same dimension of the Planck constant $q_{pl}^e = \sqrt{4\pi\varepsilon_0 \hbar c}$, $q_{pl}^m = \sqrt{4\pi\mu_0 \hbar c}$, $m_{pl} = \sqrt{4\pi\gamma_0 \hbar c}$, composed of a combination of other fundamental physical constants).

According to, equations of systems of differential equations of a single force field of polarization of a physical vacuum with vector components of a polarization field strength $\bar{E}^{pol}(r)$ and vector potential fields $\bar{A}^{W}(r)$

$$a) \quad rot\, \bar{E}^{pol} = -\sqrt{\varepsilon_0 \mu_0} \frac{\partial^2 \bar{A}^W}{\partial t^2}$$

$$b) \quad div\left(\sqrt{\varepsilon_0 \mu_0}\, \bar{E}^{pol} \right) = 0$$

$$c) \quad rot \bar{A}^W = \sqrt{\varepsilon_0 \mu_0}\, \bar{E}^{pol}$$

$$d) \quad div\left(\sqrt{\varepsilon_0 \mu_0}\, \bar{A}^W \right) = 0 \quad (28)$$

These equations are nothing but the differential equations of the polarization field of the physical vacuum.

Wave equation for $\bar{E}^{pol}$

$$rotrot\, \bar{E}^{pol} = graddiv \bar{E}^{pol} - \Delta \bar{E}^{pol} = -\sqrt{\varepsilon_0 \mu_0} \frac{\partial^2}{\partial t^2} rot \bar{A}^W =$$

$$= -\sqrt{\varepsilon_0 \mu_0} \frac{\partial^2 \bar{E}^{pol}}{\partial t^2} = -\frac{1}{c_w^2} \frac{\partial^2 \bar{E}^{pol}}{\partial t^2} \quad , \quad (29)$$

$$rotrot\,\bar{A}^{w} = graddiv\bar{A}^{w} - \Delta\bar{A}^{w} = -\sqrt{\varepsilon_0\mu_0}\,rot\,\bar{E}^{pol} =$$

$$= -\varepsilon_0\mu_0\frac{\partial^2 \bar{A}^{w}}{\partial t^2} = -\frac{1}{c_w^2}\frac{\partial^2 A^{w}}{\partial t^2}, \quad (30)$$

where A^{w} - the vector potential of the force field of vacuum polarization from the point of view of the field vector flux by induction, $C_w = \frac{1}{\sqrt{\varepsilon_0\mu_0}}$ - the speed of propagation of waves of polarization of the physical vacuum. Wave equations for polarization field components:

$$\Delta\bar{E} - \frac{1}{c_w^2}\frac{\partial^2 \bar{E}^{pol}}{\partial t^2} = 0, \quad \Delta\bar{A}^{w} - \frac{1}{c_w^2}\frac{\partial^2 A^{w}}{\partial t^2} = 0, \quad (31)$$

where $C_w = \frac{1}{\sqrt{\varepsilon_0\mu_0}} = c = 3\cdot 10^8\,{м}/{s}$.

These equations represent the equations of a single spatial field.

4 Conclusion

Based on the analysis of the physical characteristics of the forces of spatial interaction, the objectivity of the existence in nature of a single field of force interaction of bodies in the physical vacuum space due to the polarization of the vacuum medium in the presence of matter in it has been established.

It has been established that polarization waves transporting the energy of its excitation in the space of the vacuum sphere, which, when interacting with a certain physical characteristic (electrical magnetic or gravitational) of a material body, creates a dynamic response of the parameters of this body as a real energy flow. A method of representing a single spatial field with a complete system of differential and wave equations is shown.

References

1. A.A. Khalikov, O.A. Ibragimova. Scientific Almanac. **29**,3–329:230–235(2017). https://ukonf.com/doc/na.2017.03.03.pdf
2. I.K. Kolesnikov, A.A. Khalikov, J.F. Kurbanov, O.A. Ibragimova, Problems of Energy and resource conservation. **1**,2:204–220(2014).
3. A.A. Khalikov, I.K. Kolesnikov, J.F. Kurbanov, O.A. Research and development of a single spatial electromagnetic field and devices based on them. (2019).p.250.
4. V.V. Sidorenko. Irreversible processes in nature and technology. III (2011).p.215-219
5. A.A. Khalikov, I.K. Kolesnikov, J.F.Kurbanov. Fan wa technology to find out. (2019).p.250
6. A.A. Khalikov, O.A. Ibragimova. IIUM Ingenering. **14**,2:163–172(2013).
7. A.A. Khalikov, O.A. Ibragimova. VESTNIK. **1**,29:63–69(2014).
8. A.A. Khalikov, I.K. Kolesnikov, J.F. Kurbanov, O.A. Ibragimova. The magazine "Problems of Energy and Resource Saving. **1**,2:204–208(2014).
9. A.A. Khalikov, O.A. Ibragimova, I.K. Kolesnikov. Magazine "Energy saving and water treatment". Ed .: ENIV. **89**,3:9–13(2014).

10. A.A. Khalikov, O.A. Ibragimova. Center for Scientific Publications. III. (2015).p.21–25.
11. A.A. Khalikov, O.A. Ibragimova. UNIVERSUM. **11**:68(2019). https://7universum.com/ru/tech/archive/item/8231
12. A.A. Khalikov, I.K. Kolesnikov, S.S. Khalikov. The magazine "Problems of Energy and Resource Saving. **1**,2:194–198(2015).
13. A.A. Khalikov, O.A. Ibragimova. "Energy saving and water treatment". ENIV. **2,**106):14–18(2017).
14. A.A. Khalikov, O.A. Ibragimova, K.A. Musamedova. Science in the modern world. **9**:28–31(2017).

The Study of the Electrical Conductivity of Apples and Grapes as an Object of Electrical Processing

Abdurahman Radjabov, *Matkarim* Ibragimov, and *Nodir* Eshpulatov*

Tashkent Institute of Irrigation and Agricultural Mechanization Engineers, Tashkent, Uzbekistan

Abstract. Experimental studies of the electrical conductivity of whole berries, peels, and pulp of apples and grapes were carried out according to the methods that the measurement voltage should not exceed the threshold and the measurement time should be minimal, and mechanical damage to the cells during contact with the electrodes was not allowed. To study the electrical conductivity (electrical resistance) of apples, grapes, and their constituent elements, a device is developed that represents a three–electrode system. The study of the specific volume resistance of the pulp of apple and grape. Development of a methodology for a three–electrode system to determine the electrical conductivity of the pulp and skin of fruits and grapes. Three–electrode system showed sufficient accuracy for apple and grape the electrical conductivity. The volumetric resistance of the skin was much greater than the pulp. The tissue resistance of fruits and grapes with an increase in the frequency of the measuring current decreases exponentially.

Key words: Biological materials, cellular structures, electric current, electrical resistivity.

1 Introduction

The practical interest in pulsed electrical processing of biological objects is primarily associated with a non–thermal increase in the permeability of cells in an electric field [1, 2]. Dielectric breakdown [3] or called electroplasmolysis [4–6] of biological cells occurs as a result of the formation and growth of pores in cell membranes polarized by an electric field [7, 8]. In food technologies, pulsed electric processing is used for non–thermal pasteurization of liquid products [2, 9] and the intensification of mass transfer processes in plant tissues [10, 11]. It is proposed to use electroplasmolysis in industries related to the processing of plant materials as an additional method to increase the efficiency of processes such as pressing [12], separation centrifugation [13, 14], extraction [15, 16], osmotic dehydration [17], drying [18, 19]. The effectiveness of these processes is determined by the structural properties of raw materials. Therefore, a change in these properties as a result of electroplasmolysis of plant tissue is an important factor in optimizing the conditions of both the electric processing itself and its combination with other methods of processing raw materials [11, 20].

* Corresponding author: nodir_1885@mail.ru

The nature of electroplasmolysis and pore localization have not been studied yet. The generally accepted mechanism of electrical destruction of suspended cells, such asmicroorganisms and red blood cells, is the formation of pores in the lipid matrix of cell membranes [21]. Electroplasmolysis of cells of biological tissues (plant and animal) can also be explained by electroporation of cell membranes [11] and violation of the integrity of cell walls [22]. In this case, the plasmolysis of cells proper is observed, accompanied by leakage of juice from cell wall and a decrease in the volume of vacuole [23, 24]. However, the mechanism of the influence of the electric field on the cell walls is poorly understood and remains largely unclear.

The measurement of the electrical conductivity of living tissue is carried out at a minimum value of the Alternating Current (AC) potential in a wide frequency range. Moreover, at low frequency, the capacitance during the measurement is very significant. This is primarily due to electrode polarization, which manifests itself as a result of the accumulation of a space charge at the boundary of the electrode with the test substance [25].

Experimental studies of the electrical conductivity of the whole berry, peel, and pulp of apples and grapes were carried out according to existing methods [26–28], taking into account that the measurement voltage should not exceed the threshold and the measurement time should be minimal, and mechanical damage to cells upon contact with the electrodes is not allowed.

2 Material and methods

To study the electrical conductivity (electrical resistance) of apples, grapes, and their constituent elements, a device was developed that represents a three–electrode system consisting of plate potential and ring electrodes made of stainless steel (Figure 1 and Figure 2).

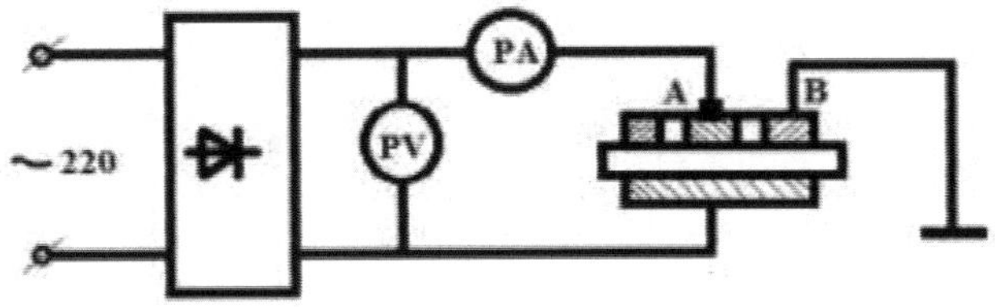

Fig. 1. Schematic diagram of the stand for measuring volumetric electrical resistance (ρ_v).

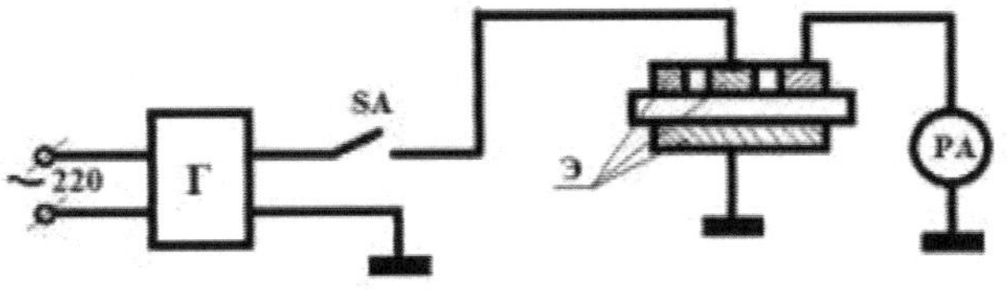

Fig. 2. Schematic diagram of the stand for measuring surface resistance (ρ_s): Г – signal generators GZSh–33; SA – switch; Э – electrodes; PA – milliammeter; O –sample.

The experiments were carried out as follows. A sample was cut from an apple into the shape of a plate and placed between the electrodes. To ensure reliable contact of the electrodes with the surface of the sample, a certain pressure was created from the outside of the electrodes. The pressure force values were determined according to the conditions of maintaining the mechanical integrity of the tissue cell (for example, for grapes – 10 000 Pa, for apples – 11 000 Pa).

After measuring the biopotential, the voltage was applied to the opposite surfaces of the sample to the electrodes, and every 30 sec, the current and voltage were recorded for 7 min. The data of the 6–fold repetition of experiments were subjected to statistical processing by a well–known method [28].

The resistivity was calculated by the Formula (1) and Formula (2)

$$\rho_v = 0{,}785 R_v \frac{d_{cp}^2}{h}, \tag{1}$$

$$\rho_s = \frac{2\pi R_s}{\ln \frac{d_2}{d_1}} \tag{2}$$

Note: $d_{cp} = \frac{d_1 - d_2}{2}$

d_1 – diameter of a measuring electrode, m;

d_2 – inner diameter of the ring electrode, m;

h – sample thickness;

ρ_v, ρ_s – bulk and surface electrical resistance of the sample.

3 Result and discussions

Initial studies were carried out at constant current according to the circuit shown in Figure 1. According to the results of measurements and calculations, dependencies are obtained that characterize the measurement of the specific volume resistance of the pulp of apple and grape. Figure 3 and Figure 4 show the patterns of change in the pulp of apples of the "Semerenko" variety from the gradient, the applied voltage, and the duration of its exposure.

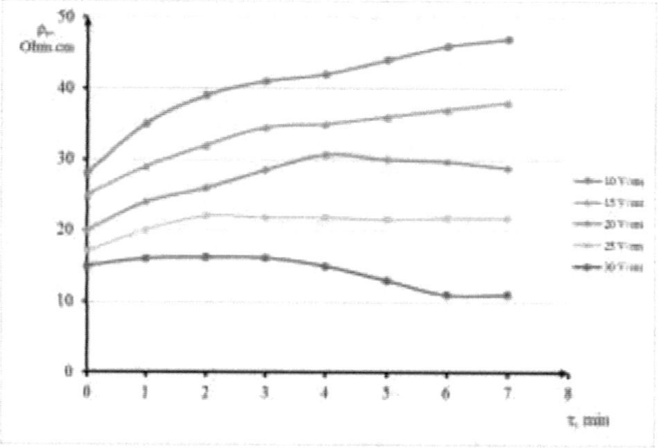

Fig. 3. The dependence ρ_v of the pulp of apples "Semerenko" on the duration and impact of direct current at various values of the gradient of the applied voltage.

The dependencies of Figure 3 showed the specific volume resistance of the pulp at gradients of the applied voltage of 10 V cm^{-1} to 20 V cm^{-1} in the initial period increases and approaches the steady–state value for 7 min, and more measurements. Such a pattern is associated with electrode polarization, which leads to a decrease in the potential at the electrodes at the initial time of measurement. At gradients of the applied voltage of

20 V cm^{-1} to 25 V cm^{-1}, the steady–state resistance value is reached faster (within 4 min to 5 min), as evidenced by a decrease in the influence of the biopotential and the phenomenon of polarization at the indicated values of the gradient of the applied measurement voltage. With a further increase in the gradient of the applied voltage of 25 V cm^{-1} or more, a constant decrease in the value of resistivity from the initial was observed due to the partial death of tissue cells from the damaging effect of the current as of the measurement duration increases.

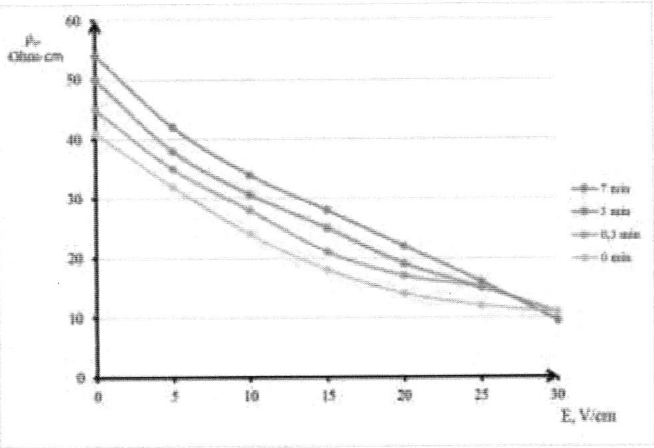

Fig. 4. The dependence ρ_v of the pulp of apples "Semerenko" on the gradient of the applied voltage for different duration of exposure.

The analysis of the dependencies in Figure 4 also indicates that measuring the volumes of resistance of the pulp of the tissue with gradients of the applied voltage of 20 V cm^{-1} to 25 V cm^{-1} gives more accurate results with a minimum measurement time.

Due to the fact that the new technology provides for the impact on the fruits and grapes of pulsed electrical discharges, and also consider the relatively long duration of measuring the resistance to direct current, the patterns of change in the electrical resistance of the electric processing object (fruits, grapes, and their constituent elements) using alternating current the circuit shown in Figure 2.

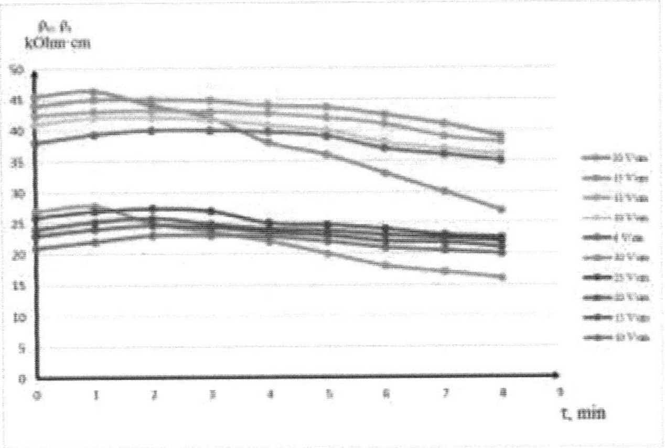

Fig. 5. Dependence ρ_v, ρ_s of apple pulp ("Rosemary") on the time of exposure to an electric current with a frequency of 1 000 Hz at various gradients of the applied voltage.

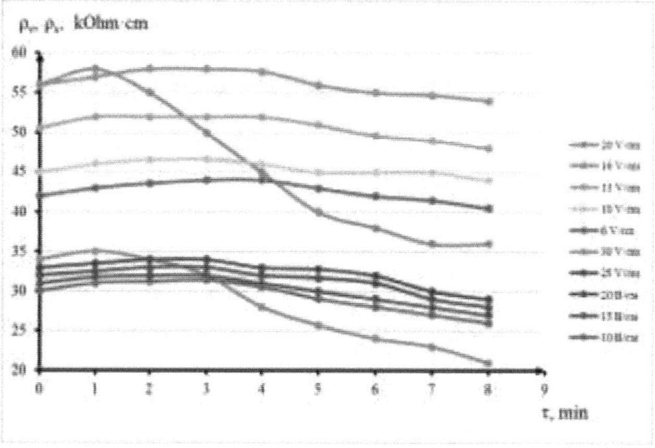

Fig. 6. Dependence ρ_v, ρ_s of grape flesh on the time of exposure to an electric current with a frequency of 1 000 Hz at different gradients of the applied voltage.

An analysis of the curves of the dependences ρ_v and ρ_s on the time of exposure to an electric current of variable frequency (f = 1 000 Hz) at various gradients of the applied voltage (in Figure 5 – the pulp of the apple, in Figure 6 – the pulp of the grape) shows that the gradient of the applied voltage at measurement ρ_v should not exceed 25 V cm^{-1}, and measurement ρ_s – 15 V cm^{-1}. an increase in the gradient of the applied voltage of the indicated values distorts the true value of the resistance.

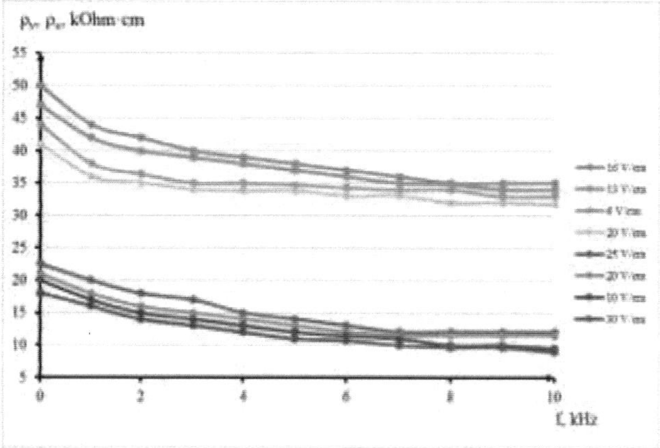

Fig. 7. Dependence ρ_v, ρ_s of the Rosemary apple on the frequency of the current at various gradients of the applied voltage.

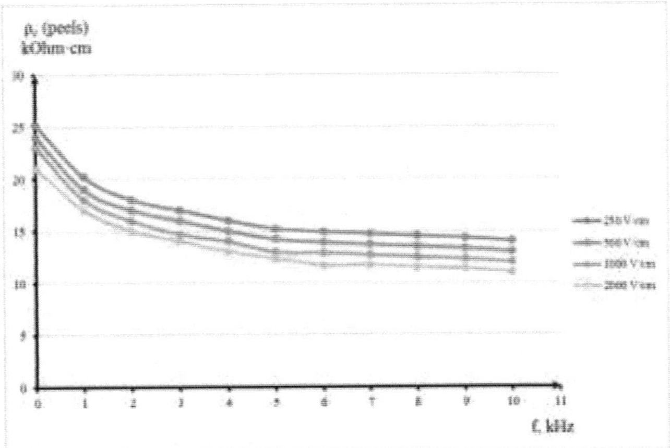

Fig. 8. Dependence ρ_v (peel) of the Rosemary apple on the current frequency at various gradients of the applied voltage.

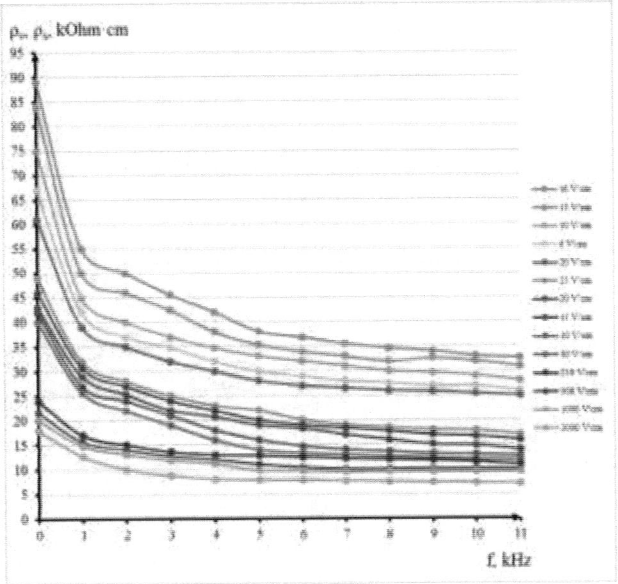

Fig. 9. Dependence ρ_v, ρ_s of pulp and ρ_v peel of "Tayfi" grape skin on current frequency at various gradients of applied voltage.

Based on the results of studying the changes ρ_v and ρ_s of the pulp of apple and grapes and ρ_v of their peel on the frequency of the measuring current for various values of the applied voltage gradient, the relationship between them in the form of graphical dependencies shown in Figure 7, Figure 8, and Figure 9. With increasing frequency of the measuring current, both volumetric ρ_v and surface resistance of the pulp ρ_s and volumetric resistance of the skin ρ_v decrease exponentially at a frequency of 5 kHz to 10 kHz and reaches a constant value. In addition, the volumetric resistance of the skin ρ_v

must be measured at gradients of the applied voltage two orders of magnitude more than when measuring ρ_v and ρ_s the pulp.

4 Conclusion

i. The developed private technique and the three–electrode system make it possible to determine with sufficient accuracy the values of the electrical conductivity of the pulp and skin of fruits and grapes.
ii. The gradient of the applied voltage to the test sample should not exceed the threshold value of the damage to living cells (when measuring ρ_v and ρ_s the pulp, respectively, no more than 20 V cm^{-1} and 10 V cm^{-1} and ρ_v the skin no more than 1 000 V cm^{-1}).
iii. The volumetric resistance of the skin is much greater than the pulp, which is due to a denser arrangement of cells of its tissue.
iv. The tissue resistance of fruits and grapes with an increase in the frequency of the measuring current decreases exponentially, which indicates its additivity with the impedance of the resistance of plant materials.

References

1. D. Knorr, A. Angersbach. Trends Food Sci. Technol. **9**,5:185–191(1998).
2. G.V. Barbosa–Canovas, M.D. Pierson, Q.H. Zhang, D.W. Schafner. Pulsed electric fields. J. Food Sci. **65**,(supplement):65–79(2000).
3. U. Zimmermann, G. Pilwat, F. Beckers, F. Rieman. Effects of external electric fields on cell membranes. Bioelectrochem. Bioenerg. **3**:58–83(1976).
4. M.R. McLellan, R.L. Kime, L.R. Lind. Electroplasmolysis and other treatments to improve apple juice yield. J. Sci. Food Agric. 57:303–306(1991).
5. M. Gudmundsson, H. Mafsteinsson. Effect of electric field pulses on microstructure of muscle foods and roes. Trends Food Sci. Technol. 12:122–128(2001).
6. ЛазаренкоБ.Р., ФурсовС.П., ЩегловЮ.А. идр. Электроплазмолиз. Кишинев, (1977).
7. M. Winterhalter, W. Helfrich. Effect of voltage on pores in membranes. Phys. Rev. A. **36**,12:5874–5876(1987).
8. J. Teissie, N. Eynard, B. Gabriel, M.P. Rols. Electropermeabilization of cell membranes. Advanced Drug Delivery Rev. 35:3–19(1999).
9. L. Barsotti, E. Dumay, T.H. Mu, M.D.F. Diaz, J.C. Cheftel. Effects of high voltage electric pulses on protein–based food constituents and structures. Trends Food Sci. Technol. 12:136–144(2001).
10. И.С. Гулый, Н.И. Лебовка, Манк В.В. и др. Научные и практические принципы электрической обработки пищевых продуктов и материалов. Киев, (1994).
11. B.I.O. Ade–Omowaye, A. Angersbah, K.A. Taiwo, D. Knorr. Use of pulsed electric field pretreatment to improve dehydration characteristics of plant based foods. Trends Food Sci. Technol. **12**,8:285–295(2001).
12. M.I. Bazhal, E.I. Vorobiev. Electrical treatment of apple slices for intensifying juice pressing. J. Sci. Food Agric. 80:1668–1674(2000).
13. T. Beveridge. Juice extraction from apples and other fruits and vegetables. Critical Rev. Food Sci. Nutr. **37**,5:449–469(1997).

14. D. Knorr, A. Angersbach, M.N. Eshtiaghi, V. Heinz, D.U. Lee. Processing concepts based on high intensity electric field pulses. Trends Food Sci. Technol. **12**,4:129–135(2001).
15. A.B. Jemai. Contribution al'étude de l'effet d'un traitementélectrique sur les cossettes de betterave a sucre. Incidence sur le procéded'extraction. Thése de Doctorat, Université de Technologie de Compiégne, France, (1997).
16. Купчик М.П. Разработка технологии сахара из свеклы с применением электрических полей низкой частоты: Автореф. дис. докт. техн. наук. М., (1991).
17. N.K. Rastogi, N.M. Eshtiaghi, Knorr D. Accelerated mass transfer during osmotic dehydration of high intensity electrical field pulse pretreated carrots. J. Food Sci. 64,6:1020–1023(1999).
18. B.I.O. Ade–Omowaye, A. Angersbah, N.M. Eshtiaghi, D. Knorr. Impact of high intensity electric field pulses on cell permeabilisation and as pre-processing step in coconut processing. Innovative Food Sci. EmergingTechnol. **1**,3:203–209(2000).
19. Б ордиян В.В., Интенсификация сушки растительного сырья электроплазмолизом: Автореф. дис. канд. техн. наук. М., (1991).
20. M.I. Bazhal, N.I. Lebovka, E.I. Vorobiev. Pulsed electric field treatment of apple tissue during compression for juice extraction. J. Food Eng. **50**,3:129–139(2001).
21. J.C. Weaver, Y.A. Chizmadzhev. Theory of electroporation: a review // Bioelectrochem. Bioenerg. 41:135–160(1996).
22. A.M. Rojas, M.A. Castro, S.M. Alzamora, L.N. Gershenson. Turgor pressure effects on textural behaviour of honeydew melon // J. Food Sci. **66**,1:111–117(2001).
23. Чебану В.Г. Повышение эффективности электроплазмолиза растительного сырья: Автореф. дис. канд. техн. наук. Киев, (1987).
24. M. Fincan, P. Dejmek. In situ visualization of the effect of a pulsed electric field on plant tissue // J. Food Eng. **55**,3:223–230(2002).
25. Капилов Ю.А., Еськов-Сосковец В.М. Основные направления в создании оборудования с использованием электрофизических методов обработки пищевых продуктов. – М., (1976). - С.39.
26. N.M. Eshpulatov. Intensification of withdrawing process of juice using of electrotechnological methods. Актуальные проблемы аграрной науки, производства и образования. Материалы IIмеждународной заочной научно-практической конференции молодых ученых и специалистов на иностранных языках. Россия, Воронеж, апрель (2016) г. 283-288 с.
27. Н.М. Эшпулатов, М.Н. Саломов. Электротехнологические методы повышения энергоэффективности в производстве соков плодов и винограда. Приоритетные направления развития современной науки молодых учёных аграриев. Материалы V-ой Международной научно-практической конференции молодых ученых, посвященные 25-летию ФГБНУ «Прикаспийский НИИ аридного земледелия». с. Соленое Займище 11-13 мая (2016) г. 784-787 с.
28. Раджабов А. Методологические основы энергосберегающей электротехнологии сушки плодов и винограда. Автореф. док. дисс. — Челябинск, (1991). — 36 с.

Improvement Generative Growth of *Coffea arabica* L. Using Plant Growth Regulators and Pruning

Ade Astri Muliasari[1,*], *Ratih* Kemala Dewi[1], *Hidayati* Fatchur Rochmah[1], *Andoniana* Rakoto Malala[2,3], and *Praptiningsih* Gamawati Adinurani[4]

[1]Department of Plantation Production Technology and Management, College of Vocational Studies, Bogor Agricultural University, Jl. Kumbang No. 14, Kampus IPB Cilibende Bogor 16151, West Java, Indonesia
[2]Graduate School of Agricultural Science, Kobe University 1–1 Rokkodai, Nada–ku, Kobe, 657–8501, Japan
[3]Centre de Formation et d`Application du Machinisme Agricole (CFAMA) Route Betafo Ambaniandrefana – B.p. 109, Antsirabe, Madagascar
[4]Department of Agrotechnology, Merdeka University of Madiun, Jl. Serayu No.79, Madiun 63133, East Java, Indonesia

Abstract. Pruning coffee plants is intended to stimulate generative growth that increases productivity to reach optimal. Plant Growth Regulator (PGR) makes it possible to exploit the production potential of plants. Pruning management, combined with the PGR in coffee plants, is expected to improve flowering and berry growth. The research was conducted for 7 mo from October 2017 to April 2018 at Gunung Gede, Bogor, West Java, Indonesia. A split–plot design with two treatment factors was used in this research. Pruning was placed as the main plot consisting of two levels, namely without (P0) and with pruning (P1). The PGR application was placed as a subplot, consisting of three dosages, namely without PGR (Z0), and concentrations of 0.3 mL L^{-1} (Z1) and 0.4 mL L^{-1} (Z2). The results showed, the pruning treatment significantly influenced microclimate, number of branches, B0, B1 and B2, number of berry sets, and berries. Pruning treatment had a very significant effect on microclimate, plant height, number of branches, number of coffee fruits, yields $crops^{-1}$, and productivity. Interaction both significantly affected plant height and B0. The best dosage had not yet been found.

Key words: Branches, coffee arabica, improve coffee flowering and berry growth, increase coffee production

1 Introduction

Indonesia is the fourth world coffee exporter after Brazil, Colombia, and Vietnam [1]. Coffee is one of the leading commodities in Indonesia, with positive impacts in socio–economic [2], health science [3, 4], and others. Coffee performance is not only in

* Corresponding author: ade.astri@apps.ipb.ac.id

beans. But it is also shown by the side product, among others by solid waste, namely husks and pulp [5–7].

In order to become the largest coffee exporter, Indonesia must increase its coffee yield by developing some technologies. The availability of technologies, such as pruning and plant growth regulators are important. Pruning activities are very important so that coffee plants remain productive continuously [8]. Pruning is useful for breaking the life cycle of pests and diseases, forming a balanced plant canopy between branches, leaves, and fruit production, preventing fruit density, and shooting death. Trimming remnants can be used for soil mulch. Mulching can prevent soil erosion. Decomposed mulch can function as organic fertilizer [8]. In addition, pruning of coffee plants is intended to stimulate generative growth, so that productivity is more optimal and continuous throughout the year.

Plant growth regulators (PGR) in low concentrations can encourage, inhibit, or modify growth and development both in quantity and quality. PGR improves the absorption of nutrients by plants [9], PGR makes it possible to exploit the production potential of plants. In addition, PGR serves to accelerate the response of plants to the environment [10]. Plants can produce natural PGR, but their quantities are limited so that synthetic PGR needs to be added. PGR acts as activators plant cell metabolism, strengthens the immune system, and reacts physiologically. In *Coffea arabica* L. plants, the number of shoots can be increased with the application of growth regulators and fertilizers [11]. Pruning management, combined with the application of PGR in coffee plants, is expected to improve flowering and berry growth.

2 Material and methods

2.1 Study site and treatments

The experiment was conducted for 7 mo from October 2017 to April 2018 at Gunung Gede, Bogor, West Java, Indonesia Teaching Farm. The planting material used was the Catimor variety of coffee plants from Catura versus Hibrido De Timor, which was 3 yr old after planting. The growth regulator used was Paclobutrazol. PGR was applied by watering through the ground instead of spraying it directly to the plant parts to reduce the negative impact.

The dosage of fertilizers was listed in Table 1.

Table 1. Fertilizer dosage

Plant age (yr)	The early rainy season (g plant^{-1})				The end of the rainy season (g plant^{-1})			
	Urea	SP 36	KCl	Kieserit	Urea	SP 36	KCl	Kieserit
1	20	25	15	10	20	25	15	10
2	50	40	40	15	50	40	40	15
3	75	50	50	25	75	50	50	25
4	100	50	70	35	100	50	70	35
5 to 10	150	80	100	50	150	80	100	50
> 10	200	100	125	70	200	100	125	70

2.2 Trial set up

A split–plot design with two treatment factors was used in the research. Pruning was placed as the main plot consisting of two levels, namely without pruning (P0) and pruning (P1). The PGR application was placed as a subplot, consisting of three dosages of PGR, namely without the application of PGR (Z0), application of concentrations of 0.3 mL L^{-1} (Z1), and 0.4 mL L^{-1} (Z2). Therefore, there were six treatment combinations, and each of them consists of three replications. There were 18 experimental units. Each experimental unit consisted of three coffee plants.

The parameters observed were microclimate, such as temperature, humidity. Vegetative growth like plant height, number of branches, and generative growth such as productive branches, number of B0 (branches not yet bearing fruit), B1 (branches that have been fruiting once), B2 (branches that have been fruiting twice), B3 (branches that have been fruiting three times), number of berry sets, number of berries, yields $crops^{-1}$, and productivity.

3 Results and discussions

3.1 Respon of microclimate to pruning treatment

Microclimate observations were carried out three times in the morning, noon, and afternoon. In general, the temperature begins to increase from morning to afternoon and then decrease in the afternoon. Pruning (P1) treatment has a significant effect on temperature at noon (Table 3) and afternoon (Table 4) but not significantly in the morning (Table 2) around the canopy coffee plant.

Table 2. Effect pruning treatment on the temperature in the morning (at 07.00 am)

Treatment	Age (months after application)					
	1	2	3	4	5	6
	Temperature (°C)					
Pruning						
P0	26.40a	26.24a	26.45a	27.26a	27.40a	27.35a
P1	26.53a	26.62a	26.40a	27.24a	27.30a	27.35a
Pr > F	No	No	No	No	No	No

Description: *: Significant difference ($p < 0.05$), **: Significant difference ($p < 0.01$), No: Not significant, P0: Without pruning, P1: Pruning

Table 3. Effect pruning treatment on the temperature at noon (at 12.00 o'clock)

Treatment	Age (months after application)					
	1	2	3	4	5	6
	Temperature (°C)					
Pruning						
P0	30.97b	31.05a	30.97b	31.11a	31.04a	31.21a
P1	31.32a	31.31a	31.34a	31.28a	31.35a	31.34a
Pr > F	*	No	*	No	No	No

Description: *: Significant difference ($p < 0.05$), **: Significant difference ($p < 0.01$), No: Not significant, P0: Without pruning, P1: Pruning

Table 4. Effect pruning treatment on the temperature in the afternoon (at 05.00 pm)

Treatment	Age (months after application)					
	1	2	3	4	5	6
	Temperature (°C)					
Pruning						
P0	26.31b	26.38b	26.44b	27.53a	27.64a	27.46a
P1	26.95a	26.94a	26.97a	27.75a	27.61a	27.84a
Pr > F	**	*	*	No	No	No

Description: *: Significant difference (p < 0.05), **: Significant difference (p < 0.01), No: Not significant, P0: Without pruning, P1: Pruning

Pruning (P1) treatment had a very significant effect on the humidity in the morning (Table 5) at the age of 5 MAP to 6 MAP (month after application) and afternoon (Table 7) at the age of 2 MAP to 4 MAP. At noon (Table 6), pruning treatment had a significant effect on the 4 MAP compared to the treatment without pruning (P0). Humidity began to decrease from morning to noon and increase in the afternoon. Pruning treatment could reduce the moisture, to reduce fruit desiccation.

Pruning techniques could improve air circulation and reduced humidity in the microclimate to reduce disease attacks. On the other hand, pruning will reduce photosynthesis competition between fruit and leaves [12].

Table 5. Effect of pruning treatment on Relative Humidity (RH) in the morning (at 7.00 am)

Treatment	Age (months after application)					
	1	2	3	4	5	6
	Relative humidity (%)					
Pruning						
P0	74.78a	76.77a	77.40a	69.77a	69.11b	69.44b
P1	76.88a	77.22a	78.22a	70.22a	70.55a	70.88a
Pr > F	No	No	No	No	**	**

Description: *: Significant difference (p < 0.05), **: Significant difference (p < 0.01), No: Not significant, P0: Without pruning, P1: Pruning

Table 6. Effect of pruning treatment on RH at noon (at 12.00 o'clock)

Treatment	Age (months after application)					
	1	2	3	4	5	6
	Relative humidity (%)					
Pruning						
P0	56.11a	57.11a	57.11a	55.77b	56.44a	56.33a
P1	56.00a	56.66a	56.66a	58.55a	58.22a	58.00a
Pr > F	No	No	No	*	No	No

Description: *: Significant difference (p < 0.05), **: Significant difference (p < 0.01), No: Not significant, P0: Without pruning, P1: Pruning

Table 7. Effect of pruning treatment on RH in afternoon (at 05.00 pm)

Treatment	Age (months after application)					
	1	2	3	4	5	6
	Relative humidity (%)					
Pruning						
P0	90.66a	91.22a	91.44a	79.44b	79.15a	80.33a
P1	87.77b	88.00b	87.88b	80.44a	79.55a	80.53a
Pr > F	*	**	**	**	No	No

Description: *: Significant difference ($p < 0.05$), **: Significant difference ($p < 0.01$), No: Not significant, P0: Without pruning, P1: Pruning

3.2 Response of vegetative growth *C. arabica* to pruning and PGR treatment

Pruning treatment had a very significant effect on plant height at the age of 2 MAP to 6 MAP (Table 8). Pruning treatment produced the lowest plant height and achieved optimal production. In addition, pruning was very useful in facilitating the harvesting of crops. The application of PGR treatment had no effect of planting height from beginning to end observation 0 MAP to 6 MAP. Interaction between pruning and PGR application had a significant effect on the plant height at the age of 2 MAP.

Table 8. Effect of pruning and PGR treatment to plant height

Treatment	Age (months after application)						
	0	1	2	3	4	5	6
	....(cm)..						
Pruning							
P0	164.37a	166.63a	177.04a	185.44a	188.55a	178.85a	183.96a
P1	166.63a	164.37a	149.00b	149.07b	149.07b	149.07b	148.70b
Pr > F	No	No	**	**	**	**	**
PGR application							
Z0	172.22a	166.39a	173.22a	170.27a	172.00a	167.66a	169.72a
Z1	159.17a	165.11a	159.16a	170.94a	172.55a	166.00a	169.44a
Z2	164.11a	157.56a	164.11a	160.55a	161.89a	158.22a	159.83a
Pr > F	No	No	No	No	No	No	No
Interaction	No	No	*	No	No	No	No

Description: *: Significant difference ($p < 0.05$), **: Significant difference ($p < 0.01$), No: Not significant, P0: Without pruning, P1: Pruning, Z0: 0 mL L^{-1}, Z1: 0.3 mL L^{-1}, Z2: 0.4 mL L^{-1}

Pruning treatment gave a very significant effect on the number of branches at the age of 1 MAP to 6 MAP (Table 9), but the application of PGR treatment did not affect the number of branches. The average number of branches was 49.85 branches on without pruning (P0) and 33.40 branches on pruning treatment (P1) in the last observation.

Pruning treatment stimulated new branches that are continuous in sufficient quantities to support the continuity of production. Pruning coffee plants facilitated the entry of light to stimulate the formation of evenly distributed flower primordia and perfect flowering [13]. The effect of pruning and PGR applications on the number of branches can be seen in Table 9.

Table 9. Effect of pruning and PGR treatment to the number of branches

Treatment	Age (months after application)						
	0	1	2	3	4	5	6
	......................................(Number of branches)...						
Pruning							
P0	38.370a	40.77a	41.22a	52.96a	53.22a	46.88a	49.85a
P1	25.04a	26.22b	33.15b	33.77b	33.85b	32.44b	33.40b
Pr > F	No	*	*	**	**	*	**
PGR application							
Z0	31.33a	34.33a	38.78a	43.72a	44.11a	39.72a	42.27a
Z1	30.11a	34.77a	38.67a	44.00a	44.11a	40.83a	42.66a
Z2	33.667a	34.11a	34.11a	42.38a	42.38a	38.44a	39.94a
Pr > F	No	No	No	No	No	No	No
Interaction	No	No	No	No	No	No	No

Description: *: Significant difference ($p < 0.05$), **: Significant difference ($p < 0.01$), No: Not significant, P0: Without pruning, P1: Pruning, Z0: 0 ml L^{-1}, Z1: 0.3 mL L^{-1}, Z2: 0.4 mL L^{-1}

3.3 The response of generative growth *C. arabica* to pruning and PGR treatment

Pruning and PGR application did not affect the productive branches formation. Unproductive branches were pruned so that the nutrients supplied could be channeled to the more productive branches. The effect of pruning and PGR treatment on productive branches were shown in Table 10.

Table 10. Effect of pruning and PGR treatment to productive branches

Treatment	Age (months after application)						
	0	1	2	3	4	5	6
	....................................(Productive branches)................................						
Pruning							
P0	23.51a	23.70a	23.77a	24.51a	24.51a	23.77a	24.18a
P1	18.88a	23.55a	23.59a	23.00a	22.59a	23.18a	23.81a
Pr > F	No	No	No	No	No	No	No
PGR application							
Z0	22.611a	23.22a	24.111a	23.27a	22.94a	22.61a	23.50a
Z1	22.611a	23.27a	23.500a	24.16a	24.16a	24.55a	25.00a
Z2	18.778a	20.27a	23.278a	23.83a	23.55a	23.27a	23.50a
Pr > F	No	No	No	No	No	No	No
Interaction	No	No	No	No	No	No	No

Description: *: Significant difference ($p < 0.05$), **: Significant difference ($p < 0.01$), No: Not significant, P0: Without pruning, P1: Pruning, Z0: 0 mL L^{-1}, Z1: 0.3 mL L^{-1}, Z2: 0.4 mL L^{-1}

The effect of pruning showed a significant effect on the amount of B0 at 1 MAP (Table 11). The number of B0 reached 16.85 branches in the treatment without pruning while in the treatment of pruning around 12.74 branches. The PGR treatment did not seem to affect the amount of B0 from the beginning to the end of the observation. Interaction of pruning treatment and PGR were evident at 2 MAP.

Table 11. Effect of pruning and PGR treatment to the number of B0

Treatment	Age (months after application)					
	1	2	3	4	5	6
	..................................(Number of B0).....................................					
Pruning						
P0	16.85a	11.11a	9.98a	10.37a	17.03a	9.51a
P1	12.74b	10.63a	8.45a	9.51a	12.18a	7.88a
Pr > F	*	No	No	No	No	No
PGR application						
Z0	15.16a	9.61a	8.92a	9.00a	15.66a	6.72a
Z1	14.72a	8.66a	8.46a	10.11a	14.05a	9.22a
Z2	14.50a	14.33a	10.27a	10.72a	14.11a	6.72a
Pr > F	No	No	No	No	No	No
Interaction	No	*	No	No	No	No

Description: *: Significant difference ($p < 0.05$), **: Significant difference ($p < 0.01$), No: Not significant, P0: Without pruning, P1: Pruning, Z0: 0 mL L^{-1}, Z1: 0.3 mL L^{-1}, Z2: 0.4 mL L^{-1}

Pruning (P1) treatment had a very significant effect (Table 12) on yields (g $plant^{-1}$), fresh and dry fruit yields ha^{-1} at the second harvest compared to without pruning (P0). The yield $plants^{-1}$ (Figure 1) in pruning treatment reached 1 668.36 g while without pruning only 958.56 g. The importance of pruning activities was to avoid cases of overbearing dieback, namely the death of coffee plants after experiencing heavy growth without being followed by proper maintenance, for example, fertilization and pruning [14]. The application of PGR and their interaction did not significantly affect the yield. Different from the result of [15], the application of a plant growth regulator increased at the number of internodes, the average number of fruits on the fourth and fifth node, and the length of reproductive branches. The productivity of coffee could be increased with the use of plant growth regulators [16].

Table 12. Effect of pruning and PGR treatment on *C. arabica* yields in the first and second harvest

Treatment application	Yield g $plant^{-1}$		Wet fruit yield (berries) kg ha^{-1}		Dry fruit yield (productivity) kg ha^{-1}	
Pruning	1	2	1	2	1	2
P0	2 381.88a	985.56b	4 763.76a	1 971.12a	952.75a	394.22a
P1	2 154.48a	1 668.36a	4 308.96a	3 336.72a	861.79a	667.34a
Pr > F	No	**	No	**	No	**
PGR application						
Z0	2 308.44a	1 526.64a	4 616.88a	3 053.28a	923.37a	610.66a
Z1	2 454.72a	1 236.00a	4 909.44a	2 472.00a	981.88a	494.40a
Z2	2 041.20a	1 218.24a	4 082.40a	2 436.48a	816.48a	487.29a
Pr > F	No	No	No	No	No	No
Interaction	No	No	No	No	No	No

Description: *: Significant difference ($p < 0.05$), **: Significant difference ($p < 0.01$), No: Not significant, P0: Without pruning, P1: Pruning, Z0: 0 mL L^{-1}, Z1: 0.3 mL L^{-1}, Z2: 0.4 mL L^{-1}

4 Conclusions

Pruning treatment significantly influenced microclimate, number of branches, B0, B1 and B2, number of berry sets, and berries. Pruning treatment had a very significant effect on microclimate, plant height, number of branches, number of coffee fruits, yields crops^{-1}, and productivity. Interaction between pruning and PGR significantly affected plant height and B0. The best dosage for PGR treatment had not yet been found.

Thank you very much to Vocational College, Bogor Agricultural University as the sponsor of this research.

References

1. [ICO] International Coffee Organization. *Trade statistics tables*. [Online] from http://www.ico.org/prices/m1-exports.pdf (2019). [Accesed on August 3 2020]
2. N. Viartasiwi, A. Trihartono. Coffee Science. **15**,e151687:1–5(2020). https://doi.org/10.25186/.v15i.1687 or http://www.coffeescience.ufla.br/index.php/Coffeescience/article/view/1687
3. I.D.A.R. Dewanti, P.E. Lestari, R. Budirahardjo, D. Setyorini, R.W.E. Yani, S. Wibisono, et al. Coffee Science. **14**,4:477–483(2019). https://doi.org/10.25186/cs.v14i4.1619 or http://www.coffeescience.ufla.br/index.php/Coffeescience/article/view/1619
4. I.D.A. Susilawati, S. Suryono, P. Purwanto, J. Burlakovs, A. Yaro. Coffee Science. **15**:e151637(2020). https://doi.org/10.25186/.v15i.1637 or http://www.coffeescience.ufla.br/index.php/Coffeescience/article/view/1637
5. R.H. Setyobudi, S.K. Wahono, P.G. Adinurani, A. Wahyudi, W. Widodo, M. Mel, et al. MATEC Web of Conference. **164**,01039:1–13(2018). https://doi.org/10.1051/matecconf/201816401039
6. R.H. Setyobudi, L. Zalizar, S.K. Wahono, W. Widodo, A. Wahyudi, M. Mel, et al. IOP Conference Series: Earth and Environmental Science. **293**,012035: 1–25(2019). https://doi.org/10.1088/1755-1315/293/1/012035 or https://iopscience.iop.org/article/10.1088/1755-1315/293/1/012035/pdf
7. D. Damat, R. Anggriani, R.H. Setyobudi, P. Soni, Coffee Science. **14**,4:493–500 (2019). https://doi.org/10.25186/cs.v14i4.1625 or http://www.coffeescience.ufla.br/index.php/Coffeescience/article/view/1625
8. W. Rademacher. J. Plant Growth Regul. **34**:845–872(2015) https://doi.org/10.1007/s00344-015-9541-6
9. D.C. Baitelle, S.D. Freitas, K.M. Vieira, C.M. Meneghelli, A.C. Verdin–Filho, D.F. Baroni, et al. Journal of Experimental Agriculture International. **21:** 1–9(2018) http://www.sciencedomain.org/abstract/23762
10. A. Calatayud, D. Roca, E. Gorbe, P.F. Martines. Scientia Horticulturae. **116**: 73–79(2008). https://doi.org/10.1016/j.scienta.2007.10.028
11. R. Cobos, R.M. Mateos, J.M. Alvarez, M.A. Olego, S. Sevillano, S.G. Garcia, et al. Applied and Environmental Microbiology. **81**:6474–6483(2015). https://aem.asm.org/content/aem/81/18/6474.full.pdf
12. V.F. Sianturi, A. Wachjar. Bul. Agrohorti. **4**,3:266–275(2016). [in Bahasa Indonesia]. https://doi.org/10.29244/agrob.v4i3.14242
13. FAO. *Coffee*. Institut Africain pour le développement économique et social. B.P. 8008, Abidjan, Côte d'Ivoire (1977). p. 1–69.

https://doi.org/10.1051/e3sconf/202122600003

http://www.fao.org/3/ad219e/AD219E06.htm#ch6.3

14. T. Whelan. *Sustainable coffee farming: Improving income and social conditions protecting water, soil and forests*. Rainforest Alliance. New York, NY (2014). p. 1–10. https://www.rainforest-alliance.org/sites/default/files/2016-08/sustainable-coffee-farming-report.pdf
15. F.S. Bacilieri, L.C. De Lima, R.M.Q. Lana, D.S.N. Guimaraes, M.A. Clemente. Biosci. J, Uberlandia. **32**,2:346–353(2016). https://doi.org/10.14393/BJ-v32n2a2016-29820
16. P.F.M.C. Filho, D.C. Baitelle, S.J. Freitas, W.S. Silva, P.C. Santos, W.P. Rodrigues. American Journal of Plant Sciences. **9**:628–636(2018) https://doi.org/10.4236/ajps.2018.94049

Improving Margins of the Indonesian Seaweed Supply Chain Upstream Players: The application of the Kaizen Approach

Agus Heri Purnomo[1,*], *Rinta* Kusumawati[1], *Asri* Pratitis[1], *Ilham* Alimin[2], *Singgih* Wibowo[1], *Mike* Rimmer[3], and *Nick* Paul[3]

[1]Research and Development Center for Marine and Fisheries Product Processing and Biotechnology, Jl. KS. Tubun Petamburan VI, Slipi, Central Jakarta 10260, Indonesia
[2]State College of Fisheries Jakarta, Jl. AUP Pasar Minggu, South Jakarta 12520, Indonesia
[3]University of the Sunshine Coast, Maroochydore, 90 Sippy Downs, Queensland 4556 Australia

Abstract. The margin obtained at the upstream in the Indonesian commercial seaweed supply chain is generally small and makes the motivation for business development low. The research aimed to identify the opportunities to overcome these problems. The research conducted with the Kaizen analysis approach that seven upstream business nodes were chosen at three locations: two *Gracilaria* sp. farmers and one *Gracilaria* sp. trader in Brebes, one Eucheumatoid farmer and one intermediary trader in Serang, also two Eucheumatoid farmers in Sumenep. The results show some inefficiency in several loci in the business lines of intermediate farmers and traders. These include a gap on moisture content (*Gracilaria* of 2 % and Eucheumatoid of 4 %), productivity of *Gracilaria* pond 0.6 t ha^{-1} yr^{-1} and its by-product is 50 %, while the productivity of Eucheumatoids was 10 %. The quality of dry *Gracilaria* is one grade only. All result loss of potential margin. Root problem analysis found several external factors outside the Kaizen domain and other internal aspects can be corrected through intervention. The Kaizen analysis determined some simple interventions including improvement of cultivation embankments and extension of maintenance periods, use of monofilament net to dry seaweed, optimization of cultivation spacing, and use of moisture checker.

Keywords: Eucheumatoid, hydrocolloid, seaweed industry

1 Introduction

Indonesia has an excellent opportunity to create massive revenue through seaweed hydrocolloid trades. Hydrocolloid markets tend to increase over time continuously and carrageenan is the most necessary type of product. Hydrocolloid trade appeared firstly in 2009, reaching USD 1.01 × 10^9, an increase of almost 60 % over the previous decade. In that period, the trade value of agar increased by 35 % while carrageenan increased by 81 %. The increase in international demand for hydrocolloid resulted in significant needs of raw materials. In the last 5 yr, it took nearly 3 500 000 t of wet carrageenophyte and almost

* Corresponding author: a_heri_p@yahoo.com or purnomo.agus88@gmail.com

1×10^6 t of wet agarophyte and now increases by 4.5 % yr^{-1}. Indonesia has the power to become an essential player in the trade or product as well as raw materials (carrageenophyte, especially *Eucheuma* sp. J.Agardh, and agarophyte, especially *Gracillaria* sp. [Greville , 1830]). As the world prominent seaweed producer, Indonesia has relatively been prosperous in mastering the market share of the raw material but plays minimal role in the processes product market (hydrocolloid). For consumption seaweed, with a 29.9 % share, Indonesia secures the second position after South Korea. Meanwhile, for non-consumption seaweed, with a share of 14.6 % Indonesia occupies the second position under Chile, outperforming China in the third. Success in the raw material market did not occur in the hydrocolloid market; in this context Indonesia caught less than 4 % of world market share, far behind China, Chile, Spain and the Philippines.

The weakness of Indonesia at the global market competition of seaweed can be traced on the micro-problems in the country. In the year of 2015, a study on Indonesian seaweed supply chain found that there is detected quality degradation, such as inconsistencies in yield, quality, price, and technologies in one part of the supply chain can affect the whole chain [1], marked with low gel strength and undesirable color. There is something wrong with managing and developing the seaweed industry. A country with high production of raw materials just becomes an insignificant player in the product market, even for intermediate products. The government wants the support of scientific data needed to make the high-value seaweed industry.

Some studies related to this problem have been carried out by previous researchers. These for example, research of the Lombok case [2], Kendari Bay case [3], the Sumenep case [4], and the Tarakan case [5]. Unfortunately, with local specific approaches adopted in these studies, the recommendations obtained are not enough to provide a basis for national policy-making [6]. The researches have come up with reasonable remedial steps including the development of seed ponds, the creation of cultivation clusters, the development of processing activities at the local level; however, local specific approaches make the resulting output insufficiently connected with supply chain nodes in other regions.

In an ACIAR 2017 research, there is a multi-location value chain analysis on the complex roles played by persons (leading producers, processors, traders, service providers) of seaweed. Problems have been identified as occurring at nodes in the commodity supply chain [7]. Among the outputs of this research is a finding the margin obtained by the person at the upstream node in the Indonesian commercial seaweed supply chain is generally small and makes the motivation for business development low.

Problems found in the 2017 research activities are still global and need to be explored through further analysis carried out in 2018, using a productivity analysis approach. Through the Kaizen technique, the productivity analysis was applied to identify the leverage actions needed to improve efficiency. The Kaizen approach itself can be briefly described as 'a change towards a better' which is done continuously. In industries, Kaizen are generally applied to formulate strategies to increase production, increase product quality, reduce operational costs, reduce waste, and increase job security. Productivity analysis has been applied generally to industrial cases in Indonesia [8], but it was reported that this approach is also being applied to seaweed processing plants in Indonesia [9].

Based on those situations, the research reported in this paper was based on the Kaizen approach and aimed at identifying opportunities to overcome problems related to the inefficiency in upstream activities in the Indonesian seaweed supply chain. In this case, focuses were given to identification of problem loci, root cause analysis, priority problem selection, and determination of intervention options.

https://doi.org/10.1051/e3sconf/202122600004

2 Methods

This research focused on productivity analysis [10], which was carried out following the Kaizen framework. Moreover, this study was a follow-up to the previous years' research results. Based on this, seven upstream business nodes, hereby referred to as Kaizen research collaborators, were chosen in the commercial seaweed supply chain at three locations: two *Gracilaria* farmers in Brebes, one *Gracilaria* trader in Brebes, one Eucheumatoid farmer in Serang, one intermediary trader in Serang, and two Eucheumatoid farmers in Sumenep, Indonesia. In each of these selected nodes, four analysis phases were carried out to identify opportunities for improving production performance i.e. i) identification of problem loci, ii) root cause analysis, iii) determination of priority problems, and iv) determination of intervention for changes. In implementing the Kaizen application, researchers must provide understanding, and get assurance that research collaborators fulfil the following requirements such as open minded, same vision to improve the work system in their business units, same mission to increase margins, and are willing to provide time/capital/energy to make observations of agreed changes. The steps of the intended Kaizen methodology shown in Figure 1.

Fig 1. Steps of Kaizen methodological approach

3 Results and discussion

Kaizen research collaborators set to represent each district/regency case, was a joint decision between the researchers and the local fisheries service department which a chairman or a group member registered in the fisheries service department. Where available, records of business activities such as those regarding seaweed ponds, buying and selling, business facilities such as warehouses, postharvest processing equipment, handling techniques, and other relevant matters, are provided by collaborators for references in discussions and consultations during implementation Kaizen activities.

The results of intensive discussion and consultations were focused on searches and the handling of shared problems, as the decisions outlined in the project charters for Kaizen improvement scenario. The results are presented in groups per district/regency case as follows. The presentation of individual district cases will start with exposure to the relevant strategic environment and proceed with the summary of specific problems identification in each Kaizen research collaborators.

3.1 General situation of the case locations

3.1.1 Brebes

Brebes Regency is one of the country's primary concentrations for *Gracilaria* sp. cultivation. *Gracilaria* sp. cultivation in this area began in 2006 and developed on the northern side of the district, especially the villages of Randusanga Wetan and Randusanga Kulon of the Brebes Sub-district, and Prapag Lor and Prapag Kidul of Losari Sub-district. Currently, businesses that are based on *Gracilaria* sp. seaweed are an important economic activity for the community in the Brebes Regency. Based on the statistics of fisheries Brebes in the year 2017, the total production of dried seaweed reached 62 057 t, with production value amounting to IDR 186 $\times$ 10^9 (conservative price assumption of *Gracilaria* sp. dried IDR 3 000).

This economic activity also creates quite a lot of employment in cultivation, handling, transportation and trade in the commodity. The seaweed cultivation to the post-harvest stage requires around 45 workers ha^{-1} [7]. The potential of seaweed aquaculture in Brebes Regency is very high, supported by 12 678 ha of ponds spread over a 53 km coastline.

At present Graciaria's cultivation is decreasing, because the payment of the products has stalled. The main factory intended for *Gracilaria* sp. dry marketing is PT. Agarindo in West Java and PT. Agarsari in East Java. PT. Agarindo is the most targeted factory because it provides high prices for high quality seaweed. Meanwhile, other factories represent only alternative destinations because they do not demand high quality even though the purchase price is lower. Within the last four months at the time of this research (February, June 2018) data on payment of products that had passed quality checking were not obtained. However, according to farmers and middlemen, there is a tendency that farmers are forced to accept the price set by the factory because there was already a lot of dried seaweed deposited in the factory, and this can be intereted as fraud by the factory.

The upstream side of the seaweed industry includes groups of cultivators and collectors, with work stages consisting of seaweed cultivation, postharvest to produce dried seaweed, and handling dried seaweed that meets factory requirements (low moisture content, clean, and uniform colour). Aside from its strategic environmental factors, there are a number of internal deficiencies in the upstream side of the seaweed supply chain in Brebes. The disadvantage is the inability of the players to control impurities i.e moss, clamps, traces, dry mud, and colours fading due to exposure to dew. Impurities occur due to poor and postharvest cultivation pond water circulation (washing and drying). Meanwhile, fading colors occur due to the night drying, which was left without cover so that it was exposed to fresh water from dew.

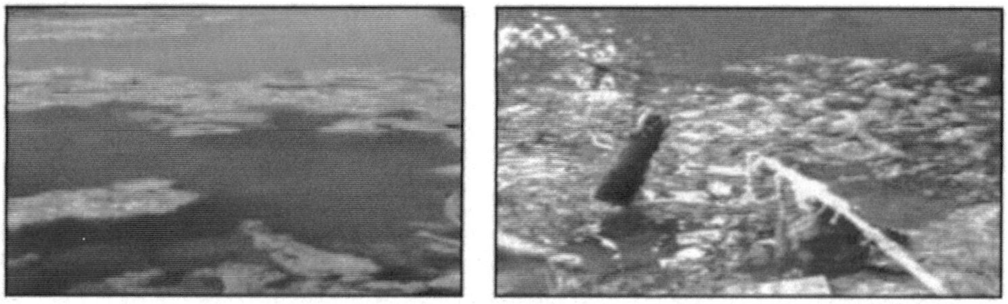

Fig 3. Barnacle and moss at the *Gracilaria* ponds

Fig 4. Distinct appearences of ponds – well maintained and poorly maintained

Fig 5. Messy and unhealthy look of harvested *Gracilaria* due to moss or barnacle

3.1.2 Serang

Seaweed production in Serang Regency comes mainly from *Euchema* cultivation, where the figure is close to 20 000 t yr^{-1}. Such high production makes Serang seaweed the biggest contributor to the statistics of commodity production at the provincial level. However, there are problems where farmers generally harvest seaweed at the age of 30 d, so the quality is poor, carrageenan content is low, and the price is cheap. Cash flow pressure is a common reason that causes farmers to harvest at less than the recommended age, which is 45 d. If harvested at the ideal age, the price of fresh dried seaweed on the market should be IDR 12 000 kg^{-1}, salted dried seaweed IDR 8 000 kg^{-1}, and wet seaweed IDR 4 000 kg^{-1}. In fact, with 30 d of harvesting, prices dropped by 40 % to 60 %.

The existence of a seaweed development policy in Serang Regency is characterized by the implementation of the Minapolitan program, which is a concept of fisheries-based area development, centralized in Domas Village, Pontang District, Indonesia. In addition, there is also a concentration of development on the area of seaweed cultivation in the Lontar waters of the Tirtayasa District. In the past, there were also intensive efforts to develop seaweed on Pulau Panjang, but at the moment it no longer continues because the waters in the region have been heavily polluted. The latest development shows that all land on the coast of Lontar, Tirtayasa Subdistrict covering an area of 100 ha is used as ponds for seaweed cultivation.

Although Serang has positive development of *Eucheuma* sp. seaweed but lately there are challenges faced for example in Lontar Village occur crop failure. Change of weather and sea water quality make difficulty for seaweed to grow, and even eventually die.

Gracilaria sp. cultivation is not as intensive as *Eucheuma*'s, but its-based economic activities can't be ignored. *Gracillaria* sp. cultivation in this area was initiated by a processing plant, which in time became the main market of dried Gracillaria in the Serang region. The cultivation began in a 36 ha area, which was projected to supply raw materials for processing plants with a capacity of 3 000 t to 6 000 t.

The general problems of the business *Gracilaria* sp. are quality. The grade of dried *Gracilaria* sp. is lower than Brebes. This problem is related to the awareness of farmers to produce and sell good quality products. A number of interventions have been tried by traders, but it did not work well.

3.1.3 Sumenep

Sumenep has a conducive area for *Eucheuma* sp. cultivation including the land protected by small islands, the calm and stable currents along year, the good of water sources. This condition also ensures the availability of seeds.

Seaweed *Eucheuma* sp. cultivation in Sumenep is high and increased in the last 4 yr (2015 to 2018). This production spread across 10 sub-districts including on several small islands, with average productivity of 3.71 t ha^{-1}. Among these locations, seaweed cultivation concentrated in Sub-districts of Bluto, Talango, and Bungkek. In Sumenep, middlemen play a major role in the development of seaweed in their area, both in terms of quality and quantity. Now, the middlemen have a stronger bargaining position that intensively carried out by improving the quality of dried seaweed and providing good quality seeds, as well as applying the ideal harvest age. The middlemen also coordinate with relevant institutions or groups such as the local fisheries office, seaweed producer associations, seaweed processing associations.

In Sumenep Regency there are eight large traders, spread in Pakandangan Sangra areas, Pakandangan Barat, Aeng Dake (Bluto). The big problem of the seaweed industry in Sumenep is unexpected results although the middlemen have tried to provide instructions to farmers. In addition, there is a low quality seaweed from Sumenep and unoptimized cultivation because of improper harvest time and not dried product. The other problem is the competition between middemens leads to unfair competition and causes a lot of poor quality seaweed on the market. Moreover, the absence of seaweed processing units also contributed to the problems.

3.2 Problems and solutions as observed in Kaizen research collabolators's cases

3.2.1 Problems as observed in Kaizen research collaborators' cases

It can be described that local businessmen have opportunities that are constrained by several external aspects. Thus, a more appropriate strategy is to identify and implement internal measures that can improve the situation. The following are the results of field observations conducted to explore such internal measures. As mentioned in the method section, the problem and the relevant potential internal measures were explored in three cases that occurred in three Kaizen research collaborators such as two *Gracilaria* sp. farmers, and one *Gracilaria* sp. trader, three euchematoid farmers, and one intermediary trades. Table 1 summarizes problems and relevant conditions facing the local businessmen as reflected by the selected respondents.

Table 1. Problems and relevant conditions

Respondent's name, type of business and general profile	Problem	Relevant condition
Warsono, *Gracilaria* farmer, Brebes, 2.7 ha ponds, intercropping *Gracilaria* + milkfish, and vannamae	i. Wastewater pollutant from the onion fields, ii. Prices are often lower because of impurities iii. Tidal flooding	i. Farmers have skill but do not practice it ii. Embankments are damaged, so cultured milk fishes are escaping iii. Drying is carried out on the streets; quality checking is done by buyers
Slamet Riyadi, *Gracilaria* middleman, Brebes, purchases seaweed from local farmers, supplies grade three dried seaweed to to agar factories in Tangerang and Malang	i. High impurities, which affects quality & price	i. Purchases made without re-drying ii. Determination of quality in the factory is very dependent on the weigher iii. Middlemen check the water content only by palpation
Abdul Latief, farmer, Brebes, 4 ha seaweed ponds intercropped with milkfish	i. Pond productivity is low so the margins obtained are low	i. Harvesting age is 8 wk; productivity is 2 t ha^{-1} yr^{-1} of dried seaweed. ii. Farmers focus on operational costs and are not aware of opportunities to increase margins
Asmawi, *Eucheuma* farmer, Serang, 100 spans of seaweed planting, each of which holds 1.2 t of seedlings	i. Distance between the nodes for the cottonii planting does not follow what is recommended ii. Demand for white cottonii increased, so that the salted dry cottonii decreased	i. Farmers receive additional knowledge of good cultivation techniques but has not applied much ii. Seaweed is sold in two types of products, i) white seaweed: not require long harvests, high price, low turnover, ii) salted seaweed: requires long harvests, low price, big turnover, continuous demand
Karim, Serang, *Gracilaria* middleman, purchases dried *Gracilaria* from local farmers, sell 90 % of it of Grade 3 to Tangerang and Malang	i. Yield is low because the material received from farmers is dried on soil ii. Extra cleaning can improve the quality from Grade 4 to three	i. Drying and cleaning are done by the employees, which are earned by him at a price of IDR 250 kg^{-1}
Junaidi, *Eucheuma* farmer and group leader, Sumenep	i. Prices of cultivated seaweed are generally IDR 1 000 lower than the average price	i. E. cottonii is harvested at 35 d; productivity of 500 kg per *ancak* in summer (4 mo yr^{-1}) and 1 t per *ancak* during the rainy season (8 mo yr^{-1})
Mr. Mukhrasam, *Cottonii* cultivator and group leader, Aeng Dake Village, Bluto District	i. The price of wet seaweed is IDR 7 000 is cheaper than the price of seaweed with 40 % water content	i. Farmers deliver harvests directly to collectors without additional handling. ii. Productivity of 500 kg per *ancak* in summer (4 mo yr^{-1}) and 1 t per *ancak* during the rainy season (8 mo yr^{-1}).

3.2.2 Potential solutions as observed in Kaizen research collaborators' cases

The problems and conditions stated in Table 1 were explored through direct and in-depth observations with business owners and staff. The problems and potential solutions are outlined in the project charters, as summarized in Table 2.

Table 2. Summary of Project Charters

Case	Project	Assumption
Warsono, *Gracilaria* farmer	Improve maintenance water circulation by embankment & waterways repairs	i. The purchase price: IDR 5 300 (highest quality) and IDR 4 700 (lowest quality) *Gracilaria* value = four harvests per location x 1.5 t per harvest x IDR 600 kg^{-1} = IDR 3.6 × 10^6 yr^{-1} per location ii. Milkfish lost when the embankment breaks =50 % Milkfish value = four harvest times x 100 kg x IDR 20 000 kg^{-1} = IDR 8 × 10^6 yr^{-1}
Slamet, *Gracilaria* middleman	Reduce the gap between product moisture content and required water content by 2 %	i. 40 wk yr^{-1} x 2 trucks wk^{-1} x 8 000 kg per truck = 640 000 kg yr^{-1} of Tangerang shipment @ IDR 7 000 ii. 40 wk yr^{-1} x 1 truck/k x 8 000 kg per truck = 320 000 kg yr^{-1} to Malang @ IDR 5 000 iii. The difference in product water content and water content required by the factory is 2 %, equivalent to a loss of IDR89.6 × 10^6 + IDR 32 × 10^6 yr^{-1} = IDR 121.6 × 10^6 yr^{-1}
Abdul Latif, *Gracilaria* farmer	Increase pond productivity from 2.0 t to 2.6 t ha yr^{-1} through a 3 wk to 4 wk harvest extension	i. Seaweed productivity averages 2.0 t ha^{-1} yr^{-1} with an average harvesting age of 8 wk ii. Production target is 2.6 t ha^{-1} yr^{-1} with an average harvesting age of 12 wk iii. Profit margin of IDR 5 300 kg^{-1} x 600 kg = IDR 3.18 × 10^6 ha^{-1} yr^{-1} iv. With a land area of 4 ha, the profit margin becomes IDR 12.7 × 10^6 yr^{-1}
Asmawi, *Eucheuma* farmer	Improve productivity by increasing the distance between nodes per span in planting	i. Seeds needed for 100 spans are 1.2 t ii. Seeds obtained from their own harvest: harvest diverted into seeds for each planting cycle/group members = 20 spans x 50 kg per span = 1 000 kg. iii. With the price of IDR 1 500 kg^{-1} of seeds, the cost of seedling/member = IDR 1 × 10^6 → total cost IDR 30 × 10^6 iv. Seeds purchased from others: costs per planting cycle/member = 1 200 kg x IDR 3 000 = IDR 3.6 × 10^6 → total cost of seeds for one group= IDR 72 × 10^6 v. Harvest weight with inter-node distances 5 cm to 7 cm is 50 t per 100 spans per planting cycle vi. Harvest weight with the distance between nodes of 10 cm is 70 t per 100 spans per planting cycle

(Continued on next page)

Table 2. Continued

Case	Project	Assumption
Karim, *Gracilaria* middlemen	Loss reduction attributed to the use of waring net mat in the drying process in ponds from 20 % to 2 % within 3 mo	i. Shipments to the factory = 40 wk yr^{-1} x 3 trucks wk^{-1} x 8 950 kg per truck = 1 074 000 kg yr^{-1} ii. Products from farmers of 20 % impurities, purchase volume = 1 070 000 kg yr^{-1} x (120/100) = 1 284 000 kg yr^{-1} iii. Drying costs = IDR 250 kg for relatively seaweed and IDR 275 kg for dirty seaweed iv. With drying with no net mat, two options: cleaning to obtain grade three or leaving it to obtain grade four v. Price of grade four = IDR 5 000 kg^{-1} and grade three = IDR 7 500 kg^{-1}. Equivalent to a loss due to unclean harvest = 1 074 000 kg x IDR (7 500 to 5 000) = IDR 2 685 × 10^6 yr^{-1} vi. With cleading, total additional costs = (1 284 000 kg x IDR 25 kg^{-1}) + (1 284 000 to 1 070 000) kg x IDR 250 kg^{-1} = IDR (32 + 53.5) × 10^6 yr^{-1} = IDR 85.5 × 10^6 yr^{-1}
Junaidi, *Eucheuma* farmer, Sumenep	Applying of Kaizen approach to improve gel strength by increasing cultivation time from 35 d to 40 d	i. Current price of dried cottonii = IDR 7 500 kg^{-1}. ii. Price of 35 d of dried cottonii = IDR 6 500 kg^{-1} iii. Harvest frequency = 5 times yr^{-1} iv. Harvest frequency = 7 and production = 500 kg of dried cottonii. Income earned by farmers from seaweed harvested at age 35 d is IDR 22.75 × 10^6 per cycle or IDR 113.75 × 10^6 yr^{-1} v. Income earned by farmers from seaweed harvested at age 40 d = IDR 26.25 × 10^6 per cycle or IDR 131 250 × 10^6 yr^{-1}. vi. Extending harvest from 35 d to 40 d will give additional revenue of IDR 3.5 × 10^6 per cycle (IDR 17.5 × 10^6 yr^{-1})
Mukhrasan, *Eucheuma* farmer, Sumenep	Applying a Kaizen approach to increase cottonii dryness from pond-wet condition to dry condition (40 % moisture content).	i. Current price: wet IDR 5 000; water content 40 % IDR 12 000; water content 38 % IDR 14 000; and water content 37 % IDR 15 000 ii. Farmers apply five cycles of cultivation per year iii. Each farmer has five cultivation ropes and production per cultivation rope of 500 kg dry cottonii; based on that: iv. The income earned by farmers from selling wet seaweed is IDR12.5 × 10^6 per cycle (5 cultivation ropes x 500 kg x IDR 5 000) or IDR. 62.5 × 10^6 yr^{-1} v. Income earned by farmers from selling dried seaweed (40 % moisture content) is IDR. 30 × 10^6 per cycle (5 cultivation ropes x 500 kg x IDR 12 000) or IDR 150 × 10^6 yr^{-1} vi. By drying up to 40 % moisture content, a farmer will get an additional income of IDR 17.5 × 10^6 per cycle or IDR 87.5 × 10^6 yr^{-1}

4 Conclusion

The results show there are inefficiencies in a number of loci in the business lines of intermediate farmers and traders include a gap (actual vs. standard value) on moisture content (2 % in *Gracilaria* and 4 % in Eucheumatoid), pond productivity (0.6 t ha^{-1} yr^{-1} of *Gracilaria*), productivity of by-products (50 % of *Gracilaria*), the productivity (10 % of Eucheumatoids). Moreover, the quality of dry *Gracilaria* is just in the same grade, thus all of the inefficiencies cause a loss of potential profit (margin). Basic problem analysis found a number of external factors outside the Kaizen domain and other internal aspects that can be corrected through intervention. The internal aspects include the objective measurement of water content and good drying method for the inter-trade business and the implementation of *Gracilaria* cultivation pond maintenance techniques and spacing on Eucheumatoid. The results of the analysis determined a number of simple interventions including improvement of cultivation embankments and extension of maintenance periods, use of monofilament net for drying seaweed, optimization of cultivation spacing, and measurement of water content using standardized gadgets. The potential benefits were IDR 13×10^6 ha yr^{-1} through the intervention of extending the 4 ha of *Gracilaria* pond life; IDR 122×10^6 yr^{-1} through intervention in measuring dry *Gracilaria* water levels; and IDR 8×10^6 yr^{-1} through *Gracilaria* by products; IDR 2.8×10^6 yr^{-1} through improvement in the way of drying *Gracilaria*; and IDR 30×10^6 yr^{-1} through the improvement of the Eucheumatoid cultivation spacing. These results indicated the opportunity to significantly increase margins through the intervention of post-harvest aspects.

References

1. H. Mulyati, J. Geldermann, Clean Technol. Environ. Policy. **19**,1:175–189 (2016) doi: 10.1007/s10098-016-1219-7
2. E. C. Wright, *The upshot of upgrading: seaweed farming and value chain development in Indonesia*. [Thesis]. University of Hawai'i at Manoa. (2017).p.112 https://scholarspace.manoa.hawaii.edu/handle/10125/62441
3. M.A. Limi, L. Sara, T.L. Ola, L. Yunus, Suriana, S.A.A. Taridala, *et al*., Bioflux. **11**,6:1927–1936 (2018) http://www.bioflux.com.ro/docs/2018.1927-1936.pdf
4. I. C. Neish, *A diagnostic analysis of seaweed value chains in Sumenep Regency, Madura Indonesia*. United Nations Industrials Development Organization. (2015)p.62 https://open.unido.org/api/documents/3370517/download/A%20diagnostic%20analysis%20of%20seaweed%20value%20chains%20in%20Madura,%20Indonesia.pdf
5. A. Wahyularassati, D. Susiloningtyas, T. Handayani. IOP Conference Series: Earth and Environmental Science. **311**:1–6(2019) doi: 10.1088/1755-1315/311/1/012082
6. R. Pomeroy, H. Navy, A.J. Ferrer, A.H. Purnomo. J. World Aqua. Soc. **48**,4:542–554(2017) doi: 10.1111/jwas.12407
7. W.E. Jamandre, H. Hasanuddin, C. Kokarkin, U. Hatch, R.B. Bolivar, R.J. Borski, *Value chain analysis of seaweed in Aceh, Indonesia*.Oregon State University. (2012)p.189 https://aquafishcrsp.oregonstate.edu/sites/aquafishcrsp.oregonstate.edu/files/09mer06nc_marketing_economic_assessment_0.pdf
8. Banyuriatiga, D.H. Darwanto, L. R. Waluyati, Agro Ekonomi. **28**,1:80–94 (2017). [in Bahasa Indonesia]. https://jurnal.ugm.ac.id/jae/article/view/24316/16904

9. H. Mulyati, *Supply chain risk management study of the Indonesian seaweed industry*.[Thesis Phd]. Georg-August University of Göttingen, Gottingen (2015)p.171 https://ediss.uni-goettingen.de/bitstream/handle/11858/00-1735-0000-0022-606A-8/Dissertation_Heti_2015_2.pdf?sequence=1
10. F.A. Hasanah, M. Nurhudah, M. Mulyono, M. Dillon, *IOP Conference Series: Earth and Environmental Science,* **414**:1–13(2020) doi:10.1088/1755-1315/414/1/012007

Determinants of Technical In-efficiencies in Swamp Rice Farming - Ciamis District, Indonesia

Agus Yuniawan Isyanto*, *Sudrajat* Sudrajat, and *Muhamad* Nurdin Yusuf

Agribusiness Department, Faculty of Agriculture, Universitas Galuh,
Jl. R.E. Martadinata No.150, Ciamis 46274, Indonesia

Abstract. Rice farming in swamps, a marginal land, has a relatively high risk related to the level of technical efficiency. This research was conducted with the aim of identifying the level of technical efficiency and the influence factors of technical inefficiency in Lakbok Subdistrict, Ciamis District, Indonesia. Simple random sampling was used with a sample size of 41 farmers. The analysis was carried out using a stochastic frontier function. The results showed that the average level of technical efficiency was 0.78. Education and family size have a significant effect on technical inefficiency.

Keywords: Farmer, marginal land, stochastic frontier

1 Introduction

Swamp rice grows in marginal land that has a relatively high risk related to the level of technical efficiency [1]. This farming is carried out in the dry season where water needs depend on rainfall [2]. Swamp rice farming usually faces serious challenges during the rainy season [3].

The main problem in the management of swamp rice farming is the occurrence of floods that inhibit plant growth and production. Farmers in swamp rice often have difficulty in predicting flood levels, so they face the risk of flooded rice plants in the vegetative growth phase [4]. Weather and climate have a direct influence on agricultural production so that weather fluctuations and climate variability play an important role in growth and yields [5].

The crucial problem of rice farming in Indonesia is the low efficiency and productivity thus the production is uncompetitive compared to other rices. Increased rice production can be done through existing technology [6]. Technical efficiency compares the level of output in relation to the level of input used [7].

Lack of skilled farmers in managing the system properly causes inefficient agricultural management leads to reduced yields and increased waste [8]. The low yield is due to several factors including agro-climatological problems and high input costs [9]. Constraints in increasing crop yields can be related to inefficient agricultural management even though inputs are used intensively [10].

The ability to allocate factors of production will affect production and the level of efficiency. The non optimal production indicates the existence of technical inefficiencies [11].

This research was conducted with the aim of identifying the level of technical efficiency and the factors that influence technical inefficiency in swamp rice farming in Lakbok Subdistrict, Ciamis District, Indonesia.

* Corresponding author: gusyun69@gmail.com

2 Methods

The research was conducted in Lakbok Subdistrict, the only sub district that has swamp lanswampland farming. The research was carried out for 3 mo in planting the first season.

The sample size was 41 farmers using simple random sampling. The study utilized stochastic production frontier which builds hypothesized efficiency determinants into the inefficiency error components. The model is defined by Equation (1):

$$\ln Y = \beta_0 + \beta_1 \ln X_1 + \beta_2 \ln X_2 + \beta_3 \ln X_3 + \beta_4 \ln X_4 + \beta_5 \ln X_5 + vi - ui \quad (1)$$

Where:
Y = output (kg)
X1 = seed (kg)
X2 = Organic fertilizer (kg)
X3 = chemical fertilizer (kg)
X4 = Pesticide (liter)
X5 = labor (man-day)
β = coefficient of regression
vi = random error
ui = technical inefficiency effects in the model.

Technical efficiency (TE) effects model developed by Battese and Coelli was employed in this study. In this model a Cobb-Douglas production function and some exogenous factors influencing technical efficiency are determined simultaneously.

Technical efficiency in the context of production relates to the level at which a farmer produces maximum feasible output from a given set of inputs (output-oriented measure), or uses a minimum level of input feasible to produce a certain level of output (a size-oriented input) [12].

Inefficiency model was defined to estimate the influence of some farmer's socio-economic variables on the technical efficiency of the farmers. Technical inefficiency effects are assumed to be distributed independently [13, 14]. The model was defined by Equation (2):

$$\mu i = \delta_0 + \delta_1 Z_1 + \delta_2 Z_2 + \delta_3 Z_3 + \delta_4 Z_4 \quad (2)$$

Where:
μi = technical inefficiency
Z1 = age (years)
Z2 = education (years)
Z3 = experience (years)
Z4 = family size (persons)
δ = regression coefficient.

3 Results and discussion

3.1 Technical efficiency

The level of technical efficiency achieved by rice farmers in swamps ranged from 0.53 to 1.00 with an average of 0.778 as presented in Table 1.

Table 1. Frequency distribution of technical efficiency

Technical Efficiency	Frequency	Percentage
0.51 to 0.60	5	12.20
0.61 to 0.70	6	14.63
0.71 to 0.80	9	21.95
0.81 to 0.90	11	26.83
0.91 to 1.00	10	24.39
minimum = 0.53; maximum = 1.00; mean = 0.78		

Table 1 showed that the average level of technical efficiency achieved was 0.78, which indicates that swamp rice farming was technically efficient. This efficiency index value implied a technical inefficiency gap of 0.22 which indicates that 22 % of higher production can be achieved by farmers without using additional resources, or the use of inputs can be reduced to achieve the same level of output. The technical efficiency will be considered as efficient if it reaches an efficiency index value of more than 0.70 [15]. The difference in the level of technical efficiency achieved by farmers shows the degree of differentiation in the application of technology [16].

3.2 The stochastic frontier production functions analysis

Analysis of factors affecting production and technical inefficiencies was carried out using the stochastic frontier production function as presented in Table 2.

Table 2. Maximum likelihood estimates and inefficiency functions

Variable	Coefficient	Standard Error	t-ratio
Production function			
Constant	3.106 7	0.651 8	4.766 6
Seed	–0.066 8	0.240 8	-0.277 4
Organic fertilizer	–0.001 3	0.013 3	-0.100 8
Chemical fertilizer	–0.045 4	0.165 1	-0.275 3
Pesticide	–0.099 4	0.091 6	-1.085 8
Labor	1.353 1	0.438 8	3.083 9*
Inefficiency function			
Constant	–0.209 2	0.068 0	3.075 6
Age	–4.868 0	0.447 1	0.108 9
Education	0.308 8	0.011 4	27.077 8*
Experience	–0.119 4	0.328 0	-0.364 1
Family size	–0.008 5	0.001 4	-5.985 7*
Sigma squared	–0.056 5	0.003 4	16.376 8
Gamma	0.999 9	0.002 4	425.254 4
Log likelihood function	18.2348		
LR Test	12.8942		

*significant at 1 % (p > 0.01)

The estimated value of the gamma parameter (γ) of 0.999 9 is statistically different from zero. This indicated that 99.99 % of the variation in the level of output in swamp rice farming is caused by technical inefficiencies in the use of inputs. The model used in this study is a linear log equation hence the value of each regression coefficient shows the production elasticity of each input. The sum of all regression coefficients is more than one (1.13) which indicates increasing returns to scale.

Table 2 showed only the labor force that has a positive and significant effect on production in swamp rice farming. The results of this study were in line with the other research [1, 17].

3.3 Technical inefficiency

Table 2 showed that education has a positive and significant effect on technical inefficiencies, which shows that improving education will reduce technical efficiency. Family size has a negative and significant effect on technical inefficiency which indicates that increasing family size will increase technical efficiency. The results of this study are in line with the results of another research [18].

Age has a negative but not significant effect on the level of technical inefficiency which indicates that the older the farmer, the more technically efficient. The results of this study are consistent with other findings [19].

Education has a positive and significant effect on technical inefficiency which shows that the more educated farmers, the lower the technical efficiency. The results of this study are in line with other findings [20].

The experience of farmers in swamp rice farming has a negative but not significant effect on the level of technical inefficiency. This shows that the more experienced farmers in carrying out swamp rice farming will increase their technical efficiency. The results of this study are consistent with other findings [21].

Family size has a negative but not significant effect on the level of technical inefficiency. This shows that the more family size will increase the technical efficiency. More family size means more workers are available to carry out rice farming activities in swamps in a timely manner thus the production process becomes more efficient [6]. Farmers who have large family sizes tend to try their best to get higher yields to meet the needs of their families. In addition, large family sizes have the workforce needed to implement agricultural management decisions [22].

4 Conclusion

The level of technical efficiency of swamp rice farming ranged from 0.53 to 1.00 with an average of 0.778 which indicated that swamp rice farming has reached a level of technical efficiency. Labor has a significant effect on production, while seeds, organic fertilizers, chemical fertilizers and pesticides have no significant effect. Education and family size have a significant effect on technical inefficiency, while age and experience have no significant effect.

References

1. M. Zakirin, E. Yurisinthae, N. Kusrini, J. Soc. Econ. Agric. **2**,1:75–84(2013). http://dx.doi.org/10.26418/j.sea.v2i1.5122
2. I. Zahri, S. Adriani, E. Wildayana, Sabaruddin, M.U. Harun, Bulg. J. Agric. Sci. **24**,2:189–198(2018). https://www.agrojournal.org/24/02-03.pdf
3. K.J. Ani, G. Maxwell, C.S. Ecoma, Int. J. Adv. Academic Res. **3**,4:1–10(2017). https://www.ijaar.org/articles/Volume3-Number4/Arts-Humanities-Education/ijaar-ahe-v3n3-m17-p4.pdf

4. Gribaldi, R.A. Suwignyo, M. Hasmeda, R. Hayati, Agrivita **38**,1:64–72(2016). http://doi.org/10.17503/agrivita.v38i1.498
5. A.V. Nkiene, N. Clement A, T. Paul, Canadian J. Trop. Geography. **3**,2:1–14(2016). https://www3.laurentian.ca/rcgt-cjtg/volume-3-issue-2/3011/?lang=en.
6. J. Mariyono, Jurnal Ilmu Administrasi dan Organisasi **21**,1:35–43(2014). [in Bahasa Indonesia]. http://journal.ui.ac.id/index.php/jbb/article/view/4042
7. M.M. Islam, D.C. Kalita, Int. J. Agric. Sci. **6**,2:938–948(2016). https://doi.org/10.46882/IJAS/1231].
8. Nurliza, E. Dolorosa, A.H.A. Yusra. Agribis. **3**,2:85–92(2017). https://doi.org/10.18196/agr.3248
9. O.M. Bamiro, J.O. Aloro. Scholarly J. Agri. Sci. **3**,1:31–37(2013). http://www.scholarly-journals.com/sjas/archive/2013/jan/pdf/Bamiro%20and%20Aloro.pdf
10. B. Ayedun, A. Adeniyi, Acta Sci. Nutr. Health. **3**,7:86–94(2019). https://actascientific.com/ASNH/pdf/ASNH-03-0325.pdf
11. T.S Azwar, T.I. Noor, Ernah, Mimbar Agribisnis, **5**,2:276–292(2019). [in Bahasa Inodensia]. https://jurnal.unigal.ac.id/index.php/mimbaragribisnis/article/view/2264
12. J.N Ugwu, G.O. Mbah, N. Chidiebere-Mark, T. Ashama, D.O. Ohajianya, M.O. Okwara, Current Res. Agric. Sci. **4**,1:1–6(2017). https://doi.org/10.18488/journal.68/2017.4.1/68.1.1.6
13. M. Bala, M.N. Shamsudin, A. Radam, I.A. Latif. CSSPO International Conference 2018, Sarawak, Malaysia 2018. E3S Web of Conferences. **52**:00030(2018). https://www.e3s-conferences.org/articles/e3sconf/pdf/2018/27/e3sconf_nrm2018_00030.pdf
14. M.F. Alam, M.A. Khan, A.S.M.A. Huq. Aquacult Int. **20**:619–634(2012). https://link.springer.com/article/10.1007/s10499-011-9491-3
15. H. Khotimah, R. Nurmalina, Forum Agribisnis **2**,2:141–160(2012).[in Bahasa Indonesia]. http://journal.ipb.ac.id/index.php/fagb/article/view/8882
16. A.Y. Isyanto, M.I. Semaoen, N. Hanani, Syafrial, J. Econ. Sustain. Dev. **4**,10:100–104(2013). https://iiste.org/Journals/index.php/JEDS/article/view/6518
17. D.L. Pudaka, Rusdarti and P.E. Prasetyo. J.Econ. Education. **7**,1:31–38(2018). https://journal.unnes.ac.id/sju/index.php/jeec/article/view/22799
18. M. Mukwalikuli. WJRR. **6**,4:60–65(2018). https://www.wjrr.org/download_data/WJRR0604025.pdf
19. Y. U. Oladimeji and Z. Abdulsalam, IOSR-JAVS. **3**,3:34–39(2013). https://doi.org/10.9790/2380-0333439
20. B.S. Balde, H. Kobayashi, M. Nohmi, A. Ishida, M. Esham, E. Tolno. J. Agric. Sci. **6**,8:179–196(2014). https://doi.org/10.5539/jas.v6n8p179
21. M.E.A. Begum, M.A.M. Miah, M.A. Rashid, M.I. Hossain. *Factors affecting the technical efficiency of turmeric farmers in the slash and burn areas of Bangladesh*. [Online] from https://doi.org/10.1007/s10457-018-0338-z. (2011) . [Accessed on 20 September 2019].
22. U. Mukhtar, Z. Mohamed, M.N. Shamsuddin, J. Sharifuddin, M. Bala. E3S Web of Conferences **52**:00049 (2018). https://doi.org/10.1051/e3sconf/20185200049

Chemical Characteristics and Viability of Starter Cultures of Freeze–Dried Sweet Potato Extract–Supplemented Synbiotic Yogurt

Agustina Intan Niken Tari[1,*], *Catur* Budi Handayani[1], *Sri* Hartati[1], *Damat* Damat[2], and *Karina* Stankeviča[3]

[1]Department of Agricultural Product Technology, Faculty of Agriculture, University of Veteran Bangun Nusantara Sukoharjo, Jl. Letjend Sujono Humardani no. 1, Sukoharjo 57521, Central Java, Indonesia
[2]Department of Food Science and Technology, Faculty of Agriculture and Animal Science, University of Muhammadiyah Malang, Jl Raya Tlogomas No. 246, Malang 65144, East Java, Indonesia
[3]Department of Environmental Science, University of Latvia, Raina Blvd.19, Riga LV-1050, Latvia

Abstract. The research aimed to determine the sucrose concentration as a *cryoprotectant* to obtain the chemical properties and the viability of Lactic Acid Bacteria (LAB) and *Lactobasillus plantarum* in the synbiotic yogurt. It adopted a one–factor Completely Randomized Design (CRD) by including sucrose in concentrations from 0 %, 2.5 %, 5 % to 7.5 %, three replications. The results showed various sucrose concentrations significantly affected the yields of the purple sweet potato extract–supplemented synbiotic yogurt, reduction–sugar level, the total of LAB, and the total of *L. plantarum* before and after the freeze–drying process, no significant impact on the moisture content and total quantity of LAB. The best treatment of the synbiotic yogurt was induced through the addition of sucrose with a concentration of 5 % as a *cryoprotectant*. The treatment signified the following characteristics: 14.797 % of yields, 7.51 % of water content, 14.59 % of reduction–sugar level, 1.98 × 10^9 CFU mL^{-1} of total LAB before the freeze–drying and 9.28 × 10^8 CFU mL^{-1} after the freeze–drying, 8.23 × 10^8 CFU mL^{-1} of total *L. plantarum* before the freeze–drying and 6.81 × 10^8 CFU mL^{-1} after the freeze–drying.

Key words: Healthy consumption, fermented drink, functional food, free radical, sucrose as cryoprotectant.

1 Introduction

Together with the increasing public awareness on the importance of healthy foods, the consumer demands related to food qualifications have begun to shift. Consumers currently look for foods that contain not only a good nutritional composition with attractive appearance and taste but also certain physiological impacts for the body, such as its benefits

*Corresponding author: intanniken@gmail.com

to maintain the balance of intestine microbiota and immune system. The healthy consumption pattern can be achieved through the addition of synbiotic yogurt. The types of symbiotic yogurt include local probiotic–based yogurt and purple sweet potato extract–supplemented yogurt.

Tari et al. [1] Reported that purple sweet potato extract–supplemented yogurt made from commercial cultures and indigenous probiotics, such as *Streptococcus thermophilus* FNCC 0040, *Lactobacillus bulgaricus* FNCC 0041, and *Lactobacillus plantarum* Dad 13 by the ratio 1: 1: 0.5 had nowbeen available completed with the physical properties (pH = 3.78, viscosity = 5.198 7 cP, chromatic color = 18.559) and chemical properties (titrated acid content = 1.273 3 %, moisture content = 85.266 4 %, ash content = 0.804 1 %, reduction sugar level = 3.327 8 %, dissolved protein content = 1.478 2 %, fat content = 0.08 %, and anthocyanin level = 8.531 5 %).

Tari et al. [2] also shared that indigenous probiotics, such as *L. plantarum* Dad 13 supplemented to the purple sweet potato extract yogurt, could act as a diarrhea–lowering agent in experimental animals. The discovery was indicated by the decreasing water content of the experimental animals' feces from 63.32 % to 62.73 %, in addition to the decreasing water content of the experimental animals' cecum from 83.31 % to 35.13 %. The indigenous probiotics are also able to reduce free radical components as indicated by the decreasing blood MDA level of the experimental animals from 4.23 mmol mL^{-1} to 1.52 mmol mL^{-1} and the decreasing liver MDA level at from 5.60 mmol mL^{-1} to 2.96 mmol mL^{-1} at the end of the study.

According to [3], fermented milk products must be stored at a temperature of less than 10 $^{\circ}$C to avoid negative environmental impacts. The storage at low temperature functions to inhibit the fermentation process thus will maintain a high number of microbes. This treatment, however, will result in high storage and distribution costs deal with the challenges, freeze–dried synbiotic yogurt could offer the solution.

Cruz [4] explained that the freeze–drying process could reduce the number of bacteria by 1 log cycle, as the process allows the microbial death (sublethal). Therefore, the addition of protective substances (*cryoprotectant*) before the freeze–drying process is required to minimize the damage.

The common use of coating material (*cryoprotectant*) derives from encapsulants, such as gum, carbohydrates, and protein to coat the core materials (bacteria) with specific purposes, such as to cover up the bad taste and odor, protect against the environmental influences, increase the stability, and prevent evaporation. This *cryoprotectant* material can be either carbohydrates or protein.

According to [5] the use of protein as a *cryoprotectant* can maintain the bacterial resistance, while the use of carbohydrates can both improve microcapsule texture and maintain the bacterial resistance. Bhat [6] stated that sucrose could function as a protective material for bacteria. It is safe for consumption and able to increase sweetness. The addition of sucrose as a *cryoprotectant* function to protect the structure and proteins in microbial cells [7].

However, there have yet been available any relevant studies that proved the function of sucrose as a *cryoprotectant* agent in lactate–fermented drinks such as indigenous probiotic yogurt (local) produced from sweet potato supplementation through a freeze–drying process to the viability of Lactic Acid Bacteria (LAB) and *L. plantarum* Dad 13, in addition to its characteristics. The notion underlies this research background.

This research aims to signify the effect of sucrose concentration as *cryoprotectant* on the yields, water content, reduction–sugar level, total LAB, and total *L. plantarum* Dad 13 of the freeze–dried lactate–fermented drinks.

2 Materials and methods

2.1 Materials

The research used a number of materials, which consisted of LAB cultures from the Universitas Gadjah Mada (UGM) Yogyakarta, Indonesia FNCC (Food and Nutrition Culture Collection). The bacterial cultures were identified in upright shapes, including *S. thermophilus* FNCC 0040, *L. bulgaricus* FNCC 0041, and *L. plantarum* Dad 13 as probiotic indigenous LAB cultures which were collected from UGM Researchers Association. The MRS (de Mann Rogosa Sharpe) Agar/Broth (Oxoid) media was utilized for the LAB culture maintenance. Meanwhile, the LPS media was utilized as the selective media for the *L. plantarum* Dad 13 growth. Other supporting chemical substances were 70 % of alcohol, methylated spirit, and distilled water that were obtained from the Biological–Chemistry and Microbiology Laboratory, Faculty of Agriculture, University of Veteran Bangun Nusantara Sukoharjo, Central Java, Indonesia. The research utilized several instruments, including glasswares (test tubes, beaker cups, Erlenmeyer flasks, and petri dish), autoclaves (All America), an incubator, an oven (binder), and a laminar airflow (LAF).

2.2 Research procedure

The research procedures ranged from the preparation of bacterial culture stocks, starter culture, purple sweet potato extract, and probiotic yogurt through supplementation of purple sweet potato extract, and freeze–drying process.

2.2.1 Preparation of bacterial culture stocks

The commercial bacteria (*S. thermophilus* FNCC 0040 and *L. bulgaricus* FNCC 0041) and *L. plantarum* Dad 13 as indigenous probiotic LAB cultures in the form of agar slants were taken one ose and grown in a 10 mL sterile liquid MRS. The cultures were then etched in a sterile agar slant and incubated at 37 °C during 24 h to 48 h. The stocks of bacterial cultures in the agar slants were stored in a refrigerator at 2 °C to 3 °C and regenerated once every 2 wk.

2.2.2 Preparation of purple sweet potato extract

The procedure of purple sweet potato extraction referred to the preliminary research conducted by [8]. The procedure was begun by dice–cutting the sweet potatoes in a 5 cm × 5 cm cube size and extracting the cuts in a juicer. The product was then left to stand for 24 h at a 4 °C temperature, so the purple sweet potato starch would settle, while the filtrate would remain. The filtrate of the purple sweet potato extract would be taken to produce theyogurt.

2.2.3 Preparation of starter cultures

This stage included the preparation of 5 mL sterile liquid MRS media in three tubes. Each tube was then inoculated with the upright–shaped bacterial cultures, including *S. thermophilus* FNCC 0040, *L. bulgaricus* FNCC 0041, and *L. plantarum* Dad 13. All of the isolated cultures were incubated at a 37 °C temperature for 24 h. To produce the starter cultures, every 0.1 mL of the cultures were then inoculated into 5 mL sterile skimmed milk and incubated at a 40 °C temperature for 7 h to 8 h or at a 37 °C temperature for 24 h. From this phase, the total production of LAB and *L. plantarum* Dad 13 as the starter cultures reached by 109 CFU mL^{-1}.

2.2.4 Preparation of purple sweet potato extract–supplemented yogurt

Fresh milk, skimmed milk (5 % w v^{-1}), and purple sweet potato extract (10 % v v^{-1}) were pasteurized at the temperature ranged from 72 °C to 80 °C for 30 min and cooled at the temperature ranged from 40 °C to 45 °C. The product was aseptically inoculated with 5 % (v v^{-1}) *S. thermophilus*, *L. bulgaricus,* and *L. plantarum* Dad 13 probiotic bacteria by the ratio 1:1:0.5 at the temperature ranging from 40 °C to 45 °C. The product was then shaken until it reached a homogeneous state and incubated at a 40 °C temperature for 8 h or at a 30 °C temperature for 20 h to produce purple sweet potato extract–supplemented yogurt.

2.2.5 Freeze–drying process for lactate–fermented drink

The yogurt produced from the supplementation of purple sweet potato extract and probiotics (*L. plantarum* Dad 13) was added with sucrose (consecutively based on the treatment) and skimmed milk (10 % v v^{-1}). The yogurt was then aseptically stirred until it reached a homogeneous state. The mixture was put into freeze–dryer glass tubes and frozen at a –20°C temperature for 12 h. The product was dried in a freeze–dryer for 10 h.

2.2.6 Research design

The research used a Completely Randomized Design (CRD) through a single treatment by including sucrose (S) as a *cryoprotectant* with the variations of concentration percentage, including S_1= 0 %, S_2= 2.5 %, S_3= 5 %, and S_4= 7.5 %. The study was conducted by three times of replications, thus 12 experimental units were produced. The observation parameters consisted of the viability of *L. plantarum* Dad 13 before and after the freeze–drying treatment, total viability of LAB before and after the freeze–drying treatment, moisture content, solubility, and reduction–sugar level. The data were analyzed using the RAL One–Way ANOVA. The treatment would proceed to the DMRT test, if the result showed a real significance [9].

2.2.7 Observation parameter

The experiment tested the fermented products and dried products in order to find out: The yields of freeze–dried purple sweet potato extract synbiotic yogurt, the result of moisture content test through a thermogravimetric method [10], the result of reduction–sugar level test through the Nelson Somogyi method [11], the total LAB before and after the freeze–drying process, and the viability of *L. plantarum* Dad 13. All of the tests utilized the Bacteriological Analytical Manual (BAM) method (2002).

3 Result and discussions

3.1 Yields of purple sweet potato extract–supplemented synbiotic yogurt

The yields of purple sweet potato extract–supplemented synbiotic yogurt (PSPE–SSY) was the final product following the freeze–drying process. The content of yields (Y) was generated by the formula (1):

$$\text{Y (Yield)}=\frac{\text{PSPE-SSY after freeze drying}}{\text{PSPE-SSY after freeze drying}}\times 100\ \% \quad (1)$$

The addition of 10 % skimmed milk as a *cryoprotectant* combined with various concentrations of sucrose gave a significant effect ($p < 0.05$) to the yields of the freeze–dried yogurt. The statistical test result was explained in Figure 1.

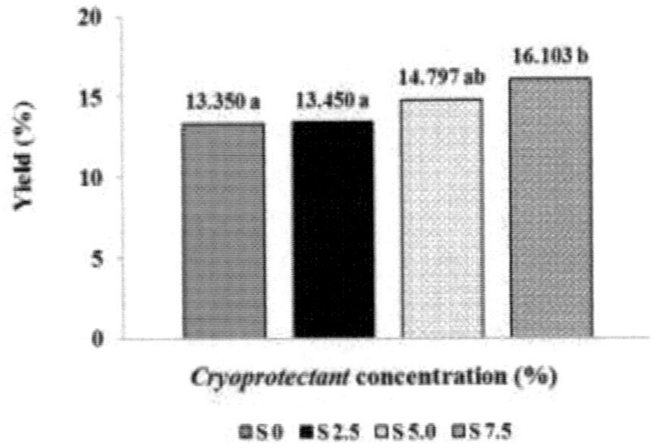

Fig. 1. Impact of various sucrose concentrations on yields of freeze–dried purple sweet potato extract–supplemented synbiotic yogurt

Figure 1 signifies that a higher sucrose concentration resulted in a higher quantity of the freeze–dried synbiotic yogurt yields. The condition was due to the combination of skimmed milk and sucrose to increase the volume and total of material solids. This research was relevant to [11] who concluded that the increase of yields was affected by the large amount of *cryoprotectant* that resulted in a higher total of material solids. A high total of material solid correlates with a large quantity of yields.

3.2 Moisture content

The moisture content of the final products became an important parameter in making instant products. The combination of 10 % skimmed milk and various concentrations of sucrose as a *cryoprotectant* indicated no impact on the moisture content of the freeze–dried purple sweet potato extract–supplemented synbiotic yogurt ($p > 0.05$) as shown in Figure 2. Such a condition was possible due to the small interval of sucrose concentration used as a *cryoprotectant*, thus offered less impact on the moisture content of the product. The use of sucrose was *cryoprotective* as a drying medium, as it can form porous structures to facilitate the release of the product moisture during the freeze–drying process and become the medium at the rehydration process [12].

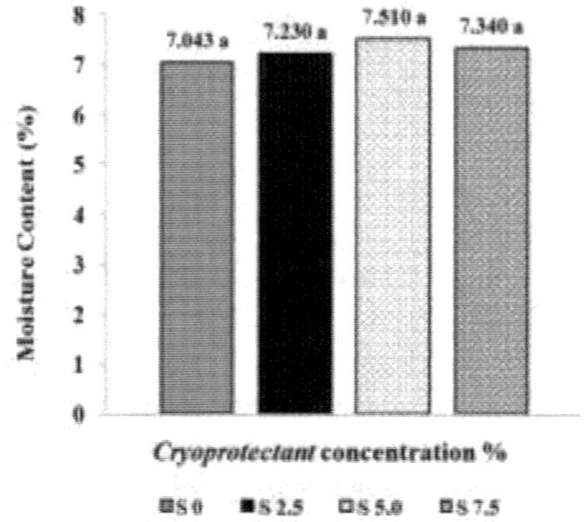

Fig. 2. Impact of various sucrose concentrations on moisture content of freeze–dried purple sweet potato extract–supplemented synbiotic yogurt.

3.3 Reduction–sugar level

A reduction–sugar is a group of monosaccharides and disaccharides that have free–reducing groups, such as glucose, fructose, lactose, galactose, and maltose [13]. Figure 3 showed that the concentration of the *cryoprotectant* made from combination of 10 % skimmed milk and various sucrose concentrations had a significantly different impact ($p < 0.05$) on the reduction–sugar of the freeze–dried purple sweet potato extract–supplemented synbiotic yogurt.

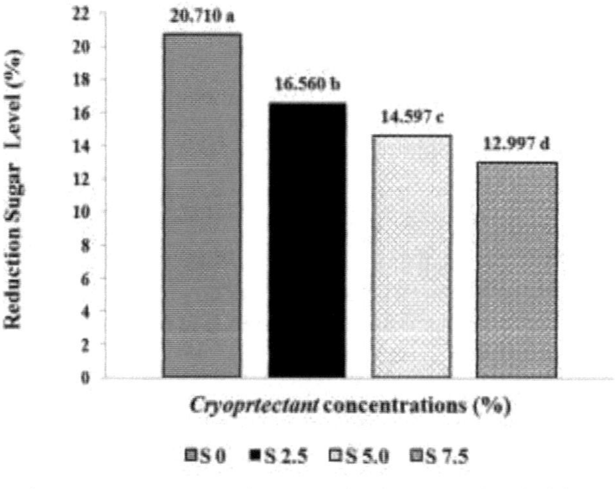

Fig. 3. Impact of various sucrose concentrations on reduction–sugar level of freeze dried purple sweet potato extract–supplemented synbiotic yogurt

The decreasing reduction–sugar level was considered due to its use as the energy source for the bacterial cell growth and propagation, in addition to the formation of bacterial metabolites during the freeze–drying process. Such a condition was concluded from the total average of LAB that ranged from 12.93×10^8 CFU mL^{-1} to 39.77×10^8 CFU mL^{-1} or around 10^9 CFU mL^{-1} before the freeze–drying process and decreased by 1 log cycle ranging from 2.75×10^8 CFU mL^{-1} to 9.28×10^8 CFU mL^{-1} after the freeze–drying. The total average of *L. plantarum* before the freeze–drying process ranged from

8.23×10^8 CFU g^{-1} to 14.73×10^8 CFU g^{-1} or around 10^8 CFU mL^{-1}, while after the freeze–drying process, the total average of *L. plantarum* ranged from 1.74×10^8 CFU mL^{-1} to 6.81×10^8 CFU mL^{-1} following the addition of *cryoprotectant*. The data confirmed the findings shared by [14] that lactic acid bacteria utilized sugar as the energy source for its growth and produced lactic acid during the fermentation process.

3.4 The total of Lactic Acid Bacteria (LAB)

The measurement of the total of LAB is an important parameter, as it closely relates to the amount of skimmed milk and sucrose as the *cryoprotectant*. Li et al. [15] stated that a good protector in the freeze–drying process ought to be *cryoprotective*, easy to dry, able to form a good matrix to maintain cell stability, and easily rehydrated. Sucrose is one of the *cryoprotectants* that meet with these criteria. The effect of sucrose concentration as *cryoprotectant* on the total LAB was explained in Figure 4.

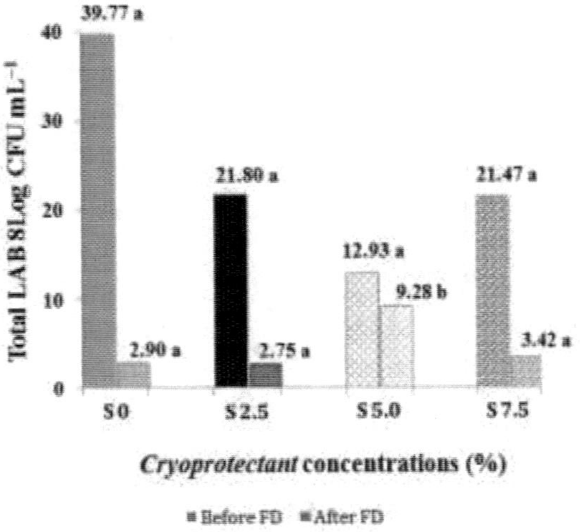

Remarks:Numbers with the same alphabetical notations signify real indifferent results in DMRT test with an α–value = 0.05

Fig. 4. The total of LAB in synbiotic yogurt before and after freeze–drying process

Figure 4 showed that the total of LAB before the freeze–drying process ranged from 12.93×10^8 CFU mL^{-1} to 39.77×10^8 CFU mL^{-1} or around 10^9 CFU mL^{-1} with no significant difference ($p > 0.05$) at various sucrose concentrations. After the freeze–drying process, the total of LAB decreased by 1 log cycle that varied from 2.75×10^8 CFU mL^{-1} to 9.28×10^8 CFU mL^{-1} with a significant difference ($p < 0.05$) at various sucrose concentrations.

The total of LAB at the 5 % sucrose concentration as the *cryoprotectant* showed a significantly different number among other treatments. The reduction in the total of LAB at the 5 % sucrose concentration before and after the freeze–drying process showed the smallest number which was less than 1 log cycle. It was suspected that sucrose as the *cryoprotectant* could protect the structure and function of microbial cell proteins [7]. Additionally, sucrose could also function to strengthen the cell resistance to the freezing condition. The mechanism of *cryoprotectant* function in the reaction of cell preservation is marked by, i) the decreasing freezing point of the *cryoprotectant* medium, ii) the protective reaction of cell membranes, and iii) the suppressing rate of a high concentration effect [15].

An over–high concentration of sucrose (7.5 %) could lead to the osmotic imbalance inside and outside the bacterial cells. The condition would potentially spur bacterial lysis and cause bacterial death. Figure 4 signifies a lower total of LAB at the 7.5 % sucrose concentration compared to the 5 % sucrose concentration.

3.5 The total of *L. plantarum* Dad 13

The enumeration of *L. plantarum* Dad 13 in the freeze–dried purple sweet potato extract–supplemented synbiotic yogurt utilized the LPSM (*L. plantarum* Selective Medium). It is a simple medium which is sensitive to *L. plantarum* and other potentially probiotic *Lactobacillus* species [15,16]. The medium contains a 4 µg mL^{-1} concentration of ciprofloxacin, as an antibiotic that functions to inhibit most of the infectious bacteria, including the endogenous acid lactic bacteria that grow in MRS agar, however, has no adverse effect on the recovery of *L. plantarum*. The effect of sucrose concentration as the *cryoprotectant* on the total of *L. plantarum* is explained in Figure 5.

Figure 5 showed that the total average of *L. plantarum* before the freeze–drying process ranged from 8.23×10^8 CFU g^{-1} to 14.73×10^8 CFU g^{-1} or around 10^8 CFU mL g^{-1} with a significant difference ($p < 0.05$) among various sucrose concentrations. After the freeze–drying process, the total average of *L. plantarum* following the addition of *cryoprotectant* at various sucrose concentrations (0 %, 2.5 %, 5 %, and 7.5 %) ranged from 1.74×10^8 CFU mL^{-1} to 6.81×10^8 CFU mL g^{-1} with a significant differences ($p < 0.05$) among various sucrose concentrations. At the control treatment (S_1= 0 %), the total average of *L. plantarum* decreased by 1 log cycle following the freeze–drying phase. The total average of *L. plantarum* after the sucrose treatment at various concentrations from 2.5 %, 5 %, to 7.5 % also decreased, however not by 1 log cycle.

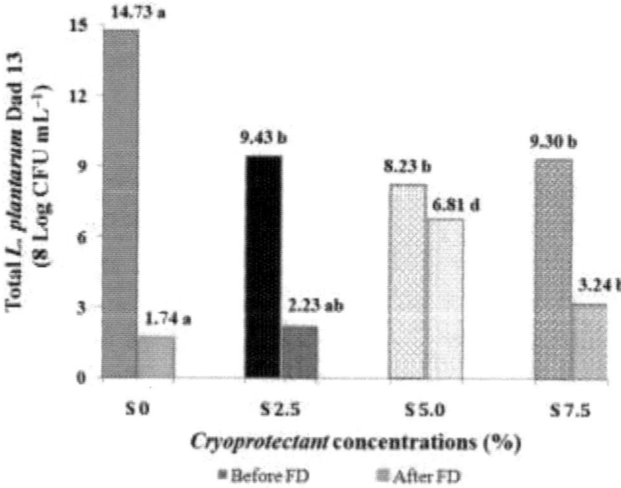

Remark: Numbers with the same alphabetical notations signify real indifferent results in DMRT test with anα–value = 0.05.

Fig. 5. The total of *L. plantarum* of synbiotic yogurt before and after freeze–drying process

Figure 5 indicated that the decreasing cell viability was probably due to the freeze–drying process. The freezing process caused the cell to lose its stability, thus it became easily damaged during the drying process. The osmotic shock was suspected as the main factor of the bacterial cell damage during the drying process, as signified by the membrane damage and displacement of hydrogen bonds that affected the hydrophilic macromolecule properties of the bacterial cells [17].

Another factor that affected the cell viability was the selection of *cryoprotectant* materials. According to [18], the use of two types of encapsulants from protein and carbohydrate matrices can produce a higher efficiency compared to the use of one type of encapsulant. It happens as the encapsulants can interact in forming granules, thus they can better overlay the encapsulated components.

The skimmed milk which contains lactose also provides good protection against the freeze–drying effects. The components of lactose in the form of glucose and galactose include simple sugars with a low molecular weight that results in easy entry intobacterial cells and protect the two sides of the cell membranes during the freeze–drying process. The presence of sucrose as a *cryoprotectant* could protect the structures and functions of microbial cell proteins [7], thus sucrose could improve cell resistance in freezingconditions.

In addition to the external factors, internal factors also affect viability. *L. plantarum* bacteria are gram–positive bacteria. According to [19], the g–positive bacteria have a cell composition that consists of 90 % peptidoglycan, lipids (1 % to 4 %), teichoic acid, and other components. The composition caused the gram–positive bacteria to be more resistant to physical and enzymatic treatments, such as the freeze–drying process than the gram–negative bacteria.

4 Conclusions

The various sucrose concentrations significantly affected the yields of the freeze–dried purple sweet potato extract–supplemented synbiotic yogurt, reduction–sugar level, the total of LAB, and the total of *L. plantarum* before and after the freeze–drying process. However, there were indicated no significant effects on the moisture content and total quantity of LAB. The best treatment was shown in the freeze–dried purple sweet potato extract–supplemented synbiotic yogurt with a 5 % of sucrose concentration as the *cryoprotectant*. The treatment using a 5 % of sucrose concentration resulted in the following characteristics: A 14.797 % of yields, a 7.51 % of water content, a 14.59 % of reduction–sugar level, a 1.98×10^9 CFU mL^{-1} of total LAB before the freeze–drying and a 9.28×10^8 CFU mL^{-1} after the freeze–drying, a 8.23×10^8 CFU mL^{-1} of total *L. plantarum* before the freeze–drying and a 6.81×10^8 CFU mL^{-1} after the freeze–drying.

The researchers acknowledge the Directorate of Research and Public Submission of the Ministry of Research, Technologies, and Higher Education Affairs due to the 2018 National Strategic Research Grants. No. SP DIPA-042.06.1.401516/2018. December 05 2017.

References

1. A.I.N. Tari, C.B. Handayani, A.M.W. Mulyono. *Bakteri probiotik terhadap mutu yogurt ekstrak ubi jalar ungudan pemanfaatannya sebagai pencegah diare dan imunomodulator* [Probiotic bacteria on the quality of yogurt of purple sweet potato extract and its use as a prevention of diarrhea and immunomodulators]. Laporan Penelitian Hibah Bersaing. Universitas Veteran Bangun Nusantara Sukoharjo (2013). p.1–95. [in Bahasa Indonesia]. http://lppm.univetbantara.ac.id/data/materi/Laporan%20PHB-I-%20Fix.pdf
2. A.I.N Tari, C.B. Handayani, Sudarmi. *Bakteri probiotik terhadap mutu yogurt ekstrak ubi jalar ungu dan pemanfaatannya sebagai pencegah diare dan imunomodulator* [Probiotic bacteria on the quality of yogurt of purple sweet potato extract and its use as a prevention of diarrhea and immunomodulators]. Laporan Kemajuan Penelitian Hibah Bersaing. Universitas Veteran Bangun Nusantara

Sukoharjo (2014). p. 1–34. [in Bahasa Indonesia]. http://lppm.univetbantara.ac.id/data/materi/Laporan%20Kemajuan%20PHB%20tahun%20ke-2%20Agustina%20Intan%20Niken%20Tari.pdf

3. M.H.B. De Almeida, S.S. Zoellner, A.G. Da Cruz, M.R.L. Moura, L.M.J. De Carvalho, M.C.J. Freitas, et al. Int. J. Dairy Technol. **61**: 178–182(2008). https://doi.org/10.1111/j.1471-0307.2008.00390.x
4. A.G. Cruz, E.H.M. Walter, R.S. Cadena, J.A.F. Faria, H.M.A. Bolini, H.P. Pinheiro, et al. Food Res. Int. **43**,5:1444–1448(2010). http://dx.doi.org/10.1016/j.foodres.2010.04.028
5. R. Heni, B.S.L. Jeni, N. Nurhidayat, C. Nurwitri. Anim. Prod. **10**,3: 179–187(2008). http://repository.ipb.ac.id/handle/123456789/9429
6. A.R. Bhat, V.U. Irorere, T. Bartlett, D. Hill, G. Kedia, M.R. Morris, et al. AMB Expr. **3**:1–9(2013). https://dx.doi.org/10.1186%2F2191-0855-3-36
7. C.A. Morgan, N. Herman, P.A White, G. Vesey. J. Microbiol. Methods. **66**,2: 183–193(2006). https://doi.org/10.1016/j.mimet.2006.02.017
8. A.I.N. Tari, C.B. Handayani, A.K. Sariri. Jurnal Teknologi Hasil Pertanian UNS. **1**:1–7(2012). [in Bahasa Indonesia]. https://jurnal.uns.ac.id/ilmupangan/article/download/13530/11276
9. J. Sarwono, H.N. Salim. *Prosedur-prosedur populer statistik untuk analisis data riset skripsi*. [Popular statistical procedures for thesis research data analysis]. Gaya Media. Yogyakarta (2017). [in Bahasa Indonesia]. https://www.semanticscholar.org/paper/Prosedur-prosedur-Populer-Statistik-untuk-Analisis-Salim/ea2d356588baa0ecbd1365dd5f12e464eceb96ea
10. H.N. Al–Mentafji. AOAC. *Official Methods of Analysis*, 18th ed. AOAC International, Maryland (2005). p. 28. https://www.researchgate.net/publication/292783651_AOAC_2005
11. S.S. Endang, Prasetyastuti. Jurnal Farmasi Kedokteran. **3**,1:353–362(2010). [in Bahasa Indonesia]. https://docplayer.info/30784494-Pengaruh-pemberian-juice-lidah-buaya-aloe-vera-l-terhadap-kadar-lipid-peroksida-mda-pada-tikus-putih-jantan-hiperlipidemia.html
12. P. Bordat, A. Lerbret, J.P. Demaret, F. Affouard, M. Descamps. EPL. **65** (2004). https://iopscience.iop.org/article/10.1209/epl/i2003-10052-0/pdf
13. V. Mougios. *Exercise Biochemistry*,2nd Ed. USA: Canada (2019). p. 1–472. https://books.google.co.id/books?id=udCMDwAAQBAJ&printsec=frontcover&dq=V.+Mougios.+Exercise+Biochemistry,+2nd+Ed.+(2019)&hl=en&sa=X&ved=2ahUKEwjyj5OC8ebqAhUzmuYKHfhdD3cQ6AEwAXoECAUQAg#v=onepage&q&f=false
14. A.S. Carvalho, J. Silva, P. Ho, P.H.P. Teixeira, F.X. Malcata, P. Gibb. Journal of INRA, EPD Science. **83**,3:203–210(2003). https://doi.org/10.1051/lait:2003010
15. P. Li, Q. Gu, X. Wu. Protein Expr. Purif. **126**:115–121(2016). https://doi.org/10.1016/j.pep.2016.06.007
16. V. Coeuret, M. GueGuen, J.P. Vermoux. J. Dairy Res. **71**,4:451–460(2004). https://doi.org/10.1017/S0022029904000469
17. P. Kaushik, K. Dowling, C.J. Barrow, B. Adhikari. J. Funct. Foods. **19**: 868–881(2015).https://doi.org/10.1016/j.jff.2014.06.029
18. A. Munin, F. Edwards–Lévy. Pharmaceutics. **3**,4:793–829(2011). https://doi.org/10.3390/pharmaceutics3040793
19. S. Brown, J.P.S. Maria, S. Walker. Annu. Rev. Microbiol. **67**:313–336(2013). https://doi.org/10.1146/annurev-micro-092412-155620

Biomass Enhancement of *Stevia rebaudiana* Bertoni Shoot Culture in Temporary Immersion System (TIS) RITA® Bioreactor Optimized in Two Different Immersion Periods

Agustine Christela Melviana[1*], *Rizkita* Rachmi Esyanti[1], *Maizirwan* Mel[2], and *Roy* Hendroko Setyobudi[3]

[1]Department of Plant Science and Biotechnology, School of Life Sciences and Technology, Bandung Institute of Technology, Jl. Ganesha No. 10, Bandung 40132, West Java, Indonesia
[2]Department of Biotechnology Engineering, Kulliyyah of Engineering, International Islamic University Malaysia, P.O. Box 10, 50728 Kuala Lumpur, Malaysia
[3]Department of Agriculture Science, Postgraduate Program, University of Muhammadiyah Malang, Jl. Raya Tlogomas No. 246, Malang, 65145, East Java, Indonesia

Abstract. Stevia plant contains steviol glycosides, which are estimated to be 300 times sweeter than sucrose. However, conventional (*in vivo*) propagation of *Stevia rebaudiana* in Indonesia was not effective due to poor results. Therefore, an alternative method to propagate stevia plants is needed. One of them is by using *in vitro* method. Multiplication with a large quantity of stevia biomass in a relatively short period can be conducted by using TIS RITA® (Recipient for Automated Temporary Immersion System). This study aimed to evaluate the effect of the immersion period of the medium on growth and the medium bioconversion into the production of shoot biomass. The bioreactors were set up with 15 min and 30 min immersions periods, scheduled every 6 h, and incubated for 21 d. The result indicated that the immersion period affected the biomass and growth rate (μ). Amount of 30 min immersion showed a greater percentage of shoot multiplication, higher biomass, percentage of leaf growth, growth rate, and productivity compared to 15 min immersion. The pattern of sucrose, mineral, and inorganic compounds consumption followed the growth of plant biomass for both systems. In conclusion, 30 min immersion gave a greater efficiency medium bioconversion to plant biomass compared with 15 min immersion.

Key words: Diabetes prevention, micropropagation, natural sweetener, sweetener plant, tissue culture

1 Introduction

Indonesia has been classified as one of the top 10 countries worldwide with a high number of who has diabetes in 2013. The same pattern is predicted to continue unless diabetes prevention and management interventions are taken [1]. The rising prevalence of diabetes is

* Corresponding author: rizkita@sith.itb.ac.id

highly caused by the high rate of foods and drink consumption that contains high sugar and artificial sweeteners, such as aspartame, saccharin, and cyclamate [2]. Therefore exploration of natural sweetener that contains low calorie is needed. Stevia (*S. rebaudiana*) is a herbaceous perennial shrub in Indonesia that contains several sweet-tasting compounds in its leaves. The main components that provide a sweet taste are stevioside, which have been estimated to be 300 times as sweet as sucrose [3]. Stevioside has been shown to possess beneficial effects on human health and can be used as a natural sweetener, especially for diabetic patients [4].

Stevia is commonly propagated using in vivo method, but this method was not effective. Seed propagation was not efficient due to low fertility [5], great variability on the stevioside level and composition [6], and poor germination rates (below 10 %) [7]. *In vitro* culture provide the best method to overcome those problems and has the potential to produce a large number of stevia plantlets in a short time.

In vitro cultivation method has been widely applied in various plants. This method requires a closed sterile container, where culture is given certain nutrients, either in solid or liquid form. Various studies on the use of in vitro culture methods for biomass propagation has been done, including in vitro culture methods in TIS (Temporary Immersion System) bioreactors [8]. The bioreactor is a modified liquid culture system where the immersion of explant happens periodically. In culture systems with temporary tissue immersion, it is clear that the immersion time is very important and decisive [9] since this factor determines nutrient uptake, which affects the biomass enhancement.

A previous study on the cultivation of stevia on TIS RITA® with immersion periods 15 min for 6 h had been conducted and proven significantly increased stevia biomass. Shoot multiplication was greatly increased, with 30 new healthy shoots produced from each bud clump, resulted in approximately 90 new stevia shoots per vessel. In comparison with semi-solid media with the same supplemented PGR, TIS produced 2.00 times more biomass production [10]. Referring to the previous research, this study was conducted to determine the potential of *S. rebaudiana* cultivation in the TIS RITA® bioreactor system, by evaluating the effect of immersion period of the medium on the growth rate, productivity, the rate of nutrient uptake by *S. rebaudiana*, and bioconversion medium into biomass during cultivation.

2 Material and methods

2. 1 Culture mmedium preparation

The medium was prepared for the initiation, subculture and acclimatization and also bioreactors. The initiation medium was consisted of MS salts, 30 g L^{-1} sucrose, and added with several plant growth regulators (PGRs) for optimization, i.e: 0.5 mg kg^{-1} Benzyl Amino Purine (BAP); 0.6 mg kg^{-1} BAP, 0.7 mg kg^{-1} BAP; 1 mg kg^{-1} Kinetin; and 2 mg kg^{-1} Kinetin. Subculture medium has a similar composition to the initiation medium but were added with 1 mg kg^{-1} Kinetin. Acclimatization and bioreactors medium as the best initiation medium consisted of half-strength MS salts, 1 mg kg^{-1} Kinetin, and 30 g L^{-1} sucrose. The pH of each medium was adjusted to 5.8 and autoclaved at 121°C for 15 min. Then, the sterilized media was added by 200 mg kg^{-1} Augmentin.

2.2 Shoot culture initiation of Stevia

Stevia rebaudiana Bertoni plants were collected from the Research Center for Tea and Quinine, Gambung, Indonesia. The tips of young, actively growing shoot tips and nodal segments were cut and washed three times with tap water to remove the dust particles and other contaminants. The explants were drenched with 3 mg kg^{-1} fungicides HEKSA (Heksakonazol 5 %) for 3 min and followed by water. In a laminar airflow chamber (LAF), the washed explants were subjected to surface sterilization with ethanol 70 % (v/v) for 1 min. Shoots were dipped in NaClO 5 % and Tween-20 for 5 min, but nodes were dipped in NaClO 10 % and Tween-20 for 10 min. Both explants were rinsed three times with sterile distilled water, placed in Petri dishes lined with filter paper, and transferred into a solid medium. The cultures were incubated at 23 °C ± 1 °C with a 16 h photoperiod. Subcultures were done after 30 d intervals. Nodal segments from the proliferated shoots were subcultures for further multiplication of shoots. The regenerated multiple shoots were cut, and individual shoots were placed in semi-solid MS medium supplemented with 1 mg kg^{-1} Kinetin.

2.3 Stevia shoots acclimatization in the liquid medium

S. rebaudiana shoots in the solid medium were sub-cultured into thin layer liquid medium culture aseptically. Approximately 10 shoots of stevia were cultivated in 5 mL of medium and incubated on a shaker and shaken at 40 rpm at room temperature (25 °C) for 3 d, under 2 000 lx to 2 500 lx ordinary white fluorescent lamp for 16/8 h daily photoperiod.

2.4 Shoot culture in RITA® bioreactor

Shoots were selected from thin layer liquid medium culture and ready to be cultivated in 250 mL media. Bioreactors were assembled and the airflow was adjusted by an automatic timer with two variations, which gave two different immersion periods: (i) 15 min immersion time; (ii) 30 min immersion time, scheduled every 6 h.

2.5 Plantlet and medium analysis

After 21 d of cultivation, plantlets were harvested, counted (number of shoots and leaves), weighed, and then dried to obtain data of the dry weight data. Final sucrose content and conductivity of the culture medium were also tested. The conductivity was measured by using a conductivity meter (Eutech Instruments Con-110) and the sucrose content in the medium was measured by using a refractometer (Milwaukee MA871).

2.6 Data analysis

Data of dry weight were transferred into the growth curve. Data of sucrose content and conductivity of medium were used to construct a mass distribution model of *Stevia rebaudiana* shoot culture in TIS RITA® bioreactor. The mass balance modeling was based on the following reaction in Formula (1) [11]:

$$0.39\ C_{12}H_{22}O_{11} + 0.23\ NH_4NO_3 + 3.4\ 3O_2 \rightarrow CH_{1.27}O_{0.43}N_{0.45} + 4.07\ H_2O + 3.64\ CO_2 \quad (1)$$

The growth and sucrose consumption curves were developed based on the following Equation (2) to Equation (4):

a. Growth kinetics
Logistic equation [10]:

$$\frac{d}{dt}[X] = \mu X\left(1 - \frac{X}{X_{max}}\right) \quad (2)$$

$$\mu = \mu_{max} * \left(\frac{[C_{12}H_{22}O_{11}]}{k_{C_{12}H_{22}O_{11}} + [C_{12}H_{22}O_{11}]}\right) \quad (3)$$

b. Sucrose consumption [12]:

$$\frac{d}{dt}[C_{12}H_{22}O_{11}] = -q_{C_{12}H_{22}O_{11}} * X$$

$$q_{C_{12}H_{22}O_{11}} = \frac{\mu}{Y_{x/C_{12}H_{22}O_{11}}} \quad (4)$$

$$Y_{x/C_{12}H_{22}O_{11}} = \Delta X / \Delta[C_{12}H_{22}O_{11}]$$

S. rebaudiana productivity in RITA® system was determined by dividing the addition of dry biomass weight (g) with the volume of medium (L) and cultivation time (d).

3 Results and discussions

The effect of various PGRs on the in vitro growth of *S. rebaudiana* was evaluated in terms of shoot number and leaf number (Figure 1 and Table 1). Kinetin 2 mg kg^{-1} produced the highest number of shoot multiplication but not followed with the highest multiplication of stevia leaves. PGR BAP produced high multiplication of shoots and leaves but the explants morphology was not normal and not healthy. The explants cultured with the addition of BAP had small brownish stems (Figure 1 (C–E)). BAP is widely used to increase the number of shoots in stevia multiplication [13]. However, a large number of shoots in the BAP medium showed abnormal phenotypes such as brownish and non-vigor stems. Meanwhile, 1.0 mg kg^{-1} Kinetin showed quite a high number of shoots and leaves multiplication with vigor stem and normal looking leaves. Kinetin 1 mg kg^{-1} was decided to become the PGR used in further experiment.

Table 1. Differences in the number of stevia shoots and leaves in various types of PGR.

PGR	∑ Number of Shoots ± SD	∑ Number of Leaves ± SD	Visual Description
1 mg kg^{-1} Kinetin	46 ± 2.243	195 ± 4.219	Greenish stems and leaves
2 mg kg^{-1} Kinetin	70 ± 2.360	156 ± 2.984	Greenish stems and leaves
0.5 mg kg^{-1} BAP	64 ± 1.569	201 ± 3.087	Brownish stems
0.6 mg kg^{-1} BAP	48 ± 1.621	210 ± 5.368	Brownish stems
0.7 mg kg^{-1} BAP	48 ± 1.581	154 ± 3.399	Brownish stems

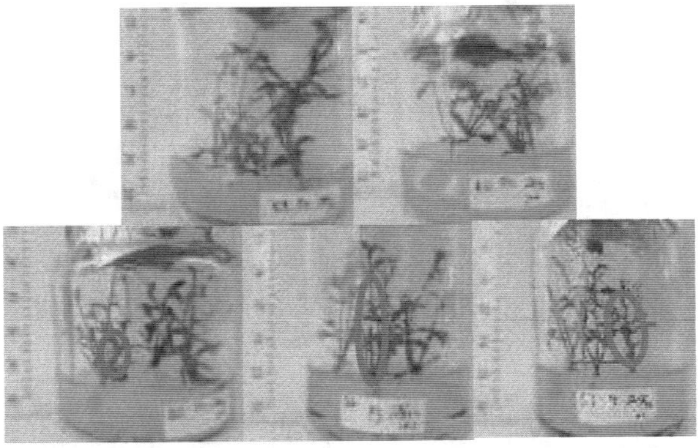

Fig. 1. Growth of Stevia shoots cultivated in a solid medium with different concentrations of PGRs: (A) 1 mg kg^{-1} Kinetin, (B) 2 mg kg^{-1} Kinetin, (C) 0.5 mg kg^{-1} BAP, (D) 0.6 mg kg^{-1} BAP, and (E) 0.7 mg kg^{-1} BAP.

The result of the initiation process showed that 70 shoots meristem (either apical shoots or lateral shoots) could proliferate into 243 shoots with 89.74 % percentage of success. Multiplication percentage of initiation reached 352.86 %, or in other words, one explant could proliferate into three new shoots. The subculture phase showed that 247 shoots could proliferate into 610 shoots with a success rate of 98.34 %. Multiplication percentage of subculture was 246.96 %, or in other words, an average of one shoot could proliferate into two new shoots. These shoots were then acclimatized in thin layer culture.

S. rebaudiana shoots were then cultivated in RITA® bioreactor for 3 wk. According to Table 2, RITA® 30 (intensity periods for 30 min) gave the highest multiplication of stevia shoots compared with RITA® 15 (intensity periods for 15 min), with an average percentage increase of shoots for RITA® 30 was 89.71 % ± 5.83 %, while RITA® 15 only 76.90 % ± 4.85 %. Percentage of leaf growth in RITA® 30 reached 85.24 % ± 5.99 % while in RITA® 15 only 79.73 % ± 7.76 %.

Table 2. The increasing number of shoots and leaves produced bioreactor RITA® with immersion periods 15 min and 30 min

TIS RITA®	Average Initial Number of Shoots ± SD	Average Number of Shoots Produced ± SD	Increasing percentage number of shoots ± SD	Average Initial Number of Leaves ± SD	Average Number of Leaves Produced ± SD	Increasing percentage number of leaves ± SD
15	1.53 ± 0.46	6.55 ± 0.64	76.90 ± 4.85	3.93 ± 1.31	19.58 ± 1.02	79.73 ± 7.76
30	1.30 ± 0.02	12.61 ± 1.88	89.71 ± 5.83	3.15 ± 0.22	21.37 ± 1.04	85.24 ± 599

TIS RITA® 30 did not only produce the highest number of shoots and leaves, but this system also gave a better culture quality (Figure 2C). After incubation of 21 d in the bioreactor, the stems in TIS RITA® 15 were purplish-brown and the leaves turned to yellow (Figure 2B). This symptom developed presumably due to RITA® 15 had shorter and less frequent immersion regime than RITA® 30, as they reduce nutrient and water uptake and consequently limited the rate of nitrogen uptake.

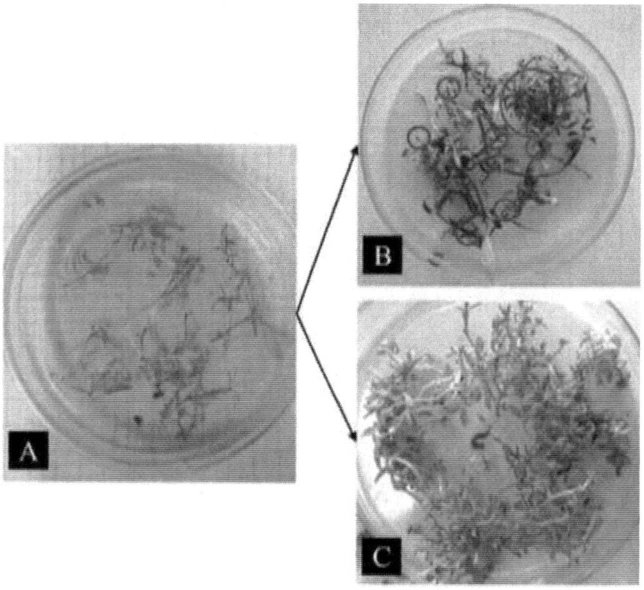

Fig. 2. Stevia shoots before and after cultivated in TIS RITA® bioreactors: (A) Day 0 for RITA® 15 min and 30 min; (B) Day 21 for RITA® 15 min; and (C) Day 21 for RITA® 30 min.

Nitrogen is the mineral element that plants require in greatest amounts. It serves as a constituent of many plant cell components, including amino acids and nucleic acids. Therefore, nitrogen deficiency rapidly inhibits plant growth. If such a deficiency persists, most species show chlorosis (yellowing of the leaves), especially in the older leaves near the base of the plant) [14]. Under severe nitrogen deficiency, these leaves become completely yellow (or tan) and died. Younger leaves may not show these symptoms initially because nitrogen can be mobilized from older leaves. Thus a nitrogen-deficient plant may have light green upper leaves and yellow or tan lower leaves [15].

When nitrogen deficiency develops slowly, plants may have markedly slender and often woody stems. This woodiness may be due to a buildup of excess carbohydrates that cannot be used in the synthesis of amino acids or other nitrogen compounds. Carbohydrates which not used in nitrogen metabolism may also be used in anthocyanin synthesis, leading to the accumulation of pigment. This condition is revealed as a purple coloration in leaves, petioles, and stems [16].

Table 3. The increasing number of shoots and leaves produced bioreactor RITA® with immersion periods 15 min and 30 min

TIS RITA®	Initial mass (g) ± SD		Final mass (g) ± SD		Growth rate (g DW d^{-1})	Productivity (g. L medium^{-1} d^{-1})	Doubling time (d)
	Fresh weight	Dry weight	Fresh weight	Dry weight			
15	0.950 ± 0.071	0.140 ± 0.014	2.490 ± 0.354	0.355± 0.007	0.045 ± 0.004	0.041	15.275 ± 1.495
30	0.900 ± 0.000	0.077 ± 0.005	3.115 ± 0.064	0.425 ± 0.021	0.081 ± 0.001	0.066	0.573 ± 8.551

The growth rate of *S.rebaudiana* cultivated in RITA® 30 min was greater than RITA® 15 min. Results in Table 3, showed that the growth rate of *S. rebaudiana* in RITA® 30 min was 0.081 g ± 0.001 g DW d^{-1} with doubling time (dt) 0.573 d ± 8.551 d while in RITA® 15 min was 0.045 g ± 0.001 g DW d^{-1} with a dt 15.275 d ± 1.495 d. The result in Table 3 also indicated two different growth patterns in shoot culture. According to Mordocco et al. [17], the immersion periods in RITA® system can determine the productivity of the systems and it has been proven by the increasing number of shoots of *Grandiflora × amelanchier* Rehder (serviceberry) up to 2.6 times with immersion 5 min every 60 min compared with 5 min every 30 min. The observation of the culture in RITA® bioreactor showed that the growth culture for RITA® 30 system was better than RITA® 15. The immersion periods is one of the critical parameters which can influence the growth of the culture, since the period of immersion related to the length of contact between the culture and medium. These conditions determine the absorption of nutrients by explants.

Conductivity value indicated the number of ions and dissolved salts contained in the bioreactor medium. The study stated that the decrease conductivity level in the culture medium showed the consumption of mineral compounds and inorganic compounds by cells [18], such as NH_4^+, NO_3^-, K^+, Ca^{2+}, Mg^{2+}, Cl^-, SO_4^{2-}, and PO_4^{3-}. All treated culture consumed minerals contained in the medium, therefore the conductivity of two bioreactors systems generally decreased. However, consumption of mineral on RITA® 30 (2.925 mS ± 0.033 mS) was relatively higher compared to RITA® 15 (2.520 mS ± 0.014 mS). Consumption of mineral compounds will be directly proportional to the addition of biomass. The higher the consumption of mineral and inorganic compounds, the higher the increase of biomass [19].

The sucrose consumption rate of RITA® 30 was higher when compared to RITA® 15. RITA® 30 had consumption rate of 0.184 g L^{-1} d^{-1} while RITA® 15 was only 0.092 g L^{-1} d^{-1}. The decrease in mass of sucrose at RITA® 15 system amounted to only 5.89 %, whereas in RITA® 30 was 11.66 %. The rate of sugar consumption would correlate with the rate of biomass enhancement. Sucrose is used as a carbon source for the cells, both for the structure and metabolites, so the addition of biomass needs more of sucrose. Conversely the lower the consumption of its substrate (sucrose), then the lower the enhancement of biomass [19].

Tables 4 and Table 5 showed the mass balance of culture in RITA® 15 and 30. The table showed a hypothetical amount of biomass and the value was different from the actual amount of biomass obtained from the measurement of plantlets weight. The hypothetical amount of biomass in RITA® 15 was 0.498 g while the actual was 0.355 g ± 0.007 g. The hypothetical amount of biomass in RITA® 30 was 0.023 g while the actual was 0.426 g ± 0.021 g. This imbalance might have resulted from the use of the root biomass formula of *Atropa belladona* Linn. for the shoot biomass formula of *S. rebaudiana*. Each cell had a different and unique substrate consumption pattern and biomass formation pattern. Another factor that might influence the construction of a mass balance model was the limiting substrate in the reaction, as this was not examined in this study. Knowledge of the limiting substrate will be very helpful in validating the model that has been built.

Table 4. Culture mass balance in TIS RITA® 15 system

Reaction Equation	0.39 $C_{12}H_{22}O_{11}$	0.23 NH_4NO_3	3.43 O_2	$CH_{1.27}O_{0.43}N_{0.45}$	4.07 H_2O	3.64 CO_2
Beginning (mol)	0.096	-	-	0.004	-	-
Reaction (mol)	0.006	0.003	0.049	0.014	0.058	0.052
Final (mol)	0.090	0.000	-	0.019	0.058	0.052
Final (g)	30.788	-	-	**0.498**	1.050	2.296

Table 5. Culture mass balance in TIS RITA® 30 system

Reaction Equation	0.39 $C_{12}H_{22}O_{11}$	0.23 NH_4NO_3	3.43 O_2	$CH_{1.27}O_{0.43}N_{0.45}$	4.07 H_2O	3.64 CO_2
Beginning (mol)	0.096			0.003		
Reaction (mol)	0.011	0.003	0.004	0.001	0.001	0.001
Final (mol)	0.084			0.001	0.001	0.001
Final (g)	28.861			**0.023**	0.011	0.064

4 Conclusions

In vitro propagation of *S. rebaudiana* cultured in TIS RITA® bioreactors was potential to be developed as an efficient method of it shoot production. Furthermore, the period of culture immersion was a critical parameter that affected the growth of stevia culture in TIS bioreactor. Ammout of 30 min immersion in medium was an optimal immersion period for growth of the *S. rebaudiana* shoots compared with 15 min immersion. Comparison between the percentage of shoot and leaf growth, growth rate, biomass acquisition, as well as productivity of *S. rebaudiana* with the consumption of sucrose and nutrition showed that the RITA® culture system with 30 min immersion period gave better results of bioconversion medium to become biomass than 15 min immersion period every 6 h.

References

1. L. Guariguata, D.R. Whiting, I. Hambleton, J. Beagley, U. Linnenkamp, J.E. Shaw, Diabetes Res Clin Pract., **103**,2:137–149(2014). https://www.diabetesresearchclinicalpractice.com/article/s0168-8227(13)00385-9/fulltext
2. A. Sanyaolu, A. Marinkovic, J. Gosse, L. Likaj, O. Ayodele, C. Okorie, et al., J Pub Heal Cat. **1**,4:86–88(2019). https://www.alliedacademies.org/articles/artificial-sweeteners-and-their-association-with-diabetes-a-review-10973.html
3. E. Gupta, S. Purwar, S. Sundaram, G.K. Rai, J Med Plants Res. **7**,46:3343–3353(2017). https://academicjournals.org/journal/JMPR/article-full-text-pdf/57B2A0342160.pdf
4. S.K. Goyal, Samsher, R.K. Goyal, Int J Food Sci Nutr., **61**,1:1–10(2010). https://www.researchgate.net/profile/Arvind_Singh56/post/How_much_do_you_know_about_Stevia/attachment/5ad97d8db53d2f63c3c6eba5/AS%3A617365834305536%401524202893114/download/Stevia.pdf
5. R. Raina, S.K. Bhandari, R. Chand, Y. Sharma, J Med Plants Res. **7**,24:1793–1799(2013). https://academicjournals.org/journal/JMPR/article-full-text-pdf/083395127660.pdf
6. M. Anbazhagan, M. Kalpana, R. Rajendran, V. Natarajan, D. Dhanavel, **22**:216–222(2010). https://www.ejfa.me/index.php/journal/article/download/275/177/
7. A. Mitra, A. Pal, Journal of Plant Biochemistry and Biotechnology, **16**,1:59–62(2007). https://www.researchgate.net/profile/Amita_Pal3/publication/257288936_In_Vitro_Regeneration_of_Stevia_rebaudiana_Bert_from_the_Nodal_Explant/links/5641adf408aec448fa61b375/In-Vitro-Regeneration-of-Stevia-rebaudiana-Bert-from-the-Nodal-Explant.pdf
8. A.C. Melviana, R.R. Esyanti, R.H. Setyobudi, M. Mel, P.G. Adinurani and J. Burlakovs, Sarhad J. Agric. **37**,1:1–8(2021). http://dx.doi.org/10.17582/journal.sja/2021/37.1.1.8

9. N. Vidal, C.Sánchez, Engineering in Life Sciences. **19**,12:896–915(2019). https://onlinelibrary.wiley.com/doi/pdf/10.1002/elsc.201900041
10. N. Noordin, R. Ibrahim, N. Sajahan, S. Nahar, N. Rashid, *Micro propagation of Stevia rebaudiana Bertoni through temporary immersion bioreactor system*. Research and Development Seminar, INIS, Malaysia. (2012), p. 1–8. https://inis.iaea.org/collection/NCLCollectionStore/_Public/44/096/44096871.pdf
11. H.I. Indra, R.R. Esyanti, A. Faizal, Int J Res Appl.**4**,3:133–140(2016). https://multisite.itb.ac.id/sith/wp-content/uploads/sites/56/2016/06/2-14-1459340628-17.applied-The-Efficiency-of-Growth-Medium-Bioconversion-into-Ahmad-Faizal.pdf
12. A. Saterbak, K. San, L. McIntire. *Bioengineering fundamentals*, New Jersey, USA: Pearson Prentice Hall (2018). https://www.worldcat.org/title/bioengineering-fundamentals/oclc/971893219?referer=di&ht=edition
13. K. Rangappa, D. Aind, Advances in Bioscience and Biotechnology. **4**,7:1–4(2013). https://www.scirp.org/html/3-7300547_34153.htm
14. P. Kumar, M. Sharma, *Nutrient deficiencies of field crops*. Boston, MA: CABI (2013). https://www.cabi.org/bookshop/book/9781780642789/
15. M. Canevari, R. Vargas, S. Wright, L. Jackson, *Small grain production Pt 14: Troubleshooting small grain production*. USA:UCANR Publications (2006). p. 1–13. http://citeseerx.ist.psu.edu/viewdoc/download?doi=10.1.1.610.3128&rep=rep1&type=pdf
16. C. Diaz, V. Saliba-Colombani, O. Loudet, P. Belluomo, L. Moreau, F. Daniel-Vedele et al. Plant and Cell Physiology, **47**,1:74–83(2006). https://academic.oup.com/pcp/article/47/1/74/1867545
17. A. Mordocco, J. Brumbley, P. Lakshmanan, In Vitro Cellular & Developmental Biology - Plant. **45**:450(2008). https://doi.org/10.1007/s11627-008-9173-7
18. D. Ryu, S. Furusaki, *Advances in Plant Biotechnology*, Amsterdam: Elsevier (2012). https://www.elsevier.com/books/advances-in-plant-biotechnology/ryu/978-0-444-89939-2
19. M. Weih, F. Pourazari, G. Vico, Scientific Reports, **6**:1–9(2016). https://www.nature.com/articles/srep35958

Physical Treatment of Oil Palm Shell for Briquette Production as Bioenergy at Remote Area

Amaliyah Rohsari Indah Utami[1,*], *Suwandi* Suwandi[1], *Yoga* Alun Mustafa[1], and *Maizirwan* Mel[2]

[1]Department of Engineering Physics, School of Electrical Engineering, Telkom University, Jl. Telekomunikasi, Terusan Buah Batu, Bandung 40257, Indonesia
[2]Department of Biotechnology Engineering, Faculty of Engineering, International Islamic University Malaysia (IIUM), Gombak, 50278 Kuala Lumpur, Malaysia

Abstract. The daily needs of energy in remote areas, Indonesia, is dominated by biomass waste. The abundant types of biomass waste for heat transfer is empty fruits shell (EFS) of oil palm as briquette. This research discussed a brief analysis of oil palm shell by using physical treatment at different crushing pressure. It was also to examine the adequate size of particle for improving the heating value of briquette. The crushing pressure had a height dimension of 20 cm, as well as the briquette mold with a diameter of 2 cm and a height of 1 cm. The oil palm shell which was applied physical treatment at different crushing pressure of 400 J, 550 J, and 700 J, that resulted in particle size is smaller than 800 μm, 800 μm to 1 700 μm, 1 700 μm to 2 000 μm, and bigger than 2 000 μm. The adequate size of particle for briquette was found to be < 800 μm, which reached the heating value of 20 042.32 J g^{-1} and ash residue of 20 %. This study showed the advantages of oil palm shell briquette by using physical treatment is suitable and sustainable alternative to daily life use in Indonesia.

Keywords: Biomass, renewable energy, waste to energy, zero waste.

1 Introduction

Increasing energy consumption accompanied by depletion of fossil fuel reserves as the main supplier of energy. It is due to an imbalance between the conversion process rate from biomass to fossil which requires millions of years compared to the rate of human energy consumption. It makes fossil fuel as an unrenewable and sustainable fuel. Therefore, a lot of researches were developed to find a sustainable source of alternative energy. Among those who attract attention is the waste of biomass that is categorized as a neutral carbon energy source [1, 2]. Thus, the use of biomass waste as a source of energy can offset the CO_2 emissions generated by fossil fuels.

Waste biomass is mainly produced by commercial plantations such as oil palm, very abundant in Indonesia. In 2017, Indonesia produced 7 071 877 t of fresh fruit palm oil with

[*] Corresponding author: amaliyahriu@telkomuniversity.ac.id

a total area of 12 307 677 ha [3]. The processing of oil palm fresh fruit into oil produces waste such as liquid waste, fiber, and empty fruit bunches [4, 5]. The amount of empty palm oil fruit bunches waste was 6.5 % of the total number of fresh fruit bunches processed, i.e. ammount 1 t [4]. Oil palm shell waste had a heating value of 18 034.56 J g^{-1} [6]. However, the oil palm shell after oil production processing remains as waste [6–8]. In addition, the increasing need for energy and decreasing reserves of fossil fuel resources, has encouraged researchers to develop renewable energy in the form of briquettes [9, 10].

Recently, briquette is one of the energy technologies that is relatively easy to develop and consume by residents in remote areas [11–13]. Some studies have reported on techniques and factors affecting the quality of briquettes [14–16]. The technique of making briquette with both low and high pressure has been conducted to improve briquette performance [17]. In addition, the performance of briquettes is influenced by factors such as ash content and ignition properties. Moreover, one major significant factor is physical properties such as the particle size of the composition briquettes that will affect on briquette utilization in energy requirement. Therefore, the present study determined to investigate the effect of the physical treatment for briquette production on the heating value and ash residue content of palm oil shell waste. In addition, this study measured ignition time and flame burning time of briquette to know the efficiency level of quality briquette. Furthermore, the use of the oil palm shell waste into briquette is an alternative way to reduce much amount of the waste.

2 Materials and methods

Lignocellulosic biomass in the form of an empty fruits shell (EFS) obtained from oil palm shell biomass located in Riau, Indonesia, was used in this study. EFS was dried under the sun for 7 d. The crushing pressure tools (Figure 1) was employed to crush 15 g of dried EFS. The tool height is 20 cm. The briquette mould has a diameter of 2 cm and a height of 1 cm.

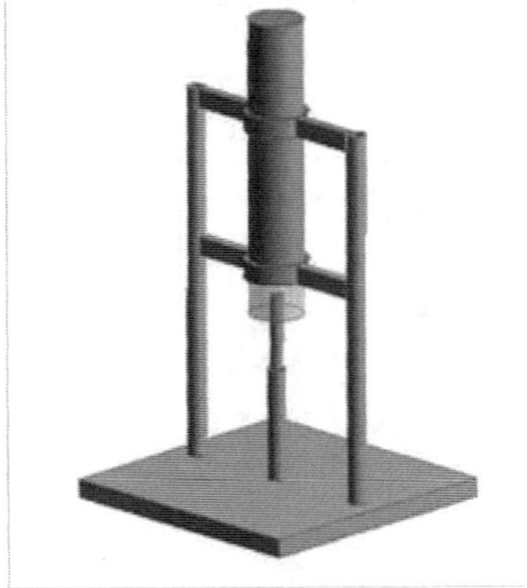

Fig. 1. Crushing pressure and briquette tools

The experimental method was explained in detail in Figure 2. The energy pressures applied to treat EFS were 0 J, 400 J, 550 J, and 700 J. Then, the treated EFS was filtered by Testing Sieve Shaker in Bogor, Indonesia. The treated EFS was mixed with water and then pressed 700 J for 1 min with the same equipment to produce briquettes. The sample was heated at 80 °C for 1.5 min in the oven. Briquette was weighed ± 15 g with a thickness of 1 cm and a diameter of 2 cm.

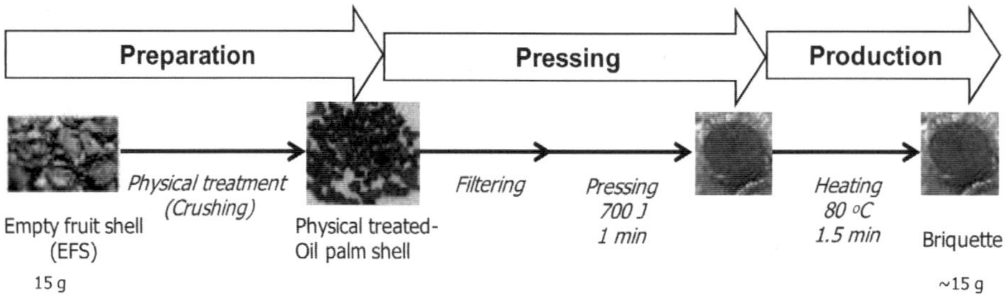

Fig. 2. The detail experimental method of briquette by using crushing pressure and briquette tools

The caloric value was analyzed by 1 261 Isoperibol Bomb Calorimeter. The ignition and flame burning time of briquette were also measured. Finally, the percentage of ash residue from combustion briquette was weighed by using the analytical balance.

3 Results and discussion

3.1 Effect of physical treatment on particle size

To evaluate the influence of EFS physical treatment to particle size, the treated EFS was sifted with CBN Testing Sieve Shaker. Figure 3 showed the physical treatment for briquette production on the percentage of particle size at crushing pressure. The physical treatments of 400 J produced 65 % particle size greater than 2 000 μm, 19.05 % had 1 700 μm to 2 000 μm, 9.14 % particle sized 800 μm to 1 700 μm, and 6.91 % smaller than 800 μm, respectively. Increasing energy press given to EFS decreased to percentage of particle size greater than 2 000 μm but increased the percentage of the smaller particles.

These results are similar to previous research that physical treatment produced the compressive energy that affected the particle size i.e. specific surface area [17]. Therefore, the physical treatment at different energy of crushing pressure was sufficient to reduce the particle size of EFS.

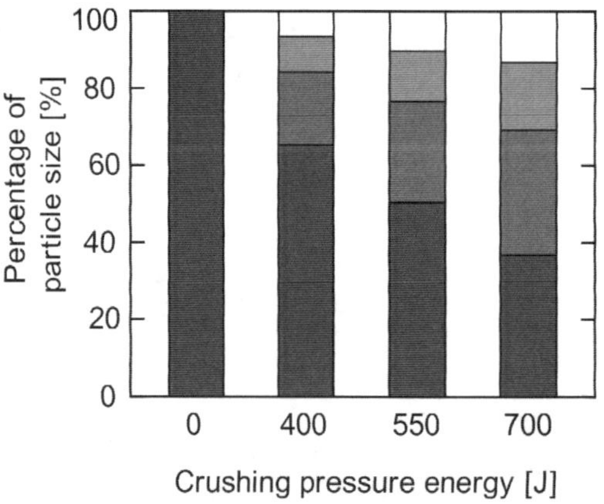

Fig. 3. Percentage of particle size at different crushing pressure of physical treatment (greater than 2 000 μm, 1 700 μm to 2 000 μm, 800 μm to 1 700 μm, and smaller than 800 μm from the bottom).

3.2 Effect of particle size on heating value

To evaluate the influence of the particle size of treated EFS in the briquette on the heating value, the sample was tested in the bomb calorimeter. The heating value of EFS without physical treatment was 18 034.56 J g^{-1}. Table 1 showed the heating value of oil palm shell waste at different particle sizes. The heating value for all EFS samples treated was higher than without physical treatment.

The results was similar to previous research that the particle size of the briquette composition affected the calory briquette value [17], because of heat transfer that occurs among the particles of the briquette. The smaller particle size, the closer of the distance between particles, then the heat transfer would occur through the convection process [18, 19]. Furthermore, these results indicated that physical treatment of oil palm shell biomass using at different crushing pressure was more sufficient to provide energy needs at a remote area.

Table 1. Heating value of palm oil briquette at a different particle size

Particle size (μm)	Heating value (J g^{-1})		
	400 (J)	550 (J)	700 (J)
< 800	18 500.98	18 382.99	20 042.32
800 to 1 700	19 834.84	19 308.28	19 803.58
1 700 to 2 000	19 215.23	19 338.16	19 668.52
> 2 000	19 042.30	18 672.77	20 039.48

3.3 Effect of particle size on ignition time

To examine the efficiency of briquette, ignition time test was applied. Figure 4 showed the ignition time of sample. For EFS not briquette sample showed the fastest ignition time i.e. 46 s. The longest ignition time occurred on the briquette sample which particle size of 1 700 μm to 2 000 μm was 122 s. For briquette sample which particle size < 800 μm, the ignition time was 60 s. The ignition time of briquette that particle size 800 μm to 1 700 μm, and > 2 000 μm were 110 s and 98 s, respectively.

Due to the ignition time was combustion reaction, therefore oxygen were required. In the case EFS not briquette, there were opened space to provide oxygen. As the result, the combustion reaction occurred quickly. It was different from briquettes that particle size > 800 μm. The presence of an empty cavity inside the briquette caused the heat transfer slowly. The pressing step might generate the absence of oxygen in the briquette and remained the loss binding among the particle [7, 16, 19]. For briquette which particle size < 800 μm, the smaller particle size, the larger specific surface area of the particles that can speed up the heat energy during the combustion process through convection. Therefore, particle size < 800 μm was the best briquette among all the samples.

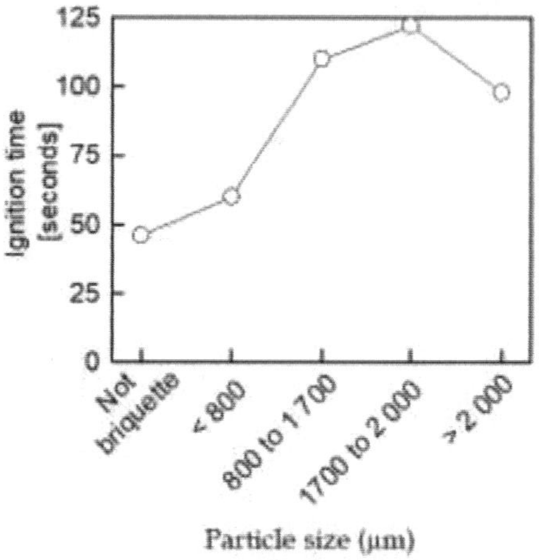

Fig. 4. Ignition time of sample

3.4 Effect of particle size on flame burning time

Determine the efficiency of briquette, the flame burning time was conducted. Figure 5 showed the flame burning time of sample. For EFS not briquette showed the fastest ignition time i.e. 8.67 min. The longest ignition time occurred on the briquette which particle size of > 2 000 μm was 16.67 min. For briquette sample which particle size < 800 μm, 800 μm to 1 700 μm, and 1 700 μm to 2 000 μm were 11.67 min, 13.5 min, 13.5 min, respectively.

These results indicated that the main factor influenced are specific surface area of the particle and the presence of cavity without oxygen content as mentioned above [7, 16, 19]. The linear trend of the results represented the supporting reason of relations between igniton time and the flame burning time due to those main factors. Compared to all treatments, briquette contained particle size < 800 μm was the best briquettes due to fast ignition and flame burning time.

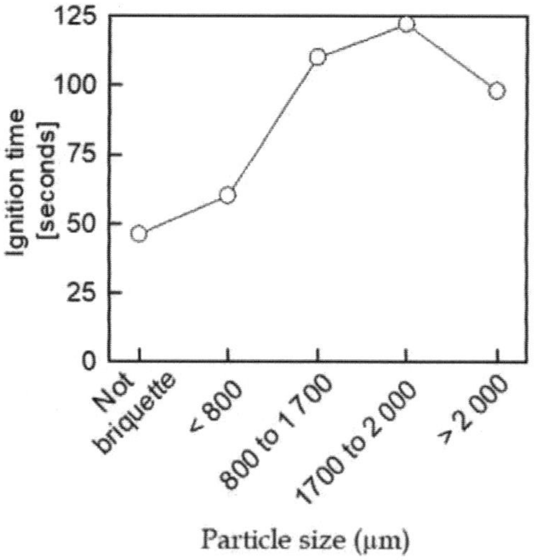

Fig. 5. Flame burning time of sample

3.5 Effect of flame burning time on ash percentage residue

To investigate the effect of flame burning time on ash content residue, the ash residue was weighed after the combustion process. Figure 6 showed the used percentage of original biomass size at the flame burning time. Almost all briquettes had more than 25 % ash residue, except briquettes with a particle size of < 800 μm only 20 % of the total weight of the briquette.

The ash percentage residue refferred that the smaller of particle size and the larger of the specific surface area that can speed up the heat transfer energy during the combustion process and the perfect combustion of briquette almost occurred [7, 16, 19]. These results supported the previous parameters such as ignition and flame burning time. From point of effiency view, the briquette contained particle size < 800 μm was the best and recommended to provide energy at the remote area.

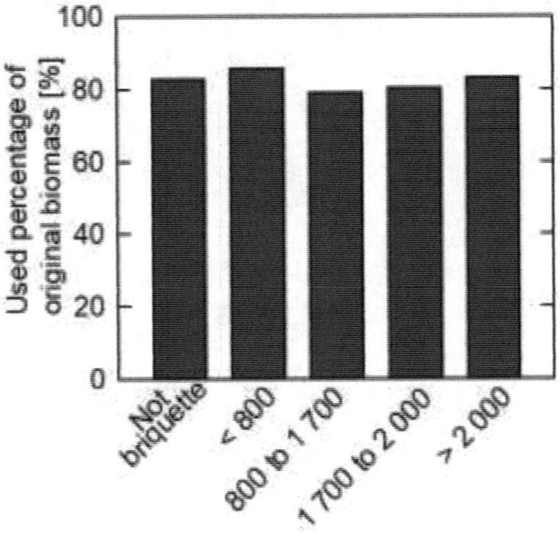

Fig. 6. Use percentage of original biomass at burning test.

4 Conclusion

The oil palm shell which was applied physical treatment at different crushing pressure of 400 J, 550 J, and 700 J, resulted in particle size was smaller than 800 μm, 800 to 1 700 μm, 17 002 000 μm, and bigger than 2 000 μm. The best briquette was found in size of particle < 800 μm, which reached the heating value 20 042.32 J g^{-1}, and ash residue 20 %. However, physical treatment with different crushing pressure showed the higher heating value than the untreated biomass.

References

1. A. Demirbas, Journal Energy Sources: Part A Recovery Utilization Environmental Effects. **31**,17:1573–1582(2009). https://doi.org/10.1080/15567030802094011
2. N. Abdullah, F. Sulaiman, H. Gerhauser, J. Phys. Sci., **22**,1:1–24(2011). http://jps.usm.my/wp-content/uploads/2014/10/22.1.1.pdf

3. I.R. Nurbahar, *Statistik Perkebunan Kelapa Sawit Indonesia 2015–2017* [Indonesian Oil Palm Plantation Statistics 2015-2017], Sawit, 81(2016) [Online] from http://ditjenbun.pertanian.go.id [Acessed on 20 May 2019].
4. N. Kamal, ITENAS, Bandung, (2014). http://lib.itenas.ac.id/kti/wp-content/uploads/2014/04/JURNAL-Netty-Kamal-ED-15
5. K. Siregar, S. Sholiati, I. Sofiah, T. Miharza, R.H. Setyobudi, O. Anne, et al. E3S Web of Conferences **190**:00021(2020). https://doi.org/10.1051/e3sconf/202019000021
6. F.B. Ahmad, Z. Zhang, W.O. Doherty, I.M. O' Hara, Renew. Sustain. Energy Rev., 109:386–411(2019). https://doi.org/10.1016/j.rser.2019.04.009
7. M. Aris, Y.S. Chin, Asian J. Sci. Res., **6**,3:537–545(2013). https://scialert.net/abstract/?doi=ajsr.2013.537.545
8. K. Siregar A. L. Machsun, S. Sholihati, R. Alamsyah, I. Ichwana, N. C. Siregar, et al. E3S Web of Conferences **188**:00018(2020). https://doi.org/10.1051/e3sconf/202018800018
9. S.K. Hoekman, Renew. Energy, **34,**1:14–22(2009). https://linkinghub.elsevier.com/retrieve/pii/S0960148108001341
10. C.S. Dees, M. Long, Contract Manag., **45**,12 :48–50(2005) https://www.worldcat.org/title/contract-management/oclc/1165262452
11. A. Budiman, Energy Proc., **68:**157–166(2015). https://doi.org/10.1016/j.egypro.2015.03.244
12. P.S. nee'Nigam, A. Pandey, *Biotechnology for Agro-Industrial residues Utilisation*. UK: Springer (2009). https://pure.ulster.ac.uk/en/publications/biotechnology-for-agro-industrial-residues-utilisation-3
13. H. Hariyanto, D. Parenden, Z. Vincēviča-Gaile, P.G. Adinurani. E3S Web of Conferences **190**:00012(2020). https://doi.org/10.1051/e3sconf/202019000012
14. Ľ. Pňakovič, L. Dzurenda, BioResources. **10,**3:5563–5572(2015). https://bioresources.cnr.ncsu.edu/resources/combustion-characteristics-of-fallen-fall-leaves-from-ornamental-trees-in-city-and-forest-parks/
15. Q.A.M.O. Arifianti, U. Anggarini, A. Nafis, E.R. Sari, E.F. Nugrahani. E3S Web of Conferences, **190**:00029(2020). https://doi.org/10.1051/e3sconf/202019000029
16. M. Lubwama, V.A. Yiga, Renewable Energy, **118**:43–55(2018). https://linkinghub.elsevier.com/retrieve/pii/S0960148117310960
17. Y.A. Mustafa, S. Suwandi, A.R.I. Utami, In: Proceeding Biology Education Conference: Biology, Science, Enviromental, and Learning, **11,**1(2014). https://sembio.fkip.uns.ac.id/prosiding-sembio-xi-2014/
18. Z. Husain, Z. Zainac, Z. Abdullah, Biomass Bioenergy, **22,**6:505–509(2002). https://doi.org/10.1016/S0961-9534(02)00022-3
19. D. Wang, L. Liu, Y. Yuan, H. Yang, Y. Zhou, R. Duan, Renew. Energy, **147,**1:1371–1379(2020). https://linkinghub.elsevier.com/retrieve/pii/S0960148119314120

The Potential of Cashew Apple Juice as Anti Hypercholesterol Agent on Whistar Rats (*Rattus norvegicus* Berkenhout, 1769)

Asmawati Asmawati[1,*], *Marianah* Marianah[1], *Abubakar* Yaro[2,3], and *Roy* Hendroko Setyobudi[4]

[1]Department of Food Processing of Technology, Faculty of Agriculture, Muhammadiyah Mataram University, Jl. KH. Ahmad Dahlan No.1, Pagesangan, Mataram 83115, Nusa Tenggara Barat, Indonesia
[2]Institute of Health Sciences & Research, AHRO Scientific Publishing, 272 Bath Street, Glasgow G2 4JR, Scotland, UK
[3]Strathclyde Institute of Pharmacy & Biomedical Sciences (SIPBS), University of Strathclyde, 161 Cathedral St, Glasgow G4 0RE, Scotland, UK
[4]Waste Laboratory, University of Muhammadiyah Malang, Jl. Raya Tlogomas No. 246, Malang, 65145, East Java, Indonesia

Abstract. Cashew (*Anacardium occidentale* L.) apple are rich in nutritional value, especially vitamine C and bioactive compounds such as tannins that are very beneficial to support health. This research aims to find out the effects of cashew apple juice of lipid profile and its potential as an anti–Hypercholesterol agent. In this study used experimental animals in the form of male Whistar Rats Type (*Rattus norvegicus* Berkenhout, 1769) as much as 21 rats are divided into three groups and each group consists of seven rats were fed with AIN '76 diet for 28 d, given the treatment of cashew apple fruit juice 2 mL, two times a day by were force–fed and as control treatment is given only drinking aquabidest. The results showed that cashew apple juice was influential for the lipid profile of the whistar rats experiment, potentially as an anti–Hypercholesterol agent by lowering total cholesterol levels 6.37 % to 7.16 %, Triglyseride (9.71 % to 10.76 %) LDL–Cholesterol (11.19 % to 13.37 %) and increased cholesterol levels of HDL (6.10 % to 6.46 %).

Keywords: *Anacardium occidentale* L., increase HDL cholesterol, reduce total cholesterol, tannin, waste to herbal

1 Introduction

Cashew plants (*Anacardium occidentale* L.) is a commodity that has an essential position because almost all of its parts can be utilized as a variety of products with high economic value and health support. This plant produces fruit consisting of two parts, namely cashew nuts as true fruit–producing cashew nuts and cashew apple as pseudo fruit. The existence of cashew nuts is only 10 % of the total weight of cashews, but the economic value is very high [1]. When compared with cashew fruit (Cashew) has an abundant potential (90 %) of the total weight of the fruit, but has not been maximally utilized, this is because (80 % to 90 %) the product is disposed of in the form of waste and animal feed. Only a small portion can be consumed directly as fresh fruit and other processed products.

* Corresponding author: asmawatiraba@gmail.com

Cashew apple has considerable potential to be developed into food and beverage products when viewed from its chemical composition and nutritional value. One that has a bright prospect is cashew apple juice. This is because the cashew apple is juicy with a moisture content of 86.3 g 100 g^{-1} [2] and rich in vitamin C which is 180 mg 100 g^{-1} [2] and 126 mg 100 mL^{-1} to 372 mg 100 mL^{-1} [3] three to six times more vitamin C than orange juice [4], and as a very good source of antioxidants [5].

The processing of cashew apples into food and beverage products has significant constraints, such as the taste of the astringency and itching (acrid). The taste of the mouthand throat, which is the characteristic of both flavors, is not so preferred by consumers, so it needs to be reduced. The presence of soluble tannins mainly causes the astringency taste in cashew apple. It is one component of polyphenols with levels of 0.35 % [6] and 147 mg 100 mL^{-1} to 164 mg 100 mL^{-1} [7] depending on variety and level of maturity.

The presence of tannin in food and drinks can have a positive and negative effect on those who consume it. The negative effects are, i) can reduce the nutritional value of macromolecules (protein, amino acids, carbohydrates, sugars), micromolecules (vitamins, minerals) and their bioavailability [5, 8]. ii) negative effect to digestive [9] by forming insoluble complexes, thereby increasing levels of nitrogenous stool, which can inhibit the growth of experimental animals including rats. The positive effect on tannin is that it can reduce total cholesterol concentration, and increase HDL cholesterol in the blood so that it can strengthen the transport back of cholesterol and inhibit absorption in the intestine, which results in the amount of cholesterol being excreted with stool. Therefore, cashew apple can be used as an anti–hypercholesterolemia [10].

Based on the description above, a research has been conducted on the potential of cashew apple juice as an anti–hyper cholesterol agent in Wistar white rats (*Rattus Norvegicus* Berkenhout, 1769). This study aims to determine the effect of giving cashew juice to the lipid profile of experimental rats and its potential as an anti–hyper cholesterol agent.

2 Material and methods

2.1 Materials

The materials used in this study are, i) Yellow–orange cashew apple obtained from Lombok Utara NTB as raw material for making cashew juice, ii) Experimental animals namely Wistar male white rats (*R. Nervegicus*) who were 6 wk old with an average body weight of 104 g $\pm$ 8 g obtained from the Experimental Animal Development Unit (UPHP) of Universitas Gadjah Mada, Yogyakarta, Indonesia who had the qualification standards for research, iii) Experimental rats feed during the study referred to the standard feed set by the American Institute of Nutrition 1976 (AIN 76) made in the nutritional laboratory of the Faculty of Agricultural Technology, Universitas Gadjah Mada, Yogyakarta, Indonesia, iv) Other materials in the form of kits for the analysis of lipid profiles of experimental rats are kit no. 1.14992.0250, 1.14210.0001 and 1.14366.0001. obtained from chemical stores in Yogyakarta, Indonesia.

2.2 Methods

2.2.1 The processing of cashew apple juice

Cashew apple juice used in this study was obtained by the following method: i) Cashew apple that have been sorted, then blanched for 1 min to soften the texture to facilitate the extraction process to produce juice, the resulting juice is called cashew apple juice control, then tannin content are analyzed (0.29 % DB), ii) The same treatment with the control of cashew apple Juice, but first soaked in a calcium hydroxide solution ($Ca(OH)_2$ 3 % for 15 min, then extracted to produce juice treatment for tannin content analysis (0.14 % DB)

2.2.2 Research procedur

The research procedure has been approved by the Ethical Commission of the Faculty of Medicine, Al-Azhar Islamic University Mataram, West Nusa Tenggara, Indonesia (Number 29/EC/FK–06/UNIZAR/VIII/2020). The stages are as follows :

i. 21 male white rats (*R.norvegicus*) were adapted for 7 d, and during the adaptation, the experimental rats were fed AIN 76 and drinking distilled water. After adjustment, all blood rats were taken to analyze the presence of lipid profiles (total cholesterol, triglycerides, LDL cholesterol, and HDL cholesterol).
ii. Experimental rats were divided into three groups, and each group consisted of seven mice.
iii. Experimental rats were treated in the form
   - Group 1 was given control juice
   - Group 2 was given drinking cashew apple juice treatment
   - Group 3 was only given aquadest (placebo group).
   - giving of drinking cashew apple juice in experimental rats and aquadest in the placebo group were given forcibly (forced feeds) each 4 mL d^{-1} given two times (morning and evening) and were still fed during the study with a standard feed from AIN 76 This treatment lasted for 28 d.
iv. At the end of the study (28 d), all rats from each group were taken blood from their eyes (Retroorbitol plexus) and their hearts to analyze their lipid profile using a microhematocrit tube. Briefly can be seen in the flow chart, as shown in the Figure 1.

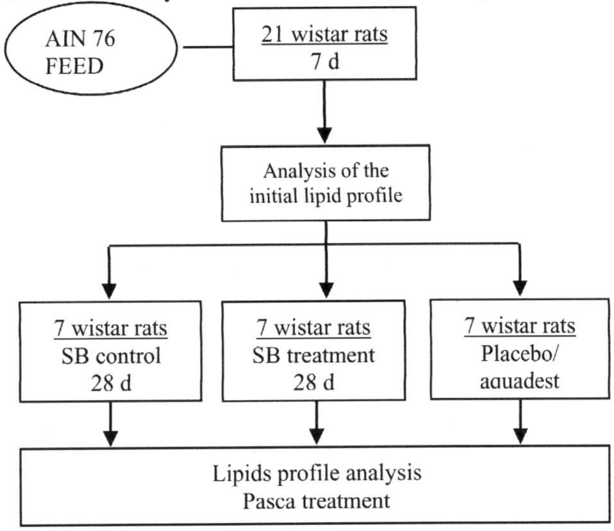

Fig. 1. Scheme of the research

2.2.3 Measurement of research parameters

i. The tannins content in Cashew apple juice were analyzed the burn method (1963)
ii. Lipid Profile of experimental rats namely total cholesterol by the CHOD–PAP method [11], TG by the GPO–PAP method [12] LDL cholesterol by the CHOD–PAP method [13] and HDL cholesterol by the CHOD–PAP method.

2.2.4 Data analysis (statistical analysis)

This research was conducted using an experimental method. The data design using a completely randomized design. Data obtained by analysis of variance, differences among samples would be tested using Least Significant Different (LSD) with the significance level set at $\alpha = 0.05$ [14].

3 Result and discussions

The results showed that the giving of cashew apple juice significantly affected the blood serum lipid protein on the experimental group of rats compared to the placebo group. These results can be seen in Table 1 below.

Table 1. The changes in blood serum lipid profile of rats experiments on various treatments given by cashew apple juice

Treatment	Cholesterol total	Triglyceride	LDL Col	HDL Col
SB control	–7.16 a	–10.76 a	–13.37 a	+6.46 a
SB treatment	–6.37 a	–9.71 a	–11.19 a	+6.10 a
Placebo	–0.60 b	–2.63 b	–0.07 b	+0.86 b

Noted: Numbers followed by the same letter in the same column are not significantly different at 5 % level. (– is down) (+ is up)

3.1 Cholesterol total

The results of an analysis of total cholesterol concentration on blood serum of the experimental rats group showed that after given of cashew apple juice was decreased. The decrease occurred respectively by 7.16 % (SB control) and 6.37 % (SB treatment). The results were not significantly different between treatments. When compared to the placebo group, there was a decrease of only 0.60 %, and it was significantly different from the other two groups. These results can be seen in the following Figure 2.

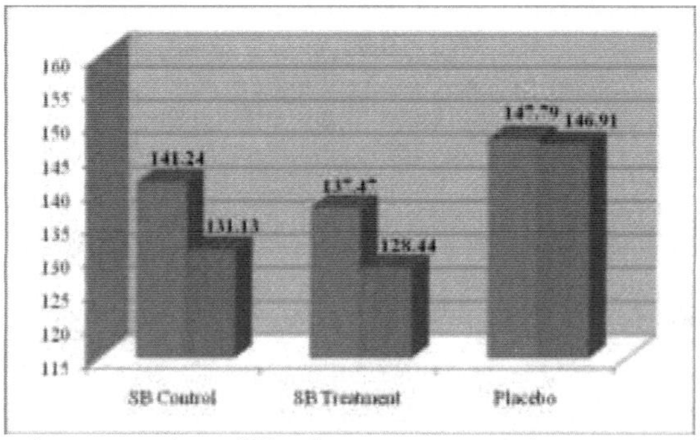

Fig. 2. Total cholesterol concentration of blood serum rats experiments on various treatments given by Cashew apple juice

https://doi.org/10.1051/e3sconf/202122600009

In Figure 2, it can be seen that there was a significant decrease in total cholesterol concentration ($p < 0.05$) after being given a cashew apple juice in rats. There was a decrease from 141.24 mg dL^{-1} after adaptation (before treatment) to 131.13 mg dL^{-1} (7.16 %) on SB control group. While in the group SB treatment, that is equal to 6.37 %, from 137.47 mg dL^{-1} to 128.44 mg dL^{-1}. As for the placebo group (only given aquadest) not significant decrease occurred, from 147.79 mg dL^{-1} to 146.91 mg dL^{-1} (0.60 %).

A decrease in the total cholesterol content in the group of rats that had been given cashew juice by 7.16 and 6.37 %. This is most likely due to the tannin content in the juice, the tannin content in SB control and SB treatment were 0.29 % db and 0.14 % db, respectively. The role of tannins in inhibiting or reducing the absorption of cholesterol in the intestine, namely by forming complexes with cholesterol, resulting in the deposition of cholesterol in the intestinal lumen resulting in the amount of cholesterol that is excreted with feces. This phenomenon is consistent with the results reported by [12].

3.2 Triglyceride

The results of triglyceride analysis showed that giving of cashew fruit juice in experimental rats could significantly reduce triglyceride concentrations in rat blood serum ($p < 0.05$) when compared to the placebo group as shown in Figure 3.

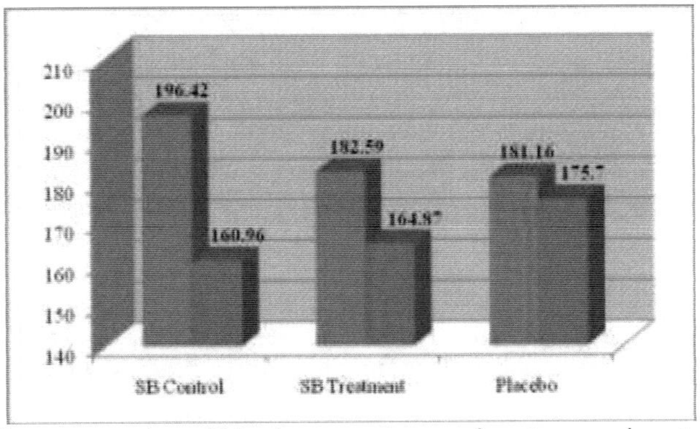

Fig. 3. Triglyceride concentration of blood serum rats experiments on various treatments given by Cashew apple juice

In Figure 3 above, it can be seen that the administration of cashew juice can decrease triglyceride concentrations from 196.42 mg dL^{-1} to 160.96 mg dL^{-1} in the SB control group (a reduction of 10.76 %). Whereas the SB treatment group could decrease the concentration from 182.59 mg dL^{-1} to 164.87 mg dL^{-1} (a reduction of 9.71 %). However, there was no significant difference between treatments compared to the placebo group which could only reduce from 181.16 mg dL^{-1} to 175.70 mg dL^{-1} (2.63 %).

The high decrease in triglycerides in SB control and SB treatment is due to the presence of tannin in cashew apple juice, which can inhibit the absorption of triglycerides so that absorption is slower or inhibited. This inhibition, according to [15], is thought to result in slower chylomicrons and remnant chylomicrons (the remaining kilo microns) formation. Besides, this inhibition also related to the chance of the lipoprotein lipase enzyme to catalyze more micron kilos to triglycerides before reaching the liver. Then triglycerides are sent to the network and cause blood triglyceride levels to decrease.

3.3 LDL (Low Density Lipoprotein)

The results showed that there was a significant decrease in LDL cholesterol as a result of giving cashew juice to rats. But the placebo group did not have a significant effect.

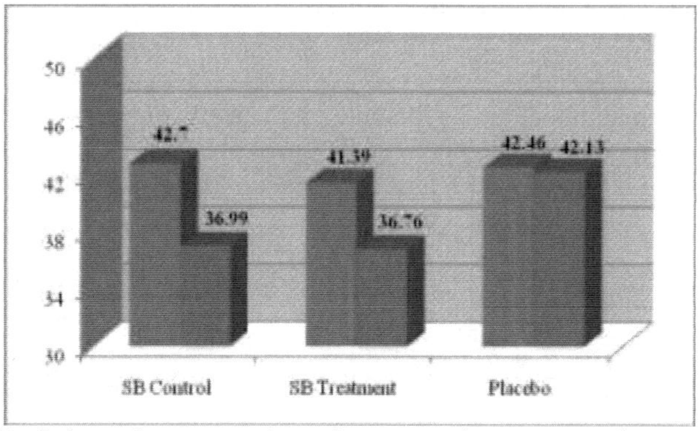

Fig. 4. Blood LDL cholesterol concentration of rats experiments on various treatments given by Cashew apple juice.

Figure 4 showed that there was a significant decrease in LDL cholesterol ($p < 0.05$) of 13.37 % in the control SB group, from 42.70 mg dL^{-1} to 36.99 mg dL^{-1} (13.37 %). In SB treatment, there was a decrease from 41.39 mg dL^{-1} to 36.76 mg dL^{-1}. They decreased LDL cholesterol along with a decrease in triglycerides. This phenomena is because LDL cholesterol is a natural result of the VLDL catabolism, which contains a lot of triglycerides. Subsequently, by VLDL, triglycerides are lifted from the liver to other tissues, especially to adipose tissue so that triglycerides in the blood are reduced [15].

3.4 HDL (High Density Lipoprotein)

The results showed that there was a significant increase ($p < 0.05$) of HDL cholesterol in the blood of experimental rats due to the administration of cashew apple juice compared with the placebo group, while the SB control and SB treatments showed did not significantly different.

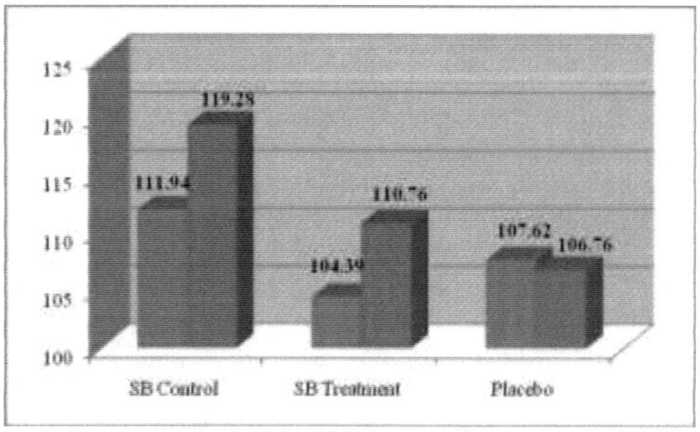

Fig. 5. Blood HDL cholesterol concentration of rats experiments on various treatments given by Cashew apple juice.

In Figure 5, it can be seen that there was an increase in HDL cholesterol in SB control and SB treatment, respectively, from 104.39 mg dL^{-1} to 110.76 mg dL^{-1} and from

111.94 mg dL^{-1} to 119.28 mg dL^{-1}. Whereas in the placebo group, there was a tendency to decrease from 107.62 mg dL^{-1} to 106.76 mg dL^{-1}.

The increasing in HDL cholesterol level in a group of rats given cashew apple juice, probably due to the presence of tannin in the juice. These results are in according to the research [12] that tannin polymer of grape seed can increase HDL cholesterol concentration and decrease plasma VLDL.

Low VLDL was causing low triglycerides in the blood of experimental rats, where a transfer of triglycerides to the liver produces more HDL cholesterol because the remaining protein and triglyceride release will cause HDL cholesterol formation [16].

4. Conclusion

i. The administration of cashew apple juice in experimental rats affected the blood serum lipid profile of the experimental rats' group compared to the placebo group.
ii. The administration of cashew apple juice with levels of 0.29 % (SB Control) and 0.14 % (SB treatment) has the same ability to reduce levels of total cholesterol, triglycerides, and LDL cholesterol and increase HDL cholesterol concentrations.
iii. Cashew apple juice has the potential as an anti–hyper cholesterol agent in rats (*R. nurvegicus*).

References

1. T.L. Honorato, M.C. Rabelo, L.R.B. Goncalves, G.A.S. Printo, S. Rodrigues, World J. Microbiol Biotechnol, **23**:1409–1415(2007). https://doi.org/10.1007/s11274-007-9381-z
2. C. Gopalan, B.V. Ramasastri, S.C. Balasubramanian. *Nutritive Value of Indian Foods*. India: National Institute of Nutrition, ICMR (2012). p.47–58. http://www.eeb.cornell.edu/biogeo/nanc/Food_Feed/table%201%20gopalan%20et%20al%201989.pdf
3. L.F. Damasceno, F.A.N. Fernandesa, M.M.A. Magalhaes, E.S. Brito. Braz. J. Chem. Eng. **25**,2:313–320(2008). https://doi.org/10.1590/S0104-66322008000200010
4. A. Marc, T.F. Achelli, M. Gbane, P.V.N. Koffi, N.G. Amani. Pakistan Journal of Nutrition. **10**,12:1109–1114(2011). http://dx.doi.org/10.3923/pjn.2011.1109.1114
5. E. Dedehou, J. Dossou, V. Anihouvi, M.M. Soumanou. African Journal of Biotecnology. **15**,47:2637–2648(2016). https://doi.org/10.5897/AJB2015.14974
6. R. Cormier. *Clarification of cashew apple juice and commercial applications*. Oxfarm Quebec, Benin, West Africa. p. 1–9(2008). https://scholar.google.com/scholar_lookup?title=Clarification%20of%20Cashew%20Apple%20Juice%20and%20Commercial%20Applications&pages=1-9&publication_year=2008&author=Cormier%2CR
7. N. Bora. Partner. **15**,2:147–152(2008). [in Bahasa Indonesia]. https://media.neliti.com/media/publications/159374-ID-pemberian-natrium-benzoat-terhadap-kuali.pdf
8. S.T.M. Sulthoniyah, T.D. Sulistiyati, E. Suprayitno. THPi STUDENT JOURNAL. **1**,1:33–45(2013). [in Bahasa Indonesia]. https://media.neliti.com/media/publications/110333-ID-none.pdf
9. A Jayanegara, H.P.S. Makkar, K. Becker. Media Peternakan. **32**,2:120–129(2009). [in Bahasa Indonesia]. http://journal.ipb.ac.id/index.php/mediapeternakan/article/download/1147/326

10. Nuradi.*Pengaruh pemberian buah sejati jambu mete (Anacardium occidentale L) sangrai terhadap kadar kolesterol total, kolesterol-LDL, dan kolesterol-HDL dalam serum tikus hiperkolesterolemik* [The effect of giving roasted cashew nuts (*Anacardium occidentale* L) on total cholesterol, LDL–cholesterol, and HDL cholesterol levels in the serum of hypercholesterolemic rats] [Thesis] Kedokteran, Universitas Airlangga (2003). [in Bahasa Indonesia]. p. 1–91. http://repository.unair.ac.id/id/eprint/35248
11. M. Harini, O.P. Astirin. Nusantara BioScience. **1**,2:53–58(2009). [in Bahasa Indonesia]. https://doi.org/10.13057/nusbiosci/n010201
12. M.W. McGowan, J.D. Artiss, D.R. Standbergh, B. Zak. J. Clin. Chem. **29**,3: 538–542(1983). https://doi.org/10.1093/clinchem/29.3.538
13. A.H.B. Wu (ed). *Tietz clinical guide to laboratory tests*, 4th ed. USA: Saunders Company (2006). p. 1856. https://books.google.co.id/books?id=aQqjBQAAQBAJ&printsec=frontcover&dq=Tietz+Clinical+Guide+to+Laboratory+Tests+4th+Edition&hl=id&sa=X&ved=2ahUKEwjnkLLQxrfrAhVNVisKHUsQAhoQ6AEwAHoECAYQAg#v=onepage&q=Tietz%20Clinical%20Guide%20to%20Laboratory%20Tests%204th%20Edition&f=false
14. K.A. Hanafiah. *Rancangan percobaan teori & aplikasi* [Design theoretical experiments & applications]. (2012). [in Bahasa Indonesia]. https://books.google.co.id/books?id=sIK3tQEACAAJ&dq=Rancangan+percobaan+teori+%26+aplikasi&hl=id&sa=X&ved=2ahUKEwiVl9fatuXqAhUUWCsKHWHVAFgQ6AEwAHoECAAQAQ
15. M. Alves–Bezerra, D.E. Cohen. Compr Physiol. **8**,1:1–44(2018). https://doi.org/10.1002/cphy.c170012
16. F.K. Welty . Curr Cardiol Rep. 15,400:1–22(2013). https://doi.org/10.1007/s11886-013-0400-4

The Ability of Water Hyacinth (*Eichhornia crasipes* Mart.) and Water Lettuce (*Pistia stratiotes* Linn.) for Reducing Pollutants in *Batik* Wastewater

Bunyamin Muchtasjar[1,4,*], *Hadiyanto* Hadiyanto[2,4], *Munifatul* Izzati[3], *Zane* Vincēviča–Gaile[5], and *Roy* Hendroko Setyobudi[6]

[1]Department of Health Science, University of Muhammadiyah Purwokerto 53181, Indonesia
[2]Department of Chemical Engineering, Diponegoro University, SH-Tembalang, Semarang 50239, Indonesia
[3]Department of Biology, Faculty of Mathematics and Natural Sciences, Diponegoro University, SH-Tembalang, Semarang 50239, Indonesia
[4]School of Postgraduate Studies, Diponegoro University, SH-Tembalang, Semarang 50239, Indonesia
[5]Department of Environmental Science, University of Latvia, Jelgavas Street 1, Room 302, Riga LV-1004, Latvia
[6]Waste Laboratory, University of Muhammadiyah Malang, Jl. Raya Tlogo Mas No.246 Malang 65144, Indonesia

Abstract. *Batik* produced by craftsmen is categorized as a small and medium industry. The problem of *Batik* production is less attention to the environment because the craftsmen lived in the middle of a residential area that has inexperience. One of less attention to the environment is no treatment of wastewater that reaches up to 95 % (61.9 L each cloth). Thus it polluted the river. This research aimed to treat the wastewater of *Batik* production by using phytoremediation. In this study, phytoremediation was carried out in three tanks in series using *Eichhornia crassipes* and *Pistia stratiotes* which are abundant in Indonesia. The treated wastewater was evaluated for its ammonia and chromium (Cr). The results showed that ammonia reduced 78.36 % by *E. crassipes*, and 73.13 % *P. stratiotes*. While *E. crassipes* reduced Cr by 63.76 % and *P. stratiotes* by 83.39 %. This research concluded that *E. crassipes* and *P. stratiotes* are potential plants to reduce the pollutants in *Batik* wastewater.

Keywords: Ammonia, chromium, phytoremediation, treatment.

1 Introduction

The role of small and medium-sized enterprises (SMEs) in economic development and growth is crucial in both developing and developed countries [1]. In Sokaraja, Indonesia there are some *Batik* SMEs that grow rapidly that collaborate as a society called 'Sawunggaling'. Those societies and other industries are located in the middle of residential

* Corresponding author: beben_ump@yahoo.com

areas. Inexperience is a main reason for less attention to the environment. None of these industries has wastewater treatment thus polluting the river.

Environmental pollution is living things, matter, energy, and other components that interfere into the environment by human activities, thus it converts environmental quality standards [2]. Environmental pollution from the batik industry can be identified from the solid color of wastewater disposal. The water usage in the *Batik* production (61.9 L each cloth) is abundant and 95 % as wastewater disposal generated from soaking, heating, and rinsing. The problem from *Batik* production involves high volumes of water and chemicals such as waxes and dyes.

These pollutants interfere with quality standards following the Indonesian Minister of Environment Regulation No. 5 of 2014, shown in Table 1.

Table 1. Comparison of wastewater batik quality and standards in Sokaraja, Indonesia

Parameters	Quality of sample (mg L^{-1})	Quality standards (mg L^{-1})
Biological Optical Density $(BOD)_5$	2.5	60
Chemical Optical Density (COD)	5.8	150
Total Suspended Solid (TSS)	10.135	50
Total phenol		0.5
Total chrome (Cr)	1.057	1.0
Total ammonia (NH_3-N)	278.681	8.0
Sulfide	381.96	0.3
Oil and fat	280	3.0
pH	9.63	6.0 to 9.0

From Table 1, it shows that some parameters have exceeded the quality standard threshold value. Even though there are regulations, the *Batik* industries ignore wastewater treatment because the punishment is too weak.

The phytoremediation model is simple, inexpensive, and easily operated [3]. Phytoremediation is considered to be a method for removing the pollutants in wastewater and recognized as a better green remediation technology using plants [4, 5] such as Water Hyacinth (*Eichhornia crassipes* Mart.) and Water Lettuce (*Pistia stratiotes* Linn.).

Water hyacinth is one of waterweed that uncontrollable growth but its nutrient absorption has provided a way for its usage in phytoremediation [4]. Moreover, Water Lettuce is a tropical plant that resists severe weather variability, fast reproduction, and simple manipulation. In addition, Water Lettuce is capable of remediating heavy metals [6]. The purpose of this study was to examine the *E. crassipes* and *P. stratiotes* plants in reducing pollutants in *Batik* industry wastewater. The pollutants indicated by ammonia and Chromium.

2 Methods

2.1 Plant material

The *E. crassipes* and *P. stratiotes* seedlings were conducted from the fish pond, in the region of Sokaraja, Banyumas, Central of Java, Indonesia. The seedlings were removed approximately 30 d growth because those plants are a perennial weed. The selection plants for this study based on their morphological character i.e green leaves and a stem length was 10 cm. The greenhouse was used to expose plants getting natural light and ambient temperature. Those situations were close to their natural environment. Plants were separated and cultivated for 14 d before the experiment was carried out with sterile water to prevent contamination with other agents and to support the growth of roots, without fertilizer or additives.

2.2 Experimental procedure

This study used a phytoremediation model with three tanks made of multilevel arranged glass with a continuous drainage system, with a flow rate of 5 L h^{-1}. The parameters observed in the treatment were ammonia and chromium.

The phytoremediation model was conducted on a laboratory scale in a greenhouse. Environmental factors were controlled by insect threats. The phytoremediation model is shown in Figure 1.

(A) (B)

Fig.1. (A) Treatment with *Eichhornia crassipes*, (B) treatment with *Pistia stratiotes*.

3 Results and discussions

E. crassipes and *P. stratiotes* reduced pollutants in *Batik* wastewater 4 % dilution. That high concentrations for *E. crassipes* and *P. stratiotes* were resistant. In the *E. crassipes* and *P. stratiotes* was used for the phytoremediation of the heavy metal, which affected leaf abnormalities, such as chlorosis and necrosis. This indicated that *E. crassipes* and *P. stratiotes* are more resistant to pollution agents proved by healthy plant and leaf quality was maintained even in the presence of high concentrations [7].

Before treatment of 4 % *Batik* wastewater, it observed ammonia of 5.73 mg L^{-1} and Cr of 596 mg L^{-1}. The *E. crassipes* and *P. stratiotes* grew in Wastewater for 15 d, with a recirculation rate of 20 L h^{-1} decreased of ammonia, as shown in Figure 2.

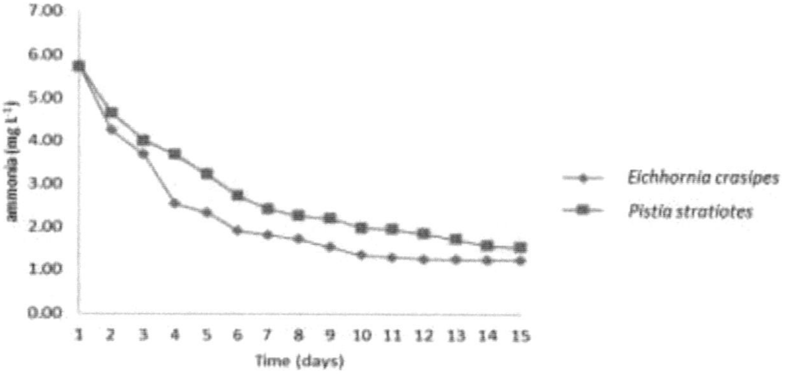

Fig.2. The decrease in ammonia parameters in *Eichhornia crassipes* and *Pistia stratiotes*.

Figure 2 showed that *E. crassipes* reduced ammonia reached up to 1.54 mg L^{-1} for 15 d, equal to 78.36 % and 1.24 mg L^{-1} or 73.13 % by *P. stratiotes.* The decrease of ammonia is an indication of absorption by plants [8].
The results of the chromium in *Batik* wastewater treatment shown in Figure 3.

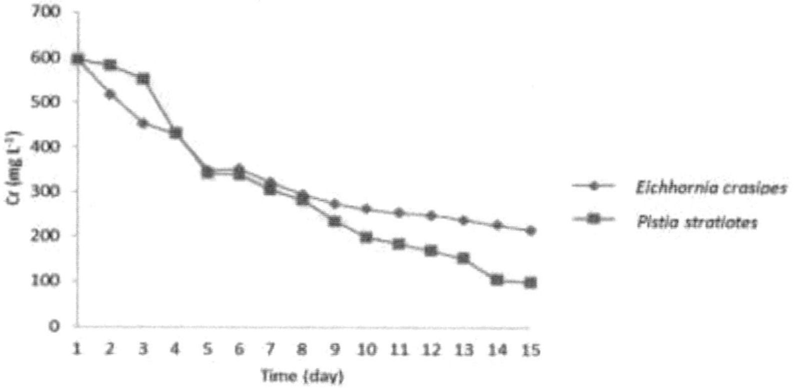

Fig.3. Decrease in Cr parameters in *Eichhornia crasipes* and *Pistia stratiotes*.

The decrease in chromium was 63.76 % in *E. crassipes* and 83.39 % in *P. stratiotes*. Both plants have hair-like roots that are used as a filter, tolerant, accumulate and eliminate environmental pollutants [6].

E. crassipes and *P. stratiotes* have shown their effectiveness in treating wastewater [9]. These plant species can sterilize higher heavy metals and help decontaminate soil, water, and others [5]. This also shows that the two plants have potential as bioindication and phytoremediation of contaminated aquatic environments [10]. In addition, Phytoremediation is related to the toxicity of heavy metals in contaminated ecosystems and bioremediation technology that is feasible, sustainable, and environmentally friendly [11]. This is in line with the results in the wastewater of his research Tewari et al. [12]. The impact of Cr contamination contained in *Batik* wastewater in physiology plants depends on metal speciation, which is responsible for mobilization, subsequent absorption and toxicity produced in plant systems [13].

4 Conclusion

This research concluded that *E. crasipes* and *P. stratiotes* plants reduced pollutants in *Batik* wastewater. The research indicated that the reduction of ammonia and Chromium were effective. Thus the two plants are potential adsorbent Ammonia and Chromium removal technology for wastewater treatment

Authors gratefully thank to the LPDP - *Lembaga Pengelola Dana Pendidikan* [Education Fund Management Agency] Number LPDP 20161141021636 of the Indonesian Ministry of Finance

References

1. R. Rahab, N. Anwar, R.E. Priyono. J. Comp. Int. Manag. **19**,1:27–45(2016) https://journals.lib.unb.ca/index.php/JCIM/article/view/24610
2. Indonesia. *Republik of Indonesia Law Number 32, 2009*, concerning environmental protection and management. (2009).[in Bahasa Indonesia] https://komisiinformasi.go.id/?p=1817
3. U. Tahir, A. Yasmin, U.H. Khan. J. King Saud Univ. Sci. **28**,2:119–130(2016). https://www.google.com/url?sa=t&rct=j&q=&esrc=s&source=web&cd=&cad=rja&uact=8&ved=2ahUKEwiY5vGsteDqAhVYfX0KHetjBNEQFjAAegQIBBAB&url=https%3A%2F%2Fcore.ac.uk%2Fdownload%2Fpdf%2F82137348.pdf&usg=AOvVaw1vO7U6daapSwNg7HDMuRzY
4. S. Rezania, M. Ponraj, A. Talaiekhozani, S.E. Mohamad, M.F.M. Din, S.M. Taib, et al. J. Environ. Manag. **163**:125–133(2015). https://www.google.com/url?sa=t&rct=j&q=&esrc=s&source=web&cd=&cad=rja&uact=8&ved=2ahUKEwi2i4HeteDqAhVBfH0KHewpCXAQFjAAegQIBRAB&url=https%3A%2F%2Fwww.ncbi.nlm.nih.gov%2Fpubmed%2F26311085&usg=AOvVaw2gApAJFEZEPyLU0BtEI-4q
5. A.R. Borker, A.V. Mane, G.D. Saratale, G.R. Pathade. Emir. J. Food Agric. **25**,6:443–456(2013). https://www.google.com/url?sa=t&rct=j&q=&esrc=s&source=web&cd=&cad=rja&uact=8&ved=2ahUKEwjE04K8tuDqAhXhmeYKHT6XBvYQFjAAegQIAxAB&url=https%3A%2F%2Fwww.researchgate.net%2Fpublication%2F271305647_Phytoremediation_potential_of_Eichhornia_crassipes_for_the_treatment_of_cadmium_in_relation_with_biochemical_and_water_parameters&usg=AOvVaw28YKypTG3v7SykliBaU0lT
6. E. Agostini, M.A. Talano, P.S. González, A.L.W. Oller, M.I. Medina. Appl. Microbiol. Biotechnol. **97**:1017–1030(2013). https://www.google.com/url?sa=t&rct=j&q=&esrc=s&source=web&cd=&cad=rja&uact=8&ved=2ahUKEwiOsMfbtuDqAhWCheYKHTR3COIQFjAAegQIBxAB&url=https%3A%2F%2Fpubmed.ncbi.nlm.nih.gov%2F23287856%2F&usg=AOvVaw1bKp0aC9TyyiipR6B1Lqhy
7. D.F. Escoto, M.C. Gayer, M.C. Bianchini, G. da C. Pereira, R. Roehrs, E.L.G. Denardin. Chemosphere, **227**:299–304(2019). https://www.google.com/url?sa=t&rct=j&q=&esrc=s&source=web&cd=&cad=rja&uact=8&ved=2ahUKEwit9-Cet-DqAhXtILcAHR8EARcQFjAAegQIBBAB&url=https%3A%2F%2Feuropepmc.org%2Farticle%2Fmed%2F30999171&usg=AOvVaw109Fo_AuTvgWUkQeCL20-q
8. K. Sung, G.J. Lee, C. Munster. Int. J. Phytoremediation. **17**,10:936–944(2015). https://doi.org/10.1080/15226514.2014.1003791

9. C.O. Akinbile, M.S. Yusoff. Int. J. Phytoremediation. **14**,3:201–211(2012). https://www.google.com/url?sa=t&rct=j&q=&esrc=s&source=web&cd=&cad=rja&uact=8&ved=2ahUKEwjbhK64t-DqAhVE9XMBHelYCPEQFjAAegQIARAB&url=https%3A%2F%2Fwww.researchgate.net%2Fpublication%2F224923917_Assessing_Water_Hyacinth_Eichhornia_crassipes_and_Lettuce_Pistia_stratiotes_Effectiveness_in_Aquaculture_Wastewater_Treatment&usg=AOvVaw3w64z5UlwvClUxS3nPpQ2C
10. F. Farnese, J. Oliveira, F. Lima, G. Leão, G. Gusman, L. Silva. Braz. J. Biol. **74**,3suppl 1:S108–S112(2014). https://www.google.com/url?sa=t&rct=j&q=&esrc=s&source=web&cd=&cad=rja&uact=8&ved=2ahUKEwiRkOzUt-DqAhUCjuYKHSREAqIQFjAAegQIBRAB&url=https%3A%2F%2Fwww.scielo.br%2Fscielo.php%3Fpid%3DS1519-69842014003000013%26script%3Dsci_arttext&usg=AOvVaw1mDg_4PDYx7QtgzVYEQ7yJ
11. D. Mani, C. Kumar. Int. J. Environ. Sci. Technol. **11**:843–872(2014). https://www.google.com/url?sa=t&rct=j&q=&esrc=s&source=web&cd=&cad=rja&uact=8&ved=2ahUKEwi5s7Trt-DqAhXq7XMBHefJCG0QFjAAegQIBBAB&url=https%3A%2F%2Flink.springer.com%2Farticle%2F10.1007%2Fs13762-013-0299-8&usg=AOvVaw31gmaoZX3pp6KMx7UhSb-u
12. A. Tewari, R. Singh, N.K. Singh, U.N. Rai. Bioresour. Technol. J., **99**:8715–8721(2008). https://www.google.com/url?sa=t&rct=j&q=&esrc=s&source=web&cd=&cad=rja&uact=8&ved=2ahUKEwjVhv-BuODqAhUl8HMBHSX1AIkQFjAAegQIBRAB&url=https%3A%2F%2Fwww.academia.edu%2F5875680&usg=AOvVaw0V2shz_eF7gc_lP0CrBu_t
13. M. Junaid, M. Adnan, N. Khan, M. U. Rahman, N. Ali. Fuuast J. Biol. **3**,2:95–103(2013). https://www.google.com/url?sa=t&rct=j&q=&esrc=s&source=web&cd=&cad=rja&uact=8&ved=2ahUKEwjM8IuxuODqAhVBIbcAHdhlCUgQFjABegQIBRAB&url=https%3A%2F%2Fwww.researchgate.net%2Fprofile%2FArvind_Singh56%2Fpost%2FMethod_for_plant_growth_measurement_for_Brassica%2Fattachment%2F59d64ebd79197b80779a8076%2FAS%253A494301905944576%25401494862166949%2Fdownload%2F95-103.pdf&usg=AOvVaw1K8PBhMbmdnrtQivoCAfl6

Potential Analysis of Low Economic Value Fish in Lamongan Regency, East Java, Indonesia

Choirul Anam[1,*], *Noor* Harini[2], *Damat* Damat[2], *Ahmad* Wahyudi[1], *Yuli* Witono[3], *Nita* Kuswardhani[3], *Moh.*Azus Shony Azar[4], *Olga* Anne[5], and *Diana* Rachmawati[6]

[1]Department of Agriculture Science, Postgraduate Program, University of Muhammadiyah Malang, Jl. Raya Tlogomas No. 246, Malang 65144, East Java, Indonesia
[2]Department of Food Science and Technology, Faculty of Agriculture and Animal Science, University of Muhamadiyah Malang
[3]Department of Agricultural Product Technology, Faculty of Agricultural Technology, University of Jember, Jl. Kalimantan no. 37, Jember 68121, East Java, Indonesia
[4]Department of Management, Faculty of Economic, Darul Ulum Islamic University, Jl. Airlangga no. 3, Lamongan 62253, East Java, Indonesia
[5]Department of Natural Sciences, Faculty of Marine Technology and Natural Sciences of Klaipeda University, H. Manto g. 84, Klaipėda 92294, Lithuania
[6]Department of Aquaculture, Faculty of Fisheries and Marine Sciences, Diponegoro University, Jl. Prof. Soedarto, SH, Tembalang, Semarang 50275, Central Java, Indonesia

Abstract. The purpose of this study was to determine the potential of low commercial value fish species in Lamongan Regency, East Java, Indonesia. The processed data included primary and secondary data. Data were processed using the LQ method and analyzed descriptively by way of compiling data into tables. The survey results showed that there are ten types of low commercial value fish in Lamongan Regency, East Java, Indonesia, namely *Sardinella fimbriata* (Tembang fish), *Leiognathus bindus (*Peperek fish), *Harpadon nehereus* (Puso fish), *Sardinella lemuru*, (Lemuru fish), *Saurida tumbil* (Beloso fish), *Gerres abbreviatus* (Kapasan fish), *Anodontostoma chacunda* (Juwi fish), and *Cynoglossus lingua* (Lidah fish). The results of the LQ method showe that there are three types of fish that have the potential to be made of high economic value products, namely *L. bindus* (11.36), *A. chacunda* (2.35), and *S. fimbriata* (2.32). Fish Auction (TPI) that has the potential in terms of fish productivity to be developed is TPI Weru, Lamongan, East Java, Indonesia. The typology of fishermen in TPI Weru has less experience working as a fisherman than other TPI but the level of education is higher and the age of fishermen is younger so that it affects the productivity of fish catches.

Keywords: Method of location quotient, trash fish, typology of fishing

1 Introduction

Lamongan is one of the regencys in East Java, Indonesia astronomically located on 6°51'54" to 7°23'6" south latitude and between 112°4'41" to 112°33'12" east longitude. Sea sector fisheries production in Lamongan regency, East Java, Indonesia has increased

* Corresponding author: choirulanam@unisda.ac.id

every year. Production in 2014, 2015, 2016 and 2017 was 71 553 t, 72 346 t, 73 142 t and 73 356 t [1]. Lamongan as one of the minapolitan areas in East Java, Indonesia [2, 3].

Fishing by fishermen is not only fish with high economic value but many fishermen accidentally catch fish or by catch with low economic value commonly called trash fish [4]. Utilization of trash fish to date has only been used as raw material for animal feed. In fact, trash fish has the potential to be used as food with higher economic value [5] because of its relatively high protein content.

Identification of trash fish potential can be done with several methodes. One method that can be used is location quotient (LQ). LQ is one of the methodes commonly used in the basic economic model that presents a relative comparison between the ability of one sector between the regions investigated with the ability of the same sector in a wider region. The LQ method is one of the methodes commonly used in the basic economic model as a first step to understanding the activity sectors that are driving growth. This method presents a relative comparison between the ability of a sector in the area being investigated with the same ability in a wider area. The unit used as a measure to produce the LQ coefficient is the amount of labor, production, or other units that can be used as criteria [6]. The LQ mathematical formula is stated as follows in Equation (1).

$$LQ = \frac{Si/S}{Ni/N} \quad (1)$$

Description:
LQ : Location quotient value
Si : Number of sectors i in an area
S : Total number of sectors in a region
Ni : Number of sectors i in the reference area
N : Total number of all sectors in the reference area

The formulation of LQ will provide alternative values ie LQ > 1, LQ < 1, and LQ = 1. If using production value as material consideration in calculating LQ, then:

i. LQ > 1, means that the commodity is a base sector. The production of the commodity concerned has exceeded consumption needs in the area where the commodity was produced and the excess can be sold abroad.
ii. LQ < 1, means that the production of these commodities is not sufficient to meet consumption needs in the area concerned and fulfillment is imported from other regions.
iii. LQ = 1, means that the production of the commodity in question is only sufficient for the needs of the local area.

Base economic theory clarifies all economic activities into two sectors, namely the base sector and the non–base sector. Explains that the basis of activity is the activity of a community whose results in the form of goods and services are intended for export out of the community or oriented outward, regionally, nationally and internationally.

While non–base activities are community activities whose results are in the form of goods or services intended for the community itself in the area of the community's economic life.

Each method of analysis has advantages and limitations as well as the LQ method. According to [7], the advantage of the LQ method in identifying the potential of superior commodities is that they are simple, easy and do not require complicated data processing programs. Completion of the analysis is enough with a spread sheet from Excel or the Lotus program, even if the data obtained is not too much you can use a calculator.

The purpose of this study was to determine the potential of low commercial value fish species in Lamongan Regency, East Java, Indonesia. Location of the research are presented in Figure 1.

Fig. 1. Location of the research in Fish Auction Sites Lamongan, East Java, Indonesia

2 Methods

2.1 Data collection techniques

This study was conducted in May 2017 to September 2017. To determine the potential of trash fish was carried out using direct observation method, mapping area, direct interviews in the field and the distribution of questionnaires to fishermen's pillars. in three Fish Auction (TPI – *Tempat Pelelangan Ikan*) Lamongan regency, East Java, Indonesia.

2.2 Data and data sources

The type of data collected consists of secondary data and primary data. Primary data obtained directly from the source through the results of observations and interviews in the field. Secondary data were obtained from the Lamongan regency fisheries service, East Java, Indonesia fisheries service.

2.3 Data analysis

Data analysis was performed in several stages, namely:

i. Insert data
   Insert data from the results of questionnaires in each TPI in Spreadsheet Ms. Excel.
ii. Calculate the amount of production of each type of fish and the total production of fish caught in each TPI.
   After the data is entered in the spreadsheet then the data is calculated by the amount of production of each type of fish and the total production of fish caught in each TPI. Each was given an S and Si notation.
iii. Calculating the amount of production of each type of fish and the total production of fish caught in Lamongan Regency, East Java, Indonesia.
   Data used to calculate the amount of production of each type of fish and the total production of fish caught in Lamongan Regency, East Java, Indonesia. Each of them is given N and Ni notations.

iv. Calculating the value of LQ
Location Quotient (LQ) is an index to compare the share of fish in each TPI in capture fisheries activities with the total share of these activities in the total activity of Lamongan Regency, East Java, Indonesia. LQ formula with the help of a spreadsheet from Excel so that its value is known. Mathematical formula LQ in Equation (2)

$$LQ = \frac{Si/S}{Ni/N} \tag{2}$$

Description:
LQ : Location quotient value
Si : Number of sectors i in an area
S : Total number of sectors in a region
Ni : Number of sectors i in the reference area
N : Total number of all sectors in the reference area

v. Interpretation of LQ
To be able to interpret the results of the LQ analysis, then:
a. if LQ > 1, it means that there is a surplus of fisheries production in the TPI and the commodity is a base sector;
b. if LQ < 1, it means that fisheries production in TPI is relatively smaller than capture fisheries activities in Lamongan Regency, East Java, Indonesia or the supply of fishery products is not sufficient;
c. if LQ = 1, it means that fisheries production in TPI is equivalent to fishery activities in Lamongan Regency, East Java, Indonesia.

3 Results and discussions

Processed data include primary and secondary data. The data obtained were processed using the LQ method and analyzed descriptively by compiling data into tables.

The survey results show that there are ten types of low commercial value fish in Lamongan Regency, East Java, Indonesia, namely *Sardinella fimbriata* (Valenciennes, 1847), *Leiognathus bindus* (Valenciennes, 1835), *Harpadon nehereus* (Hamilton, 1822), *Sardinella lemuru* (Bleeker, 1853), *Saurida tumbil* (Bloch, 1795), *Gerres abbreviatus* (Bleeker, 1850), *Anodontostoma chacunda* (Hamilton, 1822), and *Cynoglossus lingua* (Hamilton, 1822). However, every TPI is not always a landing place for all types of fish. All types of fish with low commercial value are found in TPI Weru while TPI Brondong, Lamongan, East Java, Indonesia and Lohgung, Lamongan, East Java, Indonesia only have two to three types of fish with low commercial value.

The results of interviews of 52 fishermen respondents at the Brondong Fish Auction Place are as in Table 1 below.

Table 1. Fish production (kg) in Brondong, Lamongan, East Java, Indonesia

	A. chacunda	*L. bindus*	*C. lingua*	*S. fimbriata*
N Valid	44	47	26	45
Mean	41.64	44.89	6.96	35.11
Median	40.00	35.00	6.50	30.00
Minimum	2	10	1	2
Maximum	95	100	20	90

Information: N = The number of fishermen who catch fish; and Valid = Type of descriptive analysis

Table 1 shows that the average the highest average production in Brondong is *L. bindus* as much as 44.89 kg, *A. chacunda* is 41.64 kg and *S. fimbriata* is 35.11 kg. For more details, as in Figure 2.

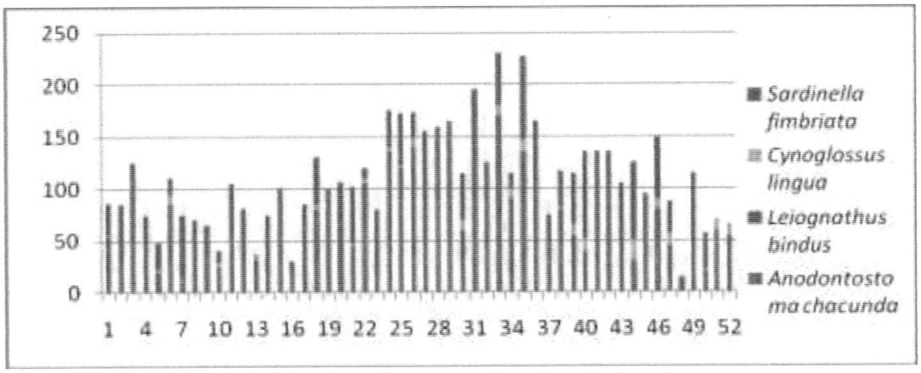

Fig. 2. Fish production (kg) in Brondong, Lamongan, East Java, Indonesia

The lowest price of potential low economic value fish in Brondong as in Table 2 below.

Table 2. Fish Prices (IDR) in Brondong, Lamongan, East Java, Indonesia

	A. chacunda	*L. bindus*	*C. lingua*	*S. fimbriata*
N Valid	44	47	27	45
Mean	3 125.00	2 297.87	10 592.59	2 300.00
Median	3 000.00	2 500.00	10 000.00	2 500.00
Minimum	2 500	2 000	10 000	1 500
Maximum	3 500	2 500	12 000	2 500
Sum	137 500	108 000	286 000	103 500

Information: N = The number of fishermen who catch fish; and Valid = Type of descriptive analysis

Table 2 shows that the lowest average price of potential fish with high production in Brondong, namely *L. bindus* is IDR 2 297.87 kg^{-1}, *A. chacunda* is IDR 3 125.00 kg^{-1} and *S. fimbriata* is IDR 2 300.00 kg^{-1}. For more details, as shown in Figure 3.

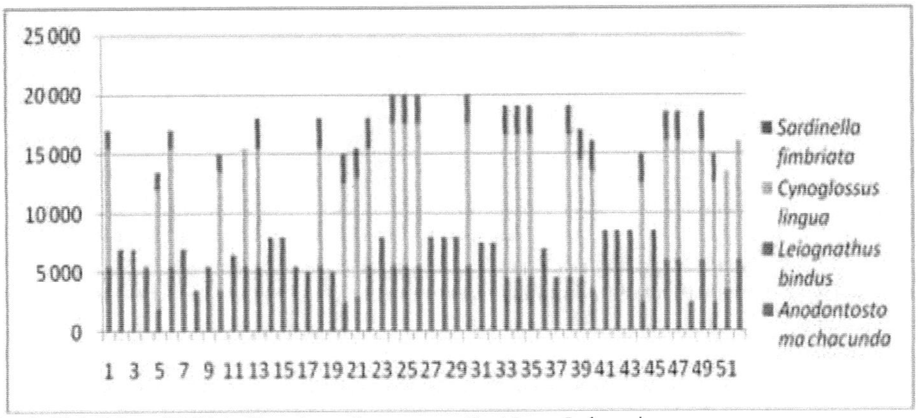

Fig. 3. Fish Prices (IDR) in Brondong, Lamongan, East Java, Indonesia

Interviews of 99 fisherman respondents at the Lohgung Fish Auction Area show that *A. chacunda* and *L. bindus* fish production, as in Table 3 below.

Table 3. Fish production (kg) in Lohgung, Lamongan, East Java, Indonesia

	A. chacunda	*L. bindus*
N Valid	51	57
Mean	8.43	7.89
Median	5.00	7.00
Minimum	2	2
Maximum	50	20
Sum	430	450

Information: N = the number of fishermen who catch fish; and Valid = type of descriptive analysis

Table 3 shows that the average fish production in Lohgung is *L. bindus* amounted to 7.89 kg and *A. chacunda* as much as 8.43 kg. For more details, as in Figure 4.

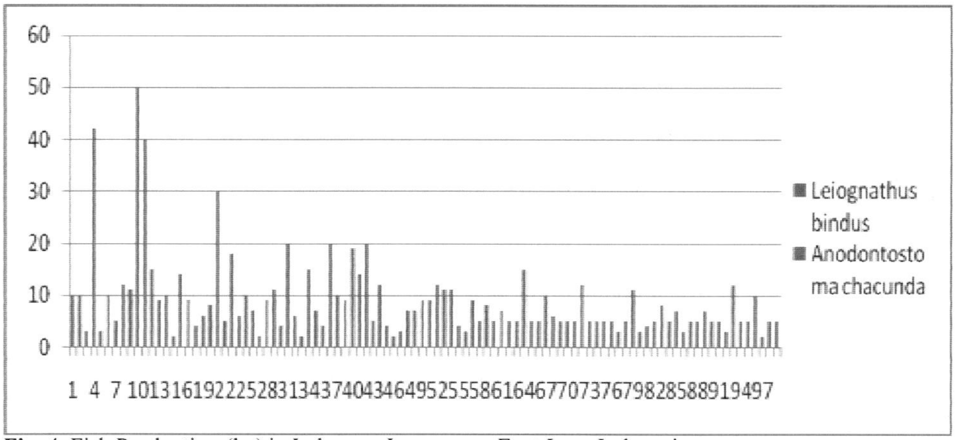

Fig. 4. Fish Production (kg) in Lohgung, Lamongan, East Java, Indonesia

The lowest price of potential low economic value fish in Lohgung as in Table 4 below.

Table 4. Fish Prices (IDR) in Lohgung, Lamongan, East Java, Indonesia

	A. chacunda	*L. bindus*
N Valid	55	84
Mean	3 500.00	2 000.00
Median	3 500.00	2 000.00
Minimum	3 500	2 000
Maximum	3 500	2 000
Sum	192 500	168 000

Information: N = The number of fishermen who catch fish; and Valid = Type of descriptive analysis

Table 4 shows that the lowest average price of fish in Lohgung is *L. bindus* i.e. IDR 2 000.00 kg^{-1}, and *A. chacunda* is IDR 3 500.00 kg^{-1}. For more details, as in Figure 5.

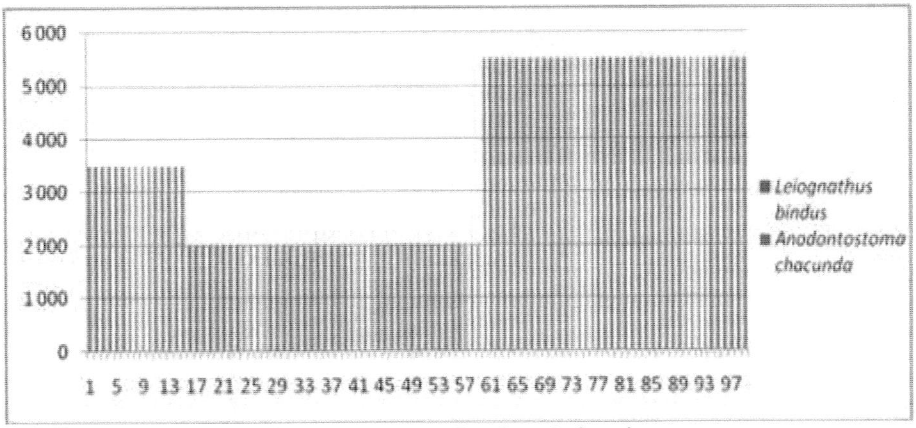

Fig. 5. Fish Prices (IDR) in Lohgung, Lamongan, East Java, Indonesia

The results of interviews of 58 fishermen respondents at the Weru Fish Auction Place show that the production of ten types of fish, as in Table 5 below.

Table 5. Fish production (kg) in Weru, Lamongan, East Java, Indonesia

	A. chacunda	*L. bindus*	*H. nehereus*	*S. lemuru*	*S. tumbil*	*G. abbreviatus*	*S. fimbriata*	*C. lingua*	*A. albimaculosus*	*P. arsius*
N Valid	28	28	19	4	2	2	12	1	1	3
Mean	35.11	38.14	12.74	8.75	16.00	16.50	17.08	30.00	30.00	12.33
Median	35.00	35.00	10.00	2.00	16.00	16.50	15.00	30.00	30.00	5.00
Minimum	2	1	5	1	2	3	5	30	30	2
Maximum	80	80	30	30	30	30	30	30	30	30
Sum	983	1068	242	35	32	33	205	30	30	37

Information: N = The number of fishermen who catch fish; and Valid = Type of descriptive analysis

Table 5 shows that the highest average fish production in Weru namely *L. bindus* by 38.14 kg, *A. chacunda* by 35 kg, and *S. fimbriata* by 17.08 kg. For more details, as in Figure 6.

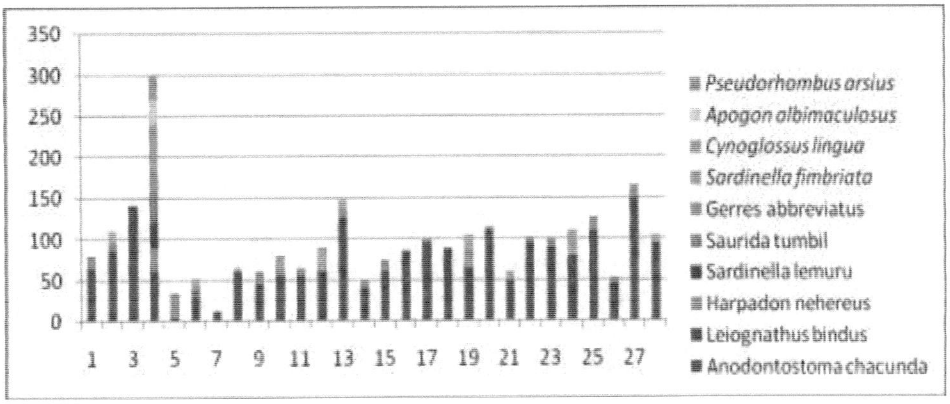

Fig. 6. Fish Production (kg) in Weru, Lamongan, East Java, Indonesia

The lowest price of potential low economic value fish in Weru as in Table 6 below.

Table 6. Fish Prices (IDR) in Weru, Lamongan, East Java, Indonesia

	A. chacunda	*L. bindus*	*H. nehereus*	*S. lemuru*	*S. tumbil*	*G. abbreviatus*	*S. fimbriata*	*C. lingua*	*A. albimaculosus*	*P. arsius*
N Valid	22	22	18	4	5	13	2	18	1	21
Mean	1 485.714	1 432.143	2 659.259	571.429	640.000	914.296	652.174	1 421.429	14.274	5 712.143
Median	1 800.000	1 800.000	4 000.000	2 000.000	4 000.000	1 800.000	7 500.000	2 200.000	2 000.000	8 000.000
Minimum	1 800.0	1 500.0	2 500.0	2 000.0	2 000.0	1 800.0	5 000.0	1 800.0	2 000.0	3 500.0
Maximum	2 500.0	2 500.0	4 500.0	10 000.0	4 000.0	3 000.0	10 000.0	3 000.0	2 000.0	20 000.0
Sum	41 600.0	40 100.0	71 800.0	16 000.0	16 000.0	25 600.0	15 000.0	39 800.0	2 000.0	188 500.0

Information: N = The number of fishermen who catch fish; and Valid = Type of descriptive analysis

Table 6 shows that the lowest average price of fish in Weru is *L. bindus* which is IDR 1 432 143 kg^{-1}, *A. chacunda* is IDR 1 485 714 kg^{-1}, and *S. fimbriata* amounting to IDR 652 174 kg^{-1}. For more details, as shown in Figure 7.

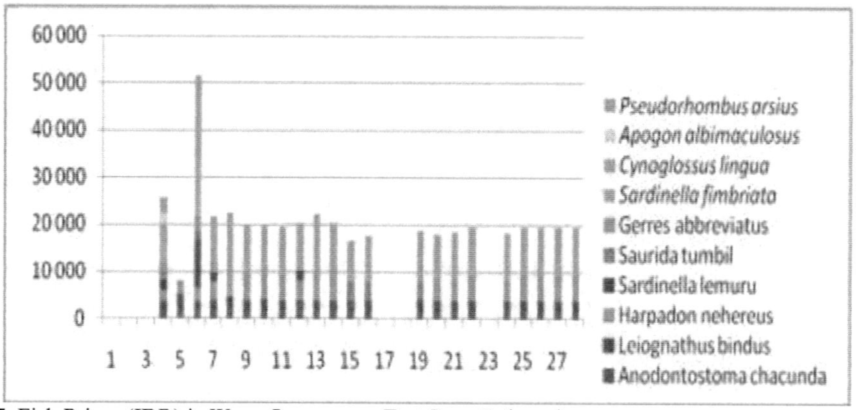

Fig. 7. Fish Prices (IDR) in Weru, Lamongan, East Java, Indonesia

From the interviews and the Lamongan regency, East Java, Indonesia fisheries service the LQ method calculation results can be calculated in Table 7.

Table 7. LQ Calculation results for each fish type in TPI Lamongan Regency, East Java, Indonesia

Type of Fish	LQ Value		
	Weru fish auction place	Brondong fish auction place	Lohgung fish auction place
A. chacunda	2.35	0.05	2.07
L. bindus	11.36	0.24	9.56
H. nehereus	0.14	–	–
S. lemuru	0.11	–	–
S. tumbil	0.03	–	–
G. abbreviatus	0.03	–	–
S. fimbriata	2.32	0.12	–
C. lingua	0.00	–	–
A. albimaculosus	0.03	–	–
P. arsius	0.20	–	–

Table 7 shows the highest LQ value or more than one for the type of fish in each TPI namely *L. bindus* (11.36) at TPI Weru and *L. bindus* (9.56) at TPI Lohgung, *A. chacunda*

(2.35) at TPI Weru and *A. chacunda* (2.07) at TPI Lohgung, and *S. fimbriata* (2.32)) in TPI Weru, while others whose value LQ < 1. [7] states that if the LQ value is greater than one then the base sector. This means that the three types of fish are the base sector in the class of fish species with low commercial value. In accordance with statement [8] which states that if the LQ value is more than one then the commodity is the base sector and the production of the commodity concerned has exceeded consumption needs in the area where the commodity was produced and the excess can be sold abroad or utilized as other products with value high economy. Besides these three types of fish are also habitats in the northern waters of Java. [9], said that the villagers of Weru, Paciran have a very high dependence on marine products. According to [10], TPI Weru has waters with broader coral reefs. The trash fish species are mostly demersal/reef fish species so that they are commonly found in the waters of TPI Weru. TPI Brondong is a high commercial fish auction center while TPI Lohgung is a center for rebon and anchovy production. Mangroves are breeding places for shrimp and rebon, while the waters in the Lohgung TPI area are overgrown with mangroves. This is why TPI Lohgung is the center of *Mysis relicta* (rebon or Zooplankton with a length of 1 cm to 1.5 cm and used for making shrimp paste) and *Stolephorus* sp. (anchovy with a body length of about 145 mm to 5 cm) production.

Based on the results of calculations using the LQ method it can be seen that the types of fish with more than one LQ value have the potential to be utilized as fishery products with high commercial value. In addition, it can be seen that the TPI that has the potential to develop fisheries results is the TPI Weru because there is more raw material supply than the others.

Fishermen's background also affects the productivity of the fish produced. The results of typology survey of fishermen can be seen in Table 8. Based on Table 8, it can be seen that fishermen in TPI Lohgung work as fishermen longer (21.60 yr) compared to other TPI. The fishermen in TPI Weru are younger (42 yr) compared to other TPI. Age greatly affects productivity. The more age a person has, his productivity will decrease. This can be seen in the LQ value of fish species in TPI Weru which is higher than the others. In addition, it is also in accordance with the statement [11, 12] which states that the age of the workforce is sufficient to determine success in doing a job, both physical and non–physical in nature. In general, older workers have weak and limited physical labor, whereas young workers have strong physical abilities.

Table 8. Fisherman Characteristics of each TPI in Lamongan Regency, East Java, Indonesia

Criteria	**Fish auction place**		
	Weru	**Brondong**	**Lohgung**
Length of work (yr)	17.44	21.32	21.60
Age (yr)	42	50	46
Education	Junior high school	Primary school	Primary school
Total of familiy	4	4	5
Total income (million)	2.18	2.60	1.01
Catching tool	payang, nets, purse seine	Fishing line	Purse seine, payang, dogol
Type of boat	Motor boat	Outboard motor	Motor boat

In addition, the educational background of fishermen in TPI Weru is higher than the others. Education level also affects productivity. [13] states that there are several factors that affect one's productivity, namely the level of education and expertise, working conditions, type of technology, health, mental attitude, incentive systems, and job satisfaction. According to [14], people who have higher formal or informal education will have broader insight. The high awareness of the importance of productivity, will encourage the relevant workforce to take productive actions.

The highest amount of income is found in fishermen in TPI Brondong (2 600 000) compared to the others. This is because TPI Brondong is the largest fish landing center in Lamongan, thus affecting the income of fishermen with the amount of capture fisheries production reaching ± 100 t d^{-1} compared to other TPI which only reaches 10 t d^{-1}. In addition, the Brondong TPI is also a fish trading center with high commercial value, thus affecting the income of the fishermen in the TPI Brondong.

The fishing gear used varies in TPI Weru using the payang, net, and purse seine. TPI Brondong uses fishing gear such as in Lampung, Indonesia bay waters [15], while TPI Lohgung uses purse seine, *payang* and *dogol* fishing gear. *Payang* fishing gear is a trawl bag used to catch hordes of surface fish (pelagic fish) where both wings are used to frighten or surprise and herd fish into the bag like on the Malabero beach in Bengkulu city, Bengkulu, Indonesia [16]. According to [16], the types of fish that are used for catching by Payang are fish that live in groups on the surface layer of water, both in groups of the same type or in different types. The catch is mainly small pelagic fish species such as *Decapterus macrosoma* (Bleeker, 1851), *Selar crumenophthalmus* (Bloch, 1793), *Thunnus alalunga* (Bonnaterre, 1788), *Sardinella fimbriata* (Valenciennes, 1847), *Anodontostoma chacunda* (Hamilton, 1822) [17]. *Dogol* is used to catch types of anchovy and Rebon fish and is widely used by fishermen from TPI Lohgung because it is a center for producing anchovy and Rebon. Purse seine is used to catch pelagic fish species [18]. The types of fish caught purse seine are *Decapterus macrosoma* (Bleeker, 1851), *Selar crumenophthalmus* (Bloch, 1793), *Sardinella fimbriata* (Valenciennes, 1847), *Thunnus alalunga* (Bonnaterre, 1788), *Anodontostoma chacunda* (Hamilton, 1822), and *Scomberomorus sp*. (Lacépède, 1801) fish such as in Fish Auction Center (PPI – *Pusat Pelelangan Ikan*) Pulolampes, Brebes in Central Java, Indonesia. Indonesia [19] and in the Java Sea, Indonesia [20].

4. Conclusions

The survey results show that there are ten types of low commercial value fish in Lamongan Regency, East Java, Indonesia, namely *A. chacunda*, *L. bindus*, *H. nehereus*, *S. lemuru*, *S. tumbil*, *G. abbreviatus*, *S. fimbriata*, and *C. lingua*. The results of the LQ method show that there are three types of fish that have avalue LQ > 1 which the haspotential to be a high economic value product, namely *L. bindus* (11.36) in Weru TPI and *L. bindus* (9.56) in Lohgung TPI, *A. chacunda* (2.35) at TPI Weru and *A. chacunda* (2.07) at TPI Lohgung, and *S. fimbriata* (2.32) at TPI Weru. While the other LQ value < 1. TPI that has the potential in terms of fish productivity to be developed is TPI Weru. The fishermen in TPI Weru have less experience working as fishermen than other TPI but the level of education is higher and the age of the fishermen is younger so that it affects the productivity of fish catches. The highest amount of income is found in fishermen in TPI Brondong because TPI is the biggest fish landing center in Lamongan Regency, East Java, Indonesia and is a center for selling fish of high economic value. The fishing gear used in catching fish are *payang*, net, purse seine, fishing line, *dogol.*

References

1. Badan Pusat Statistik Kabupaten Lamongan, *Kecamatan Lamongan dalam Angka*, BPS Kabupaten Lamongan, ISBN: 352446.1802 (2018). p. 170. [in Bahasa Indonesia]. https://lamongankab.bps.go.id/publication.html?Publikasi%5BtahunJudul%5D=&Pu

blikasi%5BkataKunci%5D=lamongan+dalam+angka+2018&Publikasi%5BcekJudul%5D=0&yt0=
2. D.R. Raissa, R.P. Setiawan, D. Rahmawati. Procedia–Social Behav. Sci. **135**: 167–171(2014). https://doi.org/10.1016/j.sbspro.2014.07.342
3. H. Martadwiprani, D. Rahmawati. Procedia–Social Behav. Sci. **135**:106–111(2014). https://doi.org/10.1016/j.sbspro.2014.07.332
4. F.K.E. Nunoo, J.O. Boateng, A.M. Ahulu, K.A. Agyekum, U.R. Sumaila. Fish. Res. **96**,2–3:167–172(2009). https://doi.org/10.1016/j.fishres.2008.10.010
5. M.D.M. Gil, M. Palmer, R. Rosselló, J. Alós, M. Cabanellas–Reboredo, A.M. Grau, et al. Fish. Res. **205**:105–114(2018). https://doi.org/10.1016/j.fishres.2018.04.011
6. S.B. Billings, E.B. Johnson. Reg. Sci. Urban Econ. **42**,4:642–647(2012). https://doi.org/10.1016/j.regsciurbeco.2012.03.003
7. M.E. Hidayat, R.R. Suprihardjo. J. Tek. POMITS. **3**,1:C16–C19(2014). [in Bahasa Indonesia]. https://scholar.google.com/scholar?hl=en&as_sdt=0%2C5&q=Identifikasi+Sub+Sektor+Unggulan+Kecamatan+di+Kabupaten+Lombok+Tengah.&btnG=
8. M. Fattah, N.U. Tiwi, A.I. Candra. J. Econ. Soc. Fish. Mar. **4**,2:135–143(2017). [in Bahasa Indonesia]. https://scholar.google.com/scholar?hl=en&as_sdt=0%2C5&q=Analisis+Potensi+dan+Peluang+Pengembangan+Sub+Sektor+Perikanan+Tangkap+Laut+di+Kabupaten+Lamongan&btnG=
9. A. Syamsuddin, A. Fauzi, A. Fahrudin, E. Anggraini.. IOP Conf. Series: Earth and Environmental Science **420** (2020) 12030. p. 7. https://doi.org/10.1088/1755-1315/420/1/012030
10. D.C. Pratiwi, D. Yona, M. Mulyanto, Y.R. Amalia. Res. J. Life Sci. **3**,2: 65–78(2017). https://doi.org/10.21776/ub.rjls.2016.003.02.1
11. K.H. Frosch. Int. J. Manag. Rev. **13**,4:414–430(2011). https://doi.org/10.1111/j.1468-2370.2011.00298.x
12. B.I.J.M. Van der Heijden, A.H. de Lange, E. Demerouti, C.M. Van der Heijde. J. Vocat. Behav. **74**,2:156–164(2009). https://doi.org/10.1016/j.jvb.2008.12.009
13. M. Soelton, N.A. Visona, I.N. Aulia, F. Rohman, Y.B. Abadi, D. Adelia. Arch. Bus. Res. **8**,5:49–62(2020). https://doi.org/10.14738/abr.85.8183
14. E.A. Okoro, M.C. Washington. J. Divers. Manag. **7**,1:57–62(2012). https://doi.org/10.19030/jdm.v7i1.6936
15. T. Hariyanto, M.S. Baskoro, J. Haluan, B.H. Iskandar. J. Saintek Perikan. **3**,2: 44–50(2008). [in Bahasa Indonesia] https://scholar.google.com/scholar?hl=en&as_sdt=0%2C5&q=Pengembangan+Teknologi+Penangkapan+Ikan+Berbasis+Komoditas+Potensial+di+Teluk+Lampung.&btnG=
16. T.D. Hapsari, A.D.P. Fitri. Aquat. Procedia. **7**:254–264(2016). https://doi.org/10.1016/j.aqpro.2016.07.036
17. B. Sambe, M. Tandstad, A.M. Caramelo, B.E. Brown. Environ. Dev. **17**,Suppl 1:105–117(2016). https://doi.org/10.1016/j.envdev.2015.11.012
18. H. Boesono, D.R. Setiawan, K.E. Prihantoko, B.B. Jayanto, A.R. Malala, Aquat. Procedia. **7**:112–117(2016). https://doi.org/10.1016/j.aqpro.2016.07.015
19. R. Rusmilyansari. Fish Sci. **2**,4:143–153(2012). [in Bahasa Indonesia]. https://scholar.google.com/scholar?hl=en&as_sdt=0%2C5&q=INVENTARISASI+ALAT+TANGKAP+BERDASARKAN+KATEGORI+STATUS+PENANGKAPAN+IKAN+YANG+BERTANGGUNGJAWAB+DI+PERAIRAN+TANAH+LAUT&btnG=

20. E.S. Wiyono, H. Hufiadi. Aquaculture, Aquarium, Conservation & Legislation International Journal of the Bioflux Society. **7**,6:475–482(2014). https://scholar.google.com/scholar?hl=en&as_sdt=0%2C5&q=Optimizing+purse+seine+fishing+operations+in+the+Java+Sea%2C+Indonesia&btnG=

Techno-Economic Analysis of Photovoltaic Utilization for Lighting and Cooling System of Ferry Ro/Ro Ship 500 GT

Danny Faturachman[1,*], *Erkata* Yandri[2,3], *Endang* Tri Pujiastuti[4], *Olga* Anne[5], *Roy* Hendroko Setyobudi[2,6], *Yahya* Yani[7], *Herry* Susanto[2,8], *Washington* Purba[2,9], and *Satriyo* Krido Wahono[10]

[1]Department Marine Engineering, Faculty of Ocean Technology, Darma Persada University, Jl. Taman Malaka Selatan, Pondok Kelapa, Jakarta 13450, Indonesia
[2]Graduate School of Renewable Energy, Darma Persada University, Jakarta, Indonesia
[3]Center of Renewable Energy Studies, Darma Persada University, Jakarta, Indonesia
[4]Department Management, Faculty of Economic, Darma Persada University, Jakarta, Indonesia
[5]Department of Natural Sciences, Faculty of Marine Technology and Natural Sciences of Klaipeda University, Lithuania
[6]Department of Agriculture Science, Postgraduate Program, University of Muhammadiyah Malang, Jl. Raya Tlogomas No. 246, Malang, 65145, Indonesia
[7]Department of Built Environment and Energy Technology, Faculty of Technology, Linnaeus University, Lückligs plats 1 (Hus M), 351 95 Växjö, Sweden
[8]Department of Mechanical Engineering, Darma Persada University, Jakarta, Indonesia
[9]JAP-Energia, Graha Mampang, Lt. 3, Suite 305, Jl. Mampang Prapatan Raya Kav. 100, Jakarta 12760, Indonesia
[10]Research Division for Natural Product Technology – Indonesian Institute of Sciences, Jl. Jogja - Wonosari, km 31.5, Gunung Kidul, Special Region Yogyakarta 55861, Indonesia

Abstract. The purpose of this study is to perform the techno-economic analysis of photovoltaic system utilization for lighting and cooling of Ferry Ro/Ro 500 GT. The world is facing a dilemma of increasing dependence on fossil energy with decreasing supply. This situation must be anticipated by all sectors by energy efficiency (EE) and utilizing renewable energy (RE). Especially for RE in the transportation sector, ships as consumers of oil energy can also take advantage of solar energy sources, for example for lighting and cooling. For that purpose, five steps must be taken. First, determine the design specifications. Second, determine the specifications of components of the PV system. Third, calculate the power required for lighting equipment. Fourth, calculate the power required for the cooling system. Fifth, make an investment comparison for propulsion systems between diesel engines and photovoltaic systems. The results show that the energy required for lighting and cooling system as well as for propulsion systems can be placed in the deck area of 148.8 m^2 for all system components, such as; PV modules, charge controllers, batteries, and inverter. This study can provide an overview of the use of PV system in designing the environmentally new or renovation ships.

Keywords: Green transportation, energy saving, solar energy, energy dilemma, ship design

* Corresponding author: fdanny30@yahoo.com

1 Introduction

According to IEA [1], the world is facing a dilemma of increasing dependence on fossil energy such as oil, gas, and coal with decreasing their supply. Especially for Indonesia, this situation must be anticipated by all sectors with implementing energy efficiency (EE) and utilizing renewable energy (RE). The RE included geothermal, biomass, hydro, wind, solar, wave and others, has the potential to be superior in comparison to fossil energy. For achieving EE, energy audits are important [2, 3]. Both EE and RE are very important to support national energy security [4].

Especially for RE in the transportation sector, ships as consumers of oil energy can also take advantage of solar energy sources, for example for lighting and cooling. In the utilization solar energy in Indonesia as equatorial and tropical areas with the land area of almost 2×10^6 km^2, endowed with irradiating the sun more than 6 h d^{-1} or about 2 400 h in a year. With a geographic location on the equator, Indonesia will always have sunlight for 10 h until 12 h a day and approximately daily irradiation of 4.5 kWh m^{-2} with the monthly variation of about 10 % [5]. Solar energy utilization for Indonesia has various advantages such as: The energy is available with large numbers in Indonesia. Strongly support the national energy policy of austerity. Verified and equitable energy. Allow built-in remote areas because it does not require the transmission of energy or transportation of energy resources.

Solar energy is an environmentally friendly energy source which can be converted to; electrical energy using solar cell or photovoltaic (PV), thermal with solar collector, or both electrical and thermal energy [6, 7]. Photovoltaic has been applied in Marine Engineering [8], such as; a passenger ferry [9, 10], cruise ship [11], bulk carrier [12], survey vessel or tanker [13]. Figure 1 shows a patented solution developed by Eco Marine Power that combines sail power (using rigid sails) with solar power. This wind-assisted propulsion (WAP) system also include marine solar power and is designed so that the practical limitations of using rigid sails and solar panels on ships are overcome [14]. As an archipelagic country, Indonesia is highly dependent on ferries for marine transportation and regional development [15]. The generator choice is specialized in idealizing systems in this role for planning because it involves a techno-economy problem. The purpose of this study is to perform the techno-economic analysis of photovoltaic system utilization for lighting and cooling of Ferry Ro/Ro 500 GT to be operated in Indonesian seas.

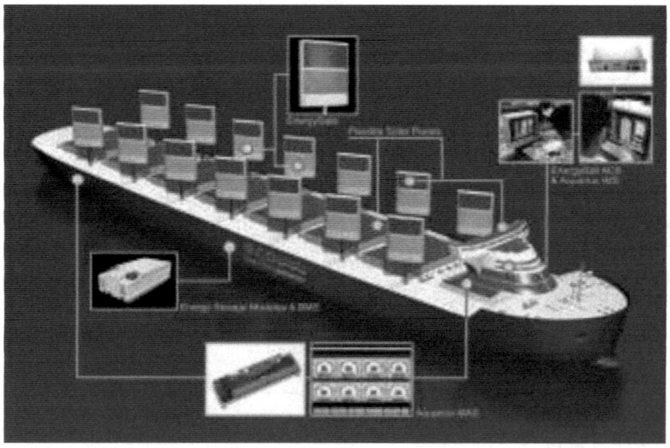

Fig 1. Sail power with solar cell in ship

2 Method

To realize the objectives of this research, there are five steps taken;
First, to determine the design specifications of the ship; Ferry Ro/Ro 500 GT, as shown in Table 1.

Table 1. Specification design of the ship

Length Over All (LOA)	45.05 m
Length between Perpendicular (LPP)	40.15 m
Length of Water Line (LWL)	42.00 m
Breadth (B)	12.00 m
Height (H)	3.20 m
Draft (T)	2.15 m
Velocity (V_s)	11 knot
Main Engine	2 × 800 HP
Auxiliary Engine	2 × 80 kVA
Gen-set emergency	25 KVa

This study focuses on Ferry Ro/Ro (Roll-on/Roll-off) with the horizontal transfer charge [4]. The power for lightings are distributed for; i). the main lighting using fluorescent and neon lamps. ii). emergency lighting lamps mounted at the steering wheel, desk maps, alleys, stairs, engine room, iii). lightings for the engine room, bathroom/toilet, kitchen and rooms open from types that are waterproof (watertight).

Second, determine the specifications of the main components for the PV system. The main components are included; PV module, inverter, charge controller, and battery. The details are shown in Table (2, 3, 4 and 5).

Table 2. Specification of PV module [16]

PV module type	PV Energy, FVG 240P – MC:
Dimension	(1650 × 990 × 35) mm
Power peak, P_{mpp}	240 W
Efficiency, η_{pv}	14.6 %
Voltage module, V_{mpp}	30.50 V
Current module, I_{mpp}	7.88 A
Open circuit voltage $V_{pv,\,oc}$	37.60 V
Short circuit current $I_{pv,\,oc}$	8.28 A

In designing this PV system, the current resulting from the PV module is the DC, while the current that is used to drive the compressor using the AC current. The DC from the PV module needs to be converted to AC using the inverter current. Table 3 shows the specification of the inverter:

Table 3. Specification of the inverter [17]

Inverter type	XANTREX model SW3024E
Power	3300 W
Voltage	24 V
Efficiency	94 %

Table 4. Specification of the charge controller [18]

Charge controller type	Marine Batteries, Rolls Series 5000
Charging current capacity, I_{cc}	60 A
Charging voltage, V_{cc}	12 V

Table 5. Specification of battery [18]

Battery type	Marine Batteries, Rolls Series 5000
Capacity, P_{bat}	370 Ah
Battery voltage, V_{bat}	12 V

The number of panels N_{pv} to meet the needs for a daily hour h of total energy E_{tot} calculated with:

$$N_{pv} = \frac{E_{tot}}{P_{pv} x h} \quad (1)$$

The number of the charge controller N_{cc} required can be calculated from;

$$N_{cc} = \frac{N_{pv}}{(I_{cc} / I_{pv})} \quad (2)$$

where, I_{cc} = 60 A, is the maximum current of the charge controller. The total output of charge controller can be calculated as;

$$I_{cc,tot} = I_{cc} \times N_{cc} \quad (3)$$

The power output of the charge controller $P_{cc,tot}$ with voltage V_{cc} (12 V) can be calculated as;

$$P_{cc,tot} = I_{cc,tot} \times V_{cc} \quad (4)$$

The energy saved by a battery E_{bat} with the capacity P_{bat} and voltage V_{bat} of 12 V can be calculated as;

$$E_{bat} = P_{bat} \times V_{bat} \quad (5)$$

The total battery required N_{bat} to support the total energy load E_{tot} can be calculated as;

$$N_{bat} = \frac{E_{tot}}{E_{bat}} \quad (6)$$

The total energy saved by the batteries $E_{bat,tot}$ can be calculated as;

$$E_{bat,tot} = E_{bat} \times N_{bat} \quad (7)$$

The number of inverters N_{inv} required can be calculated from;

https://doi.org/10.1051/e3sconf/202122600012

$$N_{inv} = \frac{P_{pv,tot}}{(P_{inv,in})} \tag{8}$$

where, $P_{pv,tot}$ and $P_{inv,in}$ are the total power of the PV module and the power input of the inverter, respectively.

Figure 2 shows the placement of the PV module on the deck wheelhouse. The installation was parallel to optimize the solar energy with proper layout [19]. The PV module of FVG 240P-MC model was considered to address the needs of load for lighting, with sufficient area on the deck of 160 m^2.

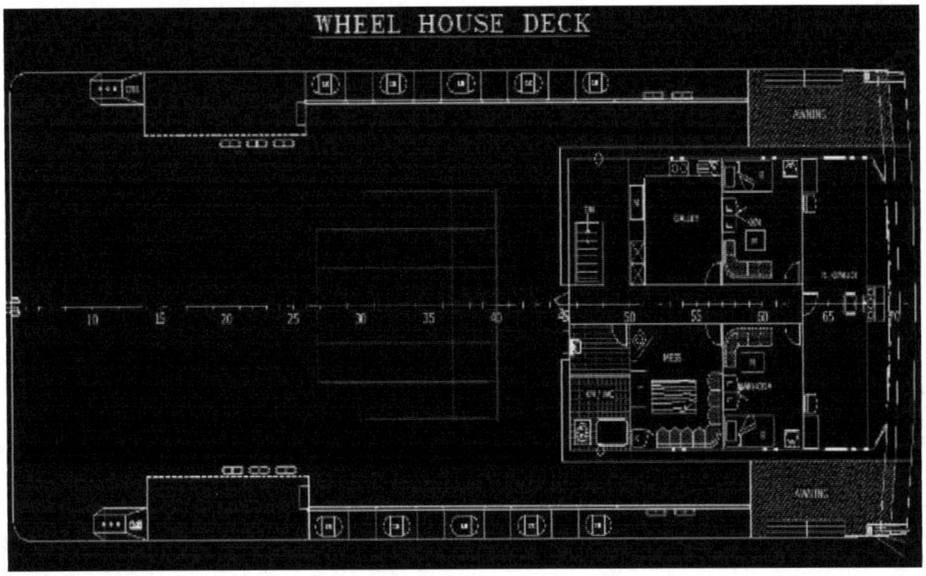

Fig. 2. The placement of solar panel in deck house

Third, calculate the power required for the lighting equipment. The summarized detail of group lighting can be seen in Table 6.

Table 6. Power required for lighting

Deck	Load range (W)	Used hour (h)	Operation	Total (pcs)	Energy (kWh)
Bottom	5 to 20	12	18:00 to 16:00	8	1.74
Vehicle	5 to 40	12	18:00 to 16:00	44	9.36
Passenger	5 to 60	12	18:00 to 16:00	53	12.72
Navigation	5 to 60	12	18:00 to 16:00	34	9.78
			Total	139	33.6

Fourth, calculate the power required for the cooling system. As shown in Table 7, the power required for the cooling system is used for Steering Room, Passenger Rooms, Cabin Crew Rooms and the Control Room, installed air conditioning (AC) machine in the form of AC Split in each room. AC Blower must be arranged so that every part of the room to get the same temperature influence. Engine/generator AC should be placed outdoors and protected from direct weather influences and the sea air or given a construction for protection against the weather. The AC generators were placed on the vehicle load space must be given a protective fender, or construction to protect the generator from the possibility of a collision with a vehicle. There are 3 factors to consider when determining the need for PK of AC power conditioners, namely AC power (BTU h^{-1}), electrical power (W), and PK of the AC compressor. Number PK on AC power is a unit on the AC compressor, not AC cooling power, so to decide on the power need, we must look from the specification of AC.

Table 7. Power required for cooling system

Room's Name	Unit	AC Power (PK)	AC Power (BTU h^{-1})	AC Power (W)
Engine Room	1	½	± 5 000	220
Captain & Engine Room Operator	1	½	± 5 000	220
Passenger Executive Room	4	11/2	± 12 000	1 560
Medical Room	1	½	± 5 000	220
Mosque Room	1	½	± 5 000	220
Mess Room	1	½	± 5 000	220
			Total	2660

Fifth, make an investment comparison for propulsion systems between diesel engines and photovoltaic systems.

3 Result and discussion

3.1 Power for lighting system

From the selection of the solar panels, it can be calculated how many pieces of solar panels needed to meet the needs of power for lighting load. For conditions in Indonesia, even though the duration of the sun shines for 8 h d^{-1} (08.00-16.00), but the effectiveness of the photon beam obtained solar panels during the day is 5 h, *h*. Thus, the number of panels N_{pv} to meet the needs of daily energy E_{tot} of 33.6 kWh (see Table 6) is 28 unit with additional 7 unit as a backup to anticipate the low irradiation less than 1000 W m^{-2}. With extensive consideration of the deck, the platform is still able to accommodate the number of solar panels, in addition to the power generated will be greater or in other words the addition of solar panels to add the amount of power generated. The amount of power generated by the 35 solar panels in 1 h : 35 panels × 240 W = 8.4 kW. The amount of power generated by the solar panels is in 5 h is 8.4 kW × 5 h = 42 kWh. It has solar panels power largest enough to area on the bridge deck 20 m × 8 m = 160 m^2 and is installed with a slope of 150° [13]. From the specification of the charge controller, the maximum current can be released is 60.0 A. The summary of the result can be seen in Table 8.

Table 8. Components and power required for lighting

Items	Eq.	Unit		Remarks
Number of PV panels, N_{pv}	(1)	pcs	28	With additional 7 pcs = 35 pcs
Number of the charge controller, N_{cc}	(2)	pcs	5	
Total current of the charge controller, $I_{cc,tot}$	(3)	A	300	
Total power of the charge controller, $P_{cc,tot}$	(4)	kW	3.6	
Energy saved by the batter, E_{bat}	(5)	kWh	4.3	
Number of battery required, N_{bat}	(6)	pcs	10	
The total energy saved by the batteries, $E_{bat,\ tot}$	(7)	kWh	43	
Number of inverters required, N_{inv}	(7)	pcs	11	

3.2 Power for cooling system

According to Table 7, The power requires for AC on board during the cruise of 10 h is 2.66 kW × 10 h = 26.6 kWh. With so many panels to meet the needs of power equal to 26.6 kW as many as the efficiency of solar panel hence: 240 × 14.6 % = 204.96 W. The number of solar panel = (26.6 kWh) / (204.96 W × 5 h) ≈26 solar panels. The amount of power generated by the solar panel in 1 h is 26 PV modules × 204.96 W PV^{-1} module = 5 328.96 kWh. The magnitude of the power generated by the PV module all over in 5 h is 5 328.96 kWh × 5 h = 26 644.8 kWh. From the existing controller charger specifications, the maximum current that can be issued charger controller is of 60 A. The current generated by a solar module with voltage of 30.5 V is 7.88 A. One charger controller can be used for 7 units of PV modules. The summary of the result can be seen in Table 9.

Table 9. Components and power required for cooling system

Items	Eq.	Unit		Remarks
Number of the PV modules, N_{pv}	(1)	pcs	7	With additional 7 pcs = 35 pcs
Number of charge controller, N_{cc}	(2)	pcs	4	
Total current of the charge controller, $I_{cc,tot}$	(3)	A	240	Max 60 A/charge controller
Total power of the charge controller, $P_{cc,tot}$	(4)	kW	5.76	
Energy saved by the batter, E_{bat}	(5)	kWh	4.4	
Number of battery required, N_{bat}	(6)	pcs	6	
Total energy saved by the batteries, $E_{bat,\ tot}$	(7)	kWh	24.5	
Number of the inverters required, N_{inv}	(7)	pcs	9	

The PV system components will be placed on the deck of the bridge or on the space under the deck of the vehicle with a total area of 12.4 m × 12 m = 148.8 m², as shown in Table 10.

Table 10. Dimensions and weight of system components for PV systems

Component	Dimension (cm)	Weight (kg unit^{-1})	Lighting System Units	Lighting System Weight (kg total)	Cooling System Units	Cooling System Weight (kg total)
PV module	165 × 99 × 3.5	21.5	28 (+7)	752.5	7	150.5
Charge controller	37 × 15 × 15	0.45	5	2.25	4	1.8
Battery	55.9 × 17.8 × 6	123.4	10	1234	6	738
Inverter	53.4 × 38.1 × 22.9	16	11	176	9	144
		Total	61	2 164.75	26	1 034.3

Lighting and cooling with the PV system provide about 1.49 % savings compared to generators.

3.2 Economic analysis for driving force system

The cost comparative analysis for driving force system between the diesel engine and PV system can be explained in Table 11 and the investment cost analysis in Table 12. From the calculation for 10 yr time, the investment cost of the PV system is much lower only IDR 553 259 000 compare to diesel system IDR 3 814 025 000 or, PV system investment is only 15 % than the diesel system. Besides, it is necessary to consider the accident factor which is experienced by the diesel-engined ships [20].

Table 11 Cost comparative analysis for ship propulsion system component

• With motor diesel: - 1unit auxiliary 80 kVA - Tool kit engine • The calculation of fuel consumption using generator power planned 80 kVA, for 10 h cruise: W_{fo}: $80 \times 210 \times 10 \times 10^{-6} \times 0.6 = 0.1008$ t The volume of fuel: $W_{fo} / \gamma_{fo} = 0.10 / 0.85 = 0.11$ m^3 = 110 L The price of diesel fuel for the total fuel shipping is 110 L and the price of 1 L of marine diesel is IDR 8 500 • The calculation for motor diesel: Investment for the purchase of diesel - Generator 1 unit: IDR 43 000 000 - Tool kit-engine 1 set: IDR 2 000 000 Operations: The fuel for the 5 trips for 1 d needs 110 L - 1 d cruise 110 L × IDR 8 500 = IDR 935 000 - For a year IDR 935 000 × 365 d = IDR 341 275 000 - For 5 yr IDR 341 275 000 × 5 = IDR 1 706 375 000 - For 10 yr = IDR 1 706 375 000	• With solar cell: - Use 26 PV modules - 4 charge controller - 6 battery - 9 inverter • For 10 yr usage performed 4 times engine maintenance and costs IDR 6 000 000 × 10: IDR 60 000 000 • Investment for the purchase of PV modules: - Using 26 units PV modules @ IDR 3 139 500 = IDR 81 627 000 - 4 pieces charger controller @ IDR 6 490 000 = IDR 25 960 000 - 6 batteries @ IDR 9 093 500 = IDR 54 561 000 - 9 inverter @ IDR 34 950 000 = IDR 314 550 000 - 1 tool kit set engine: IDR 2,000,000, - Operational battery backup 6 pieces @ IDR 9 093 500 = IDR 54 561 000 - Maintenance costs for 10 yr @ IDR 2 000 000 yr^{-1} = IDR 20 000000

Table 12. Total Investment

Year	Generator	PV System
1	IDR 341 275 000	IDR 434 329 500
5	IDR 1 706 375 000	IDR 10 000 000
10	IDR 1 706 375 000	IDR 20 000 000
After 10 yr	IDR 60 000 000	IDR 54 561 000
Total	IDR 3 814 025 000	IDR 553 259 000

Based on the above results, three technical points need to be discussed here;

First, the requirement or common rules electricity a ship for the power quality [21], such as; supply electricity to vessels needs and neutral body system of a ship grounded on may not except; zinc anode protection system must be a cathode or the outer part body of ship; system limited or local ground as system starting and starting motor in motor fuel combustion; a measuring monitor insulator instrument to the current that circulated no more than 30 mA in the worst of conditions; high voltage neutral ground to avoid dangerous areas were defined in requirements.

Second, power supply and distribution. Generator, switchboard and battery must be in a separate location from the fuel tank and oil pump, with a cofferdam or with sufficient distance. The cable that might be opened by the steam and gas needs to be protected with the proper insulation, with the possibility of reducing corrosion. Some requirements for the installation cable onboard based on the position where the cables will be placed, adapted to the structure of the ship so that the installation and buffer plate avoid the possibility of strains/stresses.

Third, when docking, ships can use the power of the land through shore connection to avoid emission by the generator [22]. If the generator is not active then the emergency source of electrical power (power source) is usually in the form of battery. Due to the nature of the emergency then only certain equipment and very important in the supply by the emergency source of electric power, for example, lights navigation, gangway lighting appliances, and others. The emergency power source will be stored automatically through the emergency switchboard, if all the generators are not active.

The results of this study can provide an overview of the use of PV system in designing the new or renovation ships, both for lighting and cooling systems as well as for propulsion systems. For environmentally friendly shipping [23], future ship design should consider the use of PV module as a technology for converting solar energy into electrical energy. According to the berthing location, the green ship concept can be integrated into the smart city concept with smart grid technology [24, 25]. For the future research direction, this analysis can be developed into a simulation tool with a variety of ship types with optimal placement on the ship [9].

4 Conclusion

The analysis of photovoltaic usage for the 500 GT Ferry-Roro has been carried out with several important points, as follows. To meet the energy needs of the lighting system with the existing component specifications; then it takes 28 PV modules, 5 charge controllers, 10 batteries and 11 inverters. As for the cooling system, it is needed 7 PV modules, 4 charge controllers, 6 batteries and 9 inverters. All system components can be placed in the provided deck area of 148.8 m^2. Lighting and cooling with the PV system provide about 1.49 % savings compared to generators. PV system investment is only 15% than the diesel system. For the propulsion, PV system investment is only 15% than the diesel system.

References

1. International Energy Agency (IEA). *World Energy Outlook 2019*. [Online] from https://www.iea.org/reports/world-energy-outlook-2019 (2019). [Accessed on 30 September 2020].
2. E. Yandri, R. Ariati, A.S. Uyun, R.H. Setyobudi, O. Anne, H. Susanto, et al., IOP Conf. Ser. Earth Environ. Sci., **2020**:490(2020). https://doi.org/10.1088/1755-1315/490/1/012005
3. E. Yandri, R. Ariati, A.S. Uyun, R.H. Setyobudi, H. Susanto, K. Abdullah et al., E3S Web of Conf. **190**:00008(2020). https://doi.org/10.1051/e3sconf/202019000008
4. E. Yandri, R. Ariati, R.F. Ibrahim, Jurnal Ketahanan Nasional, **24**,2:239–260(2018). https://doi.org/10.22146/jkn.30999
5. World Bank Group. *Solar Resource and Photovoltaic Potential of Indonesia*. [Online] from http://documents1.worldbank.org/curated/en/729411496240730378/pdf/115347-ESM-P145273-PUBLIC-IndonesiaSolarResourcePotentialWBESMAPMay.pdf (2017). [Accessed on 01 October 2020].
6. E. Yandri, Sol. Energy Mater. Sol. Cells, **201**:110066(2019). https://doi.org/10.1016/j.solmat.2019.110066
7. E. Yandri, Renew. Energy, **111**:344–352(2017). https://doi.org/10.1016/j.renene.2017.03.094
8. I. Kobougias, E. Tatakis, J. Prousalidis, Adv. Power Electron., **2013**:1–8(2013). https://doi.org/10.1155/2013/831560
9. P. Cheng, N. Liang, R. Li, H. Lan, Q. Cheng, Energies **13**,1:1(2019). https://doi.org/10.3390/en13010001

10. J. Ling-Chin, A.P. Roskilly, Appl. Energy., **181**:416–434(2016). https://doi.org/10.1016/j.apenergy.2016.08.065
11. Y. Yan, H. Zhang, Y. Long, Y. Wang, Y. Liang, X. Song, et al., J. Clean Prod., **233**:264–279(2019). https://doi.org/10.1016/j.jclepro.2019.06.047
12. W. Saidyleigh. *Investigation of auxiliary power potentials of solar photovoltaic applications on dry bulk carrier ships*. [Dissertation]. Maritime Energy Management, World Maritime University, Sweden (2017). p. 104. https://commons.wmu.se/cgi/viewcontent.cgi?article=1589&context=all_dissertations
13. H. Lan, J. Dai, S. Wen, Y.Y. Hong, D.C. Yu, Y. Bai, Energies **8**:11515–30. https://doi.org/10.3390/en81011515
14. L. Talluri, D.K. Nalianda, K.G. Kyprianidis, T. Nikolaidis, P. Pilidis, Ocean Eng., **121**:301–311(2016). https://doi.org/10.1016/j.oceaneng.2016.05.047
15. D. Faturachman, MATEC Web Conf., 177:1–8(2018). https://doi.org/10.1051/matecconf/201817701005
16. FVG Energy. *The company fvg energy - production of photovoltaic modules*. [Online] from http://www.verexzilina.sk/www/pages_template/template_doc.php?lang=sk&idp=500&fid=100&tid=1 (n.d.). [Accessed on 02 October 2020].
17. Xantrex. *FREEDOM SW 3000 Sine Wave Inverter/Charger 2010*. [Online] from http://www.xantrex.com/documents/Inverter-Chargers/Freedom-SW-3000/DS20090915FreedomSWInverterCharger3000Watts.pdf (2010). [Accessed on 03 October 2020].
18. R. Batteries, *Rolls Battery User* n.d.:46. [Online] from https://rollsbattery.com/public/docs/user_manual/Rolls_Battery_Manual.pdf (n.d.). [Accessed on 04 October 2020].
19. Ç. Karatuğ, Y. Durmuşoğlu, Sol. Energy, **207**:1259–12068(2020). https://doi.org/10.1016/j.solener.2020.07.037
20. D. Faturachman, S. Mustafa, F. Octaviany, T.D. Novita, Int. J. Serv. Sci. Manag. Eng., **1**:6–16(2014). http://www.openscienceonline.com/journal/archive2?journalId=729&paperId=316
21. Y. Qiu, C. Yuan, Y. Sun, X. Tang. Electr. Power Components Syst., 2018;46:1375–86. https://doi.org/10.1080/15325008.2018.1485185
22. D. Colarossi, P. Principi, Appl. Therm. Eng., **181**:115988(2020). https://doi.org/10.1016/j.applthermaleng.2020.115988
23. S. Fang, Y. Wang, B. Gou, Y. Xu, IEEE Trans. Veh. Technol., **69**:207–219(2020). https://doi.org/10.1109/TVT.2019.2950538
24. Z. Jin, L. Meng, J.M. Guerrero, R. Han, IEEE Trans. Ind. Informatics, **14**:703–714(2018). https://doi.org/10.1109/TII.2017.2772343
25. B. Novianto, K. Abdullah, A.S. Uyun, E. Yandri, S.M. Nur, H. Susanto, et al., E3S Web Conf., **188**:00005(2020). https://doi.org/10.1051/e3sconf/202018800005

Morphological and Physiological Responses of *Indigofera tinctoria* L. to Light Intensity

Desy Setyaningrum[1], *Maria* Theresia Sri Budiastuti[2,*], *Bambang* Pujiasmanto[2], *Djoko* Purnomo[2], and *Supriyono* Supriyono[2]

[1]Master Degree of Agronomy, Graduated School, Universitas Sebelas Maret, Surakarta 17126, Indonesia
[2]Department of Agrotechnology, Faculty of Agriculture, Universitas Sebelas Maret, Surakarta 17126, Indonesia

Abstract. Synthetic dyes can cause health and environmental impacts. Thus, there are opportunities to develop natural dyes, one of which is produced by *Indigofera tinctoria* plants. This plant is from Fabaceae that has the potential to produce a natural blue color. Natural dyes are extracted from the leaves of plants that contain indigo compounds. Indigo growth and precursors are very dependent on environmental conditions, one of which is light intensity. This study aimed to study the morphological and physiological plant responses in *I. tinctoria* to several levels of light intensity. The research was conducted in Puron Village, Sukoharjo, Indonesia with a complete randomized block design (RCBD) one factor, namely the level of light intensity (100 %, 50 %, and 25 %) with nine replications. Light intensity affected the morphology and physiology of *I. tinctoria*. Plants responded to low light intensity by increasing the leaf area index, specific leaf area and plant height. Leaf area, specific leaf area and plant height were highest at 25 % intensity. However, the number of leaves and nodes got greater at full light intensity. Higher light intensity increased the chlorophyll content a, b and total, thus, higher biomass yield which was 18.86 g at the age of 8 wk.

Keywords: Chlorophyll, leaf area, number of nodes, plant height, specific leaf area.

1 Introduction

Textile industries use synthetic dye for dying medium. Synthetic medium has some benefits in terms of characteristics; bright color, easily dissolves in water, cheaper to produce, and easier to apply on fabric [1]. Thus, the waste from these industries contains various types of synthetic dyes. The waste is considered a pollutant [2]. The waste treatment process is not yet efficient. It can cause serious environmental problems in connection with the reuse of wastewater for irrigation. Applying proper treatment process is very important in order to minimize pollution, one of which is by developing the production of the *Indigofera tinctorial* L. plant to be used as natural dyes.

* Corresponding author: mariatheresia@staff.uns.ac.id

I. tinctoria is from the Fabaceae family, which is efficacious as a natural dye industry. The plant is used as a natural dye agent in the textile industry because it contains indigo pigments that produce indigo colors [3]. *I. tinctoria* natural dyes were the main product of India and West Africa before the discovery of synthetic indigo. Natural dye compounds are extracted from plant leaves. These plant cells contain indican secondary metabolites (indoxyl-β-D-glucoside) and/or B (1H-Indol-3-yl β-D-ribo-3-hexosulopyranoside) precursors. Indican is synthesized in chloroplasts and stored in vacuoles.

Indigo growth and precursors are highly dependent on environmental conditions, one of which is light intensity. In Europe, this plant can adapt well but is not resistant to drought, and indigo content is positively influenced by the intensity of sunlight [4]. Light is the main environmental component that affects the survival rate, morphological characteristics, photosynthetic characteristics, and primary and secondary metabolism of plants [5]. Light is a vital resource for plants, and light deficiency can directly change the development of the photosynthetic apparatus of plant photosynthetic capacity [6]. Most pigments that are specifically bound to thylakoids and the photosynthetic system are strongly influenced by the availability of light. This research aimed to study the morphological and physiological responses of *I. tinctoria* to several levels of light intensity.

2 Materials and methods

The research was conducted from April to August 2019 in Puron Village, Bulu, Sukoharjo, Central Java, Indonesia. The research location was at 77.745 S and 110.83 E with an altitude of 120 m.a.s.l. Microclimatic conditions with an average precipitation of 2 289 mm yr^{-1} and the number of rainy days 105 d. Daily average temperature of 25 °C to 27 °C, humidity around 80 %. The study used a complete randomized block design (RCBD) with one factor, the light intensity with three levels; 100 %, 50 %, and 25 %. All treatment was replicated with nine replications, so there were 27 experiment units. The planting material used was *I. tinctoria* seeds, which are green in color. The tool used as an application of light intensity was a shading net with various density levels.

Chlorophyll analysis was carried out at *Laboratorium Fisiologi Tumbuhan* Universitas Sebelas Maret using the Armon [7] method with a slight modification. Chlorophyll a and b uptake was measured using a UV-Visible Double-ray Spectrophotometer (Eliso Sl-156, India) at 663 and 645 nm by using acetone 90 % as the blank. Plant physiology variables observed were chlorophyll a, b, and total, the biomass of plants aging 6 wk, 8 wk, and 10 wk. Morphological variables recorded were plant height, number of leaves, number of branches, leaf area, specific leaf area (LW/LA: LW = leaf weight, and LA=leaf area). The data were analyzed using analysis of variance with α 5 % (95 % confidence level) followed by a 5 % Duncan's Multiple Range Test (DMRT) if between the experimental units were significantly different.

Environmental variables observed were light intensity at 7 a.m., 12 p.m., and 5 p.m. (GMT+7) using lux meter (Mextech, China), temperature, and humidity using Thermo hygrometer (Shanghai, China). Observations of the environment were carried out 12 times, starting at the age of 7 d plants until harvest time. Observation of light intensity was carried out on the canopy of *I. tinctoria* ie, under the third leaf from above.

https://doi.org/10.1051/e3sconf/202122600013

Table 1. Average environmental variables in the research location

Light intensity	Variable	Time of observation (GMT+7)		
		7 a.m.	12 p.m.	5 p.m.
100 %	Light intensity (lux)	48 600	63 200	16 010
	Temperature (°C)	27.8	36.7	29.6
	Relative humidity (%)	64	46	54
50 %	Light intensity (lux)	24 600	32 400	8 430
	Temperature (°C)	27	34.2	28.9
	Relative humidity (%)	63	41	56
25 %	Light intensity (lux)	13 600	5 300	4 025
	Temperature (°C)	26.4	32	28.3
	Relative humidity (%)	66	50	57

3 Results and discussions

3.1 Morphological responses of *I. tinctoria* to light intensity

Plant morphology has manipulability and adaptation mechanisms in changing environmental conditions [8]. Light intensity significantly affected the growth of *I. tinctoria* plants at 2 wk to 12 wk old (Figure 1). Different light conditions had different effects on plant height. Plants grew higher with lower light intensity levels. This condition indicated that the stem of the *I. tinctoria* plant was etiolated. This is in accordance with the research of [9] that shading caused an increase in the height of soybean plant stems. Low light caused the internodes to grow longer [10]. Adaptation of plants to low light intensity would affect the morphology, anatomy, and physiology of plants, including an increase in leaf area and plant height as an effort to reduce the use of metabolites, and reduce the light transmitted and reflected [11].

I. tinctoria plant height at full light intensity was only 73.44 cm, while at 50 % intensity, the plant height increased by 37.34 %, which reached 117.22 cm and increased 40.72 % at 25 % intensity. Plants responded to stem elongation and inhibited the growth of stem diameter to absorb a high amount of light [12]. This was due to the low light, which activated auxin and further caused symptoms of etiolation [13]. Shading application led to carbon allocation towards stems elongation at the expense of roots and leaves development, thereby affecting the yield [14].

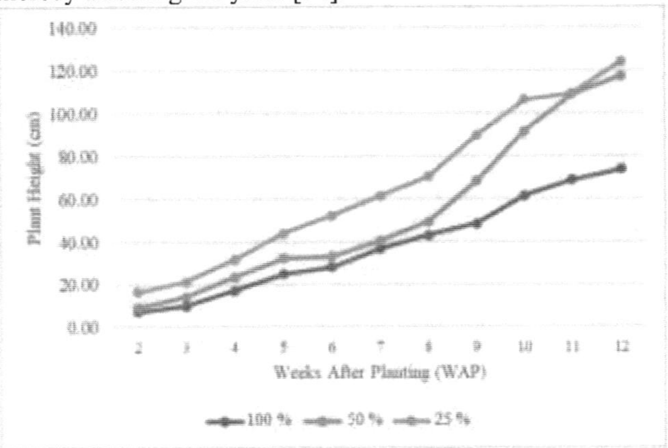

Fig. 1. Effect of light intensity on plant height

Leaves are considered as the head part of the plant. Leaf functions to collect light energy from the sun and play a role in photosynthesis to convert light energy into biochemical energy [15]. The number of leaves was considered a major factor that determined the quality of growth and the ability of plant photosynthesis. The results showed that the number of *I. tinctoria* leaves reached 368 strands in full light intensity at 12 WAT (Table 1). The number of leaves at each intensity showed significantly different results. The number of leaves at 50 % light intensity decreased by 39 % reached 225 strands and at 25 % light decreased by 51.79 %. The growth of nodes also decreased at a low light intensity. The results showed that the number of nodes at each intensity was significantly different. The number of nodes in full light reached 28 nodes. In 50 % intensity number of media is reduced by 41.69 % reached only 16 nodes. Plants responded to low light by extending internodes and leaf stalks, resulting in reduced nodes. Thus, the number of leaves was also reduced. Light intensity-modulated the content of the thylakoids as the PSII/PSI ratio, the number of reaction centers. Plants in low light exhibited lower levels of photosystem II (PSII), number of reaction centers, as well as the decreased capacity for oxygen evolution, electron transport, and limitations in electron transport between PSII and PSI [16]. An important effect of low light was the reduction in clean photosynthesis [17]. Decreased photosynthesis occurred from two main mechanisms: i) a decrease in diffusion of CO_2 into leaves, due to decreased inter-cell CO_2 and stomatal conductance, and ii) potential inhibitory metabolism for photosynthesis by inhibiting leaf growth and enlargement by controlling cell proliferation [18].

Leaf area shows the ability of plants to carry out photosynthesis because it is associated with the use of light and the use of growing facilities [19]. The results of this study indicated that light intensity significantly influenced the area of *I. tinctoria* leaves. Leaf area reached 40.22 at 25 % light; at 50 % light, the leaf area was reduced by 19 % and reduced by 61 % at 100 % light. It showed that plants responded to low light by increasing leaf area. Reduced light supply encourages an increase in leaf area [20], which is an attempt by plants to increase the area of light absorption and light capture efficiency [21]. The optimum leaf area shows that sunlight is received evenly by the leaf [22].

The specific leaf area (SLA) at 50 % light was 63.98 $cm^2\ g^{-1}$, 19 % higher than full light, and 2.1 % higher than 50 % light. Specific leaf area at light 25 % and 50 % did not show significantly different. Such a situation also appeared in soybean crop yielded in the research of [6] that SLAs were increased in low light. An increase in specific leaf area due to phototropism [23, 8]. However, an increase in SLA was not able to compensate for the decrease in leaf area caused by reduced light interception. Tolerant of intolerant species to low light would optimize the capture and utilization of the light by increasing SLAs [14], because SLA was an important parameter that reflected plant growth and carbon assimilation and was strongly influenced by changes in light conditions [24, 25].

Table 2. Light intensity on the growth of *I. tinctoria*

Light intensity (%)	Number of leaves (strand)	Leaf area	Specific leaf area ($cm^2\ g^{-1}$)	Number of nodes
100	368.56b	15.65a	46.10 a	28.78b
50	225.00a	32.24b	65.36 b	16.78ab
25	177.67a	40.22c	63.98 b	13.11a

3.2 Physiology plant response in *I. tinctoria* to light intensity

The light intensity had a significant effect on chlorophyll (Table 4). The amount of chlorophyll at a full light intensity and 50 % were not significantly different, at 25 % intensity showed significantly different and decreased by 6 %. Light intensity on chlorophyll b and total chlorophyll had no significant difference. Chlorophyll b and total decreased at a low light intensity. The decrease in chlorophyll b in light of 25 % was 2.4 %, the decrease in total chlorophyll was 4.5 %. These results indicated that low light decreased chlorophyll a, b and total in *I. tinctoria*. These results are in accordance with the results of the study of [26] that decreasing the shading from T75 to T0 increases the content of Chl a, Chl b, and Chl a + b in soybeans. Chlorophyll content also showed a decrease in water spinach planted under shading net [27]. Under changing light conditions, studies of Chl a, Chl b, and Chl a + b help as an index for sunlight absorption [28]. The amount of Chl was significantly affected by changes in light availability and decreased with the reduction in light [29]. Chlorophyll content acts as an important factor in determining plant photosynthesis [30]. Increased Chl b increases light interception under shading stress [31]. The amount of chlorophyll in shade-resistant plants increases under shading [32]. His research shows that the shade increases the content of chlorophyll a, b, and total and photosynthetic activity. This may indicate that the *I.* tinctoria plant is intolerant of low light.

Table 3. Light Intensity toward chlorophyll content of *I. tinctoria*

Light intensity (%)	**Chlorophyll**		
	A	**b**	**total (a+b)**
100	0.519 b	0.867 a	1.386 a
50	0.512 ab	0.846 a	1.358 a
25	0.487 a	0.840 a	1.327 a

I. tinctoria biomass increased as plant aged. The light intensity did not significantly affect the biomass of *I. tinctoria* at 6 WAP. Low light showed the greatest biomass at 6 WAP. Plant biomass at 6 WAP at each light intensity level was not significantly different. According to Yao et al. [32], leaf area is an important factor affecting plant light and biomass interception. The yield of plant biomass at 6 WAP had a linear relationship with the leaf area that the highest biomass resulted in light 25 %. A significant linear relationship between leaf area and biomass per plant in shady conditions [33]. This is because the reduction in leaf area directly impacts the reduction in photosynthate production. Some studies have shown that increasing the leaf area increased the rate of net photosynthesis [34]. However, this was incompatible with the *I. tinctoria* biomass at 8 WAP and 10 WAP that at full light, the leaf area was low, and biomass production was high. [19] stated that the decrease in SLA would be offset by an increase in crop dry weight. Plants with low SLA value would have narrower and thicker leaves but have a great dry plant production. This study showed that light intensity affected biomass at 8 and 10 WAP, full light could produce the greatest amount of biomass that was 18.86 at 10 WAP, 27 % greater than light 25 % and 18 % greater than biomass at 50 % light. These results indicated that the biomass of *I. tinctoria* increased in full light. This is because biomass was a product of photosynthesis that depended on the availability of light to produce carbohydrates. Plants that received low light modified the number of nodes to be very low and stimulated the plant stem internodes, causing reduced leaves to grow and reduced the number of nodes resulting in reduced biomass production. Adapting the mechanism of plants in various light environments was by changing the characteristics of photosynthesis. The rate of net photosynthesis decreases, resulting in a reduction in plant biomass in low light [35]. This is in accordance with the results of [36] research that biomass production has a linear relationship with the chlorophyll content of plants, ie the production of biomass and chlorophyll content decreases at low light. Low light intensity caused an increase in plant height, leaf area, and specific leaf area (SLA). These morphological changes allowed relatively more light interception and increased the efficiency of light use in *I. tinctoria* due to the shading effect.

Table 3. Light intensity on biomass production

Light intensity (%)	**Biomass (g)**		
	6 WAP	**8 WAP**	**10 WAP**
100	10.21 a	16.04 b	18.86 b
50	11.09 a	14.30 ab	16.37 ab
25	14.47 a	11.70 a	13.65 a

4 Conclusion

Light intensity affected the morphology and physiology of *I. tinctoria*. Plants responded to low light intensity by increasing plant height, leaf area index, specific leaf area. Leaf area, specific leaf area, and plant height were highest at 25 % intensity. An increase in leaf area had a linear relationship with biomass at 6 WAP that the highest biomass production at light 25 %. However, biomass production at 8 WAP and 10 WAP was strongly influenced by high light reaching 18.83 g. An increase in biomass was due to the number of leaves, the higher number of nodes in full light. Light intensity affected chlorophyll, the chlorophyll content increased in high light, thus, providing materials and energy base for photosynthesis.

References

1. D. Karadag, E. Akgul, S. Tok, F. Erturk. J Chem. Eng. Data. **52**,6:2436–2441(2007) Doi: 10.1021/je7003726
2. P. Sharma. Environ. Monit. Assess. 183:151–195(2011). DOI 10.1007/s10661-011-1914-
3. M.R. Hariri, T. Chikmawati, A. Hartana. J Math. Found. Sci., **49**,2:105–115(2017). DOI: 10.5614/j.math.fund.sci.2017.49.2.1
4. L.G. Angelini, S. Tozzi, N.N.o.D. Nasso. J. Agri. Food Chem., **52**,25:7541–7547(2004). https://pubmed.ncbi.nlm.nih.gov/15675801/
5. U. Niinemets, T.F. Keenan, L. Hallik. New Phytol. **205**,3:973–993(2015). https://pubmed.ncbi.nlm.nih.gov/25318596/
6. Y.S. Wu, F. Yang, W.Z. Gong, S. Ahmed, Y.F. Fan, X.L. Wu, et al. J. Integr. Agric. **16**,6:1331–1340(2017). https://doi.org/10.1016/S2095-3119(16)61525-3
7. D.I. Arnon. Plant Physiol. **24**:1–15(1949).
8. W.G. Gong, C.D. Jiang, Y.S Wu, H.H. Chen, W.Y. Liu, W.Y. Yang. Photosynthetica, **53**:259–268(2015). DOI: 10.1007/s11099-015-0103-8
9. J. Bakhshy, K.G. Golezani, S.Z. Salmasi, M.M. Vahed. Tech. J.Eng. Appl. Sci. **3**,1:30–43(2013). https://www.researchgate.net/publication/235798162_Effects_of_Water_Deficit_and_Shading_on_Morphology_and_Grain_Yield_of_Soybean_Glycine_Max_L
10. M. Lukitasari. Jurnal IKIP PGRI Madiun. 1–8(2012). https://www.academia.edu/6301530/PENGARUH_INTENSITAS_CAHAYA_MATAHARI_TERHADAP_PERTUMBUHAN_TANAMAN_KEDELAI_GLYCINE_MAX

11. M.G. Hale, D.M. Orcutt. Allelochemical stress, *in* M. G. Hale and D. M. Orcutt (eds.). The Physiology of Plants under Stress. John Wiley & Sons, New York. (1987).p.127
12. M. Kermah, A.C. Franke, S.A. Nsiah, B.D.K. Ahiabor, R.C. Abaidoo, K.E. Giller. Field Crops Res. **213**:38–50(2017). https://doi.org/10.1016/j.fcr.2017.07.008 PMID: 29104356.
13. Y. Wu, W. Gong, W. Yang. Scientific reports. 7:9259(2017). | DOI:10.1038/s41598-017-10026-5
14. C.M.M. Gommers, E. Visser, K. Onge, L. Voesenek, R. Pierik. Trends in Plant Science, 1**8,**2:65–71(2013). DOI: 10.1016/j.tplants.2012.09.008
15. F. Sukriah, L. Pranggarani. Jurnal Ilmiah Fifo. **8,**1:20-43(2016). [in Bahasa Indonesia]. http://digilib.mercubuana.ac.id/manager/t!@file_artikel_abstrak/Isi_Artikel_258872346718.pdf
16. K. Kunderlikova, M. Brectic, M. Zivcak, P. Kusniarova. J. Cent. Eur. Agric. **17**,4:950–956(2014). DOI: 10.5513/JCEA01/17.4.1797
17. C. Yang, B. Hu, N. Igbal, F. Yang, W.g. Liu, X.c. Wang, et al. J Plant Prod. Sci. **21**,3:193–202(2018). DOI: 10.1080/1343943X.2018.1484257.
18. L.Y. Feng, M.A. Raza, Z.C. Li, Y. Chen, M.H.B. Khalid, J. Du, et al. Front Plant Sci. **9**:1952(2018). https://doi.org/10.3389/fpls.2018.01952
19. S.M. Sitompul, B. Guritno. *Analisis Pertumbuhan Tanaman. Malang, Fakultas Pertanian Universitas Brawijaya*. (1995)
20. H. Poorter, K.J. Niklas, P.B. Reich, J. Oleksyn, P. Poot, L. Mommer. New Phytologist. **193,**1:30–50(2012). https://doi.org/10.1111/j.1469-8137.2011.03952.x
21. Chairudin, Efendi, Sabaruddin. Jurnal Floratek. **10,**1:26–35(2015). http://jurnal.unsyiah.ac.id/floratek/article/view/2355/2240
22. C.R. Brodersen, T.C. Vogelmann TC. Funct Plant Bio. **37**:403–413(2010). http://citeseerx.ist.psu.edu/viewdoc/download?doi=10.1.1.839.5572&rep=rep1&type=pdf
23. F. Yang, D. Liao, X. Wu, R. Gao, Y. Fan, M.A. Raba, et al. Field Crops Res. **203**:16–23(2017). DOI: 10.1016/j.fcr.2016.12.007
24. F. Yang, L. Feng, Q. Liu, X. Wu, Y. Fan, M.A. Raza, et al. Environ. Exp. Bot. **150:**79-87(2018) Https://doi.org/10.1016/j.envexpbot.2018.03.008 .
25. F. Valladares, U. Niinemets. Annu.Rev. Ecol. Evol. Syst**. 39**: 237–257(2008). DOI: 10.1146/annurev.ecolsys.39.110707.173506 or https://www.annualreviews.org/doi/abs/10.1146/annurev.ecolsys.39.110707.173506
26. M.H.B. Khalid, M.A. Raza, H.Q. Yu, F.A. Sun, Y.Y. Zhang, F.Z. Lu,et al. Appl. Ecol. Environ. Res. **17**,2:2551–2569(2019). Doi: http://Dx.Doi.Org/10.15666/Aeer/1702_25512569 or https://www.google.com/url?sa=t&rct=j&q=&esrc=s&source=web&cd=&ved=2ahUKEwjYqbbT7tfrAhXNc30KHQQ8BngQFjAAegQIARAB&url=http%3A%2F%2Faloki.hu%2Fpdf%2F1702_25512569.pdf&usg=AOvVaw2ZyLccjZ7UfY1AnDOUPowQ.
27. D. Wulandari, N. Sofiyanti, Fitmawati. Jurnal Riau Biologia. **1**,2:135-139(2016). [in Bahasa Indonesia]. https://jrb.ejournal.unri.ac.id/index.php/JRB/article/view/3773
28. Y. Fan, J. Chen, Y. Cheng, M.A. Raza, X. Wu, Z. Wang Z, et al. Plos ONE **13**,5: e0198159(2018). https://doi.org/10.1371/journal.pone.0198159 .
29. R. Li, T. Wen, Y.T. Tang, X. Sun, C. Xia, Acta Pratac. Sin. 23:198–206(2014).
30. Y. Zhou, L. Huang, X. Wei, H. Zhou, X. Chen. Plant Growth Regul. **81**:1–8(2016). DOI10.1007/s10725-016-0181-z

31. R.W. Pearcy, J.R. Seemann. Plant Physiol. **94**,2:628–633(1990). doi: 10.1104/pp.94.2.628
32. X. Yao, C. Li, S. Li, Q. Zhu, H. Zhang, H. Wang, et al. Plant Growth Regul. **83**: 1–8(2017). Doi: 10.1007/s10725-017-0307-y.
33. N. Wang, Q. Huang, J. Sun, S. Yan, C. Ding, X. Mei, et al. For. Ecol. Manag. **331**:40–49(2014). https://doi.org/10.1016/j.foreco.2014.06.046
34. S.P. Sulandjari, S. Wisnubroto, D. Indradewa. Jurnal Agrosains. **7**:71–76(2005). [in Bahasa Indonesia]. https://scholar.google.co.id/scholar?hl=en&as_sdt=0,5&cluster=18065642678899366887
35. B.Y. Su, Y.X. Song, C. Song, L. Cui, T.W. Yong, W.Y. Yang. Photosynthetica, **52**: 332–340(2014). https://link.springer.com/article/10.1007/s11099-014-0036-7
36. F. Yang, Y. Fan, X. Wu, Y. Cheng, Q. Liu, L. Feng, et al. Front. Plant Sci. **9**:56(2018). Doi: 10.3389/fpls.2018.00056

Utilization of Tofu Industry Waste and Banana Plant Waste for Growing Medium of Brown Oyster Mushrooms (*Pleurotus cystidiosus* [Jacq. Fr.] P.Kumm.)

Dian Indratmi*, *Yossy* Dian Kurniasari, *Hartawati* Hartawati, and *Ali* Ikhwan

Department of Agrotechnology, Faculty of Agriculture and Animal Science,
University of Muhammadiyah Malang, Jl. Raya Tlogo Mas no.246. Malang 65144, Indonesia

Abstract. Brown oyster mushroom is one of the consumption mushrooms with high economic value, so it is important to be cultivated commercially. Several types of plant and industrial wastes, such as dried banana leaves and tofu dregs, are available abundantly in the field. The waste has the potential to be used as medium for growing consumption mushrooms. This is because dried banana leaves and tofu dregs contain enough nutrients needed for the growth and development of oyster mushrooms. The study aimed to determine the growth response and yield of brown oyster mushrooms by giving various doses of dried banana leaves and tofu dregs flour. The research was carried out using factorial completely randomized design. First factor: dosage of tofu flour: 50 g per baglog, 150 g per baglog, and 250 g per baglog. Second factor: dosage of dried banana leaves, without dried banana leaves, 100 g dried banana leaves per baglog, and 250 g dried banana leaves per baglog. The results showed that the treatment of addition of tofu dregs flour with dried banana leaves interacted very significantly in the number of mushroom caps, diameter and thickness of the caps, the length of the mycelium, the fresh weight of the fungus, and biological efficiency.

Keywords: Biological efficiency, fungus, nutrient, waste.

1 Introduction

Brown oyster mushroom (*Pleurotus cystidiosus* [Jacq. Fr.] P.Kumm.) is a type of consumption mushroom that is popular with the public. Brown oyster mushrooms usually grow wild in nature and have not been widely cultivated. Oyster mushrooms are highly nutritious, have vitamin B, C and D, and that carry numerous medicinal benefits [1]. Brown oyster mushrooms have the advantage of thicker caps and longer shelf life [2].

Brown oyster mushroom has not been widely cultivated but the market demand for oyster mushroom continues to rise. Sawdust is a staple used as a growing medium in producing the fungus. Wood sawdust is a place to grow oyster mushrooms classified as fungi using cellulose, hemicellulose, and lignin that can break down and use it as a carbon source [3].

* Corresponding author: dian@umm.ac.id or indratmi_dian@yahoo.co.id

Kind of growing media is very important because it can affect the results of oyster mushrooms later. Brown oyster mushroom media used must contain nutrients needed for growth and production including lignin, carbohydrates (cellulose and glucose), protein, nitrogen, fiber and vitamins. Other factors that influence the growth of brown oyster mushrooms are temperature, humidity, light intensity and pH. Tofu waste and dried banana leaves are potential wastes to be used as additional growth media for consumption mushroom production.

Tofu waste is a waste from tofu processing. Tofu waste contains substances such as carbohydrates, proteins, fats, minerals and vitamins. Protein functions to stimulate mycelium growth, while fat is used as an energy source to break down carbohydrates, protein, minerals and vitamins. Tofu waste can provide earlier yields, more fruit bodies and increase fruit weight, so that at harvest time the results are better and more profitable [4].

Dried banana leaf contains hemicellulose so that it can be used as a mushroom growing media. The organic components of banana leaf (g per 100 g dry weight) are cellulose 11 g; hemicellulose 20 g; lignin 18.12 g; total C 51, and C/N ratio 29.45. Leaf waste is the best growing media for mushroom compared to other agricultural wastes such as coffee skins, reeds, and straw [5]. The addition of banana leaves waste to the wood sawdust base medium could increase the fresh weight of the white oyster mushroom fruit body. The contents of dried bananas consist of 16.0 % dry matter, 2.3 % crude protein, 3.7 % crude fiber, 6.0 % fat and 1.9 % ash content two strains *Volvariella volvacea* [(Bul.) Singer,1951] showed their highest production on banana leaves waste, with biological efficiencies of 43 % and 72 % [6]. Seeing the potential of tofu industry waste and banana plants waste, it is important to research, its use as an additional growth medium for consumption mushrooms including brown oyster mushrooms.

2 Materials and methods

The study was conducted using a Completely Randomized Design (CRD) compiled factorially with two treatment factors, repeated four times.
First factor: Addition of tofu waste flour (A) consisting of three test levels, namely:

A1: 50 g tofu dregs flour per baglog
A2: 150 g tofu dregs flour per baglog
A3: 250 g tofu dregs flour per baglog

Second factor: Addition of dried banana leaves (D) consisting of three test levels, namely:
D0: without dried banana leaves
D1: 100 g dried dried banana leaf per baglog
D2: 250 g dried dried banana leaf per baglog.

Tofu industry waste in the form of tofu dregs is dried in an oven at 105 °C for 2 d. Tofu dregs that have been dried mashed. Waste of banana plants in the form of dried banana leaves that are cut into small pieces so that it is easily inserted into plastic. The initial weight of the media grew by 800 g.

The research data included observations of the growth of brown oyster mushroom mycelium (cm), number of mushroom caps, mushroom caps diameter (cm), thickness of oyster mushrooms (mm), fresh weight of mushrooms (g), biological efficiency (%), and nutrient content analysis of tofu dregs. Harvesting is done four times. Specifically biological efficiency was measured after eigth times of harvest.

The data obtained were then analyzed statistically using analysis of variance. To find out the effect of whether or not the F test level of 5 % and 1 % is used. If the F test results show a significant effect, then proceed with the Duncan Multiple Range Test.

3 Results and discussion

3.1 Mycelium growth

The data analysis showed that there were significant interactions at (8, 16, 24, 32 and 40) d after inoculation (DAI) on mycelium growth. The mean mycelium length is shown in Table 1.

Table 1. Average length of brown oyster mushroom mycelium due to addition tofu dregs flour and dried banana leaves (8 to 40) d After Inoculation (DAI)

Treatment	**Mycelium growth (cm)**				
	8 DAI	**16 DAI**	**24 DAI**	**32 DAI**	**40 DAI**
50 g flour, 0 g leaves	0.63 bc	2.77 ab	6.16 ab	8.91 a	17.00 b
150 g flour, 0 g leaves	1.23 cd	5.26 d	8.91 d	11.94 bc	16.67 b
250 g flour, 0 g leaves	0.00 a	4.16 cd	8.71 d	13.63 c	17.00 b
50 g flour, 100 g leaves	2.30 e	7.51 e	12.60 e	17.00 d	17.00 b
150 g flour, 100 g leaves	1.48 d	4.76 cd	9.76 cd	13.06 c	17.00 b
250 g flour, 100 g leaves	0.00 ab	3.51 bc	8.42 cd	13.09 c	17.00 b
50 g flour, 250 g leaves	1.02 cd	3.63 bc	6.82 cd	9.66 ab	17.00 b
150 g flour, 250 g leaves	0.00 ab	2.30 a	5.34 a	8.79 a	15.99 a
250 g flour, 250 g leaves	0.00 ab	2.26 a	5.42 ab	10.29 ab	16.03 a

Numbers with the same letter in the same column are not significantly different according to the Duncan 0.05 test.

The 50 g tofu dregs flour and 100 g dried banana leaf treatment can accelerate the growth of fungal mycelium, so that at the age of 32 DAI mycelium fungi have filled the baglog. The addition of tofu waste flour and dried banana leaves on the mushroom growing media had a positive effect. Growth of oyster mushroom mycelium spread is influenced by nutrient content in the growing media. High protein content of 21.56 % in tofu dregs indicates that the nitrogen content is also high so that the growth of mycelium becomes thick and compact (Table 2). Tofu waste contains carbohydrates, protein, fat, minerals and vitamins [7]. added that protein functions to stimulate mycelium growth.

Crude fiber in tofu dregs can increase the rate of mycelium spread, so it can supply the nutrients needed by the growth of fungal mycelium.

Table 2. Analysis of nutrient content of tofu waste

Protein	**Crude fat**	**Crude fiber**	**Dry matter**	**Water content**	**Ash**
21.56 %	9.8 %	25.81 %	96.14 %	62.32 %	4.03 %

3.2 Number of mushroom caps

The results of the analysis of variance showed a real interaction between the treatment of the amount of tofu dregs flour with dried banana leaves on the number of mushroom caps. The average number of caps for each treatment is presented in Table 3.

Table 3. Average number of brown oyster mushroom caps due to addition of tofu dregs flour and dried banana leaves.

Treatment	Number of mushroom caps				
	Harvest I	Harvest II	Harvest III	Harvest IV	Average harvest
50 g flour, 0 g leaves	3 a	5 a	3 a	3 a	4 a
150 g flour, 0 g leaves	4 a	7 ab	5 a	4 ab	5 a
250 g flour, 0 g leaves	4 a	6 a	4 a	4 ab	4 a
50 g flour, 100 g leaves	4 a	7 ab	5 a	4 ab	5 a
150 g flour, 100 g leaves	4 a	6 a	4 a	4 ab	4 a
250 g flour, 100 g leaves	9 b	9 b	7 b	6 b	8 b
50 g flour, 250 g leaves	4 a	6 a	4 a	4 ab	4 a
150 g flour, 250 g leaves	5 a	7 ab	5 a	4 ab	5 a
250 g flour, 250 g leaves	4 a	6 a	4 a	3 a	4 a

Numbers with the same letter in the same column are not significantly different according to the Duncan 0.05 test.

The treatment of adding 250 g of tofu dregs flour and 100 g of dried banana leaves is able to provide sufficient nutrition so that it can form more primordia which will later become more fruit bodies. Nitrogen functions to form proteins, and builds enzymes for the growth of the fungus fruit caps. The crude fiber content of 25.81 % in tofu waste flour (Table 2) can increase the production of cellulose enzymes so that cell formation occurs. The dried plants leaves contain high levels of hemicellulose and lignin [8].

Low or insufficient nutrient content can cause a less optimal growth of the primordial caps, making the growth in the number of caps to be small. The use of different planting media composition affects the physical quality of the oyster mushrooms produced [9].

3.3 Mushroom caps diameter

The results of the analysis of variance showed a very real interaction on the diameter parameters of the brown oyster mushroom caps. The average yield of each treatment is presented in Table 4.

Table 4. Average diameter of brown oyster mushroom cap due to addition of tofu dregs flour and dried banana leaves

Treatment	Caps diameter (cm)				
	Harvest I	Harvest II	Harvest III	Harvest IV	Average harvest
50 g flour, 0 g leaves	8.58 c	7.42 f	6.76 c	5.60 d	7.09 e
150 g flour, 0 g leaves	6.97 abc	5.97 cde	5.77 b	5.18 bcd	5.97 cd
250 g flour, 0 g leaves	7.72 bc	6.25 de	5.06 a	4.33 a	5.84 bcd

(Continued on next page)

Table 4. Continued

Treatment	Caps diameter (cm)				
	Harvest I	**Harvest II**	**Harvest III**	**Harvest IV**	**Average harvest**
50 g flour, 100 g leaves	6.06 ab	6.08 cde	5.71 ab	5.13 bcd	5.74 bcd
150 g flour, 100 g leaves	6.01 ab	5.83 bcd	5.44 ab	4.56 abc	5.46 abc
250 g flour, 100 g leaves	5.50 a	5.32 ab	6.67 c	5.32 cd	5.70 abc
50 g flour, 250 g leaves	5.51 a	5.07 a	5.63 ab	4.75 abc	5.24 a
150 g flour, 250 g leaves	5.88 a	5.79 bc	5.37 ab	4.50 abc	5.39 ab
250 g flour, 250 g leaves	8.44 c	6.56 e	5.50 ab	4.67 abc	6.29 d

Numbers with the same letter in the same column are not significantly different according to the Duncan 0.05 test.

The treatment of adding 50 g of tofu dregs without the dry banana leaf has the highest diameter. This is allegedly because it is influenced by the number of caps formed. The 50 g tofu flour treatment without dried banana leaves has a smaller number of caps but the widest caps diameter. This shows that the growth of the caps is more maximal and the width of the caps is not crowded. The 50 g tofu dregs flour and 250 g dried banana leaves treatment has a relatively small diameter because the number of caps produced is quite large and crammed together to make the mushroom caps less optimal growth. The more mushroom caps that grow generally the smaller the diameter of the caps produced.

3.4 Thickness of mushroom caps

The results showed a significant interaction in the 1st to 4th harvest of the mushroom caps thickness. The average caps thickness of each treatment is presented in Table 5.

Table 5. Average thickness of brown oyster mushroom caps due to the addition of tofu waste flour and dried banana leaves

Treatment	The thickness of caps (mm)				
	Harvest I	**Harvest II**	**Harvest III**	**Harvest IV**	**Average harvest**
50 g flour, 0 g leaves	8.67 c	8.76 e	8.09 e	6.93 d	8.11 c
150 g flour, 0 g leaves	6.13 a	7.31 c	7.10 c	6.51 bcd	6.76 ab
250 g flour, 0 g leaves	7.33 ab	7.59 cd	6.39 cd	5.66 a	6.74 ab
50 g flour, 100 g leaves	7.11 b	7.41 cd	7.04 cd	6.46 bcd	7.01 b
150 g flour, 100 g leaves	7.07 b	7.16 bc	6.78 bc	5.90 abc	6.73 ab
250 g flour, 100 g leaves	6.56 ab	6.66 ab	8.00 ab	6.65 cd	6.97 ab
50 g flour, 250 g leaves	6.57 b	6.40 a	6.97 a	6.08 bcd	6.51 a
150 g flour, 250 g leaves	6.94 ab	7.12 bc	6.70 bc	5.82 ab	6.64 ab
250 g flour, 250 g leaves	6.66 ab	7.90 d	6.84 d	6.00 abc	6.85 ab

Numbers with the same letter in the same column are not significantly different according to the Duncan 0.05 test

The thickness parameter of the 50 g tofu dregs treatment cap without dried banana leaves has the highest caps thickness with an average total value of 8.11 mm. That is because in the growth of the number of fruit bodies, 50 g of tofu flour without tofu leaves produce a little and wider fruit caps so that the fungus can produce carbohydrates in sufficient quantities for vegetative growth. The more the number of caps, the thinner the caps thickness will be, causing the mushroom caps to be less than optimal. Mushrooms grow to form clumps, where if a large number of caps are formed in a clump, it will affect the diameter and thickness of the caps, ie the caps gets smaller.

3.5 Mushroom fresh weight and biological efficiency

The results showed a significant interaction between the addition of tofu dregs and dried banana leaves to the total fresh weight of mushrooms and biological efficiency. The average total fresh weight of mushrooms is presented in Table 6.

Table 6. Average fresh weight of brown oyster mushrooms due to addition of tofu dregs flour and dried banana leaves

Treatment	Fresh weight (g)	Biological efficiency %)
50 g flour, 0 g leaves	121.67 a	46 d
150 g flour, 0 g leaves	173.89 bc	55 ef
250 g flour, 0 g leaves	178.89 c	51 e
50 g flour, 100 g leaves	152.78 abc	48 de
150 g flour, 100 g leaves	137.78 ab	39 b
250 g flour, 100 g leaves	230.00 d	57 f
50 g flour, 250 g leaves	145.00 abc	40 b
150 g flour, 250 g leaves	168.33 bc	42 bc
250 g flour, 250 g leaves	132.78 a	31 a

Numbers with the same letter in the same column are not significantly different according to the Duncan 0.05 test.

Nutrients available in growing media that can be absorbed by the fungus will be able to increase the fresh weight of the fungus. The protein content in tofu dregs is thought to be able to add nutrients to the mushroom media. The cellulose and hemicellulose content in dried banana leaves can also stimulate tissue formation which affects the fresh weight of oyster mushrooms produced. It was suspected that mushrooms have sufficient energy reserves because the elements contained in baglog are evenly decomposed at the time of fruit body formation, thus affecting the fresh weight of mushrooms. The nutrients available in the growing media that are able to be absorbed by the fungus will be able to increase the fresh weight of the fungus. High protein content in tofu dregs increases the amount of nitrogen in the growing media so that it can spur the growth of white oyster mushroom mycelium. High nitrogen causes the growth of mycelium which is thicker and more compact [10].

The addition agricultural waste (leaves) to the base medium of wood sawdust can increase the fresh weight of the oyster mushroom body [11]. Nutritional requirements for oyster mushrooms are a source of carbon, nitrogen, and phosphate [12].

The higher the value of biological efficiency, the higher the production obtained and the more efficient use of the media by oyster mushrooms [13]. The treatment of adding 250 g flour and 100 g leaves has the best thickness of mycelium, which is dense and thick, mycelium which grows very thick can utilize the nutrients contained in the media well and efficiently. A good mycelium is mycelium that looks white like cotton, dense, and grows attached [14]. In the treatment with a higher total fresh weight of mushrooms, the biological efficiency value will also be higher.

4 Conclusion

The interaction of the treatment of the addition of tofu dregs with dried banana leaves significantly affected the growth of mycelium, the number of caps, caps diameter, caps thickness, fresh weight, and the biological efficiency of brown oyster mushrooms. The treatment of adding 250 g of tofu dregs with 100 g dried banana leaves is a better treatment than others, which is shown with the highest fresh weight and biological efficiency.

References

1. B. Chakravarty. Australian J Agric. Eng., **2**,4:102–109(2011). https://www.researchgate.net/publication/233735019_Trends_in_Mushroom_cultivation_and_breeding
2. S. Carmen. Appl. Microbiol. Biotechnol., **85**,5:1321–1337(2009). https://www.researchgate.net/publication/40444029_Cultivation_of_Pleurotus_ostreatus_and_other_edible_mushrooms
3. M. Syawal, S.A. Lasmini, Ramli. Int. J. Biol.Res., **2**,1:156–161(2019). https://www.researchgate.net/publication/331087839_The_Effect_of_Bran_and_Corn_Flour_Composition_on_Swadust_Media_Materials_Towards_the_Growth_and_Result_of_White_Oyster_Mushroom_Pleurotus_ostreatus
4. H.T. Hoa, W.Chun-Li, W. Chong-Ho. Mycobiology, **43**,4:423–434(2015). https://www.researchgate.net/publication/292205461_The_Effects_of_Different_Substrates_on_the_Growth_Yield_and_Nutritional_Composition_of_Two_Oyster_Mushrooms_Pleurotus_ostreatus_and_Pleurotus_cystidiosus
5. K.M. Alananbeh, N.A. Bouqellah, N.S.A. Kaff. Saudi Journal of Biological Sciences, **21**,6:616–625(2014). https://www.sciencedirect.com/science/article/pii/S1319562X1400093X?via%3Dihub
6. M. Obodai, J.Cleland, P.Okine, N.T. Johnson. Tropical Sci., **43**,3:121–124(2006). https://www.researchgate.net/publication/230148604_Use_of_agricultural_wastes_as_substrate_for_the_mushroom_Volvariella_volvacea
7. R. Rambey, I.D.B. Sitepu, E.B.M. Siregar. IOP Conf. Ser.: Earth Environ. Sci. 260, 012076(2019). https://iopscience.iop.org/article/10.1088/1755-1315/260/1/012076/pdf
8. S.B. Mukhopadhyay. *Oyster Mushroom Cultivation on Water Hyacinth Biomass: Assessment of Yield Performances, Nutrient, and Toxic Element Contents of Mushrooms*. (2019).p.13 https://www.researchgate.net/publication/338441296_Oyster_Mushroom_Cultivation_on_Water_Hyacinth_Biomass_Assessment_of_Yield_Performances_Nutrient_and_Toxic_Element_Contents_of_Mushrooms
9. S.S. Mkhize, J. Cloete, A.K. Basson, G.E. Zharare. Food Sci. Technol, **36**,4:598–605(2016). https://www.researchgate.net/publication/311566558_Performance_of_Pleurotus_ostreatus_mushroom_grown_on_maize_stalk_residues_supplemented_with_various_levels_of_maize_flour_and_wheat_bran
10. V. Barshteyn, T. Krupodorova. J. Microbiol. Biotech. Food Sci. **5**,6:563–577(2016). https://www.researchgate.net/publication/303702219_Utilization_of_agro-industrial_waste_by_higher_mushrooms_modern_view_and_trends
11. Widiwurjani. *Diversification of Oyster Mushroom (Pleurotus Ostreatus) Media From Agricultural Waste Materials*. Proceedings Of The International Conference Of Fossa. (2017).p.96. https://jurnal.unej.ac.id/index.php/prosiding/article/view/7950

12. Mudakir, I and U.S.Hastuti. Agrivita, **37**,1: 89–96(2015). https://www.researchgate.net/publication/277592760_Study_of_wood_sawdust_with_addition_of_plantation_wastes_as_a_growth_medium_on_yields_and_quality_of_white_oyster_mushroom
13. R.U. Priya, D.Geetha, S. Darshan. Adv. Life Sci. **5**,22:10252–10254(2016). https://www.researchgate.net/publication/315727943_Biology_and_Cultivation_of_Black_Ear_Mushroom_-_Auricularia_spp
14. T.R. Kinge, E.M. Adi, A.M. Mih, N.A.A che, T.M. Nji. African J. Biotechnol., **15**,27:1476–1486(2016). https://www.researchgate.net/publication/305452629_Effect_of_substrate_on_the_growth_nutritional_and_bioactive_components_of_Pleurotus_ostreatus_and_Pleurotus_florida

The Technical Design Concept of Hi-Tech Cook Stove for Urban Communities using Non-Wood Agricultural Waste as Fuel Sources

Erkata Yandri[1,2,*], *Bangun* Novianto[1], *Fridolini* Fridolini[3], *Roy* Hendroko Setyabudi[1,4], *Haryo* Wibowo[5], *Satriyo* Krido Wahono[6], *Kamaruddin* Abdullah[1,2], *Washington* Purba[1], and *Yogo* Adhi Nugroho[7,8]

[1]Graduate School of Renewable Energy, Darma Persada University, Jl. Radin Inten 2, Pondok Kelapa, East Jakarta 13450, Indonesia
[2]Center of Renewable Energy Studies, Darma Persada University, Jl. Radin Inten 2, Pondok Kelapa, East Jakarta 13450
[3]English Language Study Program of Vocational, Faculty of Language and Culture, Darma Persada University, Jl. Radin Inten 2, Pondok Kelapa, East Jakarta 13450, Indonesia
[4]Department of Agriculture Science, Postgraduate Program, University of Muhammadiyah Malang, Jl. Raya Tlogomas No. 246, Malang, 65145, Indonesia
[5]Institute of Energy and Power Engineering, Zhejiang University of Technology, No. 38 Zheda Road Hangzhou 310027, P.R. China.
[6]Research Division for Natural Product Technology – Indonesian Institute of Sciences, Jl. Jogja - Wonosari, km 31.5, Gunung Kidul, Special Region Yogyakarta 55861, Indonesia
[7]Individual Researcher, Alumni Graduate School of IPB University, Bogor, Indonesia
[8]Data Processing, *Rumah Paper* – Editage Services, Jl. Tokala No.1, Malang 65146, East Java, Indonesia

Abstract. The purpose of this study is to conceptualize an urban Hi-Tech Cook-Stove (HTCS) design using agricultural waste. Several steps need to be carried out. First, determine the cooking activities depend on the family size and food categories. Second, calculate the energy required for cooking. Third, determine the mass of biomass fuel required. Fourth, calculate the fuel consumption rate. Fifth, design the dimensions of the stove by considering the ergonomics, easy in manufacturing, installation, etc. Sixth, estimate the volume of each component. The result shows that the fuel supply must adjust the flow rate of fuel, air to fuel ratio controlled by a simple mechanical-electric compressor, monitor the combustion chamber visually/automatically, and the dust must be collected/disposed of automatically/ mechanically. HTCS must consider the pellets from the higher heating value and faster of biomass with a certain composition of chopped and fibre, also the safety and comfort, such as overheating control, air exhaust control, combustion control, cooling control. For the future, the HTCS technical design concept must be integrated with the electricity and hot water from solar energy by using a hybrid photovoltaic-thermal (PVT) collector and urban biogas digester in the development of smart grids and smart cities.

Keywords: Renewable energy integration, energy efficient building, briquette, pellet, rural development

* Corresponding author: erkata@gmail.com; erkata@pasca.unsada.ac.id

1 Introduction

As a tropical and agrarian country, Indonesia has a large potential of biomass from solid agriculture waste from palm oil, coconut, rubber, sugar, rice, and corn, which can be converted to solid, liquid, and gas energy [1]. Unfortunately, the utilization of biomass as energy has not been maximized yet. Technically, the yearly potential of agricultural waste in Indonesia is approximately 614.6 PJ [2]. The solid agricultural waste [3, 4], can be processed into pellets or briquettes, with uniform energy content, easily stored and transported. One of the uses of biomass solid waste is for cooking, using biomass stoves. Solid biomass gives a unique aroma and taste sensation to food, although the smoke factor must still be a serious concern. The problem is, the image of biomass stoves, especially in Indonesia, is identical to the modest tech stoves for people in low-income rural areas. Generally, rural improved cook stove programs are promoted by donors to empower women in reducing poverty [5]. Meanwhile, people in urban areas are comfortable with the ease of clean gas stoves. If there were biomass stoves made with high technology for certain urban communities, with biomass fuel from agricultural waste from pellets made by rural people, this would have significant energy, environmental and economic impact for the country, leading to national energy security [6].

The design of biomass stoves has attracted many researchers and non-government organizations around the world. There have been many efforts that have been done by various parties, both government and non-government organizations (NGOs), such as designing and promoting biomass stoves, to utilize cheap energy from agricultural waste, increase fuel efficiency, and at the same time empower the rural economy [7]. Last et al. [8] disseminated the improved traditional Haitian stove as an efficient energy stove. Raman et al. [9] evaluated the fuel flexibility, efficiency, and user acceptance with the boiling test to the three types of forced draft cookstove with wood and coconut shell fuels. Yulianto et al. [10] designed and tested a Stove for Combine Heat Power biomass using coconut shell by controlling some parameter, resulting in the flue gas and combustion chamber temperature can reach 520 °C and 840 °C, respectively. Agustina and Hafidaturrohmah [11] designed a biomass stove using corncob, coconut shell, and wood chips as biomass fuel and evaluated some parameters related to the thermal efficiency, combustion efficiency, specific fuel consumption, emission of CO and particulate emissions were evaluated. Nurhuda [12] created a biomass semi gasification stove with a pre-heating system using chopped wood, oil palm shells, hazelnut, corn cobs or pellets or briquettes from agricultural/plantation waste, thus saving 60 % to 80 % fuel compared to traditional biomass stoves.

Based on the above review, many biomass stoves have been developed for rural communities with various technical parameters to make them healthier and more efficient. However, not many biomass stoves are designed according to the needs of urban communities, especially for the upper-middle level. The challenge is how to design stoves with smaller-mass components that also have acceptable durability, affordable cost, the decisive socio-cultural, increasing the life of a cook-stove through a better choice of materials, ease of operation and maintenance, and meet user needs with economics and modern technology [13–15], by utilizing agricultural waste as fuel [12]. Based on these considerations, the purpose of this study is to conceptualize a Hi-Tech cook-stove design using pellets and briquettes as fuel from solid biomass agricultural waste non-wood produced by rural people for certain urban communities. The design of HTC supports the general concept of future household clean energy models [4] and empowering the rural economy by treating agricultural waste [12].

2 The design of hi-tech cook stove (HTCS)

Figure 1 shows the combustion chamber concept design of Hi-Tech Cook Stove (HTCS). The reason to choose the combustion chamber type stove because it seems to burn wood more effectively at varying moisture content, compared to the open fire or rocket type stoves [16]. The design consists of several main components, such as; fuel supply system (FSS) as a biomass fuel feeder, pressured air system (PAS) as an oxygen supplier during combustion, combustion chamber system (CCS) as a place for combustion, bottom ashtray system (BAS) as a bottom ash collector for combustion results, and is equipped with a combustion regulator system (CRS) button to adjust the Low-Mid-High calibrated with FSS and PAS.

Fig. 1. Concept design of Hi-Tech Cook Stove (HTCS)

3 Method

To determine the dimensions and operational parameters of each HTCS component, several steps need to be carried out as follows:

First, determine the cooking activities depend on the family size and food categories. These factors affect the energy required and times for cooking. The family size is divided into three, such as small family (three members: father, mother, and a child), middle family (five members: father, mother and three children), large family (more than six members: father, mother, and more than three children). The cooking category is divided into fast cooking (5 min to 10 min), medium cooking (10 min to 20 min) heavy cooking (more than 20 min).

Second, calculate the energy required for cooking Q_c [J]

$$Q_c = \dot{m}_c c_{p,c} \Delta T_c \quad (1)$$

where, $\dot{m}_c$ is the mass of cooked food [kg], $c_{p,c}$ is a specific heat from cooking ingredients [kJ kg^{-1} $°C^{-1}$] of selected foods [17–19], ΔT_c is the different temperature of cooking after and before [°C].

Third, determine the mass of biomass fuel required m_{bb};

$$m_{bb} = \frac{Q_c}{C_v \times Eff} \tag{2}$$

Where C_v is the heating value of biomass fuel [J kg^{-1}], and Eff is the standard combustion efficiency of biomass to heat energy [%].

Fourth, calculate the fuel consumption rate [g min^{-1}] with:

$$FCR = \frac{m_{bb}}{t} \tag{3}$$

Where, t is the total times require for cooking [min]. It refers to the amount of energy supplied by the fuel into the combustion chamber.

Fifth, design the dimensions of the stove by considering several factors, such as ergonomics for the user during cooking, easy in manufacturing, easy installation, easy to use, easy to move, easy to maintain, etc. Based on these items, the dimensions of the stove are designed 1.6 m × 0.8 m × 0.8 m for length, height and width, respectively. The volume of the stove must adjust the dimensions and position of the main stove components.

Sixth, estimating the volume of each component:

(i) FSS: calculated from Equation (2) and fuel density $\bar{V}_{bb}$ [kg m^{-3}] with assumption variables n as the frequency of refuelling in a day, Then V_{fss} [m^3] is as follows;

$$V_{fss} = n(\frac{m_{bb}}{\bar{V}_{bb}}) \tag{4}$$

(ii) PAS: calculated from Equation (1) with O_2 required during hourly cooking, then V_{pas} [m^3] is as follows [20]: when $\varepsilon = 0.3$, stoichiometric air (SA) is 4.5 kg_{air} kg_{fuel}^{-1} and air density ρ_a = 1.29 kg m^{-3} and air compression factor $f_c = 0.5$.

$$V_{pas} = \frac{\varepsilon \times FCR \times 60 \times SA}{\rho_a} f_c \tag{5}$$

(iii)CCS: calculated from Equation (2) with consideration of volume tolerance for optimal combustion, then V_{ccs} is as follows: Diameter of CCS: assuming the combustion rate (SBR) specification is 40 kg m^{-2} h^{-1}. Then D [m] is as follows:

$$D = \left(\frac{1.27 \times FCR}{SBR}\right)^{0.5} \tag{6}$$

Height of CCS: if it is assumed that the use of stove without refuelling can work for 6 h (T) and fuel density $\bar{V}_{bb}$ [kg m^{-3}]. Then the CCS height H [m] is:

$$H = \frac{SBR \times T}{\bar{V}_{bb}} \tag{7}$$

The volume of CCS;

$$V_{ccs} = \pi \times r^2 \times h \tag{8}$$

https://doi.org/10.1051/e3sconf/202122600015

where, r is the radius of the stove [m] and h is the distance between the stove and the cauldron assumed to be 3 cm. The dimension of CCS should be designed properly to prevent excess fuel loading for decreasing fuel consumption while cooking [21].

(iv) BAS: calculated from the combustion residue ash from Equation (2) with estimated residual (δ) 5% and the maximum accumulation of cooking frequency, then V_{bat} is as follows:

$$V_{bas} = n\left(\frac{\delta \times m_{bb}}{\bar{V}_{bb}}\right)10^{6} \qquad (9)$$

All aspects of the technical design must refer to internationally accepted standards and protocols, which must be considered during the development process by identifying, avoiding and overcoming errors in the design or system concept [22].

4 Result and discussion

Figure 2 illustrates the relationship between biomass needed and family size. Figure 2(a) shows the energy required for cooking in a day is proportional to the family size and cooking frequency. The energy required will also increase with increasing cooking time. Figure 2(b) shows the amount of energy required (kg) for cooking in a day with different types of biomass in pellets and briquettes. The amount of energy (in kg) depends on the caloric value (in J kg^{-1}) of the type of biomass, the higher the caloric value, the lower the biomass consumed (in kg).

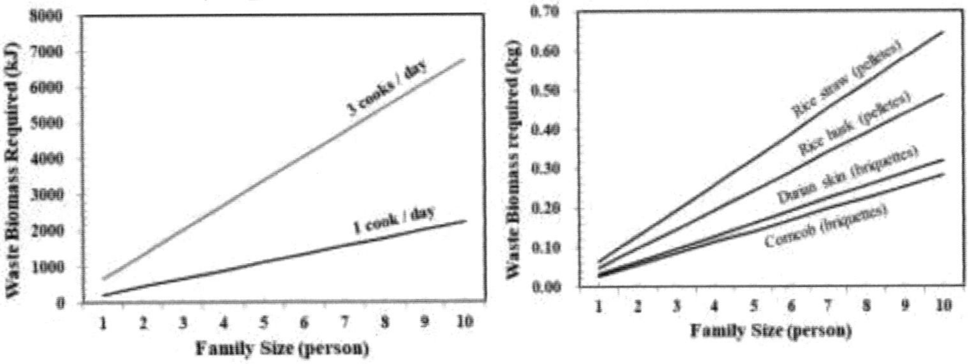

(a) (b)

Fig. 2. Waste biomass required, a). One cook d^{-1} vs. three cooks d^{-1}, b). Different biomass and types

Table 1 is the dimensions of CCS based on the high heating value,, the briquettes of corncob, and the pellets of palm oil trees, as the reference in design calculations. With higher fuel density, palm oil tree pellets provide shorter CCS heights compared to corncob briquettes.

Table 1. Specification design for CCS using briquettes and pellets

Type	Biomass Waste	Fuel Density	CCS	
			D (m)	H (m)
Briquettes	Corncob	857	0.14	0.28
Pellets	Palm oil tree	1 200	0.15	0.20

The volume and consideration for system components of HTCS are summarized in Table 2. The volume of system components is highly dependent on pellets or briquettes biomass. For ease of use, biomass pellets are considered in the HTCS technical design concept. Biomass stoves available today still cannot replace LPG stoves. This is the challenge that HTCS has to answer, how users still feel as comfortable as using an ordinary gas stove.

Table 2. Summary volume and consideration for system components

System Components	Volume (cm^3)		Design Consideration
	Briquettes	Pellets	
FSS	375	283	The fuel supply adjusts the flow rate
PAS	4 935	5 917	Simple mechanical-electric compressor application
CCS	462	530	Combustion chamber is monitored visually/automatically
BAS	48	42	Dust is collected/disposed of automatically/mechanically

Based on what has been explained before, then three things need to be discussed further. First, related to the type and composition of pellets to be used as fuel. Following HTCS specifications for immediate cooking, it is necessary to consider the type and composition of biomass for pellets [10–12]. For this reason, pellets must be produced from high heating value biomass, with a certain composition of chopped and fibre. Second, related to safety and comfort. Both of these must be an integral part of the design, prototyping and testing of HTCS. There are several important systems to be studied in more detail, such as overheating control, air exhaust control, combustion control, cooling control, etc [13–15]. To ensure the availability of fuel supply during cooking, it is necessary to design fuel supply from other sources, for example, biogas from kitchen waste digester or septic tank digester. Thus, CCS design must also be adjusted for biogas. Third, related to the standard test. HTCS must pass the following test items, such as; boiling test, time for initial start-up test, rise test, air to fuel ratio test, firing test (Low, Mid, High), etc.

In practice, HTCS must be directly applied to cook all types of cuisine, as well as the application of modern gas and electric stoves [14]. For the initial stage, HTCS was promoted for the urban middle-upper class community, which is considered easy to educate financially and operationally on this high-tech stove. In addition to contributing to utilizing non-woody biomass waste and reducing fossil energy dependency, HTCS also has an economic impact by empowering rural communities as producers of pellets and briquettes as biomass fuels [23, 24]. For future research directions, the HTCS technical design concept must be realized and tested with the special functions of the main components. In addition, with the development of smart grids, smart cities and energy efficiency in urban areas [25–27], it is also important to consider the integration of electricity and hot water from solar energy by using a hybrid photovoltaic-thermal (PVT) collector [28, 29] and future development of urban biogas digester.

The application of biogas is feasible to manage kitchen waste and excreta disposal from septic tanks in urban areas for environment friendly [30–33]. The two-stage digester system is recommended because kitchen waste tends to be acidic and float [34, 35]. Waste management is essential because it harms human health [36, 37].

5 Conclusion

A Hi-Tech Cook Stove (HTCS) concept has been designed for urban communities by utilizing non-wood agricultural waste into pellets as fuel. With dimensions of length = 1.6 m, width = 0.8 m, and height = 0.8 m, all important components of HTCS, such as CCS (combustion chamber system), FSS (fuel supply system), PAS (pressurized air system), and BAS (bottom ashtray system) are designed inside the stove. To meet operational standards, the HTCS must refer to; high heating value pellet, high levels of safety and comfort and pass certain standard tests. Further development integrates HTCS with other renewable energy sources such as electricity and heat from the PVT collector and biogas from the urban biogas digester in the development of smart grids and smart cities.

References

1. N.L. Panwar, R. Kothari, V.V. Tyagi, Renew. Sustain. Energy Rev., **16**,4:1801–1816(2012). https://www.sciencedirect.com/science/article/pii/S1364032112000251
2. IRENA, *Renewable Energy Prospects in Indonesia*, Abu Dhabi: IRENA(2017). https://www.irena.org/-/media/Files/IRENA/Agency/Publication/2017/Mar/IRENA_REmap_Indonesia_report_2017.pdf
3. S.S. Harsono, B. Prayogo, M. Mel, F. Ridha, Journal of Advanced Research in Fluid Mechanics and Thermal Sciences, **2**,2:215–226(2018). http://www.akademiabaru.com/doc/ARFMTSV51_N2_P215_226.pdf
4. A.K. Quinn, N. Bruce, E. Puzzolo, K. Dickinson, R. Sturke, D.W. Jack, et al., Energy Sustain. Dev., **46**:1–10(2018). https://www.sciencedirect.com/science/article/pii/S0973082618302679
5. M. Kees, L. Feldmann, Energy Policy, **39**,12:7595–7599(2011). https://www.sciencedirect.com/science/article/pii/S0301421511002102
6. E. Yandri, R. Ariati, R.F. Ibrahim, J. Ketahanan Nas., **24**,2:239–260(2018). [in Bahasa Indonesia]. https://dev.jurnal.ugm.ac.id/jkn/article/view/30999
7. S.A. Mehetre, N.L. Panwar, D. Sharma, H. Kumar, Renew. Sustain. Energy Rev., **73**:672–687(2017). https://www.sciencedirect.com/science/article/pii/S1364032117301648
8. K. Lask, K. Booker, T. Han, J. Granderson, N. Yang, C. Ceballos, et al., Energy Sustain. Dev., **26**:79–86(2015). https://www.sciencedirect.com/science/article/pii/S0973082615000101
9. P. Raman, J. Murali, D. Sakthivadivel, V.S. Vigneswaran, Biomass and Bioenergy, **49**:333–340(2013). https://www.sciencedirect.com/science/article/pii/S0961953412005260
10. M. Yulianto, S.E. Agustina, E. Hartulistiyoso, L. Oscar, AIP Conference Proceedings, **1826**:020031(2017); https://doi.org/10.1063/1.4979247
11. N. Agustina, H. Hafidaturrohmah, E3S Web Conf., **67**,02034:1–8.(2018). https://www.e3s-conferences.org/articles/e3sconf/abs/2018/42/e3sconf_i-trec2018_02034/e3sconf_i-trec2018_02034.html
12. M. Nurhuda, Majalah Energi dan Mineral, **9**,4:4–10(2011). [in Bahasa Indonesia]. https://litbang.esdm.go.id/publikasi/5908-majalah-energi-dan-mineral/8965-majalah-

mineral-dan-energi-desember-2011
13. J.K. Jetter, P. Kariher, Biomass and Bioenergy, **33**,2:294–305(2009). https://www.sciencedirect.com/science/article/pii/S0961953408001505
14. M.P. Kshirsagar, V.R. Kalamkar, Renew. Sustain. Energy Rev., **30**:580–603(2014). https://www.sciencedirect.com/science/article/pii/S136403211300748X
15. K.B. Sutar, S. Kohli, M.R. Ravi, A. Ray, Renew. Sustain. Energy Rev., **41**:1128–1166(2015). https://www.sciencedirect.com/science/article/pii/S136403211400776X
16. E.A.T. Yuntenwi, N. MacCarty, D. Still, J. Ertel, Energy Sustain. Dev., **12**,2:66–77(2008). https://www.sciencedirect.com/science/article/pii/S0973082608604305
17. E. Contreras-Gallegos, F.A. Domínguez-Pacheco, C. Hernández-Aguilar, J.A. Salazar-Montoya, E.G. Ramos-Ramírez, A. Cruz-Orea, J. Therm. Anal. Calorim., **128**,1:523–531(2017). https://link.springer.com/content/pdf/10.1007/s10973-016-5864-1.pdf
18. F.S. Carrillo, L. Saucier, C. Ratti, Int. J. Food Prop., **20**,3:573–584(2017). https://www.tandfonline.com/doi/abs/10.1080/10942912.2016.1171776
19. K.D. Barine, N. Victor, Journal of Food Technology Research, **3**,1:23–272016). http://www.conscientiabeam.com/pdf-files/agr/58/JFTR-2016-3(1)-23-27.pdf
20. A.T. Belonio, *Rice husk gas stove handbook*. Philippines: Central Philippine University, (2005). http://bioenergylists.org/stovesdoc/Belonio/Belonio_gasifier.pdf
21. S. Bentson, D. Still, R. Thompson, K. Grabow, Energy Sustain. Dev., **17**,2:153–157(2013). https://www.sciencedirect.com/science/article/pii/S0973082612000804
22. Z. Zhang, Y. Zhang, Y. Zhou, R. Ahmad, C. Pemberton-Pigottc, H. Annegarn, et al., Renew. Sustain. Energy Rev., **72**:1343–1354(2016). https://www.sciencedirect.com/science/article/pii/S136403211630692X
23. E. García-Frapolli, A. Schilmann, V.M. Berrueta, H. Riojas-Rodríguez, R.D. Edwards, M. Johnson, et al., Ecol. Econ., **69**,12:2598–2605(2010). https://www.sciencedirect.com/science/article/pii/S0921800910003289
24. B. Novianto, K. Abdullah, A.S. Uyun, E. Yandri, S.M. Nur, H. Susanto, et al., E3S Web of Conferences **188**:00005(2020) https://doi.org/10.1051/e3sconf/202018800005
25. E. Yandri, R.H. Setyobudi, H. Susanto, K. Abdullah, Y.A. Nugroho, S.K. Wahono, et al., E3S Web of Conferences **188**:00007(2020). https://doi.org/10.1051/e3sconf/202018800007
26. E. Yandri, R. Ariati, A.S. Uyun, R.H. Setyobudi, H. Susanto, K. Abdullah, et al., E3S Web of Conferences **190**:0008(2020) https://doi.org/10.1051/e3sconf/202019000008
27. E. Yandri, Renew. Energy, **111**:344–352(2017). https://www.sciencedirect.com/science/article/pii/S0960148117302860
28. S.S. Harsono, R. Dila, M. Mel, J. Environ. Sci. Curr. Res. **2**:004(2019). http://www.heraldopenaccess.us/openaccess/coffee-husk-biopellet-characteristics-as-solid-fuel-for-combustion-stove
29. E. Yandri, Sol. Energy Mater. Sol. Cells, **201**:10066(2019). https://www.sciencedirect.com/science/article/pii/S0927024819303952
30. R. Hendroko, T. Liwang, Salafudin, L.O. Nelwan, G.A. Praptiningsih, S.K. Wahono, *Biorefinery skala rumah tangga: integrasi energi terbarukan - biogas dan pangan* [Household scale biorefinery: integration of renewable energy - biogas and food]. Seminar Hasil Penelitian Semester Ganjil 2011/2012, Lembaga Penelitian, Pemberdayaan Masyarakat dan Kemitraan Universitas Darma Persada, (Jakarta, Indonesia, 2012), pp. 1–13. [in Bahasa Indonesia] http://lp2mk.unsada.ac.id/wp-content/uploads/2016/04/PROSIDING-GJL-11-12.pdf
31. R. Hendroko, T. Liwang, Salafudin, G.A. Praptiningsih, L.O. Nelwan, Y. Sakri, et al., *Sinergi bio-metana berbahan baku limbah* Jatropha curcas *L., dan pangan dalam penerapan program kawasan rumah pangan lestari* [Synergy of bio-methane made

from *Jatropha curcas* L. waste, and food in the implementation of sustainable food home area program]. Prosiding Simposium dan Seminar Bersama PERAGI-PERHORTI-PERIPI-HIGI, (Bogor, Indonesia, 2012), pp. 437–443. [in Bahasa Indonesia] http://lipi.go.id/publikasi/sinergi-bio-metana-berbahan-baku-limbah-jatropha-curcas-l-dan-pangan-dalam-penerapan-program-kawasan-rumah-pangan-lestari/20594

32. K. Abdullah, A.S. Uyun, R. Soegeng, E. Suherman, H. Susanto, R.H. Setyobudi, et al., E3S Web of Conferences **188**:00016(2020). https://doi.org/10.1051/e3sconf/202018800016
33. H. Susanto, R.H. Setyobudi, D. Sugiyanto, S.M. Nur, E. Yandri, H. Herianto, et al., E3S Web of Conferences **188**:00010(2020). https://doi.org/10.1051/e3sconf/202018800010
34. P.G. Adinurani, T. Liwang, Salafudin, L.O. Nelwan, Y. Sakri, S.K. Wahono, et al., Energy Procedia, **32**:84–89(2013). https://www.sciencedirect.com/science/article/pii/S1876610213000131
35. P.G. Adinurani, R.H. Setyobudi, S.K. Wahono, A. Sasmitod, L.O. Nelwan, A. Nindita, et al., International Journal of Renewable Energy Development, **3**,1:73–78(2014). https://doi.org/10.14710/ijred.3.1.73-78.
36. K. Khoiron, A.N. Probandari, W. Setyaningsih, H.S. Kasjono, R.H. Setyobudi, O. Anne, Ann Trop & Public Health S463, **23**,3(A):60–67(2020). http://doi.org/10.36295/ASRO.2020.23316
37. M. Soleh, H. Hadiyanto, J. Windarta, O. Anne, R.H. Setyobudi, M. Mel . E3S Web of Conferences **190**:00027(2020). https://doi.org/10.1051/e3sconf/202019000027

Detergents Effect on Egg Hatchability, Morphometry and Larval Bone Structure of Native Indonesian Fish: Wader Pari (*Rasbora lateristriata* Bleeker, 1854)

Farahsani Umi Abida[1], *Parvez* Alam[2] and *Bambang* Retnoaji[1,*]

[1]Faculty of Biology, Universitas Gadjah Mada, Jl. Teknika Selatan, Senolowo, Sinduadi, Yogyakarta 55281, Indonesia
[2]School of Engineering, Institute of Material and Processes, The University of Edinburgh, Edinburgh EH9 3FB, United Kingdom

Abstract. This study was conducted to determine the effect of detergents on eggs hatchability, survival rates, morphometry and bone structure of Wader Pari fish (*Rasbora lateristriata* Bleeker, 1854). The fish were treated with detergent solution, with a concentration of 0 mg L^{-1}, 3 mg L^{-1}, 6 mg L^{-1}, 9 mg L^{-1}, 18 mg L^{-1}, 27 mg L^{-1}, and 50 mg L^{-1}, respectively. Each test consisted of one aquarium filled with 30 eggs. The number of eggs hatched was counted and these were treated continually for 1.5 mo to determine fish survival rates, morphology, growth rates and bone structure assays. The bone structure assay was prepared with Alizarin's Red-Alcian Blue staining. The results show that detergent treatments at 0 mg L^{-1}, 3 mg L^{-1}, 6 mg L^{-1}, 9 mg L^{-1} did not significantly affect hatchability, survival rate, or morphometry of either egg or fish ($p > 0.05$). However, a higher concentration at 18 mg L^{-1}, 27 mg L^{-1}, and 50 mg L^{-1}, significantly affected the fish egg hatchability and survivability. Moreover, fish vertebral structures were normal at treatments of 0 mg L^{-1}, 3 mg L^{-1}, 6 mg L^{-1}, whereas at 9 mg L^{-1}, was caused abnormal vertebral structures.

Keywords: Fish, larvae, water polution, survival rate, vertebrae

1 Introduction

Detergents are cleaning products that are a cause for environmental concern as they may be leached into rivers, especially where water pollution control is not well regulated. The laundry industry is a sector that utilizes detergents in high quantities, and the rates of utility are on an upward trajectory. The number of licensed and supervised laundry services in the Yogyakarta region, Indonesia are currently only 96 % [1]. However, its estimated that this number will multiply rapidly, especially as home-scale laundry outlets gain popularity. Most home-scale laundry services lack effective waste disposal facilities for the treatment and disposal of their waste. As such, laundry waste is typically released directly into local open water sources such as rivers and streams [1].

* Corresponding author: bambang.retnoaji@ugm.ac.id

Important constituents in detergents include surfactants, builders, organic additives, solvents, enzymes, and stains. Surfactant is the most important and active component in detergents. One example of the most widely used surfactant is LAS (Linear Alkylbenzene Sulfonate). LAS, contains a long chain of carbon atoms, which cannot be degraded by microorganisms [2–5]. Detergent enzymes usually account for 2 % of their total weight. LAS decreases the stability of detergent proteases [3].

Detergents in water are absorbed by fish through their gills or skin, affecting proteins, fats and carbohydrates content [6]. The fish egg and larval phase is the most sensitive stage for water pollution.

Detergent can destroy fish eggs at a specific concentration, before the gastrulation stage of the embryo is more susceptible to pollutant stress compared to the embryo which has completed the gastrulation [7, 8]. The increasing concentration of surfactants in the waters can also result in a decrease in surface tension of the egg membrane so that the surfactant easily penetrates the egg thus disturbing the egg metabolisms. The detergent is also able to inhibit several enzyme systems by suppressing egg respiration, which results in oxygen starvation, leading to egg mortality. Moreover, the detergent can interfere with the pattern of axis formation, and head and eye development, respectively [2].

Detergent can also affect adult fish. Amount 3 mg L^{-1} detergent can reduce the blood protein content of *Cirrhinus mrigala* [Hamilton, 1822] fish from 3.14 mg dL^{-1} to 1.28 mg dL^{-1} within the space of 72 h. This decrease may inhibit protein synthesis to the detriment of the formation of bone matrices, which in turn could lead to vertebral column abnormalities. Detergent and surfactant exposure may furthermore cause liver cell defects and have locomotory effects on fishes [2, 9–11]. Given that wader pari fish may be adversely affected by the growing concentrations of detergents in water, research to determine the specific effects of detergents on eggs in terms of hatchability, survival rate, morphometry and bone structure, is urgently needed.

2 Materials and methods

2.1 Fish reproduction and egg collection

Female and male fish with mature gonad were selected and maintained at the laboratory, with adequate food, as well as controlled environmental conditions of 16 h bright photoperiod cycle and 8 h dark, 28 °C to 29 °C of temperature, and at a pH of 7.0 to 7.5. Soon after spawning was conducted, the fish were returned to the initial aquarium, and eggs were collected, cleaned and maintained on egg water for further treatment.

2.2 Egg treatment

Eggs were put in aquariums using 30 eggs for each treatment. The setup of media were 0 mg L^{-1}, 3 mg L^{-1}, 6 mg L^{-1}, 9 mg L^{-1}, 18 mg L^{-1}, 27 mg L^{-1}, and 50 mg L^{-1} of detergent, with three replications at each concentration.

2.3 Eggs hatchability

After 24 h of treatment, eggs were observed and the number of hatched eggs was recorded in each treatment. Furthermore, the larvae were kept for about 2 mo.

2.4 Morphometry of the fish

During 2 mo the larvae were observed for survivability and growth and these were also morphologically documented.

2.5 Skeletal staining

Fish bone staining was performed to detect the possible bone defects resulting from the treatment. Three fish from each treatment were randomly chosen for bone staining for Alizarin's Red-Alcian Blue following Inouye [12], with modifications. Fish were fixed with 96 % ethanol for 3 d and internal organs were removed. After fixation, embryos were treated with acetone for 24 h and then incubated in staining solution (0.015 % Alcian Blue, 0.015 % Alizarin Red in 70 % ethanol) at 37 °C for 3 d to 4 d. After being washed with water, the embryos were cleared using 1 % KOH for 48 h to 72 h and successively incubated in 1 % KOH-20 % glycerol. Stained embryos were stored and observed in 50 % glycerol.

2.6 Data analysis

Quantitative data obtained from this study, such as percent of eggs hatchability, survival rates, and fish body ratios, were statistical analyzed with ANOVA.

3 Results and discussions

3.1 Environmental parameter

The water quality parameters were measured initially 24 h after treatment and the routinely each week, which included the temperature, pH and dissolved oxygen (Table 1). Overall, the water properties were similar for both the control and the treatment save the pH value during the first 24 h of treatment using 9 mg L^{-1}, which has a pH value of 8.

Table 1. The parameters of environmental condition measured during the experiment

Water quality	**First 24 h**		**Weekly Average**	
	control	**treatment**	**control**	**treatment**
pH	7	8	7 to 7.5	7 to 8
Temperature (°C)	28	28	27 to 30	27 to 30
Dissolved oxygen	8	8	7 to 8	7 to 8

The overall value of the whole period of the experiment was in a good range for egg hatching and larval development.

3.2 Fish eggs hatchability

The hatching rate of eggs treated with detergents in several concentrations is presented in Figure 1. The hatching rate of the eggs in all treatments, ranged from 93.33 % to 100 %. The lowest hatching percentage was on 3 mg L^{-1}, while the others reached 100 % hatchability. However, overall treatment was not significantly different ($p = 0.05$).

 https://doi.org/10.1051/e3sconf/202122600016

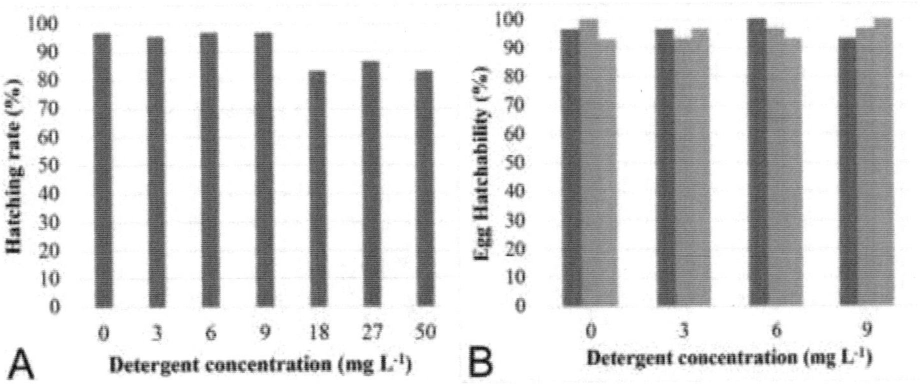

Fig. 1. Graph of egg hatching rate, A) Percentage of egg hatching rate all treatments, showing the high hatching rate on control and concentration of 3 mg L^{-1}, 6 mg L^{-1}, and 7 mg L^{-1} respectively, and less on concentration of detergent of 18 mg L^{-1}, 27 mg L^{-1} and 50 mg L^{-1}, respectively, B) The detail of egg percentage hatchability on each replication.

Referring to the result, this study conducted testing at the next exposure level of detergent concentrations of 18 mg L^{-1}, 27 mg L^{-1}, 50 mg L^{-1}. The testing was carried out to ensure the effect of very high concentration of detergent on eggs hatchability. The results showed that the range percentage of egg hatchability after exposure to a detergent concentration of 18 mg L^{-1}, 27 mg L^{-1} and 50 mg L^{-1}, were as follows; 83.33 %, 86.67 %, and 83.33 %, respectively. These suggest a very high concentration of detergent lowers the percentage of fish egg hatchability.

Temperature could induce slow and rapid choriolytic enzyme production. At low temperatures (< 24 °C), enzyme production is slow to cause chorion softening, such that the embryo is difficult to penetrate. At high temperatures (> 38 °C), rapid enzyme production causes premature hatching of embryos, which is for the most part, fatal to the embryo [13].

3.3 Survival rates of the larvae

The treatment on fish larvae decreased the fish larva survival rate for up to 1.5 mo (age). In the control, the survival rate was 98.85 %, 96.63 %, 95.48 %, and 90.84 %, from the start to the end of treatment. On the other hand, 3 mg L^{-1}, 6 mg L^{-1} and 9 mg L^{-1} of detergent treatment resulted in reductions in the numbers of surviving fish, in that respective order (Figure 2).

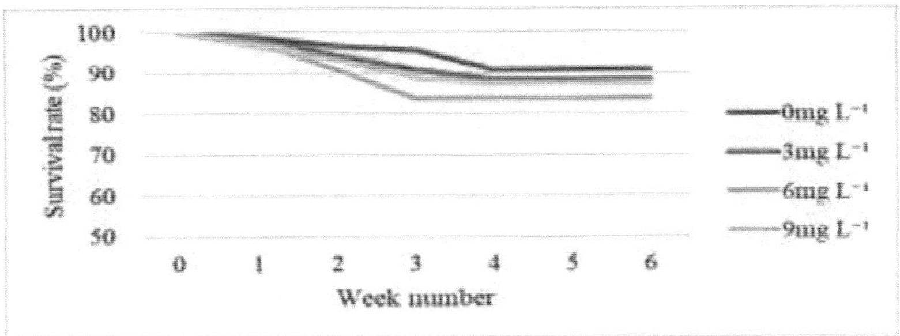

Fig. 2. Fish survival rate for 6 wk.

 https://doi.org/10.1051/e3sconf/202122600016

The treatment showed no significant ($p > 0.05$) effect on fish survival rate, which suggested that the detergent concentration up to 9 mg L^{-1} was not strong enough to interfere fish metabolism. It was reported that moderate concentration of detergent caused severe effect on fish. However, it needs higher concentration (1 000 mg L^{-1}) to severe impact on the Anabas fish respiratory performance [14–18].

3.4 The growth rate of the fish Larvae

The growth rate of fish was measured based on total body length, and the ratio of head, body, and tail length. The variation of fish length each of the experiments were showed on Figure 3.

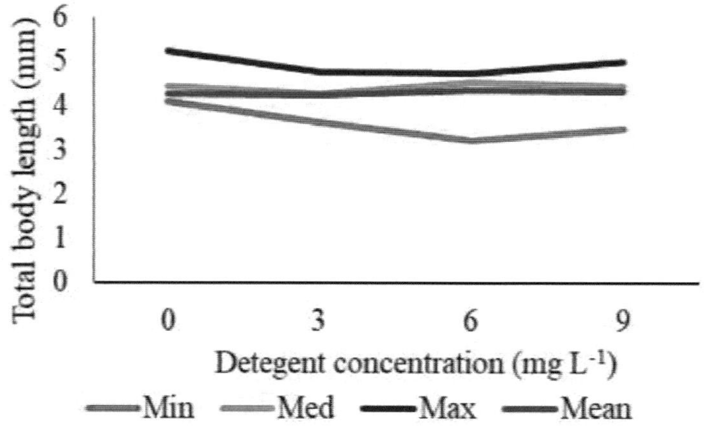

Fig. 3. Fish total body length for 6 wk.

The total body length of fish showed that control was longest body size of 4.45 cm, which followed by 9 mg L^{-1} (4.37 cm), 6 mg L^{-1} (4.33 cm) and 3 mg L^{-1} (4.24 cm).

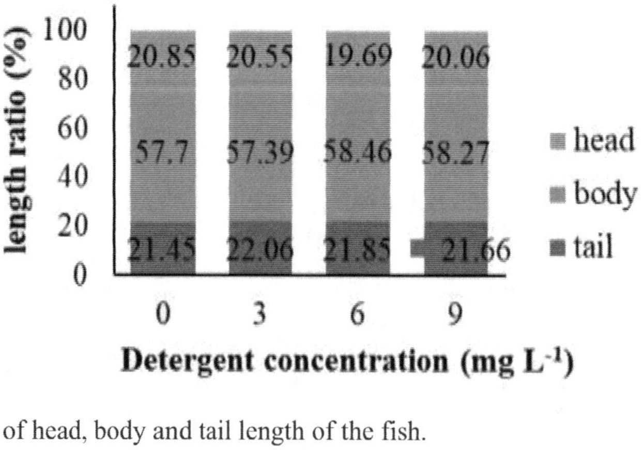

Fig. 4. The ratio of head, body and tail length of the fish.

The average head, body and tail length ratio showed the different results of different concentration of 0 mg L^{-1}, 3 mg L^{-1}, 6 mg L^{-1} and 9 mg L^{-1} (Figure 4). The average length percentage of head and to total body length ratio of 20.85 %, 20.55 %, 19.69 % and

20.06 %, respectively. The body length to total body length ratio of 0 mg L^{-1}, 3 mg L^{-1}, 6 mg L^{-1}, and 9 mg L^{-1}, were 57.7 %, 57.39 %, 58.46 % and 58.27 %, respectively. The tail length to total body length ratio of 0 mg L^{-1}, 3 mg L^{-1}, 6 mg L^{-1}, and 9 mg L^{-1}, were 21.49 %, 22.06 %, 21.85 %, and 21.66 %, respectively. The result showed that detergent concentration treatment of 0 mg L^{-1}, 3 mg L^{-1}, 6 mg L^{-1}, and 9 mg L^{-1} did not significantly affect fish growth, and the ratio of head, body, and tail ($p > 0.05$).

3.5 Vertebrae structure of the fish

The vertebrae structure of all the fish were stained and observed for number and morphology. Vertebrae number of fishes were consisted of 33 vertebral bone in total, which consisted of three cervical, 11 precaudal, 16 caudal, and three caudal fin vertebrae (Figure 5).

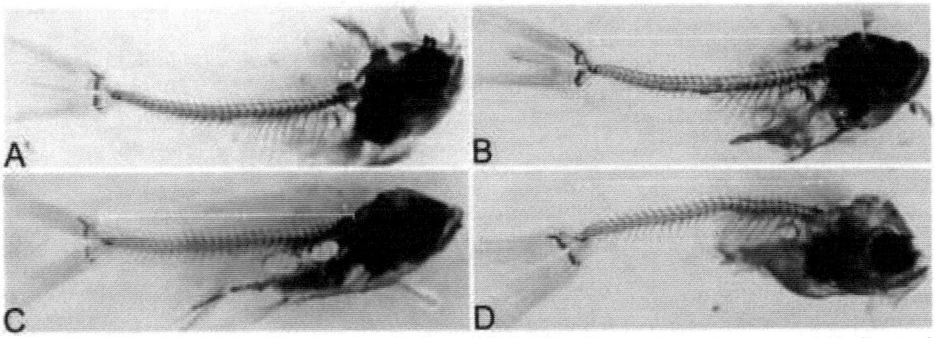

Fig. 5. Vertebrae feature of control and treated fish after 2-month maintenance: A) Control; B) 3 mg L^{-1}; C) 6 mg L^{-1}; D) 9 mg L^{-1}. Showing the vertebrae arrangement of i) Caudal Fin Vertebrae; ii) Caudal Vertebrae; iii) Precaudal Vertebrae; iv) Cervical Vertebrae

The result showed that all concentrations of detergent treatment did not affect the vertebral segment number of fish. However, the 9 mg L^{-1} of detergent caused morphological abnormality of the vertebrae arrangement, especially in the area of precaudal bone.

According to Vashanti et al., detergents at a concentration of 3 mg L^{-1} within the space of 72 h exposures, were reduced in protein content from 3.14 mg dL^{-1} to 1.28 mg dL^{-1} in *C. mrigala* fish blood [2]. As a consequence, can inhibit the process of protein synthesis in the production of amino acids, one of which can be used for bone matrix formation, resulting in abnormal ossification.

4 Conclusion

It can be concluded in this study on wader pari (*R. lateristriata*) that detergent exposure to a concentration up to 9 mg L^{-1} does not affect egg hatchability, fish survival rate, larval growth rate, of the number of fish vertebrae. However, 9 mg L^{-1} of detergent concentration was observed to affect the vertebral morphology of fish larvae.

References

1. M. Syaifullah, Pembuangan Limbah Laundry di Yogyakarta Ditertibkan [Control Laundry Waste Disposal in Yogyakarta]. [Online] from https://nasional.tempo.co/read/249971/pembuangan-limbah-laundry-di-yogyakarta-ditertibkan/full&view=ok (2010) [Acessed on 20 May 2019].
2. J. Vasanthi, S. Binukumari and N. Saradhamani, J. Pharm. Biol. Scie. **8**,5:19–22(2013). https://pdfs.semanticscholar.org/a2b3/08497549fd693b932c3fde7bb18aef6f25be.pdf
3. U. Zoller, P. Sosis, *Handbook of detergents, part F: Production*, CRC Press, (2008).p.624. https://www.routledge.com/Handbook-of-Detergents-Part-FProduction/Zoller-Sosis/p/book/9780824703493
4. P. S. Topale, A. U. Ghaware, R. G. Jadhao, Int. Res. J. Sci. Eng. **1**:10–12(2013). http://oaji.net/articles/2014/731-1396631673.pdf
5. U. Zoller, *Handbook of detergents, Part B: Environmental impact*, CRC Press, (2004).p.838. https://www.taylorfrancis.com/books/9780429215230
6. K. A. Najam, W. Bhowate, The Bioscan. **5**,2:321–322(2010). https://pdfs.semanticscholar.org/c59f/37756b579c332cef4f224b45843bb10a07c6.pdf
7. B. Isyaku, J. Solomon, DRJAFS. **4**,12:351–360(2016). https://directresearchpublisher.org/drjafs/abstract/effect-of-detergent-on-the-growth-of-african-catfish-clarias-gariepinus/
8. B. Jezierska, K. Ługowska, M. Witeska, Fish physiol. Biochem. **35**: 625–640(2009). https://doi.org/10.1007/s10695-008-9284-4
9. S. Santanu, Int. J. Pure Appl. Biosci. **1**:36–41(2013). http://www.ijpab.com/form/2013%20Volume%201,%20issue%205/IJPAB-2013-1-5-36-41.pdf
10. C. A. Zimring, W. L. Rathje, *Encyclopedia of consumption and waste: the social science of garbage*, Sage Publications, Inc., 2012. https://sk.sagepub.com/reference/consumption-waste
11. Y. Wang, Y. Zhang, X. Li, M. Sun, Z. Wei, Y. Wang, et al., Scientific Reports , **5**:10107(2015). https://doi.org/10.1038/srep10107
12. M. Inouye, OJCARAJ. **16**,3:171–173(1976). https://doi.org/10.24540/cgafa.16.3_171
13. A. El-Gamal, Global Veterinaria, **3**:80–90(2009). http://www.idosi.org/gv/gv3(2)09/3.pdf
14. E. Mathew, P. Sunitha, P. L. Thomas, JESTFT. **5**,3:1–3(2013). http://www.indiaenvironmentportal.org.in/files/file/Phosphate%20in%20detergents%20Kuttanad%20Kerala.pdf
15. C. N. Eknath, Res. J. Anim. Vet. Fishery Sci. **1**,9:14–19(2013). http://www.isca.in/AVFS/Archive/v1/i9/3.ISCA-RJAVFS-2013-052.pdf
16. N. Chandanshive, *International Research Journal of Environment Sciences*, **3**:1–5, (2014). http://www.isca.in/IJENS/Archive/v3/i8/1.ISCA-IRJEvS-2013-210.pdf
17. E. Oyoroko, E. N. Ogamba, Biotechnol. Res. **3**,59–64(2017). https://www.futuredatum.com/wp-content/uploads/2019/11/59-64.pdf

 https://doi.org/10.1051/e3sconf/202122600016

18. E. Idowu, A. Adewumi, A. Babalola. IOSR-JAVS. **10**,67–75(2017). https://pdfs.semanticscholar.org/0021/59170d687ce18543e28ffde7fc879d6a7e06.pdf?_ga=2.224593310.522309668.1595752936-677940155.1595752936

Modelling Biodiesel Supply Chain: Current State and Opportunities for Future Research

Fitriani Tupa Ronauli Silalahi[1,2,*], *Togar* Mangihut Simatupang[3], *Manahan* Parlindungan Siallagan[3], and *Rizal* Horas Manahan Sinaga[4]

[1]School of Business and Management, Bandung Institute of Technology, Jl. Ganesha no 10, Bandung 40132, West Java, Indonesia
[2]Faculty of Industrial Technology, Del Institute of Technology, Laguboti 22381, Toba, North Sumatera, Indonesia
[3]School of Business and Management, Bandung Institute of Technology, Jl. Ganesha no 10, Bandung 40132, West Java, Indonesia
[4]Faculty of Engineering, University of Nottingham, University Park, Nottingham, NG7 2 RD, United Kingdom

Abstract. The use of renewable energy is an important issue that is being promoted internationally. Considering the supply of fossil energy is increasingly depleting and requires a very long time to be renewed. One of the uses of renewable energy that is being promoted is the use of biodiesel as a substitute for diesel fuel. This utilization is carried out by mixing biodiesel with diesel fuel. The quantity of biodiesel blended with diesel fuel is following the rules in each country. That is why the biodiesel supply chain must be described to obtain minimum economic costs and environmental costs. This paper presents a comprehensive review of current biodiesel supply chain models. A study of the biodiesel supply chain's current state concludes by highlighting the unaddressed area or the gaps existing in the existing literature by suggesting future research in the biodiesel supply chain, specifically in the area case study in Indonesia.

Keywords: Biodiesel energy, environmental impact, systematic literature review, renewable energy, research gaps.

1. Introduction

Over the last decade, research on renewable energy use is increasingly carried out to reduce dependence on fossil energy and develop renewable and environmentally friendly energy sources. Unfortunately, humans still rely heavily on non-renewable energy sources. Until the end of 2017, the use of renewable energy globally is estimated at 26.5 %. This consumption increased by 5 % compared to renewable energy consumption in 2016 by 20.5 % (Figure 1) [1]. The use of renewable energy is still very little compared to the fossil energy that feared to run out. This dependency on traditional sources is causing adverse effects on human health and the environment due to global deforestation and greenhouse emissions [2].

* Corresponding author: fitriani_tupa@sbm-itb.ac.id

 https://doi.org/10.1051/e3sconf/202122600017

Renewable energy resources can be classified into two types: i) static renewable energy and ii) biofuels. Static resources like wind, solar, geothermal heat, tides, and waves should be consumed just after generation. In contrast, biofuels can be transported in order to use in far districts for consumption. At the same time, biofuels have shown a great potential to be a promising alternative for fluid fossil fuel consumption and can reduce both fossil fuel dependency and greenhouse gas (GHG) emissions. Currently, biofuel is divided into bioethanol and biodiesel. Biodiesel is a fuel with low viscosity, non-toxic, biodegradable, and cleaner than fossil diesel. There are more than 350 oil-bearing crops known in the world that are suitable for biodiesel production [3].

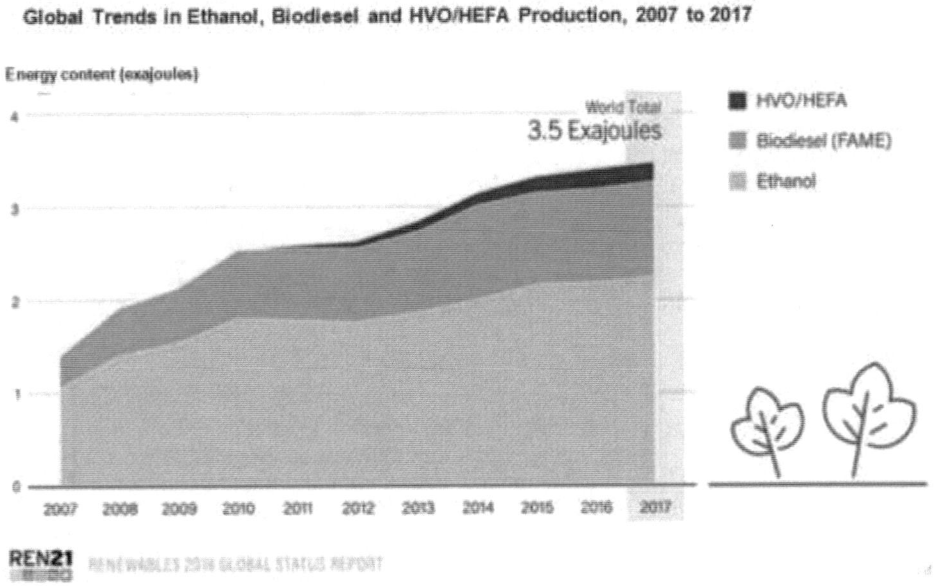

Fig.1. Global trends in biofuel production [1]

The use of biodiesel fuel as a substitute for diesel fuel has been widely promoted as part of a government program in the context of utilizing renewable energy. The utilization of biodiesel is done by increasing the percentage of biodiesel mixed with diesel fuel in accordance with the targets set. Countries that carry out the mandate for the use of biodiesel include Argentina (B10), Colombia (B10), Australia (B2), India (B5), Indonesia (B20), Malaysia (B10), and other countries [4]. Therefore, research on the development of biodiesel technology and the optimization of biodiesel supply chains are increasingly being carried out.

The flow of biomass from the land to its end use for producing biodiesel is called the biodiesel supply chain [5]. The biomass supply chains include different types of activities, such as cultivation, production, blending biodiesel with petroleum fuel, transporting, storing, and distributing to consumers. In this supply chain, at least three actors are involved, namely the biodiesel industry, the distributor of fuel oil, which distributes to consumers, and farmers of biodiesel raw materials. If there is government intervention, then there are four actors involved, including the government.

Biodiesel in the supply chain has become a very interesting topic to be investigated because of the times that require every country to look for alternative energy from renewable sources. Many researchers have produced models for biodiesel supply chains that address several issues, such as adding profits, adding to the supply chain costs, operating efficiency, or combining them as their goals. A systematic literature review that can point out exploration of the biodiesel supply chain, especially in Indonesia, is required to develop a better understanding of the biodiesel supply chain. This paper provides a literature review on the state of the art in biodiesel supply chain and research gaps in the biodiesel supply chain by classifying the previous literature into several categories. This review's objectives are i) an analysis of the state of the art of biodiesel supply chain research ii) a perspective from which this study suggest recommendations for future research that would expedite biodiesel usage and commercialization.

2. Method

To see the research trend in renewable energy study, this study conducted a systematic literature review. The first step is searching for articles from three Academic Databases, namely Scopus, Science Direct, and EBSCO. A literature review is done by looking at highly reputable papers. The keywords used in the literature review process are "biodiesel supply chain" in titles, and result in 286 papers. In the second step, elimination of duplicates articles was done and reduce articles to 266. In the third step, analysis of articles titles and abstracts was performed. Based on the analysis, 135 of the 266 articles were selected. In the fourth step, in depth screening of full texts was done by reading abstract, introduction, methodology, conclusion and future research suggestion. The articles that did not contain the relevant topic were therefore excluded. The systematic process of literature review is shown by Figure 2.

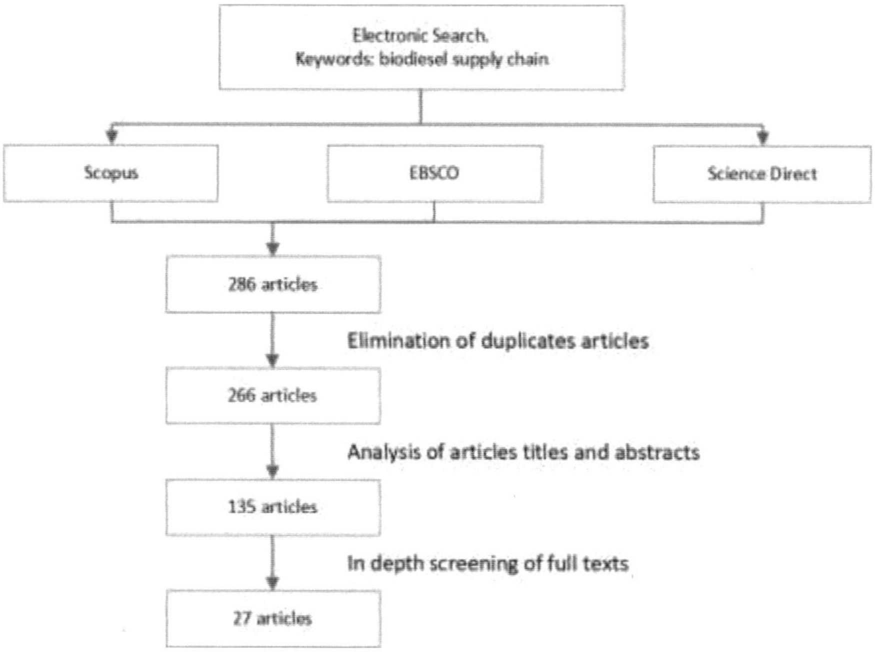

Fig.2. Systematic literature review

The classification framework for the biodiesel supply chain's literature review is described in six dimensions based on literature analysis. The six dimensions are resources, objectives, output decisions, addition issues, decision-making levels, and solution methodology. The classification can be seen in Figure 3.

The first dimensions describe the distinction between single resources and multiple biomass resources that are used for biodiesel fuel. The number of resources and the type of resources studied is very dependent on the geographical conditions and potential of the country being studied. The second objective describes the objectives of the paper on sustainable development. There are three types of objectives from literature i.e., objectives in the economic aspect, objectives in the social aspect, objectives in the environmental aspect, and objectives that are integrated between economic, social, and environmental aspects. The third dimensions describe the distinction between the type of decision to solve the problem. There are decisions in the plantation stage, production stage, blending stage, distribution stage, and consumer decision. The fourth dimension is additional issues added to the biodiesel supply chain. Through the literature study, three main issues that are important and widely discussed in the biodiesel supply chain, namely uncertainty, government intervention, and food security, are discussed.

Based on the analysis of literature works, the biodiesel supply chain also employed the planning decision level depending on the time horizon. Therefore the fifth dimension is based on the planning decision level, such as strategic, tactical, and operational planning level. The sixth dimension is used to review and classify the literature according to each biodiesel supply chain model's appropriate solution method. Amount two methods used to optimize and solve biodiesel supply chain problems that are simulation and optimization.

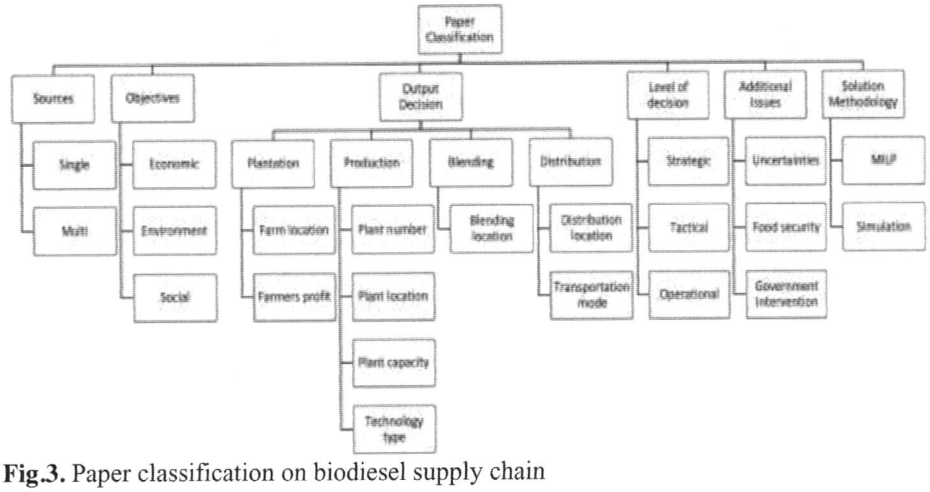

Fig.3. Paper classification on biodiesel supply chain

3. Data analysis and result

Based on the analysis of the literature, a classification framework for the literature review of the biofuel supply chain is done. The classification can be seen in Figure 3. This review of 27 paper [3, 6–31] gathered from science direct, EBSCO, and Scopus (Table 1). This 27 articles reviewed show the optimization of the biodiesel supply chain from the economic, social or environmental objective.

Tabel 1. State of the art table

No	Author	Year	RE		Objectives			Output decision					Additional Issues	Method
			S	M	O1	O2	O3	OD1	OD2	OD3	OD4	OD5		
1	[6]	2011	√		√				√		√		-	O
2	[7]	2012		√	√			√			√		-	O
3	[8]	2011		√	√						√		-	O
4	[3]	2012		√	√				√		√		-	S
5	[9]	2012		√	√			√	√		√		-	O
6	[10]	2013		√	√				√		√	√	-	O
7	[11]	2014	√		√	√			√		√	√	-	O
8	[12]	2014	√		√			√	√		√	√	-	O
9	[13]	2014		√	√				√		√		-	O
10	[14]	2015		√	√				√		√	√	-	O
11	[15]	2015	√		√	√			√		√	√	GI	
12	[16]	2016			√				√		√	√	GI	S
13	[17]	2016			√				√		√		-	O
14	[18]	2016			√				√	√	√	√	-	O
15	[19]	2016			√				√		√	√	-	O
16	[20]	2017		√	√	√					√	√	-	O
17	[21]	2017		√	√			√			√		GI	O
18	[22]	2017			√				√		√		-	S
19	[23]	2017		√	√				√		√		-	S
20	[24]	2017	√		√				√		√	√	-	S
21	[25]	2018	√		√			√	√		√		-	O
22	[26]	2018	√		√				√		√		UN	O
23	[27]	2018		√	√			√			√		-	O
24	[28]	2018		√	√			√	√		√	√	GI	O
25	[29]	2018		√	√				√		√	√	GI	O
26	[30]	2018	√		√				√		√	√	-	O
27	[31]	2018	√		√	√					√	√	-	O

S : Simulation
RE : Re sources
O1 : Economic
O2 : Environment
O3 : Social

OD1 : Plantation
OD2 : Production
OD3 : Blending
OD4 : Distribution
OD5 : Transportation Issue

S : Single source
M : Multi source
O : Optimization

3.1 Biofuel resources

The first dimension describes the distinction between single resources and multi resources. Biofuel can be determined as fuel from organic resources, including plants and animals.

Biofuel has a specific characteristic that is renewable, whereas it can be produced by using raw material that can be grown or developed. There are a variety of biofuels potentially available, but the main biofuels being considered globally are biodiesel and bioethanol. However, first-generation biofuel technologies are often criticized for their implications on food prices and production. Then the biofuel supply chains using second-generation biofuel technologies is not compete with food security. These supply chains use crop residues, wood residues, or dedicated energy crops as feedstocks, thus avoid the potential impact on food markets.

Through a systematic literature review, it was found that based on the source of the raw material studied, research in biodiesel supply chain was categorized into two categories: research with one source of biomass raw materials and research with some sources of biomass feedstock. For research that falls into one biomass source category, the study will optimize the supply chain by optimizing each supply chain process. For research that falls into several biomass sources, optimization in the supply chain is carried out by selecting existing resources and then selecting biomass sources or combining several biomass sources to produce the maximum amount of biodiesel at minimum cost.

3.2 Research objectives

The research objectives in biodiesel supply chains are classified into three types: economic objectives, environmental objectives, and social objectives. For economic objectives, the study will try to meet the biodiesel quantity target produced at the minimum cost by optimizing the biodiesel supply chain. For economic objectives, [12] minimizes biodiesel supply chain's complete costs from feedstock supply to biodiesel delivery at the gas stations to determine the number, optimal location, size, and configurations of biodiesel production plants in Finland. Andersen et al. [7] minimizes supply chain cost by determining alternative locations and capacities for farms, biomass plantations, biodiesel production plants, and storage facilities, as well as distribution centers to internal and external markets [7]. Akram [9] minimize the total discounted costs of the supply chain involves the whole supply system's operating cost and the annualized capital costs. Reference [16] developed a stochastic mixed-integer programming model to make optimal decision include location of open biorefineries, amount of biomass transported to the biorefineries, amount of biodiesel production in biorefineries, amount of biodiesel transported to the markets, amount of Inventory and shortage of biodiesel in markets. Babazadeh [20] proposed model that is capable of determining the optimum numbers, locations, capacity of facilities, suitable transportation modes, and appropriate technology at bio-refinery, material flow, and production planning in different periods with a real case in Iran.

Social objectives are assessed from several things. But many studies have only evaluated social objectives by measuring job creation in the supply chain because it is easily quantified while other social measuring indicators are not easily measured. The social objective was quantified by the number of jobs generated. Leao et al. [6] determine the social aspect from calculating a minimum number of families to be hired in a selected production zone in a particular year or during the whole period. Quintero et al. [3] identified social aspects from the participation of associations of smallholders. Santibañez-Aguilar et al. [32] was quantified social aspects by the number of jobs generated. If concluded, all research on social aspects in the biodiesel supply chain is as far as the author's observation only assessed social objectives by measuring job creation.

Some studies carried out research with more than one goal or so-called multi-objective optimization. "Multi-objective optimization" is required when conflicting objectives are considered simultaneously. This often involves the tradeoff between economic, environmental, and social performances of the supply chain as well as the risks associated with design and operation.

3.3 Output decision

Through literature studies, the output of biodiesel supply chains is categorized into five major stages: the stages of the plantation, the stages of production, the stages of blending, the stages of distribution, and the consumer stages.

3.3.1 Plantation

The fundamental source of biofuel is the biomass feedstock, which is made of renewable biological materials. Production of biomass feedstocks includes all of the operations required for farmers to generate the feedstock, including plant breeding, planting, managing, and harvesting the crops from farmland, field, or forest.

3.3.2 Production

In production, the harvested biomass from regional farms is first collected and staged at the nearby shipping terminal or elevators and then subsequently shipped to a nearby biorefinery. The biomass conversion occurs at the biorefineries through bio-decomposition, fermentation, and distillation, yielding marketplace products. The problem of determining the best location for production zones and crushing units is to find an optimum point between the two extremes, seeking to balance logistical costs with investment costs. At one extreme, only one very large crushing unit would be built (minimum investment costs); at the other, a large number of very small units would be built (minimum logistical costs).

3.3.3 Blending

Several studies discuss the mixing of diesel and biodiesel. Andersen et al. [7] conducted research on the biodiesel supply chain's optimal design using the B10 policy scenario and determined the biodiesel mixing plant and transportation to be used. Hombach et al. [29] conducted a political policy scenario about the combination of the percentage of biodiesel mixed with diesel to combine ecological and social aspects with economic considerations.

3.3.4 Distribution and transportation

The final stage of a biofuel supply chain is the distribution and sales of fuels to end-consumers. Most of the biofuels in liquid form are shipped to retail outlets through existing fuel distribution systems to the transportation fuel market. These retail market outlets are typically gas stations owned by large oil corporates such as Pertamina in Indonesia, who also often own the biofuel blending facilities. High transportation cost, especially biodiesel transport, was a decisive factor in determining the numbers, capacities, and locations of new biodiesel plants. Neither increasing transportation cost nor decreasing plant construction cost changed the optimized supply chain configuration because the transportation cost was already high enough to dominate the supply chain optimization [30].

Supply chain optimization by paying attention to the transportation route is done by adjusting the area's spatial conditions being studied. For example, Jeong et al. [30] uses transportation routes despite the GIS network analysis by combining transportation modes.

Andersen uses three modes of transportation (truck, rail, and ship) in the biodiesel supply chain, but does not use transportation routes (based only on the distance between the 2 points). Babazadeh et al. [21] uses two modes of transportation (truck and rail) for supply chain optimization.

In most studies, transportation costs or impacts were determined by assuming truck transportation and simulating the shortest distances over a road network. Transportation costs or impacts were determined by weighting the distances by feedstock and biofuel flows and unit-cost or impact factors. However, the shortest distances simulated over a single-mode do not guarantee economic routes. For example, Andersen et al. [7] uses three modes of transportation (truck, rail, and ship) in the biodiesel supply chain, but does not use transportation routes (based only on the distance between the two points). Babazadeh et al. [21] uses two modes of transportation (truck and rail) for supply chain optimization. To obtain a real transportation cost calculation, the optimization model with a geographical information system (GIS) must be applied. Several previous studies have done this. For example, Jeong et al. [30] uses transportation routes despite the GIS network analysis by combining transportation modes and uses the network analysis using OD cost matrix was used to determine minimum cost routes over regional multimodal transportation network, which is different from other studies in terms of analysis criteria, and multimodal network consideration

3.4 Additional issues

In this study, additional issues are categorized into three parts, namely food security, uncertainties, and government intervention.

i. Government intervention

Another critical factor in shaping the supply and demand of biofuel is government policy. To accelerate the development of the biofuel industry, the government is establishing various policies and imposing necessary instruments [9, 3, 12, 15]. The major legislation promoting biodiesel is the government target to achieve BX blending in biodiesel. This policy aimed to stimulate the biodiesel industry by pressuring biofuel producers and distributors, to meet a biodiesel quota. Another instrument for promoting the biodiesel industry are subsidy [23] that granted to the farmer for growing rapeseed in German and incentives for diesel fuel mixture product and biodiesel in the form of product subsidy of IDR 1×10^6 per kL [24] in Indonesia.

Government subsidy to biodiesel production is another policy which decreases the biodiesel price [16]. The energy tax act specifies the biofuel policy defined an annual increase in biodiesel's tax rate [22]. Policies related to carbon was described by Maruffazaman et al. [33] and define carbon regulation into carbon cap, carbon tax, carbon cap and trade, and carbon offset.

ii. Uncertainties

Biofuel supply chains are subject to many uncertainties, which include, but are not limited to, seasonality and random yield of biomass, demand and price fluctuations, and production and logistical uncertainty. Other external factors, such as unpredictable regulatory policy changes and technological breakthroughs, also become challenges in biofuel supply chain management. Many biofuel supply chain models focus on the supply side of uncertainties associated with crop yield and seasonality. For example, Senna et al. [19] explain uncertainties in crop prices due to weather conditions, such as rainfall that impact seed production. Ghelichi et al. [26] describe uncertainty in biodiesel demand and raw material supply.

iii. Food security

The series of direct and indirect land-use changes have raised global concern. First, farmland competition between the food and biomass has led to the so-called 'food vs. energy' dilemma. Indeed, accompanying the rapid expansion of the biofuel industry, the food price worldwide has increased significantly. For example, the US's corn grain price increased from as low as USD 1.5 per bushel in 2000 to as high as USD 7.1 per bushel in 2013 (1 US bushel = 35.239 07016 688 L), during the period in which the corn was the primary feedstock to bioethanol production [34]. Current research focuses on producing biodiesel from second-generation biofuel resources to avoid issues on food vs. fuel competition.

3.5 Level of decisions

There are three main decision-making processes in supply chain management: strategic, tactical, and operational decisions

i. Strategic decisions in biofuel supply chain

The strategic decisions are those decisions that influence over years and decades. Once a strategic decision is made, it is very unlikely to be altered in the short term. Strategic decisions in the biofuel supply chain include but are not limited to; i) selection of energy production technologies, ii) network configuration, iii) supply and demand contracts, and iv) ensuring sustainability, v) selection of plant location [35].

ii. Tactical decision in the biodiesel supply chain

Tactical decisions are medium-term decisions that involve sourcing decisions, production decisions, scheduling, transportation and logistical contracts, and planning process definition. Inventory decisions such as location, quality, and quantity of inventory are also considered. Decisions taken at the tactical decision level are planned to achieve and execute the strategic decisions [35].

Biomass sourcing decisions are crucial in the Biofuel Supply Chain (BSC) to minimize the geographical distance and increase accessibility to the raw material sources, among other factors. This ensures that the relatively isolated geographical allotment of significant biomass can raise researchers' interest in identifying the available biomass quantities over a region and subsequently proceeding with the selection of the optimal biomass sources.

iii. Operational decisions in biofuel supply chain

Operational decisions are short-term decisions that ensure the plants' continuous operation and other processes in the supply chain. These decisions are made daily or weekly, and sometimes several times to make sure that products are manufactured, moved, and sold in a timely and cost-effective manner. Some of the operational level decisions are detailed production scheduling, daily fleet management, and daily or weekly inventory review. The focus here is geared towards achieving the plan or framework set by the tactical supply chain decisions [35]. In the biofuel supply chain, this involves daily activities and planning such as transportation and logistics scheduling, demand forecasting, and review to meet the monthly targets. The plants' manufacturing planning and the detailed production and material requirements planning are usually reviewed at this decision level.

3.6 Solution methodology

Since multiple conflicting objectives are often involved when optimizing biofuel supply chains' sustainability, several multi-objective optimization techniques can be applied, such as the epsilon-constraint method, goal programming method, etc. In the literature study, two standard methods were used that is mathematical modeling methods for optimization and simulation to see interactions that occur in the system. Many studies use the optimization method, as stated in Table 1. It is used to demonstrate the effects of maximizing the economic gain as well as minimizing the environmental impacts, with the help of appropriate presentation methods and optimization techniques. Integer variables are used in the model to carry out the selection of the geographic location(s) of the biomass and product storage and distribution facilities and the biofuel production plants.

4. Discussions

This review of 27 papers is gathered from science direct, EBSCO, and Scopus. Amount 27 articles reviewed show the biodiesel supply chain's optimization from the economic, social, or environmental objective. Although all of the articles point out the biodiesel supply chain, there are still some limitations of the previous research.

The first research gaps are transportation costs by determining infrastructure factors. In terms of objectives, economic optimization is the most objective type in biodiesel supply chain research. However, there are several limitations to the current literature. In transportation costs, many studies that calculate the cost of transportation only through transportation cost or impact are determined by transportation and simulating the shortest distances over a road network. Meanwhile, the shortest distance does not always guarantee an economic value that is smaller than another distance. Other factors, such as infrastructure, also determine transportation values. Research involving transportation by paying attention to minimum cost routes was only done by [30].

The second research gaps are uncertainties in biodiesel price. Due to the nature of the biofuel supply chain, many uncertainties exist. The uncertainties include but are not limited to: raw material supply and price uncertainties, finished goods demand and price uncertainties, pre-treatment uncertainties, production and yield uncertainties, and transportation uncertainties. To achieve optimal performance, the decisions of biofuel supply chain management should incorporate these uncertainties. Furthermore, biodiesel supply chain uncertainties will impact supply chains' performance and should be incorporated in many decision-making. Then [26] conducts research on biodiesel supply chains with uncertainty on supply and demand. Hombach et al. [17] see uncertainty in terms of demand uncertainty. Seeing that biodiesel prices are constantly changing, research with price uncertainty is important and is the second gap research of this study.

The third research gap is determining land use for food security in the biodiesel supply chain. The search for productive land for palm cultivation can lead to the conversion of natural ecosystems (tradeoff between cost and emissions) or the replacement of food crops (tradeoff between cost and food security). In these cases, there is a direct change in land use, which can be beneficial or harmful to the environment depending on whether the new crop absorbs more CO_2. On the other hand, if the biofuel sources replace food crops and the latter move to natural areas such as unprotected forests, unmanaged pastures (scrubland), or extensive grazing areas (meadows), an indirect change of land use is generated. This replacement may lead to an increase in emissions (tradeoff between emissions and food security) [31]. This issue makes us must pay attention to land use for food security when thinking of biodiesel production from biomass. This is the third research gap in this research.

The fourth research gaps are unique in the Indonesian case study. Many transportation problems were carried out in previous studies, and transportation used was limited to trucks and trains adapted to state conditions. Previous research has been carried out in continental Europe. However, the problem of transportation in the biodiesel supply chain in island countries that must participate in ships (barge) has never been studied. That became the fourth research gap from this study.

Since multiple conflicting objectives are often involved when optimizing biofuel supply chains' sustainability, several multi-objective optimization techniques can be applied, such as epsilon-constraint method, goal programming method, etc. In the literature study, two common methods were used that is mathematical modeling methods for optimization and simulation to see interactions that occur in the system.

5. Conclusions and recommendations

Amount 27 articles have been reviewed and classify based on six categories. The first category is based on the type of biodiesel resources: single or multi-resources. The second category is based on the objective of biodiesel supply chain: economic objective, environment objective and social objective. The third category is based on output decisions from each model. The fourth category is based on the level of decision: strategic, tactical, operational. The fifth category is based on additional issues founded: uncertainties, food security and government intervention. The sixth category is based on the solution methodology.

Future research on biodiesel supply chain should be focused on distribution type (location of plantation, production, blending, transportation, and flow amount in each stage) by determining infrastructure factors using a Geographical Information System (GIS). Besides it will optimize economic cost, the optimize distribution will minimize environmental impact by calculating GHG emissions in each stage and transportation used. Future research can also focus on determining land use to minimize conflict with food security and conduct research in island nations that require complicated modes of transportation, such as Indonesia.

References

1. J.L. Sawin, F. Sverrisson, J. Rutovitz, S. Dwyer, S. Teske, H.E. Murdock, et al. *Renewables 2018-Global status report. A comprehensive annual overview of the state of renewable energy*. Energy Planning, Policy and Economy. (2018).p.542 https://inis.iaea.org/search/search.aspx?orig_q=RN:49053208
2. Kumar, A., B. Sah, A.R. Singh, Y. Deng, X. He, P. Kumar, R.C. Bansal. Renew. Sust. Energ. Rev. **69**,596–609(2017). https://doi.org/https://doi.org/10.1016/j.rser.2016.11.191
3. J.A. Quintero, E.R. Felix, L.E. Rincón, M. Crisspín, J.F. Baca, Y. Khwaja, et al. Energy Policy, **43**:427–435(2012). https://doi.org/10.1016/j.enpol.2012.01.029
4. Lane, J. Biofuels Mandates Around the World 2019. 2019 [cited 2019 5 November]; Available from: https://www.biofuelsdigest.com/bdigest/2019/01/01/biofuels-mandates-around-the-world-2019/.
5. N.Z. Atashbar, N. Labadie, C. Prins. IFAC-PapersOnLine. **49**,12:604–615(2016). https://doi.org/10.1016/j.ifacol.2016.07.742
6. R.R.d.C.C. Leão, S. Hamacher, F. Oliveira. Bioresour. Technol. **102**,19:8958–8963(2011). https://doi.org/10.1016/j.biortech.2011.07.002

7. F. Andersen, F. Iturmendi, S. Espinosa, M.S. Diaz. Comput. Chem. Eng. **47:**170–182(2012). https://doi.org/10.1016/j.compchemeng.2012.06.044
8. C. Papapostolou, E. Kondili, J.K. Kaldellis. Energy, **36**,10:6019–6026(2011). https://doi.org/10.1016/j.energy.2011.08.013
9. A. Akram. Renew. Sust. Energ. Rev. **16**,6:4196–4203(2012). https://doi.org/10.1016/j.rser.2012.03.023
10. B. Ivanov, B. Dimitrova, D. Dobrudzhaliev. J. Chem. Technol. Metall. **48**,5:513–523(2013). https://dl.uctm.edu/journal/node/j2013-5/13-Bojan%20Ivanov-513-523.pdf
11. M. Marufuzzaman, S.D. Ekşsioğlu. Transp. Sci. **48**,4:555–574(2014). https://doi.org/10.1287/trsc.2013.0505
12. K. Natarajan, S. Leduc, P. Pelkonen, E. Tomppo, E. Dotzauer. Renew. Energ. **62**,319–330(2014). https://doi.org/10.1016/j.renene.2013.07.013
13. J.E. Santibañez-aguilar, G. Guillen-gosálbez, R. Morales-rodriguez, L. Jiménez-esteller, A.J. Castro-montoya, J.M. Ponce-ortega. Bioenergy Res. **9**,4:1053–1069(2016). https://doi.org/http://dx.doi.org/10.1007/s12155-016-9743-1
14. J. Ren, S. Tan, L. Yang, M.E. Goodsite, C. Pang, L. Dong. Energy Conversion and Management, **92:**312–321(2015). https://doi.org/10.1016/j.enconman.2014.12.066
15. L.E. incón, M.J. Valencia, V. Hernández, L.G. Matallana, C.A. Cardona. Energy Econ. **47:**154–167(2015). https://doi.org/10.1016/j.eneco.2014.10.018
16. A. Azadeh, H.V. Arani. Renew. Energy, **93**:383–403(2016). https://doi.org/10.1016/j.renene.2016.02.070
17. L.E. Hombach, C. Cambero, T. Sowlati, G. Walther. J. Clean. Prod. **133:**565–575(2016). https://doi.org/10.1016/j.jclepro.2016.05.107
18. B. Ivanov, S. Stoyanov. Energy. **99**:221–236(2016). https://doi.org/10.1016/j.energy.2016.01.038
19. P. Senna, D. Pinha, R. Ahluwalia, J.C. Guimarães, E. Severo, A. Reis. Production, **26**,3:501–515(2016). https://doi.org/10.1590/0103-6513.200015
20. R. Babazadeh. Renew. Sust. Energ. Rev. **75**:1089–1100(2017). https://doi.org/10.1016/j.rser.2016.11.088
21. R. Babazadeh, J. Razmi, M. Rabbanu, M.S. Pishvaee. J. Clean. Prod. **147**:694-707(2017). https://doi.org/10.1016/j.jclepro.2015.09.038
22. J.A. Moncada, M. Junginger, Z. Lukszo, A. Faaij, M. Weijnen. Appl. Energy, **195**:370–381(2017). https://doi.org/10.1016/j.apenergy.2017.03.047
23. J.A. Moncada, Z. Lukszo, M. Junginger, A. Faaij, M. Weijnen. Appl. Energy, **185**:895–915(2017). https://doi.org/10.1016/j.apenergy.2016.10.070
24. H. Wijaya, Y. Arkeman, E. Hambali. Agric. Eng. Int: CIGR J. 268–282(2017). http://cigrjournal.org/index.php/Ejounral/article/download/4554/2671
25. I.A. Emara, M. Gadalla, F. Ashour. Chem. Eng. Trans. **70**:433–438(2018). https://doi.org/10.3303/CET1870073
26. Z. Ghelichi, M. Saidi-Mehrabad, M.S. Pishvaee. Energy, **156**,1:661-687(2018). https://doi.org/10.1016/j.energy.2018.05.103
27. F. Ezzati, R. Babazadeh, A. Donyavi. Renew. Energy Focus, **26**:81–92(2018). https://doi.org/10.1016/j.ref.2018.07.005
28. M. Leila, J. Whalen, J. Bergthorson. Energy, **156**:181–195(2018). https://doi.org/10.1016/j.energy.2018.04.196
29. L.E. Hombach, C. Büsing, G. Walther. Eur. J. Oper. Res. **269**,1:302–312(2018). https://doi.org/10.1016/j.ejor.2017.07.015
30. H. Jeong, H.L. Sieverding, J.J. Stone. Bioenergy Res. **12:**229–240(2018). https://doi.org/10.1007/s12155-018-9943-y

31. J.A. Orjuela-Castro, J.A. Aranda-Pinilla, C.E. Moreno-Mantilla. Comput. Electron. Agric. **161:**162–169(2018). https://doi.org/10.1016/j.compag.2018.03.009
32. J.E. Santibañez-Aguilar, J.B. González-Campos, J.M. Ponce-Ortega, M. Serna-González, M.M. El-Halwagi. J. Clean. Prod. **65**:270–294(2014). https://doi.org/10.1016/j.jclepro.2013.08.004
33. M. Marufuzzaman, S.D. Eksioglu, Y. Huang. Comput. Oper. Res. **49**:1–17(2014). https://doi.org/10.1016/j.cor.2014.03.010
34. M.K. Lim, Y. Ouyang. Springer. 143–162(2016). http://publish.illinois.edu/mlim/files/2012/12/Lim_Ouyang_Biofuel-SCN_Final.pdf
35. I. Awudu, J. Zhang. Renewable and Sustainable Energy Reviews. **16,**2:1359–1368(2012). https://doi.org/https://doi.org/10.1016/j.rser.2011.10.016

Vinasse as Cultivation Medium of *Chlorella* sp. to Produce Poly–Hydroxy Butyrate in Various Limited Low–Cost Primary Nutrient

Gregorius Prima Indra Budianto[1,*], *Yari* Mukti Wibowo[2], *Hadiyanto* Hadiyanto[3], *Widayat* Widayat[3], and *Wisnu* Arfian Anditya Sudjarwo[4]

[1]Department of Chemical Engineering, Faculty of Engineering, Setia Budi University, Jl. Letjend Sutoyo Mojosongo, Solo 57127, Central Java, Indonesia
[2]Department of Chemical Analyst, Faculty of Engineering, Setia Budi University, Jl. Letjend Sutoyo Mojosongo, Solo 57127, Central Java, Indonesia
[3]Department of Chemical Engineering, Faculty of Engineering, Diponegoro University, Jl. Prof. Soedarto, Tembalang, Semarang 50275, Central Java, Indonesia
[4]Department of Analytical Chemistry, Faculty of Chemistry, University of Vienna, Universitätsring 1, 1010 Wien, Austria

Abstract. Vinasse is ethanol wastewater that still contains nutrients. It can be medium cultivation for *Chlorella* sp. Cultivation *Chlorella* sp. in vinasse did not only minimize its COD content but also potentially produce Poly–Hydroxy Butyrate (PHB) in a limited nutrient. This paper presents a cultivation process of *Chlorella* sp. in vinasse on the various limited nutrient. *Chlorella* sp. was cultivated in vinasse by adding complete nutrient (urea and TSP), TSP (limited N), and urea (limited P). Experimental data was optimized by a mathematical model to predict the behavior of the *Chlorella* sp. in limited nutrient systematically. The study confirmed that the best condition of the cultivation medium of *Chlorella* sp. to minimize COD content in vinasse by addition phosphate into the reactor. However, PHB could be best accumulated in the cell on limited phosphate.

Key words: Biological plastic, cultivation medium, environment friendly, stillage, wastewater utilization

1 Introduction

The cultivation process of microalgae (*Chlorella* sp.) is a costly process when developed by a specific medium. Alternatively, the cultivation process can be done in wastewater. One wastewater that has the potential to cultivate *Chlorella* sp. is vinasse. The contents of vinasse are simple nutrients such as total carbon (COD), total nitrogen, and PO_4^{3-} [1, 2]. The utilization of vinasse as a cultivation medium does not only decrease production cost but also promise to remove the COD [3]. Besides, it is potential to produce algae–based plastics in a specific condition. The characteristic of algae–based plastics is degradable in nature due to containing Poly–Hydroxy Butyrate (PHB). PHB is produced in a microalgae cell as a stress response of limited nutrients [4, 5]. In addition, to improve productivity,

* Corresponding author: gregoriusjoseph87@gmail.com

Chlorella sp. has a special characteristic; it can grow in autotroph and heterotroph condition [6]. In this case, CO_2 in the air as an inorganic carbon source to support autotroph condition and vinasse as an organic carbon source to support heterotroph condition.

The aim of this research is to analyze the influence of limited primary nutrient difference into kinetic parameters on *Chlorella* sp. cultivation process in vinasse, especially on PHB production. The quantitative analysis developed to predict the rate of COD degradation (k_L), biomass productivity (μ), lag phase (λ), and yield PHB to COD ($Y_{PHB/COD}$). The assumptions taken to quantify the rate of substrate degradation is quasi single substrate expressed as COD. The rate of COD degradation was approached by Equation (1):

$$-\frac{dS}{dt} = k_L S^n \tag{1}$$

where k_L is the rate of COD degradation constant (L mg^{-1} d^{-1}), S is concentration COD at t time (mg L^{-1}) and n is kinetic order (suitable value was obtained by trial–error method).

The rate of PHB forming, modeled by Yield concept, as depict in Equation (2)

$$-\frac{dS}{dt} = \frac{1}{Y_{PHB/COD}}\frac{dP}{dt} \tag{2}$$

where $Y_{PHB/COD}$ is yield of PHB formation per unit COD consumed.

Cultivation *Chlorella* sp. in vinasse is approached by Modified Gompertz to know the correlation between biomass productivity and lag phase [7], as depict in Equation (3)

$$A(t) = A_\infty \exp\left\{-\exp\left[\frac{2.718282\,\mu}{A_\infty}(\lambda - t) + 1\right]\right\} \tag{3}$$

where, $A(t)$ is dry mass at t time (mg L^{-1}), A_∞ is dry mass at stationer phase (mg L^{-1}), μ is biomass productivity (mg L^{-1} d^{-1}) and λ lag phase (d).

The data (COD, PHB, and dry mass) is fitted to Equation (1) (2) and (3), respectively, and the coefficient determination (R^2) is used as a fitting validation. The obtained constants (k_L, $Y_{PHB/COD}$, μ and λ) are used to quantitative analysis the performance of the reactor in the different limited primary nutrients.

2 Materials and methods

2.1 Materials

Vinasse was used for cultivation medium. It was collected from one of the ethanol industry in Sukoharjo, Central Java, Indonesia. Vinasse was sterilized to eliminate bacterial content then diluted in tap water until 5 mg L^{-1} (COD = 1 435 mg L^{-1}). *Chlorella* sp. was provided by CV. Algae Park, Sukoharjo, Central Java, Indonesia. Both of them were mixed and added low–cost primary nutrients. Four variations were tested at low–cost primary nutrient of (R1) no nutrient added, (R2) 40 mg L^{-1} urea and 10 mg L^{-1} TSP, (R3) 10 mg L^{-1} TSP, and (R4) 40 mg L^{-1} urea.

2.2 Experimental setup

Reactor artificial with a fluid capacity of 1 000 mL used 1 L glass. The reactors were equipped by 54 Watt neon lighting and conventional aerator (4 L min^{-1}). The process was conducted in 10 d on batch mode. A sample was taken every day to be assayed for its dry mass, COD, and PHB contents. Dry mass and COD followed by APHA Method [8] while the PHB contents followed by Senior Method [9].

3 Results and discussion

Vinasse is potential to utilize as microalgae cultivation medium especially *Chlorellla* sp. because vinasse contains 33.60 mg L^{-1} N and 0.028 mg L^{-1} P. However, vinasse also contains toxic material (0.146 mg L^{-1} phenol) so it needs intensified in utilization as cultivation medium. Based on the calculation, kinetic parameter was obtained and it is shown in Table 1. These parameters are use to compare the performance among the reactors.

Table 1. The value of k_L, $Y_{PHB/COD}$, μ and λ in each reactor

Reactor	COD		Dry mass			PHB	
	k_L*	R^2	μ	λ	R^2	$Y_{PHB/COD}$	R^2
R1	0.493E–4	0.965	0.643	13.843 0	0.976	0.022 2	0.944
R2	1.847E–4	0.865	0.158	0.105 8	0.999	0.204 7	0.804
R3	0.865E–4	0.912	0.343	9.655 7	0.971	0.030 3	0.646
R4	1.491E–4	0.825	0.096	0.691 0	0.991	0.265 5	0.824

* The suitable parameter n (Equation 2) that gave the best–fitting of the data was n = 2 for all reactors

3.1 COD degradation

COD is a suitable parameter to approach amount of organic material utilized by *Chlorella* sp. for growing. COD degradation in each reactor best fitted by mathematical model (R^2 closer to 1) as shown in Figure 1, it was also proved that nutrient content in vinasse could support *Chlorella* sp. to degrade COD.

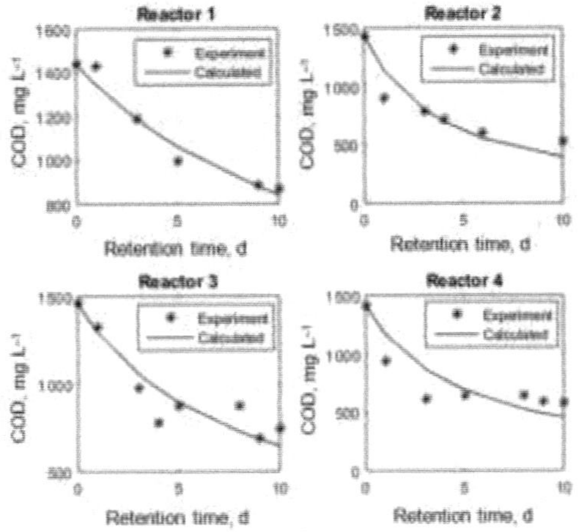

Fig. 1. COD best–fitting on the model proposed in Equation (1)

In vinasse treatment by microalgae (*Chlorella* sp.), the efficiency of COD removal becomes a primary parameter to measure the performance of process, the best performance can be reached when the highest of COD efficiency. Figure 1 and Table 2 showed that the nutrient addition could make the process of COD degradation faster than without nutrient addition. This phenomenon was correlated with the growth of *Chlorella* sp. can grow in a mixotroph condition. In the autotroph growth, *Chlorella* sp. could make its own nutrient by photosynthesis process. The other way, the heterotroph growth force *Chlorella* sp. convert organic material for growing. However, to obtain value products especially PHB, the limited nutrient should be treated as well as the previous research.

Limiting nutrients, especially limiting nitrogen (N), made *Chlorella* sp. consume organic material as their nutrient in heterotroph condition. On the other hand, limited N was triggered by organic material degradation to be faster [3, 10]. However, in this research was obtained the rate of COD degradation (*kL*) in the reactor without N and P addition (R1) on the lowest value was 0.493E–4, followed by reactor without N addition (R3), reactor without P addition (R4) and reactor with N and P addition (R2) with the value 0.865E–4 1.491E–4 dan 1.847E–4 respectively. It was because of constant lighting since the process. Based on the previous research, the heterotrophic condition occurs without lighting, and the function of aeration was just to help *Chlorella* sp. consumed organic material [11]. The lighting was only needed by *Chlorella* sp. as a catalyst in the photosynthesis process (autotrophic condition). In this condition, *Chlorella* sp. needed macronutrient such as C, N and P in big amount [12], so R1 and R3 were in situation when heterotrophic condition could not be reached, but losing big amount of nutrient since the process as effect the presence of aeration and lighting. It was equal to the previous research. The weakness of the heterotrophic condition was lighting, organic material, and CO_2 [13]. On the other hand, in R2 and R4 could grow well because the need of nutrient and environment condition support to photosynthesis process (autotrophic condition) and could obtain the high efficiency of COD removal because of constant aeration since the process. The higher oxygen in the system can give an advantage in COD degradation [14].

Table 2. Percentage of COD removal in each reactor

COD	**Reactor**			
	R1	**R2**	**R3**	**R4**
Influent (mg L^{-1})	1 437	1 435	1 459	1 413
Effluent (mg L^{-1})	868	533	746	578
Efficiency removal (%)	39.60	62.86	48.87	59.09

Chlorella sp. was expected to grow well in a mixotroph condition compared to either autotrophic or heterotrophic. It was proven by k_d parameter from previous research which had equal meaning with k_L in this research, but the value of k_d was higher than k_L. It was due to a lighting scheme so that *Chlorella* sp. could make its own nutrient by autotrophic condition (light) and degraded vinasse in heterotroph condition (dark) [15].

3.2 Biomass productivity and lag phase

The other parameter in order to approach the biological treatment process was the lag phase. The lag phase is the first phase of the growth curve that reflects the adaptation of *Chlorella* sp. in their environment. In general, the microorganism growth curve was sigmoidally shaped (Figure 2), which had similarity to microorganism dry mass [7].

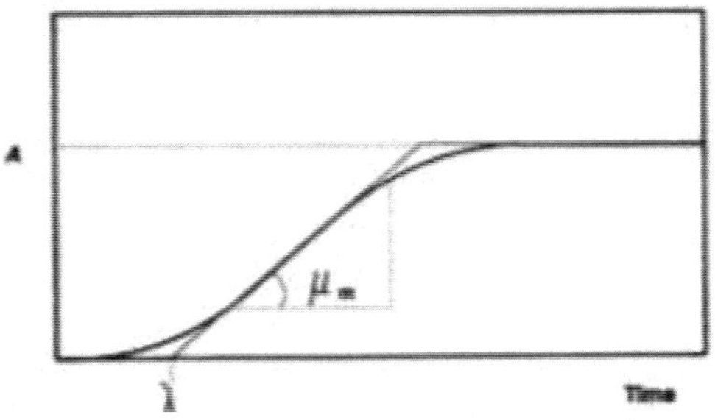

Fig. 2. Microorganism growth curve

In Figure 3 showed that the profile of dry mass in each reactor. R2 and R4 curves had a similar shape to the sigmoidal curve. It was due to the short lag phase. Based on the microorganism growth curve (Figure 2), the shorter lag phase showed that *Chlorella* sp. already grew well (exponential phase).

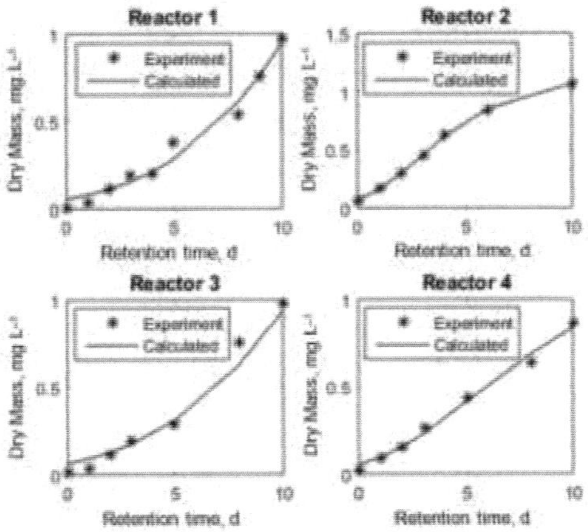

Fig. 3. Dry Mass best–fitting on the model proposed in Equation (3)

Quantitatively, the value of lag phase in each reactor was shown by Table 1, it was shown that *Chlorella* sp. in R1 and R3 undergo longer adaptation phase compared by R2 and R4. It was equal with previous research, which said that in limited N condition made the lag phase of *Chlorella* sp. to be longer [3]. On the other hand, the addition of nitrogen in the cultivation process of *Chlorella* sp. in vinasse could cut the lag phase to be shorter.

Limited N also affected on biomass productivity. Theoretically, biomass productivity is linear, with the rate of COD degradation in heterotrophic conditions [16]. In this research was obtained different scheme where the profile of biomass productivity (μ) in each reactor is R1 > R3 > R2 > R4 and the rate of COD degradation (kL) in each reactor is R2 > R4 > R3 > R1. It was shown that the lower N condition, the higher biomass productivity. However, the high biomass productivity in R1 and R3 were not accompanied by the rate of COD degradation and only occurred during the lag phase, so it could be called the biomass productivity during lag phase. It was just a response of *Chlorella* sp. self–defense.

3.3 Poly–Hydroxy Butyrate (PHB) formation

The profile of PHB in each reactor was shown in Figure 4. The R1 could form PHB. It was pointed out that vinasse could be utilized as a cultivation medium, especially in order to form PHB without nutrient addition. However, the yield of PHB to COD ($Y_{PHB/COD}$) was low, so that it needed nutrient addition to increasing the $Y_{PHB/COD}$. Based on previous research, limiting primary nutrient (N or P) could increase $Y_{PHB/COD}$ [17]. Quantitatively, biomass productivity (μ) and $Y_{PHB/COD}$ in each reactor was shown in Table 1.

Biomass productivity represents the rate of growth of *Chlorella* sp. in vinasse, and the higher value of biomass productivity reflects the higher value of PHB. The same idea with previous research that said accumulated PHB reaches 80 % by their dry mass [18]. In contrary to these statements, in this research, especially in R1 and R3, which obtained higher biomass productivity but the lower yield of PHB. It was due to *Chlorella* sp. undergo deficiency of N, which affect to the lack of metabolic energy so that the *Chlorella* sp. cell decrease in endurance. On the other hand, vinasse contains a phenolic compound, which was inhibitor substance for *Chlorella* sp. [19]. The impact of decreasing cell endurance and the presence of phenolic compounds caused R1 and R3 poisoned. Meanwhile, R2 and R4 could form PHB well because *Chlorella* sp. could grow well in autotroph condition. Nevertheless, to obtain the PHB, one of the efforts was limiting P [17].

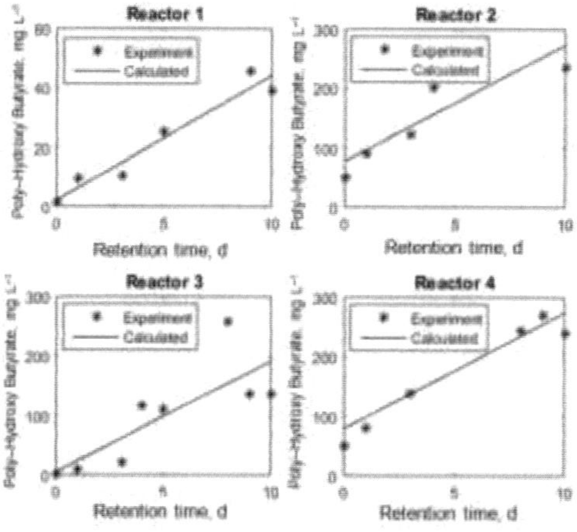

Fig. 4. PHB best–fitting on the model proposed in Equation (2)

4 Conclusions

The reactor with complete nutrients can give the best performance in treating COD (efficiency COD removal = 62.86 %). On the other hand, the reactor without nutrient N addition was undergone N deficiency and toxicity as an impact on the presence of phenol. It was indicated from the highest value of biomass productivity (μ = 0.643 mg L^{-1} d^{-1}). The longest lag phase (λ = 13.8430 d) and the shortest value of the rate of COD degrade (k_L = 0.493E–4 L mg^{-1} d^{-1}). N nutrient addition into the cultivation process of *Chlorella* sp. in vinasse could cut the lag phase and made it shorter (λ = 0.158 d). However, the yield of PHB formation in the reactor with limited phosphate medium was the highest ($Y_{PHB/COD}$ = 0.2655). The insignificant difference among the parameters (k_L and $Y_{PHB/COD}$) in each reactor show that the cultivation of *Chlorella* sp. in vinasse especially to produce PHB was not only influenced by limited nutrient, but also there was some aspect such as substrate ratio, lighting scheme and aeration correlated by CO_2 existence.

This research was funded by the national competitive research grant PKPT from the Higher Education Directorate, Ministry of Research, Technology, and Higher Education, Indonesia (Number of Contract 002/LPPM–USB/Pekerti/V/2019).

References

1. C. Candido, A.T. Lombardi. Journal of Applied Phycology. **29**:45–53(2017). https://doi.org/10.1007/s10811-016-0940-2
2. C.E.R. Reis, B. Hu. Frontiers in Energy Research. **5**:1–7(2017). https://doi.org/10.3389/fenrg.2017.00007
3. L.F.A. de Mattos, R.G. Bastos. Desalination and Water Treatment. **57**,20: 9465–9473(2016). https://doi.org/10.1080/19443994.2015.1028454
4. R. Carpine, F. Raganati, G. Olivieri, K.J. Hellingwerf, A. Pollio, P. Salatino, et al. Algal Research. **29**:49–60(2018). https://doi.org/10.1016/j.algal.2017.11.011
5. S.K. Das, A. Sathish, J. Stanley. *Materials Today: Proceedings*. **5**:1674–1678(2018). https://doi.org/10.1016/j.matpr.2018.06.020
6. M.M.A. Nur, H. Hadiyanto. Journal of Engineering and Technological Sciences. **47**,5:487–497(2015). http://dx.doi.org/10.5614%2Fj.eng.technol.sci.2015.47.5.2
7. L. Frunzo, R. Garra, A. Giusti, V. Luongo. Communications in Nonlinear Science and Numerical Simulation. **74**:260–267(2019). https://doi.org/10.1016/j.cnsns.2019.03.024
8. R. Baird, A.D. Eaton, E.W. Rice. Standard Methods for Examination of Water and Wastewater. 23rd ed. Washington: American Public Health Association. Part 2540 68–69 and Part 5220 21–22 (2017), p. 1360. https://doi.org/10.2105/SMWW.2882.030 and https://doi.org/10.2105/SMWW.2882.103
9. P.J. Senior, G.A. Beech, G.A. Ritchie, E.A. Dawes. The Biochemical Journal. **128**,5:1193–1201(1972). https://doi.org/10.1042/bj1281193
10. M.I. Khan, J.H. Shin, J.D. Kim. Microbial Cell Factories. **17**,36:1–21(2018). https://doi.org/10.1186/s12934-018-0879-x
11. D. Morales–Sánchez, O.A. Martinez–Rodriguez, J. Kyndt, A. Martinez. World Journal of Microbiology & Biotechnology. **31**,1:1–9(2015). https://doi.org/10.1007/s11274-014-1773-2
12. G. Markou, D. Vandamme, K. Muylaert. Water Research. **65**:186–202(2014). https://doi.org/10.1016/j.watres.2014.07.025
13. J. Lowrey, M.S. Brooks, P.J. McGinn. Journal of Applied Phycology. **27**,4: 1485–1498(2015). https://doi.org/10.1007/s10811-014-0459-3
14. H. Lu, G. Zhang, T. Wan, Y. Lu. Bioresource Technology. **102**,20:9503–9508(2011). https://doi.org/10.1016/j.biortech.2011.07.114
15. S. Gupta, R.A. Pandey, S.B. Pawar. Bioremediation Journal. **21**,1:38–51(2017). https://doi.org/10.1080/10889868.2017.1282936
16. H. Wang, H. Xiong, Z. Hui, X. Zeng. Bioresource Technology. **104**:215–220(2012). http://doi.org/10.1016/j.biortech.2011.11.020
17. D.M. Arias, E. Uggetti, M.J. García–Galán, J. García. New Biotechnology. **42**: 1–11(2018). https://doi.org/10.1016/j.nbt.2018.01.001
18. E. Koutra, C.N. Economou, P. Tsafrakidou, M. Kornaros. Trends in Biotechnology. **36**,8:819–833(2018). https://doi.org/10.1016/j.tibtech.2018.02.015
19. A.H. Scragg. Enzyme and Microbial Technology. **39**:296–799(2006). https://doi.org/10.1016/j.enzmictec.2005.12.018

Relationship of Group Dynamics and Fishermen Independence in Fisheries Agribusiness Attaining Maximum Sustainable Yield

Hamdi Rosyidi[1,*], *Jabal* Tarik Ibrahim[2], *Sutawi* Sutawi[3], *Olga* Anne[4], and *Diana* Rachmawati[5]

[1]Master's Degree, Department of Agriculture Science, Postgraduate Program, University of Muhammadiyah Malang, Jl. Raya Tlogomas No. 246, Malang 65145, East Java, Indonesia
[2]Department of Agribusiness, Faculty of Agriculture and Animal Sciences, University of Muhammadiyah Malang
[3]Department of Animal Sciences, Faculty of Agriculture and Animal Sciences, University of Muhammadiyah Malang
[4]Department of Natural Sciences, Faculty of Marine Technology and Natural Sciences of Klaipeda University, H. Manto g. 84, Klaipėda 92294, Lithuania
[5]Department of Aquaculture, Faculty of Fisheries and Marine Sciences, Diponegoro University, Jl. Prof. Soedarto, SH, Tembalang, Semarang 50275, Central Java, Indonesia

Abstract. This study aimed to examine the factors and the influence of group dynamics and the independence of fishing group members in catch fisheries agribusiness activities. This research was carried out using a survey method in the *Rukun Nelayan Blimbing* in Paciran District, Lamongan Regency, East Java Province, Indonesia from February to March 2019. The result showed the group dynamics and independence of fishermen in catch fisheries was high, but group dynamics did not influence the independence of fishermen in catching fisheries agribusiness. The factors that influence group dynamics were structuring, coaching and development, atmosphere, effectiveness. The factors affected the independence of the group were the addition of production facilities, prices, market, and fishermen. Those factors are needed to attain the good management in maximum sustainable yield (MSY).

Keywords: Development fisheries sector, fishermenhood, increase welfare, social interaction, sustainability

1 Introduction

Indonesia is an archipelagic country with an abundant landscape of marine wealth. In its theoretical study, Lasabuda states that Indonesia has been recognized internationally through UNCLOS 1982 that the total area of Indonesia's sea area reaches 5 900 000 km^2

* Corresponding author: hamdiorzams@gmail.com

[1]. International recognition was later ratified by Indonesia in Law No. 17 of 1985. There are 3 200 000 km^2 of territorial waters that connect the entire archipelago. This condition makes Indonesia the largest archipelago country in the world.

PPN–*Pelabuhan Perikanan Nusantara* [Archipelago Fisheries Port] of Brondong Lamongan is the largest docked fisheries base in the northern region of East Java Province, Indonesia whose production reaches an average of 100 ton d^{-1}. PPN Brondong activities involved 12 160 fishermen, 21 695 fishing gear, 1 385 motorboats, 312 traders/distributors and 170 fish processors. The north sea of the island of Java has various types of fish both pelagic and demersal. The fishing areas in Brondong District are in 712 Fisheries Management Areas (WPP 712) covering the areas of Masalembu, Matasiri, Kramean, and around Bawean, Indonesia.

The fisheries sector has a major role to play at the level of PAD- *Pendapatan Asli Daerah* [Local Government Revenue] of Lamongan Regency, Indonesia. Data Distribution of Percentage PDRB- *Produk Dosmetik Regional Bruto* [Gross Regional Domestic Product] of the fisheries sector in Lamongan Regency reached 17.49 %, outperforming the food crop sector contributed 13.09 % [2]. While the contribution of the fisheries sub-sector to the agriculture, forestry and fisheries sector in Lamongan in 2016 reached 44.39 % [3].

BPS-*Badan Pusat Statistik* Lamongan mentioned that the capture fisheries activities in 2015 employed 17 892 workers, then increased 6.36 % up to 19 030 in 2016 [2]. The large number of workers in the fisheries sector makes the emergence of fellow fishermen community groups to help each other solve common problems.

The fisheries sector can be developed through a group approach. Triwahyuni mentioned the development of dairy goat businesses driven by the role of livestock farmer groups to increase added value and efficient management [4]. Farmers can interact with one another in their group to increase knowledge and ability to manage agribusiness systems. Sondakh mentioned that social interaction of fishermen groups is the key to success and organizational goals undertaken by members consciously to build interaction between fellow members and administrators of fishing groups [5].

Research aims to determine the level of group dynamics, level of independence and the relationship between the two variables in the fishermen group in terms of catch fisheries agribusiness. Research provides an overview related to group dynamics and the independence of fishermen in catch fisheries agribusiness as scientific development material.

2 Methods

The group approach to fishermen is carried out through the RN- *Rukun Nelayan* (fishermenhood/Community Association). The study was conducted on RN Blimbing. RN Blimbing is one of the 17 RN in Lamongan Regency, Indonesia. RN Blimbing consists of 5 321 fishermen who work on 410 ships operating in the Java Sea. This research was conducted using a quantitative survey method. That aimed to explain the relationship between variables (explanatory). The sampling used simple random sampling. The indicator of group dynamic i.e. group goals, structure, function and task, coaching and development, compactness, atmosphere, pressure, effectiveness, hidden agenda, group dynamic. The research instrument has been declared valid and reliable in the test of validity and reliability. Research data have also been tested for normality, linearity, and homogeneity, then proceed to path analysis.

3 Results and discussions

3.1 Level of group dynamics

Table 1. Results of indicators assessment of group dynamics (%)

Indicator	Low	Medium	High	Very high
Group goals	22.22	37.78	28.89	11.11
Structure	4.44	24.44	46.67	24.44
Function and tasks	11.11	22.22	37.78	28.89
Coaching and development	8.89	17.78	48.89	24.44
Compactness	15.56	20.00	26.67	37.78
Atmosphere	11.11	37.78	42.22	8.89
Pressure	24.44	40.00	28.89	6.67
Effectiveness	17.78	28.89	40.00	13.33
Hidden Agenda	15.56	37.78	26.67	20.00

Based on Table 1, group dynamics were in the high category as well as the supporting indicators structure such as functions and tasks, coaching and development, group atmosphere and effectiveness and cohesiveness. As for the group goals, the pressure and hidden intent were in the medium category. This result indicated that RN Blimbing has group activities that always accommodate the interaction between its members so that it brings up dynamics. Febriani et al. states that dynamics implied the interaction and interdependence between members in own group and other groups [6].

Groups make positive contributions to the organization such as creating collaboration between group members, build effective communication, increase job satisfaction and motivation, and a culture of competition. On the other hand, conflict can have negative impacts such as decline poor productivity and competition. To prevent it, organization need to be aware of group functions, group dynamics, individual boundaries, as well correct communication methods and techniques [7].

Based on their work activities, the fishing profession has social and economic stratification. Resilience status of ship owners as a means of livelihood has the strongest position, followed by the captain and the weakest, namely ABK [8]. Zain et al. revealed that the low quality of life of fishermen is caused by competition between fishermen. Fishermen can compete with each other for big profits [9].

The higher resilience of certain groups in fishing communities contributes to increased social differentiation, which in turn undermines community resilience. Resilience at the individual, household and group level will encourage persistence, adaptability, and the ability to transform in work so as to encourage complementarity, not conflict [10]. In this case, fishermen groups have a strategic position to suppress socio-economic differentiation and build a spirit of cooperation and collaboration for the welfare of all fishermen members.

In addition, Blimbing RN also has an open relationship with members outside groups such as the government in the fisheries sector. This interaction is more responsive and critical in responding to issues outside the group and becomes a conversation within the group. Social interaction carried out by fishermen can be a strength in achieving group goals together. Communication with related local governments has a strategic role and even has an impact on fishermen's income. Moento, et al. states that the Merauke Regency

Fisheries Department plays a role in increasing the income of the fishing community in Pelita Samkai Village [11].

Sondakh et al. states that social interaction of fishermen groups is the key to success and organizational goals are carried out on the awareness of members to consciously establish communication or interaction between one another, between members and management of fishermen groups [5]. Squires et al. adding aspects of social and institutional sustainability can be seen from how group members resolve conflicts and continue to work together [12]. Conversely, social exclusion influences interest and affiliation behavior in the group [13].

Fishermen of RN Blimbing on average go fishing for 1 wk to 3 wk. This potent to become an obstacle for the group in carrying out activities to gather whole members. Group member interactions have problems such as no free time because the members have to go fishing [14]. By high dynamics, the management and members of the Blimbing RN have had the right strategy in dealing with these problems. They utilize free time to carry out group activities when bad weather, so that the majority of members can be involved.

3.2 Level of fishermen independence

Table 2. Result of indicators assessment of fishermen independence (%)

Indicators	Low	Medium	High	Very high
Independence determines the catching area	22.22	37.78	28.89	11.11
Independence determines the addition of production facilities	4.44	24.44	46.67	24.44
Independence determines prices	11.11	22.22	37.78	28.89
Independence of market	8.89	17.78	48.89	24.44
Independence in occupational health and safety	15.56	20.00	26.67	37.78
Independence of fishermen	11.11	37.78	42.22	8.89

The fishermen's independence reached a high category 42 %. Indicators of independence that are in the high category are the means of production, pricing, marketing, and occupational health and safety in the very high category. The independence in determining the area of capture is in the medium category.

The high level of independence illustrates that so far fishermen have been able to face their own problems so that they do not have excessive dependence on other parties' assistance. Government assistance was not a priority for fishermen to keep running their fishing activities. Fishermen become resilient due to the pressure of the necessities of life that requires them to continue to exist and survive in their work.

In India, fishermen groups are said to have a role in increasing the maximum yield from fishery resources so as to increase the income and welfare of their members [15]. Education or learning in fishing group activities can improve the socio-economic status of fishermen. Basavakumar et al. states that the socio-economic status of fishing communities can be improved by providing education such as the application of better fishing methods [16]. Good fishing methods can improve the quality of fish caught. Fish quality is a factor that has a significant effect on fish prices [17].

Richmon et al. explained that the framework for the sustainability of fishing communities includes four categories, namely economy and markets; social and community; physical infrastructure and critical services; environment and regulation. The sustainability of this category is very important for the long-term survival of fishing communities and strategic plans to encourage improvement, expansion, or investment in all areas. Each category is connected and dependent on each other in the fishing community [18].

A well-managed fishing group can encourage professional investment collaborations. Additional capital for investment and operational costs will encourage participation and institutional strengthening. Group members have the values of helping each other, caring, open communication, and collaboration in groups. They develop mutual understanding and can manage conflict peacefully [19].

4 Conclusion

The results of this study are taken into consideration and become input in developing and fostering fishermen groups. As an effort to optimize the potential of fishermen groups in conducting capture fisheries agribusiness activities, especially in Lamongan Regency, East Java Province, Indonesia. Both the group dynamics and independence of fishermen in catch fisheries was high, but group dynamics did not influence the independence of fishermen in catching fisheries agribusiness. The factors that influence group dynamics were structuring, coaching and development, atmosphere, effectiveness. The factors affected the independence of the group were the addition of production facilities, prices, market, and fishermen. Those factors are needed to attain the good management in maximum sustainable yield (MSY).

References

1. R. Lasabuda. JIP. **1,** 2:92–101(2013). [in Bahasa Indonesia]. https://ejournal.unsrat.ac.id/index.php/platax/article/view/1251
2. Badan Pusat Statistik Kabupaten Lamongan, *Distribusi Persentase Produk Domestik Regional Bruto Kabupaten Lamongan Atas Dasar Harga Konstan 2010-2016* [Distribution of Percentage of Gross Regional Domestic Product in Lamongan Regency on the basis of Constant Price 2010-2016] [Online] from https://lamongankab.bps.go.id/statictable/2017/07/04/671/distribusi-persentase-produk-domestik-regional-bruto-lamongan-atas-dasar-harga-konstan-2010-2016.html (2019).p.1. [Acessed on 20 May 2019). [in Bahasa Indonesia]
3. Badan Pusat Statistik Kabupaten Lamongan. *Produk Domestik Regional Bruto Kabupaten Lamongan Menurut Lapangan Usaha 2013-2017* [Lamongan Regency Gross Regional Domestic Product by Business Field 2013-2017] [Online] from https://lamongankab.bps.go.id/publication/2018/08/15/e37aa39888d81f987e570e32/produk-domestik-regional-bruto-kabupaten-lamongan-menurut-lapangan-usaha-2013-2017.html (2017).p.109. [in Bahasa Indonesia]

4. A. Triwahyuni, M. Sulistyati, L. Nurlina, Students e-Journals. **5**,3:1–9(2016). [in Bahasa Indonesia]. http://jurnal.unpad.ac.id/ejournal/article/view/9623
5. V. B. Sondakh, J. A. Andaki, M. P. Wasak, J. Akulturasi, **5**,9:589–602(2017). [in Bahasa Indonesia]. https://ejournal.unsrat.ac.id/index.php/akulturasi/article/view/16984
6. E. M. Febriani, Saharudin, J. Sosiol. Pedesaan, **2**,1:17–29(2014).[in Bahasa Indonesia]. https://journal.ipb.ac.id/index.php/sodality/article/download/9409/7373
7. H. Gençer, Univers. J. Educ. Res. **7**,1:223–229(2019). https://www.researchgate.net/publication/331482821_Group_Dynamics_and_Behaviour
8. B. Vatria, B. Wiryawan, E. S. Wiyono, M. S. Baskoro, AACL Bioflux, **12**,1:310–319(2019). http://www.bioflux.com.ro/docs/2019.310-319.pdf
9. R. M. Zain, M. K. A. Kamarudin, M. H. M. Saad. Int. J. Acad. Res. Bus. Soc. Sci., **8**,10:640–650(2018). https://hrmars.com/papers_submitted/4770/Assessment_of_Quality_of_Life_on_Fishermen_Community_in_Kuala_Terengganu,_Malaysia_A_Review.pdf
10. M. Leite, H. Ross, F. Berkes. Ecol. Soc. **24**,3:1–13(2019). https://www.jstor.org/stable/26796970?seq=1#metadata_info_tab_contents
11. P. A. Moento, A. P. Yusuf, A. F. Adam, E. E. Maturbongs, A. P. Tljlien, M. Yunus. IOP Conference Series: Earth and Environmental Science. **473**:1–7(2019). https://iopscience.iop.org/article/10.1088/1755-1315/473/1/012033/pdf
12. K. Squires, M. G. Wiber. Ecol. Soc., **23**,2:1–12(2018). https://www.ecologyandsociety.org/vol23/iss2/art25/ES-2018-10137.pdf+&cd=12&hl=id&ct=clnk&gl=id
13. D. F. Sacco, M. J. Bernstein, Am. Psychol. Assoc. **19**,4:211–224(2015). https://www.apa.org/pubs/journals/features/gdn-gdn0000035.pdf
14. D. K. Kusnani, P. Muljono, A. Saleh, J. Penyuluhan, **11**,2:129–142(2015). [in Bahasa Indonesia]. https://journal.ipb.ac.id/index.php/jupe/article/view/10578
15. L. K. Tyagi, A. S. Bisht, A. Pal, K. K. Lal. J. Community Mobilization Sustain. Dev., **8**,1:90–93(2013). https://krishi.icar.gov.in/jspui/bitstream/123456789/4749/1/Functioning%20of%20fishing%20cooperative%20societies%20paper.pdf
16. K. V. Basavakumar, S. Devendrappa, S. T. Srenivas. J. Agric. Sci. **24**,5:684–687(2011). https://www.researchgate.net/publication/266458070_A_study_on_profile_of_fishing_community_of_a_village_in_Karnataka/link/556c31fc08aefcb861d62277/download
17. M. P. Alapan, E. L. I. Arpilleda, K. J. R. Altizo, G. K. R. Frias, J. D. Ravelo. J. Environ. Ecol. **7**,2:34–41(2016). https://www.researchgate.net/publication/312237781_Factors_Affecting_the_Market_Price_of_Fish_in_the_Northern_Part_of_Surigao_Del_Sur_Philippines
18. L. Richmond, R Cumoucel, H. Pontarelli, L. Casali, W. Smith, K. Gillick, et al. Sustain. **11**,7:1–22(2019). https://ideas.repec.org/a/gam/jsusta/v11y2019i7p1904-d218457.html
19. H. Saediman, F. F. Syah, T. La Ola, M. Tufaila, Bahari, L. Daud, J. Humanit. Soc. Sci. **24**,8:5–12(2019). http://www.iosrjournals.org/iosr-jhss/papers/Vol.%2024%20Issue8/Series-6/B2408060512.pdf

Degradation of Phorbol Esters on the *Jatropha curcas* Linn. Seed by Biological Detoxification

Hany Handajani[1,*], *Riza* Rahman Hakim[1], *Ganjar* Adhywirawan Sutaro[1], *Boy* Ronald Mavuso[2], *Zhong–Wen* Chang[3], and *Soni* Andriawan[1]

[1]Department of Fisheries, Faculty of Agriculture and Animal Science, Universitas Muhammadiyah Malang, Malang 65144, Indonesia
[2]Ministry of Agriculture, P.O Box 162, Mbabane, Eswatini, Africa
[3]Department of Aquaculture, National Pingtung University Science of Technology, Pingtung 91201, Taiwan

Abstract. The application of fermentation is one of methods to increase food quality biologically. Availability of *Jatropha curcas* residual from oil factory be focused as a soybean meal or fish meal replacement. On the other hand, *J. curcas* residuals possess a toxic compound as well. This study aimed to examine the effect of *Aspergillus niger* on the nutrition and harmful content of *J. curcas* as a potential ingredient of feed. In brief, *J. curcas* residual was fermented with a detoxification method at 3 d, 5 d, and 7 d. Crude protein, fat, and crude fiber content were assessed to discover the biological responses of *J. curcas* post–fermentation while phorbol ester was evaluated to toxic content post–detoxification. The results showed that crude protein and fat content were highest on 7 d post–fermentation but it was no significant difference ($p > 0.05$). While crude fiber content showed significant difference which the 3 d fermentation had the highest content of fiber. For phorbol ester content, 3 d fermentation showed a better result than the control group ($p < 0.05$). The present findings suggest that *A. niger* is recommendable as starter to reduce fiber and toxic content of *J. curcas* residual at 3 d fermentation.

Keywords: Fermentation, nutrient, protein substitution, residual, waste to feed

1 Introduction

The feed is one of the factors that influence the growth of farmed fish. Intensive cultivation relying on commercial feed. Sources of protein in the diet comes from animal protein and vegetable protein [1, 2]. The primary source of vegetable protein feed is generally taken from soy flour, which has about 34.39 % protein content [3]. According to [4], the use of

* Corresponding author: handajani@umm.ac.id

soy flour in commercial diets ranged between 10 % to 25 %. However, soybean prices are high now and still be imported [5, 6]. According to [7], that the cost of imported soybean in Indonesia about IDR 11 342 kg^{-1}. The increase in soybean prices has resulted in an increased cost of commercial feed [8]. It is not followed by the value of fish, affecting the sustainability of fish farming in Indonesia. One alternative to reduce the use of soy flour that is to take advantage of other feed ingredients such as *Jatropha curcas* seed.

The availability of *J. curcas* seed meals in many countries, especially Indonesia, is very abundant due to the increased production of JCO – *J. curcas* oil as an energy source instead of petroleum [9–13]. Amount of 1 t of dry beans will produce 200 L to 300 L of oil, the residue in the form of a *J. curcas* seed meal at 700 kg to 800 kg [14]. It can be utilized as feed ingredients because of the availability in abundant amounts, relatively cheap, easily digested by fish, have good nutritional content (protein), and do not compete with humans needs [15, 16].

According to [17], the crude protein content within skinned *J. curcas* seed meal in Indonesia is 19 % to 21 %, while the crude protein in the *Jatropha* seed meal without the husk is around 45 % to 50 %. Additionally, composition and percentage of amino acids, and minerals almost the same as other grains. The nutrient content makes *J. curcas* seed meals is considerable potential as a partial replacement of local feed ingredients soy flour. However, *J. curcas* seed meal is not commonly used because it has a toxic substance that is phorbol ester [18–20]. It causes lysis of fish erythrocytes that interfere with the formation of red blood cells when it is consumed by fish [21, 22]. However, the toxic substances can be minimized by detoxification. Several methods of detoxification has done, both physically (heating using an autoclave for 30 min, 121 °C), chemically (extraction using methanol), and biologically (fermentation using various molds) [22–25].

Further research is needed to biologically detoxification by fermentation using *A. niger* because it is grow fast saprophytic mold and harmless due to not produce mycotoxins. The results of the study on *J. curcas* detoxification is expected to be applied to an alternative feed ingredient safety in fish farming.

The purpose of this study is as follows: i) To determine the effect of the biological detoxification by fermentation using *A. niger* on the content of toxic substances and nutrients in the seed meal of *J. curcas* L. ii) To get the best fermentation time on seed meal of *J. curcas* L. which gives the best results in the biological detoxification using *A. niger*.

2 Material and Methods

2.1 Experimental design

The present study used an experimental approach with a Completely Randomized Designed (CRD). This research applied three treatments and three replications.
T1: 3 d fermentation of *J. curcas* seed meals by *A. niger*
T2: 5 d fermentation of *J. curcas* seed meals by *A. niger*
T3: 7 d fermentation of *J. curcas* seed meals by *A. niger*

2.2 *A. niger* preparation

The preparation of *A. niger* followed [26] with some modification. The fungus

A. niger was inoculated to Erlenmeyer contained 100 mL media of sterile bean sprouts. Inoculum of *A. niger* was incubated in the shaker incubation for 72 h at room temperature.

2.3 Fermentation process and nutrient analysis

Fermentation of *J. curcas* seed meals by *A. niger* followed [20, 27, 28]. The fermentation used plastic bags. *J. curcas* seed meals were analyzed for phorbol ester and nutrients content before used. A 250 g of *J. curcas* seed meals were steamed for 1 h to reduce moisture content to 40 %. Inoculum *A. niger* as much as 6 g was added for 1 kg *J. curcas* seed meals and stirred as that of the substrate after that of the substrate temperature down about 40 °C. Substrate was put into plastic bags and placed on a plastic basket. On the next day, plastic bags were hollowed on both sides. Plastic bags were reversed daily, and fermented for 3 d, 5 d, and 7 d. Results of fermentation were dried in the oven with 60 °C and powdered. The sample was analyzed for *Phorbol ester* and nutrient contents.

Measurement of the phorbol ester based on [29, 30]. Briefly, 50 g of fermented *J. curcas* seed meals extracted by methanol 100 mL, used magnetic stirrer on the temperature 60 °C for 60 min. Sample was then filtered using paper filter to separate solution with drain mater. A dark solution was stored in a bottle. Sample was measured using liquid chromatograpy mass spectroscopy (LC–MS) to separate or mixes compounds based on its polarity.

2.4 Data analysis

Analysis of Variance (ANOVA) was applied to determine the differences among treatments. If it had a real impact then done by test advanced, election follow-on test would be used depending on the value of the coefficients diversity from the data.

3 Results and discussions

J. curcas seeds used in this study were obtained from the collection of the Laboratory of the Vegetable Oils University of Muhammadiyah Malang (UMM). *J. curcas* seeds were prepared as much as 10 kg peeled off the shell to get the kernel. Amounted 10 kg of *J. curcas* seeds produced 5 kg kernel, then pressed to produce the oil, and the residual pressed to shaped plates and then dried with aerated for 1 d. *J. curcas* seed meal used in the study analyzed at UMM Nutrition Laboratory has protein and fat of 20.89 % and 54.56 % respectively. The results of *J. curcas* seeds meal proximate analyses could be seen in **Error! Reference source not found.**.

Fermentation is one of food processing technology biologically using microorganisms that increase the quality of nutrition of feed such as carbohydrates, fat, protein, fiber and other organic matter [31, 32]. *A. niger*, one of the fermentation microorganisms, because can break up tannic acid into glucose and gallic acid, and have the ability producing a different type of acid i.e. oxalic acid, 2–hydroxypropanoic–1,2,3–tricarboxylic acid, gluconic acid and some types of enzymes [28, 33].

The study revealed that *A. niger* reduced Phorbol ester concentration using fermentation proces. Results showed that three days fermentation come out with the optimum time fermentation to reduce Phorbol ester concentration.

Table 1. Nutrient and phorbol ester content of fermented *J. curcas* seeds meal

Treatments (Fermentation)	Protein (%)	Fat (%)	Crude fiber (%)	Phorbol ester (μg g^{-1})
Raw material	20.89	54.56	3.84	2.97
A : 3 d	20.18	55.58	3.59^{b}	0.625^{a}
B : 5 d	20.03	56.18	2.58^{ab}	4.382^{c}
C : 7 d	20.71	57.00	2.92^{a}	2.102^{b}

Phorbol ester (phorbol–12–myristate 13–acetate) is a significant poison in *J. curcas* [34, 35]. Phorbol ester contained in the seeds and roots of Jatropha. Phorbol esters heat resistant and are present in the oil remaining on the cake, as much as ± 11 % [19]. The content of phorbol ester before detoxification of 2.979 ug g^{-1}, after going through rehab biologically fermented with *A. niger* for three days could reduce the content of phorbol ester on *J. curcas* seeds meal. Similar study by [36], *A. niger* could aflatoxin B^1 by 58.2 % after fermentation.

4 Conclusion

Based on the results of the study concluded, i) Fermentation with *A. niger* significantly effect phorbol ester and crude fiber, but not to crude protein and fat in *J. curcas* seeds meal. ii) 3 d fermentation give the best result referring to the decrease of the phorbol ester and crude fiber in *J. curcas* seeds meal.

The authors gratefully acknowledge for Faculty of Agriculture and Animal Science, Universitas Muhammadiyah Malang research grant aid Block Grant No: E.2.a/121/FPP-UMM/II/2017

Reference

1. M. Song, T.T. Fung, F.B. Hu, W.C. Willett, V.D. Longo, A.T. Chan, et al., JAMA Internal Medicine, **176**,10:1453–1463(2016). https://pubmed.ncbi.nlm.nih.gov/27479196/
2. P.J. Tuso, M.H. Ismail, B.P. Ha, C. Bartolotto. Perm. J. **17**,2:61–66(2013). https://doi.org/10.7812/TPP/12-085
3. P. Singh, R. Kumar, S.N. Sabapathy, A.S. Bawa, Compr. Rev. Food Sci. Food Saf. **7**:14–28(2008). https://doi.org/10.1111/j.1541-4337.2007.00025.x
4. A.G.J. Tacon, M.R. Hasan, M. Metian, FAO Fisheries and Aquaculture technical paper, 564:1–87(2011). http://www.fao.org/docrep/015/ba0002e/ba0002e.pdf
5. N. Hasan, E. Suryani, R. Hendrawan, Procedia Computer Science, **72**:605–612(2015). https://doi.org/10.1016/j.procs.2015.12.169
6. I.W. Rusastra, T.A. Napitupulu, R. Bourgeois, The impact of support for imports on food security in Indonesia, Bogor: UNESCAP–CAPSA (2008). p.1–58. http://dx.doi.org/10.22004/ag.econ.96656
7. M.M. Adie, A. Krisnawati, Int. J. Renew. Energy Dev. **3**,1:37–43(2014). http://dx.doi.org/10.14710/ijred.3.1.37-43

8. F. Asche, A. Oglend, S. Tveteras, J. Agric. Econ. **64**,1:97–111(2013). https://doi.org/10.1111/j.1477-9552.2012.00357.x
9. J. Bergmann, D. Tupinambá, O. Costa, J. Almeida, C. Barreto, B. Quirino, Renew. Sust. Energy Rev. **21**:411–420(2013). https://doi.org/10.1016/j.rser.2012.12.058
10. H.S. Dillon, T. Laan, H.S. Dillon, Biofuels–At What Cost? Government Support for Ethanol and Biodiesel in Indonesia. The Global Subsidies Initiative of the International Institute for Sustainable Development, Jakarta. 2008. p.1–76. http://citeseerx.ist.psu.edu/viewdoc/download?doi=10.1.1.438.3249&rep=rep1&type=pdf
11. H.P.S.A. Khalil, N.A.S. Aprilia, A.H. Bhat, M. Jawaid, M.T. Paridah, D. Rudi, Renew. Sust. Energy Rev. **22**:667–685(2013). https://doi.org/10.1016/j.rser.2012.12.036
12. P.G. Adinurani, R.H. Setyobudi, S.K. Wahono, M. Mel, A. Nindita, E. Purbajanti, et al., Proc. Pakistan Acad. Sc **54,**1:47–57(2017). http://www.paspk.org/wp-content/uploads/2017/03/Proceedings-B.-Life-Sciences-541-March-2017.pdf
13. R.H. Setyobudi, A. Sasmito, P.G. Adinurani, A. Nindita, A.S. Yudhanto, Y.A. Nugroho, et al. Energy Procedia. **65**:300–308(2015). doi: 10.1016/j.egypro.2015.01.056
14. S. Nopitasari, T. Widiyastuti, T.R. Sutardi, Jurnal Ilmiah Peternakan, **1**,2:446–454(2013). [in Bahasa Indonesia]. https://scholar.google.co.id/scholar?hl=id&as_sdt=0%2C5&q=Pengujian+Kecernaan+Bungkil+Biji+Jarak+Fermentasi+ditinjau+dari+Produksi+VFA+dan+NH3+secara+In+Vitro&btnG=
15. M. Ahmad, S. Ibrahim, Int. J. Fish. Aquat. Stud. **4**,1:276–279(2016). http://www.fisheriesjournal.com/archives/2016/vol4issue1/PartD/3-3-1.pdf
16. T. Widiyastuti, T.R. Sutardi, R.H. Setyobudi. Energy Procedia. **65:**362–367(2015) . doi: 10.1016/j.egypro.2015.01.069
17. R. Rahmasari, *Pengaruh perlakuan kimiawi dan biologis terhadap penyusutan bahan, kandungan antitripsin, lektin dan nutrien bungkil biji Jarak Pagar* (*Jatropha curcas* L.) [The effect of chemical and biological treatment on material shrinkage, antitrypsin content, lectins and nutrients for Jatropha curcas (*Jatropha curcas* L.) seed meal.] [Skripsi] Ilmu Nutrisi dan Teknologi Pakan. Fakultas Peternakan. Institut Pertanian Bogor, (2009) p.1–45. [in Bahasa Indonesia]. http://repository.ipb.ac.id/handle/123456789/12763
18. C.Y. Li, R.K. Devappa, J.X. Liu, J.M. Lv, H.P.S. Makkar, K. Becker, Food and Chemical Toxicology, **48**,2:620–625(2010). https://doi.org/10.1016/j.fct.2009.11.042
19. G. Goel, H.P.S. Makkar, G. Francis, K. Becker, Int. J. Toxicol. **26**:279–288(2007). https://doi.org/10.1080/10915810701464641
20. H.P.S. Makkar, G. Francis, K. Becker, J. Sci. Food Agric. **88**:1542–1548(2008). https://doi.org/10.1002/jsfa.3248
21. V. Kumar, H.P.S. Makkar, W. Amselgruber, K. Becker, Food Chem. Toxicol. **48**,8–9:2063–2072(2010). https://doi.org/10.1016/j.fct.2010.05.007
22. H.P.S. Makkar, Anim. Feed Sci. Technol. **222**:87–99(2016). https://doi.org/10.1016/j.anifeedsci.2016.09.013

23. B.S. Sharath, B.V. Mohankumar, D. Somashekar, Appl. Biochem. Biotechnol. **172**:2747–2757(2014). https://doi.org/10.1007/s12010-013-0698-9
24. T.T. Sumiati, E. Wina, Y. Yusriani, *Pemanfaatan Bungkil Biji Jarak Pagar (Jatropha curcas) yang mengalami detoxifikasi sebagai sumber protein substitusi bungkil kedelai 45 % pada ayam kampung* [Utilization of detoxified Jatropha Curcas (*Jatropha curcas*) as a protein source of 45 % soybean meal substitution in native chickens]. Laporan Hasil Penelitian Hibah Departemen Pertanian. Institut Pertanian Bogor, Bogor, (2010). [in Bahasa Indonesia]. http://repository.ipb.ac.id/handle/123456789/46419
25. T. Widiyastuti, M. Indraji, A. Wibowo, R.H. Setyobudi. Energy Procedia. **47**:290–294(2014) doi: 10.1016/j.egypro.2014.01.227
26. D.K. Wisatya, P.R. Sarjono, N.S. Mulyani, Jurnal Kimia Sains dan Aplikasi, **13**,2:46–50(2010). [in Bahasa Indonesia] https://doi.org/10.14710/jksa.13.2.46-50
27. T. Ncube, R.L. Howard, E.K. Abotsi, E.L.J. van Rensburg, I. Ncube, Ind. Crop. Prod. **37**,1:118–123(2012). https://doi.org/10.1016/j.indcrop.2011.11.024
28. C. Joshi, P. Mathur, S.K. Khare, Bioresource Technology, **102**,7:4815–4819(2011). https://doi.org/10.1016/j.biortech.2011.01.039
29. U. Nishshanka, H. Jayasuriya, C. Chattopadhaya, P.J. Kijak, P.S. Chu, R. Reimschuessel, et al., J. Chromatogr. B, **1020**:90–95(2016). https://doi.org/10.1016/j.jchromb.2016.03.023
30. K. Ichihashi, D. Yuki, H. Kurokawa, A. Igarashi, T. Yajima, M. Fujiwara, et al., J. Am. Oil Chem. Soc. **88**:851–861(2010). https://doi.org/10.1007/s11746-010-1741-4
31. F. Bourdichon, S. Casaregola, C. Farrokh, J.C. Frisvad, M.L. Gerds, W.P. Hammes, et al., Int. J. Food Microbiol. **154**,3:87–97(2012). https://doi.org/10.1016/j.ijfoodmicro.2011.12.030
32. P. Chelule, M. Mokoena, N. Gqaleni, Current Research, Technology and Education Topics in Applied Microbiology and Microbial Biotechnology, **2**:1160–1167(2010). https://www.researchgate.net/profile/Mduduzi_Mokoena/publication/268186803_Advantages_of_traditional_lactic_acid_bacteria_fermentation_of_food_in_Africa/links/56f2a4e008aee034d8c63cb8.pdf
33. M. Papagianni, Biotechnology Advances, **25**,3:244–263(2007). https://doi.org/10.1016/j.biotechadv.2007.01.002
34. J. Langrand, C. Médernach, C. Schmitt, I. Blanc–Brisset, A.F. Villa, L. de Haro, et al., Bulletin de la Societe de Pathologie Exotique, **108**:139–143(2015). https://doi.org/10.1007/s13149-014-0412-3
35. R.G. Menezes, N.G. Rao, S.S. Karanth, A. Kamath, S. Manipady, V.V. Pillay, Indian J. Pediatr., **73**:634(2006). https://doi.org/10.1007/BF02759934
36. W. Zhang, B. Xue, M. Li, Y. Mu, Z. Chen, J. Li, et al., Toxins, **6**,11:3157–3172(2014). https://doi.org/10.3390/toxins6113157

Durian Rind Micro Composter Model: A Case of Kampung Durian, Ngrogung, Ponorogo, Indonesia

Haris Setyaningrum[1,*], *Atika* Rukminastiti Masrifah[2], *Adib* Susilo[2,3], and *Imam* Haryadi[2]

[1]Department of Agrotechnology, University of Darussalam Gontor, Siman Road Km 5, Ponorogo, East Java, Indonesia 63471.
[2]Department of Islamic Economic, University of Darussalam Gontor, Siman Road Km 5, Ponorogo, East Java, Indonesia 63471.
[3]Kulliyah of Economics and Management Sciences, International Islamic University Malaysia, Gombak, Selangor, 553100 Malaysia.

Abstract. The case of waste from durian rind in Ngrogung, Ngebel, Ponorogo, Indonesia increased significantly per year. Agglomerations and decomposition of durian rind surrounding agro park caused air pollution and bad view in park areas. The decomposition of durian rind is very important to reduce those impacts. The micro composter model in Kampung Durian consisted of three main components. The first component was drying area with cemented ground. The drying area was set to accommodate 1 t fresh durian rind per day. The materials come from surrounding areas of agro park. The second was a machine house with one ton material capacity per day. The third was a composter house consisted of composter tube and packing area. The result of the micro composter house showed a cleaning agglomeration of durian rind from Ngebel lake and surrounding areas. In addition, the model produced liquid organic fertilizer and compost. A cycle of composting was 30 d in the dry season. Amount 1 t of durian rind decomposition resulted of 100 L liquid organic fertilizers and 50 kg compost. The result of composter had a direct impact to control waste problems in Ngebel and surrounding areas.

Keywords: Agro park, compost durian, *Durio zibethinus*, micro composter house, organic fertilizer

1 Introduction

Durian is an iconic product of Ponorogo, East Java, Indonesia. It is produced in Ngebel. Production in the last 2 yr rises significantly from 1 969.2 t in 2016 to 9 871.9 t in 2017 [1]. The Ponorogo region, located in 111° 17' –111° 52'E and 07° 49' – 08° 20'S, consisted of two different levels of geographic, i.e. lowland and upland [1]. Ngebel is the new tourist destination that is integrated with an agro park named *Kampung Durian*. It is located 1.5 Km from Ngebel lake. The rind composter is potential to develop in Kampung Durian.

The higher produced durian the higher its rind. The durian composed of 75 % of rind and the remains are seed and flesh [2]. For example, the durian rind in 2017 was 6 910.33 t.

* Corresponding author: abuharist@gmail.com or haris.setyaningrum@unida.gontor.ac.id

These huge numbers of rind caused several problems in Ngebel as a tourism destination, including air pollution by chemical and biological decomposition process and health problems which coming from insects [3]. Reference [4] stated waste has the potential to pollute the environment, among others, because it produces leachate and gas. The waste can pollute three environmental media: the atmosphere, lithosphere, and hydrosphere. The local government which responsible to overcome this rising issue does not have any specific composter. Design and build a specific composter is very important.

The rind of durian is sources of lignin, cellulose, and hemicelluloses [4, 5]. Otherwise, rind of durian are sources of cellulose with high economic value such as textile fiber, biopesticide, and organic compost [6–8].

The objective of this study was to make a role model of micro composter house of rind durian which applicable to other communities.

2 Material and method

2.1 House components

The experiment was conducted from September 2019 to August 2019. The house of micro composter located in *Kampung Durian*, Ngrogung in the upland area where easy to water access and distribution. The house was separated into three different parts: i) Shelter for drying rind. ii) Machinery room. iii) Fermentation room. To make easy for moving and handling, those three parts were connected.

2.2 Durian rind sources

The rind of durian in the house composter were obtained from two places, which were *Kampung Durian* Ngrogung and Ngebel lake. *Kampung Durian* only supplied a small amount of durian rind. While, the Ngebel lake as tourism destination provided continuously even was not durian season.

2.3 Composting process

The rind was deposit in the shelter to get dried rind material. A cycle of composting process was designed for one ton of fresh durian rind. Amount 1 t of fresh rind dried for 3 d to 4 d in dry season. The dried rind was cut into small pieces used cutter machine. The small pieces of dried rind was fermented with probiotic and tap water, put in 100 L container under anaerobe condition. Fermentation process was conducted for 30 d with weekly observation especially for fresh weight material. The total container used in this research was eight containers, the materials were distributed equally.

3 Results and discussions

A cycle of composting process since drying to fermented compose was 30 d to 35 d depend on water content of durian rind. This result inline with Kinantan et.al [3]. The details of a cycle composting process of durian rind was presented in Figure 1.

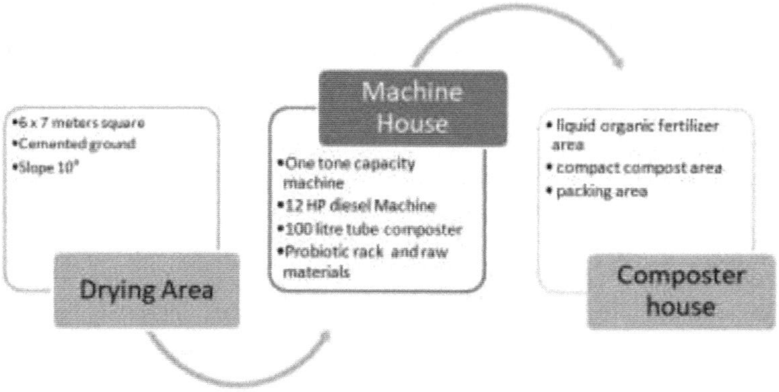

Fig 1. Scheme of processing durian rind decomposition in Micro Composter House

The last product of composting were liquid compost and solid compost which reach up to 80 % of total product. The rest of 20 % product was waste material that used for another purposed with specific processed, included biopesticides [9].

Amount 1 t material resulted 100 L of liquid organic compost (fertilizer) and ± 50 kg solid material. These current result was not categorized as a solid organic fertilizer and final waste, but as solid organic that has function as organic fiber. This type of fiber useful for reinforcement of many materials such as concrete. While as a natural source of cellulose, the durian rind has not been explored yet [5, 10]. Other product of composting was not observed yet. The other product may useful for pesticide, as stated by Kusumaningtyas et.al [7] and edible material for many purposed as stated by Amaliyah [11].

The cycle of composting in Ngrogung had several benefits such as for reducing agglomeration of durian rind as waste in Ngebel lake and surrounding areas. It was calculated when a cycle of composting in Ngebel take a ton fresh material of durian rind and can be increased depend on the availability of drying space.

Another benefit of this composting were availability of organic fertilizer for durian in such area which increased durian production through optimized and maximized of organic fertilizer. The pesticide as vital component in cultivation of durian, also be covered from composting product with some treatments [12].

4 Conclusion

The micro composter model in *Kampung Durian* Ngrogung was helpful for farmers in Ngrogung and surrounding areas of Ngebel lake, Indoensia. The benefits were reducing agglomeration of durian rind and providing organic fertilizer. Future studies of micro composter efficiency was needed to get a higher impact for durian farmers.

References

1. P. Badan Pusat Statistik. Ponorogo Dalam Angka 2018. (2018).[in Bahasa Indonesia].https://ponorogokab.bps.go.id/publication/2018/08/16/7400511f763983ffbab97edf/kabupaten-ponorogo-dalam-angka-2018.html
2. P. Damanik, Volmer; Musa, Lahuddin; Marbun. Agroekoteknologi. **2,**1:455–461 (2013). [in Bahasa Indonesia]. https://media.neliti.com/media/publications/97671-ID-pengaruh-pemberian-kompos-kulit-durian-d.pdf
3. B. Kinantan, A. Rahim Matondang, J. Hidayati. IOP Conf. Ser. Mater. Sci. Eng. **505,**1:1–9(2019). https://doi.org/10.1088/1757-899X/505/1/012106

4. K. Khoiron, A.N. Probandari, W. Setyaningsih, H.S. Kasjono, R.H. Setyobudi, O. Anne, Ann Trop & Public Health S463, **23**,3(A):60–67(2020). http://doi.org/10.36295/ASRO.2020.23316
5. P. Penjumras, R. B. A. Rahman, R. A. Talib, K. Abdan. Agric. Sci. Procedia, **2**:237–243(2014). https://doi.org/10.1016/j.aaspro.2014.11.034
6. R. Lubis, B. Wirjosentono. IOP Conf.Series Journal Physic Conf. Ser. **1232,**012017:1–9(2019) . https://doi.org/10.1088/1742-6596/1232/1/012017
7. R.D. Kusumaningtyas, H. Suyitno, R. Wulansarie. Jurnal Penerapan Teknologi dan Pembelajaran, **15,**1:38–43(2018).[in Bahasa Indonesia] https://doi.org/10.15294/rekayasa.v15i1.12576
8. S.N.B.B. Khalib, I.A.B.Zakarya, T.N.B.T. Izhar. IOP Conf. Ser. Mater. Sci. Eng. **551,**012101:1–7(2019). https://doi.org/10.1088/1757-899x/551/1/012101
9. T. Arrizqiyani, R. Hidana, M.A. Sopian, R. Nurpalah, Meri. J. Phys. Conf. Ser., **1179,**012174:1–5(2019). https://doi.org/10.1088/1742-6596/1179/1/01217
10. S. R. Masrol, M. H. I. Ibrahim, S. Adnan. Procedia Manuf., **2**:171–180(2015). http://dx.doi.org/10.1016/j.promfg.2015.07.030
11. D. M. Amaliyah, J. Ris. Ind. Has. Hutan. **6**,1:27–34(2014).[in Bahasa Indonesia]. https://doi.org/10.24111/jrihh.v6i1.1222
12. N. H. Hasem, S. F. Z. Mohamad Fuzi, F. Kormin, M. F. Abu Bakar, S. F. Sabran. IOP Conf. Ser. Earth Environ. Sci. **269,**012019 (2019). https://doi.org/10.1088/1755-1315/269/1/012019

Exploration and Effectiveness of *Trichoderma* sp. from Jember and Trenggalek, East Java, Indonesia Cacao Plantation as A Biological Control of *Phytophthora palmivora*

Henik Sukorini[1], *Feby* Wirasdenty Aigahayunindy [1], *Erfan* Dani Septia[2,*], and *Netnapis* Khewkhom[3]

[1]Department of Agrotechnology, Faculty of Agriculture and Animal Science, University of Muhammadiyah Malang, Jl. Raya Tlogomas no 246 Malang 65145, East Java, Indonesia
[2]The Center of Biotechnology Development, University of Muhammadiyah Malang, Jl. Raya Tlogomas no 246 Malang 65145, East Java, Indonesia
[3]Department of Plant Pathology, Faculty of Agriculture, Kasetsart University, 50 Thanon Ngamwongwan, Lat Yao, Chatuchak, Bangkok 10900, Thailand

Abstract. Fruit rot disease is very damaging to cacao pods, which is caused by *Phytoptora palmivora*. The attack rate of *P. palmivora* varies. In Java, losses due to this disease reduce yields by 90 %. *P. palmivora* is a soil–borne pathogen. It is currently included in the Kingdom Chromista. Control with fungicides is not successful at this time, the alternatives is biologis control with *Trichoderma* sp. This research used a Completely Randomized Design (CRD) which was arranged in factorial with two factors. The first factor was *Trichoderma* sp. the second factor was *P. palmivora*. All treatment combinations were repeated three times. *Trichoderma* sp. antagonist test to *P. palmivora* was analyzed using Analysis of Variance (ANOVA) and then further tested using a 5 % BNJ. *Trichoderma* sp. origin from Jember and Trenggalek districts, East Java, Indonesia were able to act as antagonists against *P. palmivora* with the highest inhibitory of 78 %. In comparison, the lowest inhibitory was 70 % of isolates from Jember district, East java, Indoe. Characteristics of *Trichoderma* sp. The origin of Trenggalek Regency and Jember Regency, East Java, Indonesia in inhibiting the growth of *P. palmivora* has the same species, namely *Trichoderma harzianum*.

Key words: Biologis control, environmentally friendly, fruit rot, increase cacao tree productivity, *Theobroma cacao* L.

1 Introduction

Cacao (*Theobroma cacao* L.) is one of the primary export commodities of Indonesia where over the past 5 yr, Indonesia has provided about 1 951 270 ha area of cacao plants. Unfortunately, its productivity level in the same period results in only around 655 kg ha^{-1}. In comparison, the productivity of cacao in Malaysia reaches about 1 800 kg ha^{-1}, around 800 kg ha^{-1} in Ivory coast, while in Ghana it reaches about 360 kg ha^{-1}.

Indonesia is the third–largest cacao producing country in the world after Ivory Coast and Ghana. Total cacao land of 1 400 000 ha in 2008 could contribute to the state of IDR 1 800 × 10^6 [1]. Currently, cacao development areas in Indonesia include South

* Corresponding author: erfandani@umm.ac.id

Sulawesi, West Sulawesi, Southeast Sulawesi, Central Sulawesi, West Papua, East Java, Lampung, West Sumatra, North Sumatra, and Aceh. Nationally, domestic cacao production reaches 700 000 t yr^{-1}, which is contributed the most by Sulawesi at around 70 % [9]. However, increasing productivity and quality of cacao have experienced many obstacles, including pests and disease attacks on cacao plants.

One of the critical diseases in cacao plants is cacao pod rot caused by *Phytophthora palmivora* [(E. J. Butler) E. J. Butler]. The disease can reduce cacao production by 32.6 % to 99 %. The distribution of this disease is widespread; its pathogenic diversity is also a threat to the decline in cacao production [14] Rates of *P. palmivora* attacks vary, more than 10 % occur in Peninsular Malaysia and 80 % to 90 % in Cameroon. In Java, losses due to this disease reduce yields 33 % to 50 % [4]. The threat of rapid fruit rot disease develops in the cacao planting area, so first precautions need to be taken. In areas that have been infected with the disease can be controlled with biological agents that are environmentally friendly. The use of biological agents is very useful and positive in controlling fruit rot disease caused by *P. palmivora* [10]. According to [2], good biological agents are agents originating from areas where the disease in question is found. Therefore, in this study *Trichoderma* sp. indigenous East Java, Indonesia which is used as a biofungicide against pathogens, especially against *P. Palmivora* by *in vitro* testing.

Up to now, research used *Trichoderma* from the collection but not from the same location as the pathogen, so there is a possibility that there is a mismatch of environmental conditions that will affect the antagonistic ability of *Trichoderma* sp. itself in controlling *P. palmivora*. [12] Used *Trichoderm*a sp. isolates from the collection; then it could suppress pathogens. [4] In vitro test of the three types of fungus *Trichoderma* were found on cacao plantations that *Trichoderma* sp., *Trichoderma harzianum* Rifai, and *Trichoderma viride* Pers., showed that *T. viride* was the most capable of inhibiting the growth of the fungus *P. palmivora* with the highest percentage of 71.95 %. Four antagonists [*T. asperellum* (SF04), *T. virens* (255C1), *T. harzianum* (THP) and *T. longibrachiatum* (4088)] were tested, All Trichoderma significantly ($p \leq 0.05$) reduced the incidence and severity of disease. The 4088 (*T. longibrachiatum*) isolate was the best controller agent of *P. palmivora* in postharvest [7]. The purpose of this study was to determine the effectiveness of *Trichoderma* sp. isolates from Jember and Trenggalek districts, East Java, Indonesia against causes of cacao pod rot (*P. palmivora*).

2 Material and methods

This study began in December 2018 until March 2019. This study used a Completely Randomized Design (CRD). CRD was arranged factorial with two factors. The first factor was the type of *Trichoderma* sp., the second factor was the type of *P. palmivora*, all treatment combinations were repeated three times. The treatments given were:
T: *Trichoderma* sp. (TR1: *Trichoderma* sp. origin of Suruh cacao plantation, Trenggalek, East Java, Indonesia; TR2: *Trichoderma* sp. origin of Karangan cacao plantation Trenggalek, East Java, Indonesia; JB1: *Trichoderma* sp. origin of PUSLITKAKAO cacao plantation, Jember, East Java, Indonesia; JB2: *Trichoderma* sp. origin of Banjarsari cacao plantation, Jember, East Java, Indonesia).
P: *P. palmivora* (P1: *P. palmivora* from Suruh cacao plantation, Trenggalek, East Java, Indonesia; P2: *P. palmivora* from Karangan cacao plantation, Trenggalek, East Java, Indonesia; P3: *P. palmivora* from Puslitkoka (*Pusat Penelitian Kopi dan Kakao Indonesia* – Indonesian Coffee and Cacao Research Institute) cacao plantation, Jember, East Java, Indonesia; P4: *P. palmivora* from Banjarsari cacao plantation, Jember, East Java, Indonesia).

2.1 The exploration of the fungus *Trichoderma* sp.

Fungus *Trichoderma* sp. explored from cacao plantations in the Trenggalek and Jember districts, East Java, Indonesia. The point of taking soil was determined by the criteria for healthy cacao growing conditions (without symptoms of disease affecting the cacao plant). After the cacao plant was determined, it then digs up the soil as deep as 15 cm [3]. The isolation method of *Trichoderma* sp. used the dilution plate method.

2.2 Exploration and Isolation of *P. palmivora* pathogen

P. palmivora was isolated from diseased cacao with blackish brown spots. The area of the pathogen exploration same as *Trichoderma* sp. exploration. The surface of the cacao fruit was sterilized with tissue with 70 % alcohol. Approximately 0.5 cm of flesh was taken from the tissue, then planted into the Water Agar media and incubated for 4 d at room temperature. After 4 d, the mycelia was purified on V8 agar medium (200 mL of V8 juice, 800 mL of distilled water, 1 g of $CaCO_3$ and 20 g of Agar)

2.3 Identify fungi

The fungus identification was carried out after the fungus grew the colony, the fungus *Trichoderma* sp. and *P. palmivora* were identified using a microscope so that the type of *Trichoderma* and *P. palmivora* was known by means of a colony in the observation of the microscope observed based on a journal or book so that the type of *Trichoderma* sp. as well as the identification of pathogens *P. palmivora*.

2.4 Koch postulate test

Healthy cacao pods were sterilized with sterile water, 0.5 % Clorox, and then washed with sterile water. The surface of the cacao fruit is then injured using a needle and then inoculated with *P. palmivora*. Covered using a cotton cloth and taped. The disease incidence was determined based on the symptoms that appear. The results of the postula koch test were then isolated using V8 media and compared with origin *P. palmivora* isolate from exploration. If they showed the same characteristics, the postula koch test can prove the symptoms *P. palmivora* attack.

2.5 *Trichoderma* sp. against pathogenic *P. Palmivora*

Trichoderma sp. against *P. palmivora* cacao pod rot using a dual method on PDA media. Pieces of *P. Palmivora* isolates aged sevendai was placed with a distance of 2 cm from the edge of the petri dish then on the opposite side, was placed isolates of *Trichoderma* sp. with a range of 2 cm from the edge of the petri dish. The observation of the growth pathogen of *P. Palmivora* was carried out from one day to 7 d after the test.

The growth radius of *P. palmivora* was observed by measuring the growth of colonies of each fungus. Inhibition presentation using the Formula (1):

$$Z = (R1 - R2) / R1 \times 100\ \% \quad (1)$$

where Z: Percentage of inhibition, R1: Radius of *P. palmivora* without *Trichoderma* sp. (Control), R2: *P. palmivora* radius with *Trichoderma* sp.

The data were obtained then analyzed using the Analysis of Variance (ANOVA), if there were significantly different effects, further tests would be carried out using a 5 % Tukeys.

3 Result and discussions

3.1 Exploration *Trichoderma* sp. results from plantation in Trenggalek Regency and Jember Regency, East Java, Indonesia

Based on the exploration results of *Trichoderma* sp. from the cacao plantation area in Suruh District and Karangan District, Trenggalek Regency, East Java, Indonesia. Likewise, from the cacao plantations in PUSLITKAKAO and Banjaarsari District, Jember Regency, East Java, Indonesia each site was *Trichoderma* sp.

Microscopic appearance of *Trichoderma* sp. isolates, namely TR1, TR2 (Trenggalek), and JB1, JB2 (Jember) were green hyphae, short phyalid, greenish conidia, oval–shaped and there was also some conidia formed in clusters of light green on the surface of the conidiophore. Phyalid had a length of ± 11.1 μ and conidiophore branches of ± 13.4 μ. There were many upright, branched conidiophores, vertically arranged branches, short thick phyalids, colony on potato Dextrose Agar media was dark green and round. The diameter of the colony reaches more than 9 cm within 5 d. The characters from several isolates origin namely TR1, TR2, JB1, and JB2 showed the characteristics of *T. harzianum* [15, 6]

T. harzianum is generally found in hot climates, the optimum temperature for growth of *T. harzianum* is 15 °C to 30 °C, but the optimum growth is at 30 °C and for maximum temperatures of 30 °C to 36 °C. Normal growth of *T. harzianum* is at pH 3.7 to 4.7 [13]

The benefits of *T. harzianum* was most often used in biological control because it has several comparative advantages compared to other organisms. A wide range of environments, mycoparasitic and able to compete in gaining space and producing antibiotics and enzymes that harm to the pathogens [13].

3.2 *P. palmivora* exploration

P. palmivora was isolated from diseased cacao with blackish–brown spots. Fruit exploration was carried out in cacao plantation areas where in the same area with *Trichoderma* sp. All isolates that were obtained characterized as *P. palmivora*. The characteristic of *P. palmivora* isolates a a rounded edge colony. There were four forms of ovoid sporangia, limoniform, obturbinate, and obpyriform. Sporangium length 40 μm to 62 μm and width 28 μm to 43 μm, have papillae, short pedicels, and simple simpodiabranching models. This criterion was as described by Stamps [13].

3.2.1 Koch's postulates for identification test

Tests were carried out on healthy cacao pods, then given *P. palmivora* pathogens (exploration results) and observed the symptoms of the pathogen as presented in Figure 1.

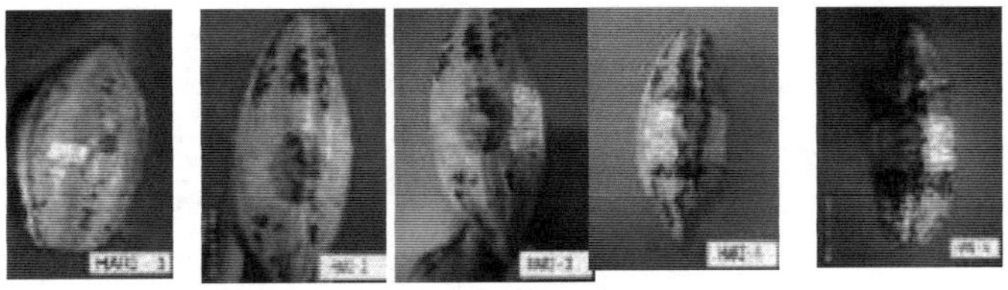

Fig. 1. Postulate test of fungus identification on healthy cacao

Based on Figure 1, *P. palmivora* disease attack on healthy cacao fruits showed the same symptoms as cacao pods affected by the disease. Koch postulates was used to prove the isolates obtained the causative agents of the observed disease symptoms.

Koch's postulatewas also used as a method to identify fungi that cause signs of cacao pod rot (*P. palmivora*).

3.3 Antagonistic test of *Trichoderma* sp. against *P. palmivora*

An in vitro antagonist test was a way to evaluate the ability of antagonists (biological control agents) in a narrower scope and controlled environmental conditions (in vitro). The aim was to determine the potential or effectiveness of biological control agents in inhibiting the growth and development of pathogens [3]. The results of the antagonist test were presented in Table 1.

Based on the Tukey's test, the type of *Trichoderma* sp. TR1P1 showed the highest antagonicity (78 %) on 7 d observations against *P palmivora* on PDA medium, It was significantly different from *Trichoderma* sp. JB2P3 (70 %) on 7 d observations. The origin of *T. harzianum* isolate did not significantly influence the inhibition of *P. palmivora* from two different regions. The result of research was not in line with the results of the research of [2], that good controlling agents to control the disease were those from the same area with pathogen isolates. Antagonistic of all *T. harzianum* was no significant difference. It was likely due to the origin of the isolates being the same from the Java region, which has almost the same ambient temperature and humidity.

The difference in antagonism of *Trichoderma* sp. against *P. palmivora*, indicates the diversity of *Trichoderma* sp. who have been tested for their abilities as antagonists. This occurs a significant difference in the sporulation ability of each treatment type *Trichoderma* sp. following the results of Darmono's research [5], six isolates of *Trichoderma* sp. from the tested nature shows a difference in the ability to suppress *P. palmivora*.

According to [12] in general the mechanism of *Trichoderma* spp. in suppressing pathogens, namely as a micoparasitic and aggressive competitor. Initially, hypha *Trichoderma* spp. grow elongated, then convolve and penetrate the host fungal hyphae so that the host hyphae are vacuated, lysis, and finally destroyed. According to Harman [8] *Trichoderma* spp. penetrate the host cell wall with the help of cell wall degrading enzymes, namely chitinase, glucanase, and protease, then use the contents of the host hyphae as a food source. *Trichoderma* spp. also produce antibiotics such as gliotoxins and viridian when they penetrate cell wall.

Table 1. Antagonist test *Trichoderma* sp. to *P. palmivora*

Treatment	Disease intensity % in day after inoculation (d)						
	1	2	3	4	5	6	7
TR1P1	55 a	62 a	65 abc	66 ab	69 abc	72 ab	78 b
TR1P2	44 a	53 a	68 c	69 b	72 c	76 b	76 ab
TR1P3	59 b	62 a	65 abc	65 ab	70 bc	77 b	77 ab
TR1P4	58 b	58 a	66 abc	66 ab	68 abc	70ab	75 ab
TR2P1	48 a	57 a	65 abc	67 ab	69 abc	72 ab	75 ab
TR2P2	52 a	59 a	67 bc	67 ab	70 bc	74 ab	76 ab
TR2P3	47 a	56 a	65 abc	65 ab	68 abc	73 ab	75 ab
TR2P4	46 a	55 a	65 abc	66 ab	66 abc	72 ab	72 ab
JB1P1	52 a	62 a	66 abc	68 ab	70 bc	74 ab	77 ab
JB1P2	48 a	56 a	63 abc	66 ab	66 abc	75 ab	75 ab
JB1P3	45 a	55 a	62 abc	65 ab	65 abc	72 ab	74 ab
JB1P4	42 a	58 a	60 abc	63 ab	64 abc	74 ab	74 ab
JB2P1	44 a	55 a	58 a	62 a	64 abc	73 ab	75 ab

(Continued on next page)

Table 1. Continued

Treatment	Disease intensity % in day after inoculation (d)						
	1	2	3	4	5	6	7
JB2P2	55 a	56 a	58 a	62 a	62 ab	70 ab	72 ab
JB2P3	53 a	55 a	63 abc	63 a	63 ab	69 a	70 a
JB2P4	55 a	56 a	65 abc	65 ab	65 a	69 a	74 ab
BNJ 5 %	6.021	6.35	6.98	7.046	7.24	7.62	7.93
	**	**	**	**	**	**	**

Note: Values followed by the same letter in the same column in the Tukey's test at a level of 5 % show very significant different

TR1: Isolate *Trichoderma* sp. Trenggalek Regency in Suruh District, East Java, Indonesia; TR2: Isolate *Trichoderma* sp. Trenggalek Regency in Karangan District, East Java, Indonesia; JB1: Isolate *Trichoderma* sp. Jember Regency in PUSLITKAKAO Jember, East Java, Indonesia; JB2: *Trichoderma* sp. Jember Regency in the District of Banjarsari, East Java, Indonesia; P1: *P. palmivora* from the Suruh cacao plantation in Trenggalek Regency, East Java, Indonesia; P2: *P. palmivora* from Karangan cacao plantation in Trenggalek Regency, East Java, Indonesia; P3: *P. palmivora* from PUSLITKAKAO cacao plantation in Jember Regency, East Java, Indonesia; P4: *P. palmivora* from Banjarsari cacao plantation in Jember Regency, East Java, Indonesia.

Control mechanisms with biological agents against plant pathogenic fungi are generally divided into three types, firstly competition for growth sites and nutrition, secondly antibiosis, and thirdly parasitism [8]. The mechanism of parasitism is an interesting phenomenon that plays an essential role in the process of biological control. *Trichoderma* spp. usually use this mechanism with other, namely competition and antibiosis.

4 Conclusions

Antagonistic test results, isolates of *Trichoderma* sp. very effective in inhibiting the growth of *P. palmivora*. Isolates from Trenggalek district, Suruh sub–district (TR1P1) were the most effective isolates because they can inhibit the growth of *P. palmivora* by 78 %, and a significant difference from *Trichoderma* sp. JB2P3 of 70 %

References

1. B. Scanu, B.T. Linaldeddu, A. Deidda, T. Jung. PLoS One. **10**,12:1–24(2015). https://doi.org/10.1371/journal.pone.0143234
2. Alfizar, Marlina, F. Susanti. Floratek. **8**,1:45–51(2013). [in Bahasa Indonesia]. http://www.jurnal.unsyiah.ac.id/index.php/floratek/article/download/860/799
3. U.S. Hastuti, I. Rahmawati. J. Pure Appl. Chem. Res. **5**,3:178–181(2016). https://doi.org/10.21776/ub.jpacr.2016.005.03.260
4. R. Sriwati, T. Chamzurni, L. Soesanto, M. Munazhirah. Agrivita. **41**,1:175–182(2019). https://doi.org/10.17503/agrivita.v41i1.2146
5. T.W. Darmono. Menara Perkebunan. **62**,2:2530(1994). [in Bahasa Indonesia]. https://scholar.google.co.id/scholar?hl=id&as_sdt=0%2C5&q=Kemampuan+beberapaisolate+Trichoderma+spp.+dalam+menekan+inokulum+Phytophthora+sp.+di+dalam+jaringan+buah+kakao&btnG=
6. D.L. Taylor, R.L. Sinsabaugh. *The Soil Fungi: Occurrence, Phylogeny, and Ecology*, Elsevier, USA (2015). p.77–109. https://doi.org/10.1016/b978-0-12-415955-6.00004-9
7. T.A.S. de Oliveira, L.E.B. Blum, E.A.A. Duarte, E.D.M.N. Luz. Biosci. J. **34**,6: 1513–1521(2018). https://doi.org/10.14393/bj-v34n6a2018-39548
8. G.E. Harman. The American Phytopathological Society. **96**,2:190–194(2006). https://doi.org/10.1094/PHYTO-96-0190

9. A.G. Maulana, A. Permana. *Atasi impor kakao, pemerintah didesak remajakan tanaman*. [in Bahasa Indonesia]. [Online] from http://www.m.bisnis.com/industri/read/20150313/99/411481/atasi-imporpemerintah-didesak-remajakan-tanaman.com (2016) [Acessed on 10 August 2020].
10. I.Y. Opoku, A.Y. Akrofi, A.A. Appiah. Eur. J. Plant Pathol. **117**: 167–175(2007). https://doi.org/10.1007/s10658-006-9082-8
11. A. Singh, M. Shahid, M. Srivastava, S. Pandey, A. Sharma, V. Kumar. Virol. Mycol. **3**,1:1–7(2014). https://www.researchgate.net/publication/270564856_Optimal_Physical_Parameters_for_Growth_of_Trichoderma_species_at_Varying_pH_Temperature_and_Agitation
12. A.O. Nwankiti, VI. Gwa. Trends Tech. Sci. Res. **1**,1(2018). https://scholar.google.co.id/scholar?hl=id&as_sdt=0%2C5&q=Evaluation+of+Antagonistic+Effect+of+Trichoderma+Harzianum+against+Fusarium+oxysporum+causal+Agent+of+White+Yam+%28Dioscorearotundata+poir%29+Tuber+Rot&btnG=
13. H.H. Ho. J. Bacteriol. Mycol. Open Access. **6**,1:40–45(2018). https://doi.org/10.15406/jbmoa.2018.06.00174
14. K. Nidhina, K.M. Sharadraj, V.H. Prathibha, V. Hegde, K.P. Gangaraj. Vegetos. **29**,2:19–26(2016). https://scholar.google.co.id/scholar?hl=id&as_sdt=0%2C5&q=Antagonistic+activity+of+Trichoderma+Spp.+to+Phytophthora+infecting+plantation+crops+and+its+beneficial+effect+on+germination+and+plant+growth+promotion&btnG=
15. T. Watanabe. *Pictorial Atlas of Soil and Seed Fungi*, 3rd ed. Boca Raton: CRC Press. (2010). p. 426. https://doi.org/10.1201/ebk1439804193

Variability of *Fusarium oxysporum* f. sp. *lycopersici* from different altitudes in East Java, Indonesia

Henik Sukorini[1], *Erfan* Dani Septia[2*], and *Netnapis* Khewkhom[3]

[1]Department of Agrotechnology, Faculty of Agriculture and Animal Science, University of Muhammadiyah Malang, Jl. Raya Tlogomas no 246 Malang 65145, East Java, Indonesia
[2]The Center of Biotechnology Development, University of Muhammadiyah Malang
[3]Department of Plant Pathology, Faculty of Agriculture, Kasetsart University, 50 Thanon Ngamwongwan, Lat Yao, Chatuchak, Bangkok 10900, Thailand

Abstract. Fusarium wilt disease attacks the tomato plants, and Solanaceae plants with losses can reach 90 %. Thus, this disease was tough to control. The morphology and pathogenicity variation caused failure disease control. Both variations could be affected by the altitude of the planting site. Therefore, it also required special handling of management as well. The specific objectives of this study to investigate the Fusarium isolate from different altitude response of physical properties (acid-base pH, low-high temperature, and fungicides of other active ingredients) and growth rate of Fusarium. All Fusarium sp. were isolated from the leaves, soil, roots, and stem of infected tomato plants. Before a series of tests were carried out, all of the isolates were tested for pathogenicity, then tested for the influence of abiotic factors. This study obtained four isolates from high altitude, four isolates from medium altitude, and five lowland isolates. In morphological observations, there were striking differences in colony color and growth rate among isolates on several media and pathogenicity tests. Each isolate had different pathogenicity, although from the same location but other parts of the plant. Knowing the characteristic morphology and Fusarium sp. response of the three areas can be used as references in controlling.

Keywords: Isolates, pathogenicity variation, solanaceae, wilt disease.

1 Introduction

Fusarium oxysporum f. sp. *lycopersici* (Sacc.) W.C. Snyder and H.N. Hans. is a cosmopolitan species that can be found in all soil types in the world [1]. *Fusarium* is an economically very important species because it causes wilt and root rot disease in some plants [2]. Distinguishing Fusarium pathogens is not easy [3]. The identification of Fusarium is based primarily on differences in the shape and size of macro and microconidia, presence or absence of chlamydospores, and appearance of colonies pigmentation, growth rates on agar mediums [4]. However, some forma species and races

* Corresponding author: erfandani@umm.ac.id

of *F. oxysporum* was tough to distinguish in this way [5]. A report reveals that there are cultural and morphological variations in *Fusarium oxysporum* f. sp. *psidii* isolate and *F. solani* when tested in vitro on growth and physiological parameters [6].

Some *Fusarium* species have been reported to grow and produce spores in pH 5.0 to pH 6.0 [7]. Arcía et al. [8] reported that *Fusarium* spp., *F. oxysporum* and *F. solani* require pH 4.5 and 6.0, while *F. graminearum* and *F. equiseti* at pH 3.5 and pH 6.5. That pH 5.5 to pH 7.0 was the optimum pH for growth and sporulation of *F. solani* [9]. Besides, Hsuan et al. [10] tested *Gibberella fujikuroi* at various temperatures, suitable temperatures for growth and sporulation was a temperature of 30 °C followed by 25 °C, 20 °C, and 35 °C. Tyagi 2014 [11] studied the effect of temperature on growth and sporulation of *F. oxysporum* f. sp. *lines*.

This research was a follow-up from previous research controlling *Fusarium* sp., a wilt disease on tomato plants. From the previous research result, *Fusarium oxysporum* isolates of the same low altitudes site plant, but different plant parts had the same character of shape and size of spore, but growth characteristics were different, as well as pathogenicity and virulence. So that the response to bio-pesticide from bio-diesel waste of *Jatropha* seeds also different levels of success. Other studies have also revealed that in the presence of a specific form of the *Fusarium* sp., it causes difficulties in controlling it [12]. Amandeep 2015 [13] studied the variability of isolate *F. oxysporum* f. sp. *ciceri*. As a result, it is difficult to control the disease [14]. So, it needs to be revealed more about the difference of isolated characters from many altitudes.

2 Material and methods

2.1 Pathogen isolates

Fusarium sp. isolates obtained from the isolation of the three regions that have different altitudes, at low (140 m to 200 m above sea level, m asl), medium (678 m asl to 700 m asl), and high altitudes (1 100 m asl to 1 200 m asl). Isolation was done by culturing 0.5 cm of the infected plant part (root, stem, leaves) and 1 g of soil from around the root on Potato Dextrose Agar media, then purifying for testing [15].

2.2 Pathogenicity test

The pathogenicity test was done by deep root method. Plants that have been sown for 30 d were removed and cleaned from the soil. The suspense was made from 7 d old isolate. The population density of the suspense was 1×10^8. The root ends were cut as many as ten pieces, then depth in each inoculum's suspense for 20 min. Then, the inoculated plants were transferred into polybags containing sterile soil and sand at a ratio of 1:1, then incubated for 2 wk to 4 wk at 27 °C to 30 °C. Isolates that could show disease symptoms were considered to have the ability to cause disease. To ensure the result of the pathogenicity test, it was done twice [16].

2.3 Characterization of growth

Fusarium diameter and sporulation were observed by culturing the isolate on Potato Dextrose agar (PDA) medium, Czapek's Dox Agar medium (CDA), Cooke's Rose Bengal Agar medium (CRBA) and Oatmeal Agar medium (OA) [17]. Each Petri dish was filled with 20 mL of media then 5 mm diameter of *Fusarium* isolate was placed in the middle of the Petri dish, then incubated at a temperature of 28 ° C ± 2° C for 9 d. Measurement of growth diameter and sporulation was done every 3 d after incubation. Total spore was calculated by cutting 5 mm the edge of isolate culture, then added 10 mL sterile water, shaken, then observed with haemocytometer. Moreover, growth characterizations such as colony thickness and colony color were observed. The growth rate of isolate was also studied by using liquid medium Potato Dextrose Broth (PDB) medium, Czapek's Dox Broth medium (CDB), Cooke's Rose Bengal Broth medium (CRBB), and Oatmeal Broth medium (OB). Isolates were grown in 100 mL of liquid medium and stored for seven days at a temperature of 28 ° C ± 2 °C. The mycelium was then filtered with Whatman filter paper number 42, dried at 60 °C., then measured.

2.4 In vitro isolates resistance test to temperature, pH, and fungicide

The influence of temperature on growth and sporulation of *Fusarium* sp isolates was observed by growing 5 mm colonies diameter in the middle of Potato Dextrose agar (PDA) medium, then grew at (10, 16, 22, 28, 34, and 40) ° C ± 1 °C. Growth diameter and sporulation were observed at 7 d after inoculation [9]. The growth rate on pH (5, 6, 7 and 8) was observed after 7 d cultured on PDB, but resistance to Fungicides was observed by measuring the colony diameter on PDA at 7 d after inoculation.

3 Result and discussions

There were thirteen isolates from three different altitudes, four from high, four from a medium, and five isolates from low altitudes. In the lowlands, two isolates from the leaves were obtained, which looked morphologically different.

3.1 Pathogenicity test

In pathogenicity test, results indicated that most of *F. oxysporum* were pathogenic. These isolates caused wilt symptoms on at least one plant on the tomato plant, stem cross-section also done to ensure the pathogenicity. The result of pathogenicity was presented in Table 1.

Table 1. Pathogenicity test of *Fusarium* sp isolates from three different altitudes

Altitude	Sample Source	1 st test	2nd test
High	Soil	+	-
	Root	+	-
	Stem	+	+
	Leaves	-	+
Medium	Soil	+	+
	Root	-	+
	Stem	+	+
	Leaves	+	+
Low	Soil	+	+
	Root	+	+
	Stem	+	+
	Leaves 1	+	+
	Leaves 2	+	+

+ Appeared symptom, No symptom appeared

Pathogenicity tests was done twice. The distance between the first and second tests is 2 wk in the greenhouse. Several isolates were not consistent to show wilt even symptoms on leaves. Isolates that always appeared symptom were isolates from the lowland.

3.2 Characterization of growth

Four solid media (OA, CRBA, PDA, and CDA) were used for cultural studies. The colony color varies from white to purplish white, but all of the *Fusarium* isolates were white on CDA medium, and the thickness of the colony also varies greatly (Table 2).

Table 2. The thickness and colony color of isolates from three different altitudes on OA, CRBA, PDA, and CDA medium

		Thickness				Colony color			
Altitude	Sample Source	OA	CRBA	PDA	CDA	OA	CRBA	PDA	CDA
High	Soil	++	+	++	++	White purplish	white purplish	white	white
	Root	+	+	++	++	white	white	white	white
	Stem	+ ++	+	+++	+	white	white	white	white
	Leaves	+	+	+++	+++	white	white	white	white
Medium	Soil	+	+	++	++	White purplish	White purplish	White purplish	white
	Root	+	+	++	++	white	white	white	white
	Stem	+ ++	+	+	+	white	white	white	white
	Leaves	+	+++	+++	+++	white	white purplish	white purplish	white
Low	Soil	++	+	++	++	purple	white	white	white
	Root	++ ++	++	++	++	white	white	white	white
	Stem	+ ++	++	+	+++	white	white	white	white
	Leaves 1	+ ++	+++ +++	+++ +++	++ +++	white	white	white	white
	Leaves 2	+	+	+	+	white	white	white	white

Knowing another morphological character, colony diameter, sporulation, and growth rate of each isolate was culture on four different media. All isolates had the same growth in all four media, except for isolates from high and low altitudes, respectively, isolates from leaves from the highlands and isolates from stems from lowlands (Figure 1). While for sporulation, isolates originating from the lowlands, namely isolates from roots and soil, can produce spores that were much stronger than other isolates. In isolate No. 1 originating from leaves in the lowlands, sporulation in CRBA media was much lower than that of other media (Figure 3). The growth rate, observed by growing *Fusarium* sp. grown on four types of liquid media, then filtered, dried, then weighed. The results of the weighing showed that the isolates originating from the plain medium could grow faster than the isolates from the high and lowlands (Figure 3).

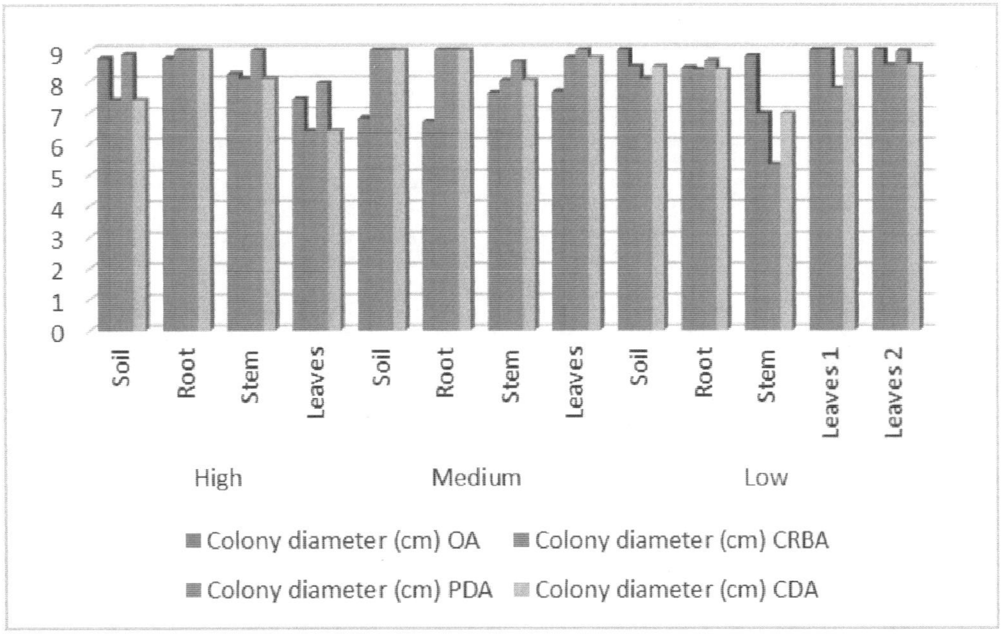

Fig. 1. Colony diameter (cm) of *Fusarium* sp. isolates on OA, CRBA, PDA, and CDA medium, 7 d after inoculation

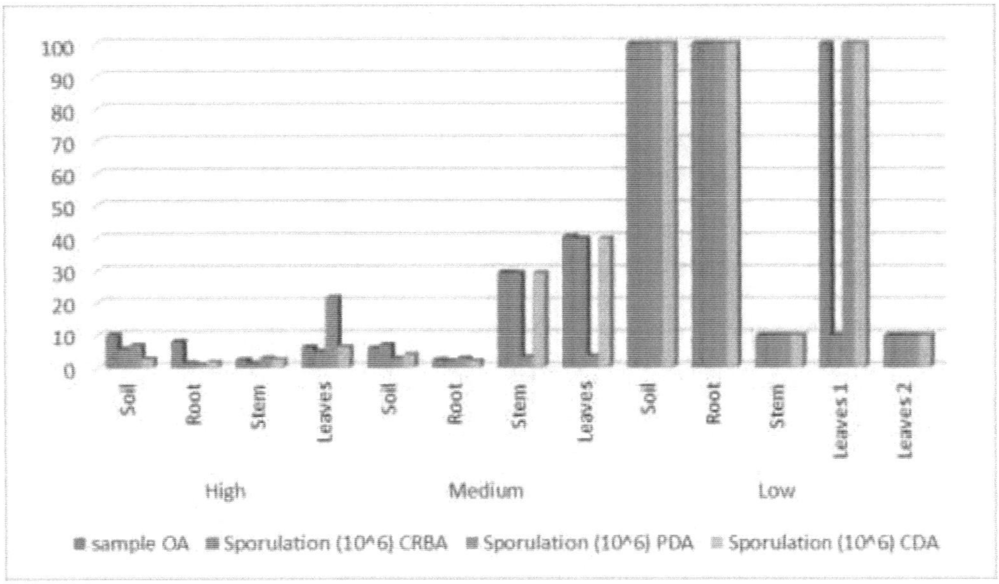

Fig. 2. Sporulation (x10^6) of *Fusarium* sp. isolates on OA, CRBA, PDA, and CDA medium, 9 d after inoculation

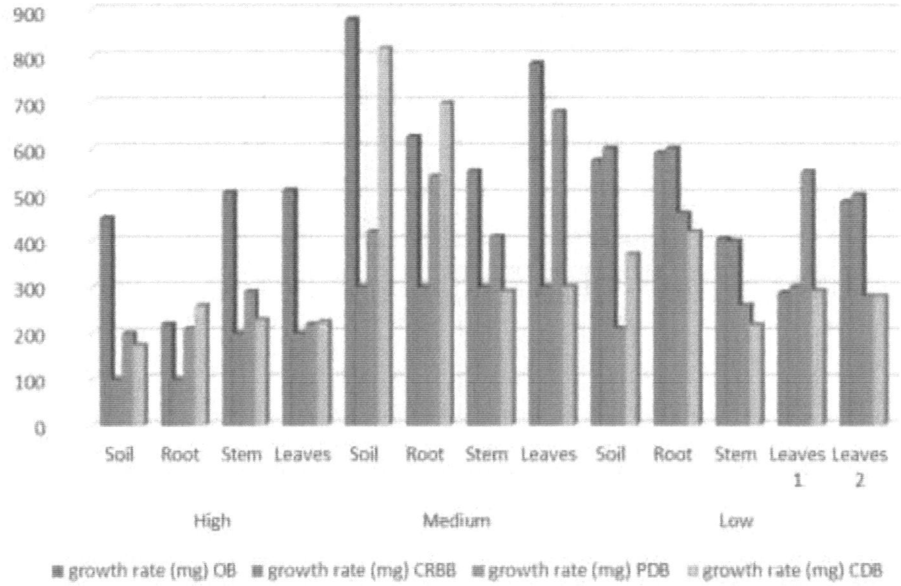

Fig. 3. Growth rate (mg) of *Fusarium* sp. isolates on OB, CRBB, PDB, and CDB liquid media, 7 d after inoculation

3.3 Effect of Growth temperature, pH, fungicide on *Fusarium* sp growth

The growth pattern of all isolates from three heights was the same, all isolates continued to increase to a temperature of 22 °C, but then the growth of all isolates decreased to 34 °C, except for isolates from the lowlands, which were leaves (no 2), roots, and stems (Figure 4). The sporulation trend was exciting because only a few isolates increased the number of spores as the temperature rises. Isolates from high land were isolated from roots and soil respectively, isolates from temperate plains which originated from roots and soil, and isolates from lowland which isolates from the soil, starting at 16 °C the number of spores produced increased continuously to 34 °C, whereas other isolates despite an increase but not too significant (Figure 5).

Testing the sensitivity of isolates to pH was carried out at pH 5 to pH.7. The test results showed that all isolates grew optimum at pH seven and would decrease if the pH of the media was raised to pH 8 (Figure 6). this result was slightly different from the last research [1]. Most of the *Fusarium oxysporum* isolates favored the pH of 7.5, slightly alkaline conditions. Mezzomo et al. [9] found that mycelia of *F. oxysporum, which* grew within the range of pH 2 to pH 12, pH 0 to pH 6, was the most suitable for the growth of all *Fusarium* species, while a highly acidic medium was unsuitable for the sporulation of all *Fusarium* species.

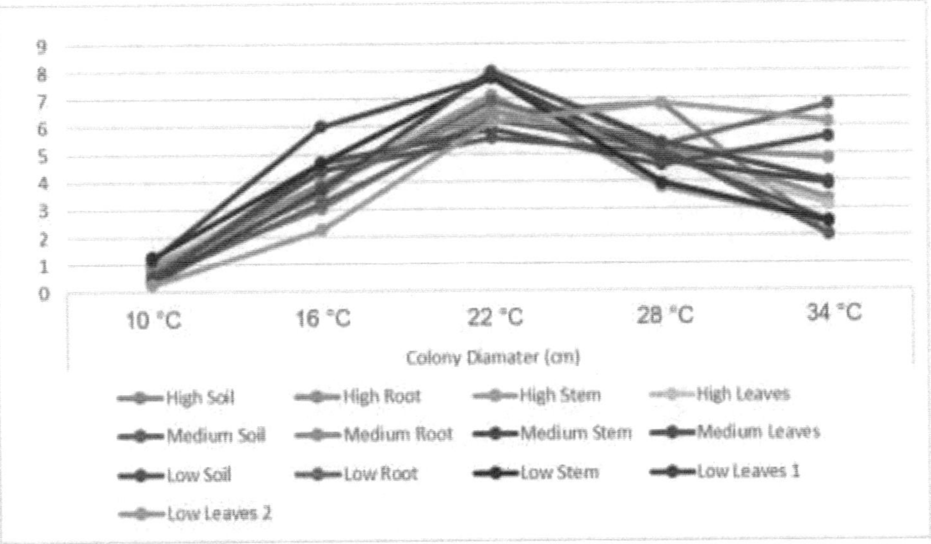

Fig. 4. Colony diameter of *Fusarium* sp. from three different altitudes that have grown on (10, 16, 22, 28, and 34) °C

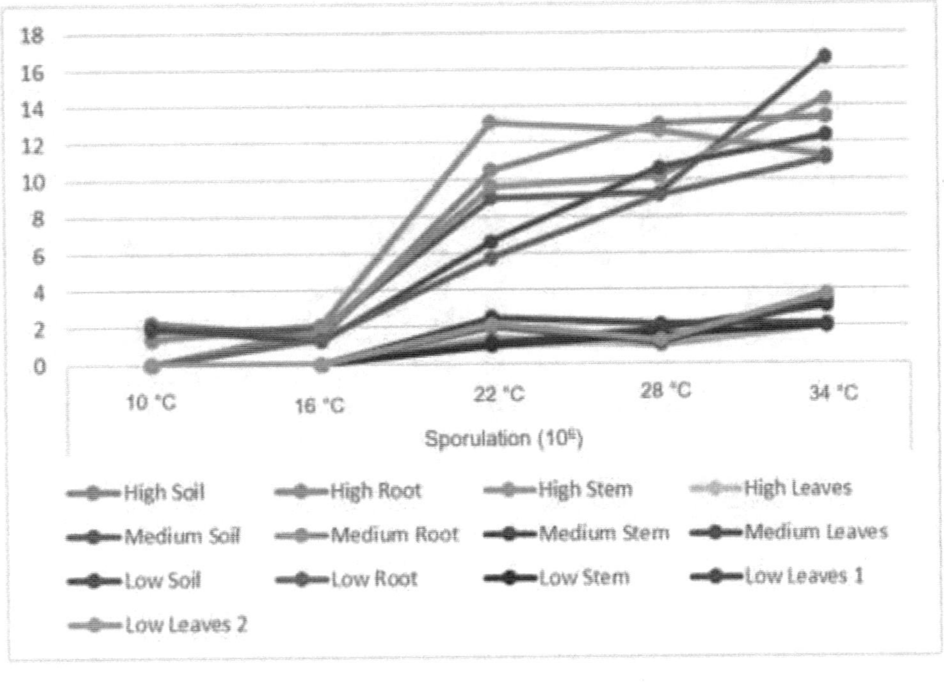

Fig. 5. Sporulation of *Fusarium* sp. from three different altitudes that grown on (10, 16, 22, 28, and 34) °C

 https://doi.org/10.1051/e3sconf/202122600023

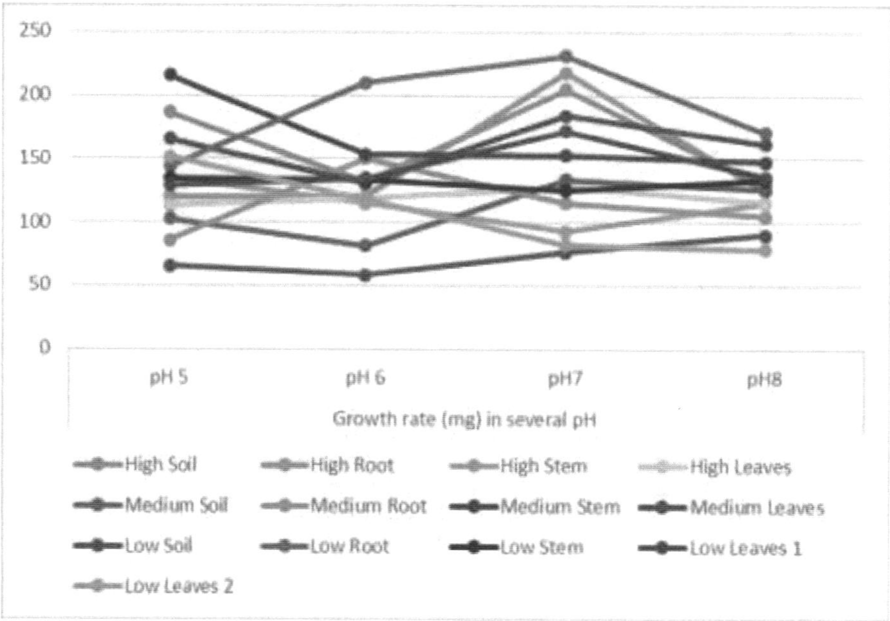

Fig. 6. Colony diameter of *Fusarium* sp. from three different altitudes that grown on (10, 16, 22, 28, and 34) °C

Testing the sensitivity of the active ingredients of fungicides is done by adding fungicides commonly used by farmers at recommended dosages. The results showed that the fungicide, which had an active ingredient of Mancozeb 64 % + Metalaxyl 8 %, inhibited the growth of *Fusarium* sp. by 60 % (Figure 7).

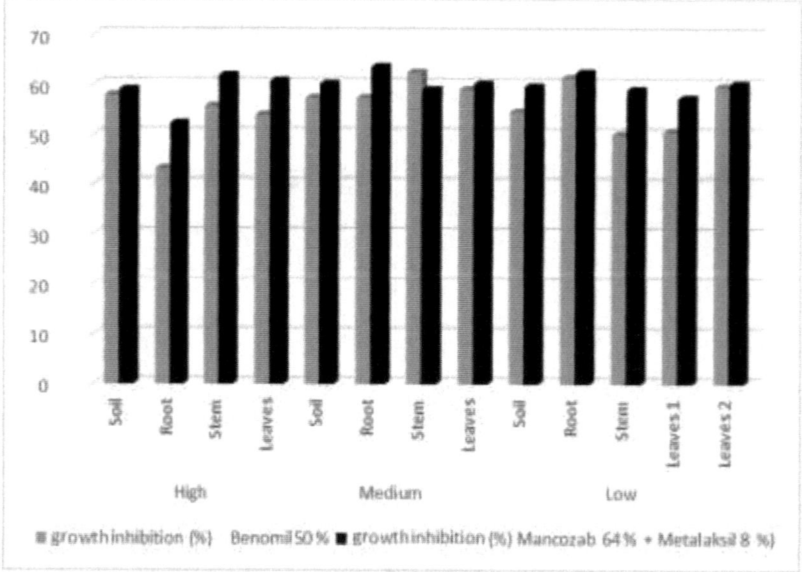

Fig. 7. Inhibition percentage of Fungicide to Fusarium growth

4 Conclusions

The character of the isolates varied in several media, i.e., the character of growth and color of the isolate. The optimum temperature for growth and sporulation is at 22 °C, while the optimum pH was 7. Fungicide with Mancozeb 64 % + Metalaxyl 8 % only could inhibit not more than 60 %.

References

1. M. Joshi, R. Srivastava, A.K. Sharma, A. Prakash, J. Appl. Nat. Sci., **5**:108–117(2013). https://doi.org/10.31018/jans.v5i1.290
2. S. Srivastava, C. Kadooka, J.Y. Uchida, Microbiol. Res., **207**,2017:188–195(2018). https://doi.org/10.1016/j.micres.2017.12.002
3. S.N. Rampersad, Pathogens, **9**,5:340(2020). https://doi.org/10.3390/pathogens9050340
4. A. Robles-Carrión, M. Leiva-Mora, L. Herrera-Isla, A. Sánchez-Rodríguez, R. Torres-Gutiérrez, Rev. Protección Veg., **31**,3:184–193(2016). http://scielo.sld.cu/scielo.php?script=sci_abstract&pid=S1010-27522016000300005
5. V. Edel-Hermann, C. Lecomte, Phytopathology, **109**,4:512–530(2019). https://doi.org/10.1094/PHYTO-08-18-0320-RVW
6. G. Sharma, R. R. Pandey, J. Yeast Fungal Res., **1**,8:157–164(2010). https://doi.org/10.5897/JYFR.9000029
7. M.Z. Hussain, M. Rahman, M.N. Islam, M. Latif, M. Bashar, Bangladesh J. Bot, **41**,1:49–54(2012). https://doi.org/10.3329/bjb.v41i1.11082
8. J. Adame-García, R. Rodríguez-Guerra, L.G. Iglesias-Andreu, J.M. Ramos-Prado, M. Luna-Rodríguez, Bot. Sci., **93**,3:669–678(2015). https://doi.org/10.17129/botsci.142
9. R. Mezzomo, J.M. Rolim, T. Poletto, M.B. De Oliveira, M. Lazarotto, M.F.B. Muniz, Sci. Agrar., 19,1:14–19(2018). https://doi.org/10.5380/rsa.v19i1.55844 d
10. H. M. Hsuan, B. Salleh, L. Zakaria, Int. J. Mol. Sci., **12**,10:6722–6732(2011). https://doi.org/10.3390/ijms12106722
11. S. Tyagi, R. Paudel, Int. J. Basic Appl. Biol., **2**,1:103–106(2014). https://www.krishisanskriti.org/ijbab.html
12. I.M. Singha, Y. Kakoty, B.G. Unni, M.C. Kalita, J. Das, A. Naglot, et al., World J. Microbiol. Biotechnol., **27**,11:2583–2589(2011). https://doi.org/10.1007/s11274-011-0730-6
13. K. Amandeep, K.S. Vineet, S. Asmita, K. Jaspal, S. Gursahib, K. Pardeep, African J. Microbiol. Res., **9**,15:1089–1097(2015). https://doi.org/10.5897/ajmr2014.7267
14. K. Sasaki, K. Nakahara, S. Tanaka, M. Shigyo, S. I. Ito, Phytopathology, **105**,4:525–532(2015). https://doi.org/10.1094/PHYTO-06-14-0164-R
15. M. Ignjatov, D. Milošević, Z. Nikolić, J. Gvozdanović-varga, Pesticidi i fitomedicina, **27**,1:25–31(2012). https://doi.org/10.2298/PIF1201025I
16. C. Srinivas, D.N., Devi, K.N., Murthy, C.D., Mohan, T.R., Lakshmeesha, B. Singh, Saudi J. Biol. Sci., **26**,7:1315–1324(2019). https://doi.org/10.1016/j.sjbs.2019.06.002
17. S.P. Gautam, P.S. Bundela, A.K. Pandey, M.K. Awasthi, S. Sarsaiya, Ann. Environ. Sci.,**5**:23–34(2011). http://openjournals.neu.edu/aes/journal/article/view/v5art4

Design of Rotary Dryer for Sand Drying using Biomass Energy Sources

Herry Susanto[1,2], *Roy* Hendroko Setyobudi[2,3*], *Didik* Sugiyanto[1], *Yefri* Chan[1], *Erkata* Yandri[2], *Satriyo* Krido Wahono[4], *Kamaruddin* Abdullah[2], *Juris* Burlakovs[5], *Wahyu* Widodo[3], *Yogo* Adhi Nugroho[6,7], and *Abubakar* Yaro[8,9]

[1]Department of Mechanical Engineering, Darma Persada University, Jl. Taman Malaka Selatan No.22, Jakarta 13450, Indonesia
[2]Graduate School of Renewable Energy, Darma Persada University, Jakarta, Indonesia
[3]Department of Agriculture Science, Postgraduate Program, University of Muhammadiyah Malang, Jl. Raya Tlogomas No.246, Malang 65145, Indonesia
[4]Research Division for Natural Product Technology – Indonesian Institute of Sciences, Jl. Jogja - Wonosari, km 31.5, Gunung Kidul, Special Region Yogyakarta 55861, Indonesia
[5]Department of Water Management, Estonian University of Life Sciences, Tartu, Estonia; Friedrich Reinhold Kreutzwaldi 1a, 51014 Tartu, Estonia
[6]Individual Researcher, Alumni Graduate School of IPB University, Bogor, Indonesia
[7]Data Processing, *Rumah Paper* – Editage Services, Jl. Tokala No.1, Malang 65146, East Java, Indonesia
[8]Institute of Health Sciences & Research, AHRO Scientific Publishing, 272 Bath Street, Glasgow G2 4JR, Scotland, UK
[9]Strathclyde Institute of Pharmacy & Biomedical Sciences (SIPBS), University of Strathclyde, SPBBS, 161 Cathedral St, Glasgow G4 0RE, Scotland, UK

Abstract. Hebel brick is used in making the walls of the house. One of the materials used in making hebel bricks is sand. The water content of the sand will affect the quality of the hebel bricks. The less water content in the sand, the better the quality of hebel bricks. In order to get sand with a little water content, it is necessary to do the drying process. To overcome this problem, a rotary dryer for sand drying using renewable energy, particular biomass energy sources was designed. It can be placed in a sand quarry near a river or near a beach. It is designed to have the ability to dry the wet sand by putting it in a rotary dryer, making this drying system usable both in sunny and rainy conditions. The drying process can take place continuously using heat from biomass energy. The heat is produced from the combustion process in the biomass furnace and then distributed to the rotary dryer. The advantage of this rotary dryer system could be done continuously with a little human power. Moreover, from these calculations can be analyzed the performance of this rotary dryer machine, to determine the suitability between design calculations and test results.

Keywords: Drying, dry sand, hebel brick, heat, rotary dryer, renewable energy.

* Corresponding author: roy_hendroko@hotmail.com

1 Introduction

Hebel brick or lightweight brick is a type of building material replacement for red brick, has a lightweight compared to other bricks. Because of its various advantages, hebel bricks are increasingly sought after and in demand, especially by housing developers. Hebel brick is used as the main foundation in making the walls of the house. This is because hebel bricks are economical, lightweight, efficient, and fast enough to be used in the construction of housing projects, apartments, hotels, and other property projects. One of the materials used in making hebel bricks is sand. The water content of the sand will affect the quality of the hebel bricks. The less water content in the sand, the better the quality of hebel bricks. In order to get sand with a little water content, it is necessary to do the drying process. Some do the drying process by utilizing direct solar heat. And there are also some who do the drying process using a drying machine. The drying process uses a drying machine, in the use of energy for drying, some use electricity, some use fuel oil and some use renewable energy. Currently more and more renewable energy-based drying systems are being developed. Some developed a foodstuff drying system [1], elaborated a cocoa beans drying system [2], evolved a wood drying system [3, 4], promoted a shoe drying system [5], extended a biomass-energized drying system [6, 7], unfolded a cloth drying system [8], developed the radio frequency and microwave heating treatment to disinfest "kutu beras" [*Sitophilus oryzae* (Linnaeus, 1763)] [9], development of building rehabilitation with dry and wet systems [10], promoted of heating zone temperature control for the drying process [11], and development of the solar-energized drying system [13]. Hence the development of a drying system using this renewable energy source, becomes a challenge and also a solution to improve economy and prepare for facing increasingly fierce competition.

2 Research methodology

2.1 Research flow diagram

The initial step of this research is to collect the data needed in designing this rotary dryer and analyze these data, such as data on heat transfer, thermodynamics and machine elements. From the analysis of these data, the optimal design parameters will be obtained. After that, proceed with making a working drawing of this rotary dryer machine. And after the working drawing has been completed, the fabrication process for this rotary dryer will be carried out. The research steps are shown in the following research flow diagram in Figure 1.

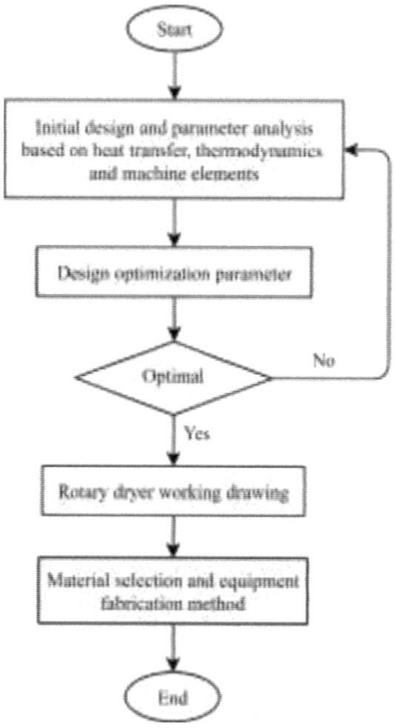

Fig. 1. Research flow diagram.

2.2 Rotary dryer of biomass-energized drying system

Design of the rotary dryer of biomass-energized drying system can be seen in Figure 2 and Figure 3, the advantage of this biomass-energized drying system is the drying process can take place continuously using heat from biomass energy. The heat from the biomass energy source is channeled to the drying machine after going through the energy conversion process in the biomass furnace chamber. The main components of the rotary dryer shown as the Figure (2):

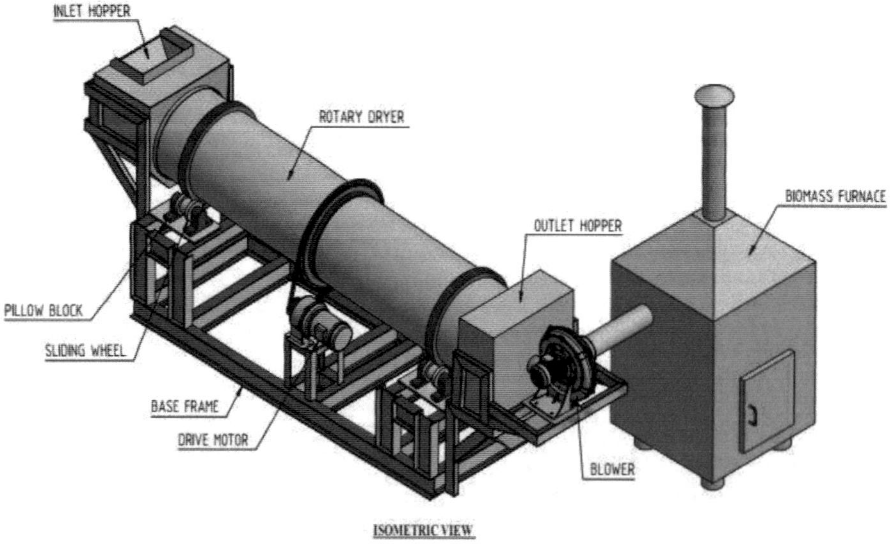

Fig. 2. The rotary dryer components.

The way this rotary dryer works is as follows, namely: The wet sand material to be dried is put into the rotary dryer through the inlet hopper. After the wet sand material reaches the rotary dryer shell chamber, the rotary dryer shell chamber will be rotated, hence the wet sand material in the rotary dryer shell chamber will be stirred using baffles attached to the wall of the rotary dryer shell. The stirring process of the wet sand in the rotary dryer shell chamber is carried out continuously, starting from the wet sand material is introduced until the dry sand material is removed from the rotary dryer shell chamber. The dry sand material will later be removed through the outlet hopper.

Meanwhile, the heat energy needed to dry the wet sand is produced by the biomass furnace which is used to burn biomass wastes. The heat energy produced from the biomass furnace is then channeled into the rotary dryer shell chamber using the blower that sucks the heat energy from the biomass furnace. The heat energy that is channeled will release the water content in the wet sand and then remove the water content from the rotary dryer shell chamber, hence resulting in the dry sand that will be used to make hebel bricks. The dimensions of the rotary dryer can be seen in Figure 3.

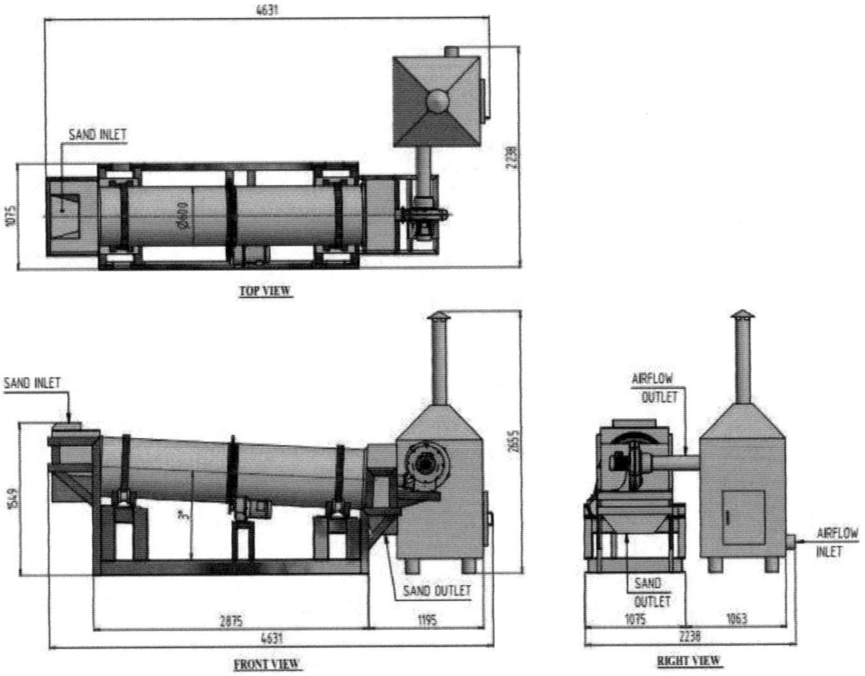

Fig. 3. The rotary dryer dimensions.

The initial design parameters of the rotary dryer can be seen in Table 1.

Table 1. Initial design parameter of the rotary dryer.

No	Input Parameters	Value	Unit
1	Material	Sand	
2	Transfer distance	3	m
3	Drying system type	Rotary dryer	
4	Gearmotor Power	0.75	kW

Tabel 1. Continue to the next page

5	Motor rotation (n)	1 400	rpm*
6	Gearbox ratio (r)	1:20	
7	Motor rotation output (n_0)	70	rpm*
8	Outer diameter of the rotary dryer shell (OD)	600	mm
9	Inner diameter of the rotary dryer shell (ID)	588	mm
10	Maximum frequency of the inverter (f_t)	50	Hz
11	The number of the sprocket teeth on the motor (z_m)	20	th
12	The number of the sprocket teeth on the rotary dryer (z_r)	106	th
13	The tilt angle of the rotary dryer (α)	3	°
14	The sand mass is put into the rotary dryer (m_i)	100	kg

*Note: 1 rpm = 1/60 Hz

This equipment will be used in areas that are still not covered by electricity. Hence, to operate the rotating motor of the rotary dryer shell and hot air distributing blower in the rotary dryer shell, an alternative source of electrical energy is needed that is not dependent on the supply of electrical energy from the State Electricity Company (PLN = *Perusahaan Listrik Negara*). Therefore, the solution is to provide a generator or power plant that uses renewable energy.

Currently, many researchers are conducting research to develop equipment that can be operated using electrical energy sourced from renewable energy. Among them are those who developed the Organic Rankine Cycle (ORC) turbine design, which is devoted to producing electrical energy from heat energy sources with a temperature of less than 100 ºC [12]. There are also researchers who develop drying machine equipment that can be operated using electrical energy generated from solar energy [13], and also develop livestock feed processing equipment that operates using electrical energy generated from biogas energy [14], and there are also researchers who develop equipment for moving goods that are operated using electrical energy generated from other renewable energy sources [15, 16]. Even now, a smart microgrid system equipment has been developed that can utilize solar energy and wind energy to produce electrical energy combined in a hybrid manner [17].

If the sand is taken from the river, it is possible to use renewable energy power plants, namely pico-hydro power to drive the rotary dryer. Reference [18–20] describes this technology, supported by several other researchers [21–23]. Bio-oil can also be used to drive rotary dryers. This vegetable oil can be produced from agricultural waste in villages around the river, including corn cobs [24]. Another renewable energy that can be utilized is biogas. Reference [25] has calculated the economic analysis of biogas production from agricultural biomass.

The quality of sea sand is considered lower as a building material. However, sea sand is needed for areas without volcanoes, with some technologies [26–29]. The rotary dryer on the beach can be operated with bio-oil. Reference [30, 31] describes bio-oil technology from fish waste. Another source of bio-oil is the use of cooking oil waste [32–34].

Reference [14–17] has stated biogas that drives several types of agricultural equipment. Precisely [15] recommends fish waste at the Fishing Ports (*Pelabuhan Perikanan*), Fish Auctions (TPI -*Tempat Pelelangan Ikan*), or Fish Auctions Center (PPI – *Pusat Pelelangan Ikan*) as feedstocks for the biogas digester. Intending to increase biogas production and also improve stakeholder fishery healthy, [15] suggests using multi feedstocks, among others are mixed with latrines. The application of a two-stage digester may overcome several difficulties that arise due to the unevenness of the feedstocks

[35, 36]. This biogas and bio-oil will be able to operate the rotary dryer for sand drying on the beach.

3 Mathematical equation

3.1 Rotary dryer performance

The performance of this rotary dryer, can be known through the following formulas [12–15]. The motor rotation output calculated using the Equation (1).

$$n_m = \frac{f_m}{f_t} . n_0 \tag{1}$$

Then calculate the rotary dryer rotation using the Equation (2):

$$n_r = \frac{z_m}{z_r} . n_m \tag{2}$$

Calculate the transfer distance of the sand material to be dried in each rotation of rotary dryer using the Equation (3):

$$L_x = 2 . D . \tan\alpha \tag{3}$$

Calculate the maximum rotations number of the rotary dryer when transferring the sand from entry to exit using the Equation (4):

$$n_{max} = \frac{L}{L_x} \tag{4}$$

Calculate the traveling time of the sand material from entry to exit using the Equation (5)

$$t = \frac{n_{max}}{n_r} \tag{5}$$

Calculate the transfer speed of the sand material in the rotary dryer using the Equation (6):

$$v = \frac{L}{1\,000 . t} \tag{6}$$

Calculate the drying capacity of the sand material in the rotary dryer using the Equation (7):

$$Q_r = \frac{3 . m_i}{50 . t} \tag{7}$$

Calculate the amount of torque that occurs in the rotary dryer using the Equation (8):

$$T = \frac{m_i . g . D}{2\,000} \tag{8}$$

Calculate the amount of power needed by the rotary dryer using the Equation (9):

$$P = \frac{2 . \pi . n_r . T}{60} \tag{9}$$

Calculate the amount of motor power needed by the rotary dryer using the Equation (10)

$$P_m = \frac{P}{0.85} \qquad (10)$$

Where:
f_m = Operational frequency of the inverter (Hz)
f_t = Maximum frequency of the inverter (Hz)
n_0 = Maximum output rotation of the motor (rpm) (1 rpm=1/60Hz)
n_m = Operational output rotation of the motor (rpm)
n_r = Output rotation of the rotary dryer (rpm)
z_m = The number of the sprocket teeth on the motor (th)
z_r = The number of the sprocket teeth on the rotary dryer (th)
L_x = The transfer distance in each rotation of the rotary dryer (mm rev^{-1})
D = Inner diameter of the rotary dryer (mm)
α = The tilt angle of the rotary dryer (°)
n_{max} = Maximum rotations number of the rotary dryer (revolution)
L = Length of the rotary dryer chamber (mm)
t = The traveling time of the sand material in the rotary dryer (min)
v = The transfer speed of the sand material in the rotary dryer (m min^{-1})
Q_r = The drying capacity of the sand material in the rotary dryer (t h^{-1})
m_i = The sand mass is put into the rotary dryer (kg)
T = The amount of torque that occurs in the rotary dryer (Nm)
g = Earth's gravity acceleration = 9.81 m s^{-2}
P = The amount of power needed by the rotary dryer (W)
P_m = The amount of motor power needed by the rotary dryer (W)

4 Result and discussion

4.1 Calculation and analysis of the rotary dryer performance

The rotary dryer perfomance can be seen in Table 2. The data in Table 1 be used as a reference to calculate the rotary dryer performance using the Equations (1) to the Equation (10) as shown in Table 2.

Table 2. Calculation result of the rotary dryer performance.

f_m (Hz)	n_m (rpm)	n_r (rpm)	L_x (mm rev^{-1})	n_{max} (rev)	t (min)	v (m min^{-1})	Q_r (t h^{-1})	T (Nm)	P (W)	P_m (W)
25	35	6.6	61.6	49	7.4	0.41	0.81	288.4	199.5	234.6
30	42	7.9	61.6	49	6.1	0.49	0.98	288.4	239.3	281.6
35	49	9.2	61.6	49	5.3	0.57	1.14	288.4	279.2	328.5
40	56	10.6	61.6	49	4.6	0.65	1.30	288.4	319.1	375.4
45	63	11.9	61.6	49	4.1	0.73	1.47	288.4	359.0	422.4
50	70	13.2	61.6	49	3.7	0.81	1.63	288.4	398.9	469.3

The graph in Figure 4, Figure 5 and Figure 6 are shown of the rotary dryer performance.

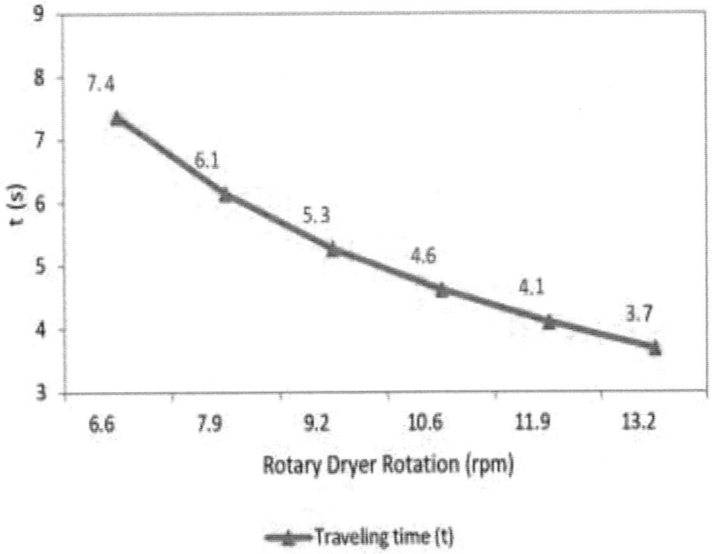

Fig. 4. Graph of traveling time (t) vs rotary dryer rotation (rpm).

Figure 4 illustrates the relationship between traveling time and rotary dryer rotation. This figure shows the decrease in traveling time from 7.4 s at the rotary dryer rotation of 6.6 rpm, to 3.7 s at the rotary dryer rotation of 13.2 rpm. This shows that by increasing the rotation of the rotary dryer, the traveling time of the dry sand will be shorter.

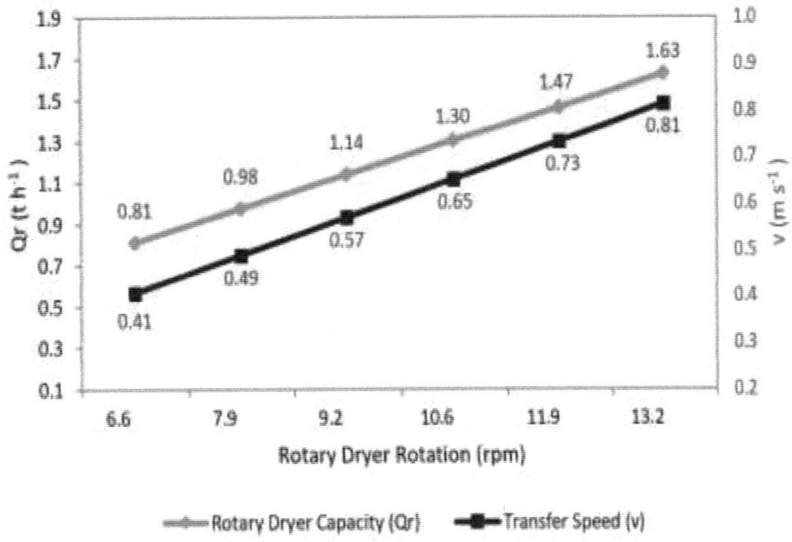

Fig. 5. Graph of drying capacity (Q_r) & transfer speed (v) vs rotary dryer rotation (rpm).

Figure 5 illustrates the relationship between drying capacity and transfer speed against the rotary dryer rotation. This figure shows an increase in the rotary dryer capacity of 0.81 t h^{-1} and the transfer speed of 0.41 m s^{-1} at the rotary dryer rotation of 6.6 rpm and become the rotary dryer capacity of 1.63 t h^{-1} and the transfer speed of 0.81 m s^{-1} at the rotary dryer rotation of 13.2 rpm. This shows that by increasing the rotation of the rotary dryer, the capacity and the transfer speed of dry sand will increase.

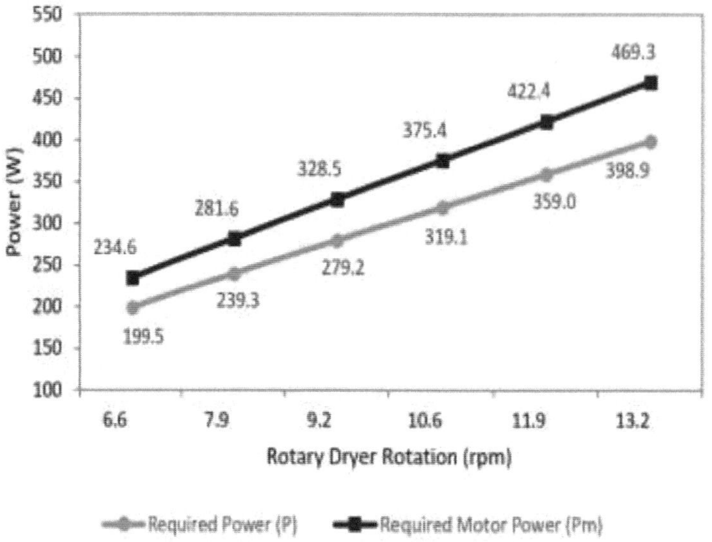

Fig. 6. Graph of required power (P) & required motor power (P_m) vs rotary dryer rotation (rpm).

Figure 6 illustrates the relationship between the required motor power and the rotary dryer rotation. This figure shows an increase in the required motor power of 234.6 W at the rotary dryer rotation of 6.6 rpm and become the required motor power of 469.3 W at the rotary dryer rotation of 13.2 rpm. This shows that the maximum required motor power is 469.3 W, hence it can use the motor that is available in the market with the motor power of 0.55 kW.

4.2 Discussion of the rotary dryer performance

In the design of the rotary dryer of biomass-energized drying system, there are several things that must be considered properly, that is: This equipment will be used in sand mining sites close to rivers or the sea that have high salt content and allows corrosion, then all components, especially in the frame, must be coated with an anti-corrosion coating through the galvanic or hotdipe process. Moreover, because the dried material is sand, it must also be taken into account regarding the possibility of erosion on the shell of the equipment. Hence the thickness and lifetime can be determined. The maintenance time of this equipment need to pay attention, hence it can more durable and be used longer [12–15]. Determination of the number of baffles, size and position of placement in the rotary drayer shell chamber determine the success of the drying process. The speed of hot air distributed into the rotary dryer shell chamber must be regulated through the blower by adjusting the opening and closing of the inlet. And for safety during operation, it is necessary to add a safety cover on the chain and the sprocket of the rotary dryer.

5 Conclusion

From the calculation results of the rotary dryer performance, by varying the rotary dryer rotation from a minimum rotation of 6.6 rpm to a maximum rotation of 13.2 rpm. The traveling time of the material reduced by 50 %. The rotary dryer capacity and the required motor power increase by 50 %. Moreover, the required motor power at maximum conditions is 469.3 W. Therefore, it is known that the faster the rotary dryer rotation, the rotary dryer capacity, and the required motor power will increase, but the traveling time will decrease.

References

1. T. Estiasih, K. Ahmadi. *Teknologi Pengolahan Pangan.* [Food Processing Technology], Malang: Bumi Aksara (2011). [in Bahasa Indonesia]. http://pustaka.poltekkes-pdg.ac.id/index.php?p=show_detail&id=1153
2. M.P. Sandra. *Rancang Bangun Alat Pengering Tipe Konveyor Otomatis Untuk Peningkatan Mutu Biji Kakao Hasil Pengeringan.* [Design of Automatic Conveyor Dryers for Quality Improvement of Cocoa Beans from Drying]. Scribd. (2009). [in Bahasa Indonesia]. https://www.scribd.com/document/393784168/Rancang-Bangun-Alat-Pengering-Tipe-Konveyor-Otomatis-Untuk-Peningkatan-Mutu-Biji-Kakao-Hasil-Pengeringan-2009-Art
3. T. Hussain, M. Shamraiz, S.U. Abrar, Int. J. Technol., **5**:795–809(2016). http://www.ijtech.eng.ui.ac.id/old/index.php/journal/article/view/2135
4. A. Latifah, H. Aulawi, M.A. Ramdhani. MATEC Web of Conferences **197**:11009(2018). https://www.matec-conferences.org/articles/matecconf/abs/2018/56/matecconf_aasec2018_11009/matecconf_aasec2018_11009.html
5. S.M. Atnaw, C.K.M. Faizal, A.J. Oumer. MATEC Web of Conferences **97**:01081(2017). https://www.matec-conferences.org/articles/matecconf/abs/2017/11/matecconf_etic2017_01081/mateconf_etic2017_01081.html
6. D.S. Kirar, H.S. Bhadoria, R. Pandey. MATEC Web of Conferences **57**:01019(2016). https://www.matec-conferences.org/articles/matecconf/abs/2016/20/matecconf_icaet2016_01019/matecconf_icaet2016_01019.html
7. S.Z.A. Nyzam, R.A. Rahim, H.A. Rahim. MATEC Web of Conferences **140**:01023(2017). https://www.matec-conferences.org/articles/matecconf/abs/2017/54/matecconf_iceesi2017_01023/matecconf_iceesi2017_01023.html
8. M. Macieira, P. Mendonca. MATEC Web of Conferences **68**:13009(2016). https://www.matec-conferences.org/articles/matecconf/abs/2016/31/matecconf_iciea2016_13009/matecconf_iciea2016_13009.html
9. Z.A. Rabidin, G.K. Seng, M.J. Abdul Wahab. MATEC Web of Conferences **108**:10001(2017). https://www.matec-conferences.org/articles/matecconf/abs/2017/22/matecconf_icmaa2017_10001/matecconf_icmaa2017_10001.html
10. C. Wutthithanyawat, N. Srisiriwat. MATEC Web of Conferences **65**:03002(2016). https://www.matec-conferences.org/articles/matecconf/abs/2016/28/matecconf_icnscm2016_03002/mateconf_icnscm2016_03002.html

11. T.A. Yassen, H.H. Al-Kayiem, K. Habib., MATEC Web of Conferences **13**:06007(2014). https://www.matec-conferences.org/articles/matecconf/abs/2014/04/matecconf_icper2014_06007/matecconf_icper2014_06007.html
12. H. Susanto, K. Abdullah, A.S. Uyun, S.M. Nur, T.M.I. Mahlia., MATEC Web of Conferences **164**:01012(2018). https://www.matec-conferences.org/articles/matecconf/abs/2018/23/matecconf_icesti2018_01012/matecconf_icesti2018_01012.html
13. H. Susanto, R.H. Setyobudi, Y. Chan, S.M. Nur, E. Yandri, J. Burlakovs, et al., IOP Conf. Series: Earth and Environmental Science, **490**:012009(2020). https://iopscience.iop.org/article/10.1088/1755-1315/490/1/012009/pdf
14. H Susanto, R.H. Setyobudi, D. Sugiyanto, S.M. Nur, E. Yandri, H. Herianto, et al., E3S Web of Conferences **188**:00010(2020). https://doi.org/10.1051/e3sconf/202018800010
15. H. Susanto, R.H. Setyobudi, S.M. Nur, E. Yandri, A.S. Uyun, A. Yaro, et al., E3S Web of Conferences **190**:00014(2020). https://doi.org/10.1051/e3sconf/202019000014
16. K. Abdullah, A.S. Uyun, R. Soegeng, E. Suherman, H. Susanto, R.H. Setyobudi, et al., E3S Web of Conferences **188**:00016(2020). https://doi.org/10.1051/e3sconf/202018800016
17. B. Novianto, K. Abdullah, A.S. Uyun, E. Yandri, S.M. Nur, H. Susanto, et al., E3S Web of Conferences **188**:00005(2020). https://doi.org/10.1051/e3sconf/202018800005
18. R.A. Subekti, A. Susatyo, P. Irasari. Perancangan dan analisis prototip unit turbin-generator tipe submersible skala piko hidro untuk aplikasi pada aliran sungai datar. [Design and analysis of the pico hydro scale submersible type turbine-generator prototype for applications in flat river flow] (2011). [in Bahasa Indonesia]. http://www.telimek.lipi.go.id/xdata/docs/ELDA02.pdf
19. A. Zuhud, S.I. Cahyono, D.D.D.P. Tjahyana. Redesign runner turbin pembangkit listrik tenaga pico hydro dengan metode reverse engineering melalui pendekatan teoritis. [Redesign of the pico hydro power plant turbine runner with the reverse engineering method through a theoretical approach]. Jurnal Teknik Mesin Indonesia, **11**(2):61–65(2016). [in Bahasa Indonesia]. https://doi.org/10.36289/jtmi.v11i2.54
20. H.P. Dewanto, D.A. Himawanto, S.I. Cahyono. Pembuatan dan pengujian turbin propeller dalam pengembangan teknologi pembangkit listrik tenaga air piko hidro (PLTA-PH) dengan variasi debit aliran. [Manufacturing and testing of propeller turbines in the development of the pico-hydro power plant technology (PLTA-PH) with variations in flow rates]. Jurnal Teknik Mesin Indonesia **12**(2):54–62(2017). [in Bahasa Indonesia]. https://doi.org/10.36289/jtmi.v12i2.72
21. S.J. Williamson, B.H. Stark, J.D. Booker. Renew. Energy **61**:43–50(2014). https://doi.org/10.1016/j.renene.2012.06.020
22. Budiarso, Warjito, M.N. Lubis, D. Adanta. Energy Proc., **156**:447–451(2019). https://doi.org/10.1016/j.egypro.2018.11.087
23. E.Y. Setyawan, Y.I. Nakhoda, A.U. Krismanto, L. Mustiadi, E. Yandri, J. Burlakovs. E3S Web of Conference **188**:00006(2020). https://doi.org/10.1051/e3sconf/202018800006
24. C.A. Mullen, A.A.Boateng, N.M.Goldberg, I.M. Lima, D.A. Laird, K.B. Hicks. Biomass and Bioenergy **34**(1):67–74(2010). https://doi.org/10.1016/j.biombioe.2009.09.012

25. M. Mel, A.S.H. Yong, Avicenna, S.I. Ihsan, R.H. Setyobudi. Energy Procedia **65**:204 –214(2015). https://doi: 10.1016/j.egypro.2015.01.026
26. J. Xiao, C. Qiang, A. Nanni, K. Zhang. Constr Build Mater., **155**:1101–1111(2017). https://doi.org/10.1016/j.conbuildmat.2017.08.130
27. A. Verma. Evaluation of sea sand and river sand properties and their comparison. RG, 1–33(2015). https://doi.org/10.13140/RG.2.1.4906.6327
28. V.M. Gnanasundar, S. Suvetha, S. Preethy. AIP Conference Proceedings **2128**:020011(2019). https://doi.org/10.1063/1.5117923
29. S.J. Jiao, & S. Wang, Adv Mat Res, **838–841**:1806–1809(2013). https://doi.org/10.4028/www.scientific.net/amr.838-841.1806
30. A. Wisniewski Jr, V.R.Wiggers, E.L. Simionatto, H.F. Meier, A.A.C. Barros, L.A.S. Madureira. Fuel, **89**,3:563–568. https://doi.org/10.1016/j.fuel.2009.07.017
31. V.R.Wiggers, A. Wisniewski Jr., L.A.S. Madureira, A.A. Chivang Barros, H.F.Meier. Fuel, **88**,11:2135–2141(2009). https://doi.org/10.1016/j.fuel.2009.02.006
32. A.B.H. Trabelsi, K. Zaafouri, W. Baghdadi, S. Naoui, A. Ouerghi. Renewable Energy **126**:888–896(2018). https://doi.org/10.1016/j.renene.2018.04.002
33. S.S. Prabu, M.A. Asokan, R. Roy, S. Francis, M.K. Sreelekh. Energy **122**:638–648(2017). https://doi.org/10.1016/j.energy.2017.01.119
34. T. Kraiem, A.B. Hassen, H. Belayouni, M. Jeguirim. Production and characterization of bio-oil from the pyrolysis of waste frying oil. Environ Sci Pollut Res **24**:9951–9961(2017). https://doi.org/10.1007/s11356-016-7704-z
35. P.G. Adinurani, R.H. Setyobudi, S.K. Wahono, M. Mel, A. Nindita, E. Purbajanti, et al., Proc. Pakistan Acad. Sc. Part B **54**,1:47–57(2017). http://www.paspk.org/wp-content/uploads/2017/03/Proceedings-B.-Life-Sciences-541-March-2017.pdf
36. R.H. Setyobudi, A. Sasmito, P.G. Adinurani, A. Nindita, A.S. Yudhanto, Y.A. Nugroho, et al, Energy Procedia, **65**,2015:300–308(2015). https://doi.org/10.1016/j.egypro.2015.01.056

The Use of Probiotic and Antioxidants to Improve Welfare and Production of Layer Duck at Commercial Farms for Global Warming Mitigation

Imam Suswoyo[1,*], *Ismoyowati* Ismoyowati[1], *Wahyu* Widodo[2], and *Zane* Vincēviča–Gaile[3]

[1]Department of Animal Production, Faculty of Animal Science, Jenderal Soedirman University, Jl. Profesor DR. HR Boenyamin No.708, Banyumas 53122, Central Java, Indonesia
[2]Department of Animal Science, Faculty of Agriculture and Animal Science, University of Muhammadiyah Malang, Jl. Raya Tlogo Mas no.246. Malang 65144, East Java, Indonesia
[3]Department of Environmental Science, University of Latvia, Jelgavas Street 1, Room 302, Riga LV-1004, Latvia

Abstract. Global warming affected increasing the risk of ducks to be more succeptable to heat stress which leads to decrease welfare and production. This research aimed to study duck welfare and productivity under probiotic and antioxidants administration at commercial farms. The method used was experiment with Completely Randomized Designed (CRD) based on factorial pattern. The treatment was dose of natural/homemade and commercial probiotics combined with vitamin C at 400 mg kg^{-1} and 600 mg kg^{-1} feed. Thus there were four treatment combinations. Each treatment was replicated five times, totally were 20 flocks of duck. Each flock had 50 laying females so there were 1 000 ducks. The treatment was conduced for 2 mo. The parameters observed included, i) duck welfare based on Heterophyl/Lymphocyte (H/L) ratio; ii) egg production consisted of (a) duck day production, and (b) egg weight. This study concluded that administration of combination between homemade probiotic and vitamin C at 600 mg kg^{-1} feed significantly ($p < 0.05$) increased duck welfare and egg production but did not affect egg weight.

Key words: Heat stress, increase duck welfare and egg production, local ducks, vitamin C

1 Introduction

High ambient temperature is a serious obstacle in developing poultry production in tropical areas, especially in recent climate change phenomenon. Climate change now is one of big issues concerns to the world. It is understood since climate change has severe impact not only for human being but also animal. Indonesian National Aviation and Space Agency reported that the climate in Indonesia has become warmer during the 20th century with an average annual temperature has increased by about 0.3 °C since 1900 [1]. Ducks, as a homeotermic animal, is succeptable to heat stress which leads to decreasing its welfare and productivity. Previous study reported that under dry intensive system, local ducks suffered from heat stress as indicated by rectal temperature, behaviour, and body and plumage

* Corresponding author: suswoyo_01@yahoo.com

condition [2]. Therefore, releasing heat stress is an essential factor to improve duck welfare. Recently it has been proven that oxidative stress is a molecular mechanism of many different stresses, including environmental stresses (e.g. heat stress), nutritional stresses (e.g. mycotoxins) and internal stresses (e.g. bacterial, viral, etc.) [3]. Manipulation on daily management is required to protect the ducks from suffering from heat stress, among others by using vitamins as functional feed.

Several studies have been conducted to investigate the use of probiotic and vitamins for alleviation of heat stress in poultry. In broiler chickens, probiotic *Lactobacillus* strains have proven to be able to restore the microbial balance and maintaine the natural stability of indigeneous bacterial microbiota following heat stress–induced changes [4]. Under heat stress conditions, chickens fed with diets containing *Lactobacillus* cultures have higher antibody production than those on control diet [5]. Supplementation of probiotic in diet significantly increases white blood cells count and decreases Heterophyl/Lymphocyte (H/L) ratio which is important in reduction of stress effect in poultry [6]. Relating to heat stress, probiotic might be useful for ameliorating the adverse influence of heat on the egg production and the gut health of laying hens [7].

The most important and well–characterised natural antioxidants in the animal body are vitamins E and C. All antioxidants are responsible for prevention of the damaging effects of free radicals and toxic products of their metabolism [8]. [9] Stated that in broiler chicks, optimal antioxidant supplementation is shown to be important to maintain growth rate, immunocompetence and meat quality. Whereas in broiler chickens, supplementation of 250 mg kg^{-1} vitamin C was the most effective in reducing heat stress [10]. Recently it has been proven that vitamin C (ascorbic acid) – an important antioxidant synthesised in chickens and its dietary supplementation is shown to be effective in stress conditions, when its requirement substantially increased. [9]. Some studies have shown that vitamin C has been reported to enhance immune response by modifying corticosteroid synthesis in adrenal glands [11]. Most previous studies, however, concerned on vitamins use for chickens. Thus, study on effect of probiotic and vitamin combination on duck welfare is lacking, especially that under commercial farm management.

Keeping this in mind, this study was conducted to examine the effect of probiotic and vitamin C combination on welfare and egg production of local ducks kept in commercial farms.

2 Materials and methods

2.1 Methods

This study used experimental method that conducted in collaboration with '*Berkah Abadi*' duck farmer group and ducks reared under dry system intensively. Under intensive system, the ducks are mostly kept in sheds with rice straw bedding and solid floor ranch infront of the sheds [12]. The study was located at Tegal City, Central Java, Indonesia which is located in the northern coastal area of Java Island. The City and its surrounding areas is one of the most important duck centers in Indonesia with its very famous local layer duck namely *Tegal Duck*. It is believed that the duck belongs to *Indian Runner* family.

2.2 Materials and methods

The materials used was local layer ducks which the age of 13.4 mo ± 1.3 mo. Intensive system refers to the conventional reraing by the farmers in which ducks were confined around the farmer's village by closed fence. Hence ducks had no access to the outside area, and the amount of feed provided could be controlled and measured. Drinking water was provided *ad libitum* three times a day i.e. in the morning, noon, afternoon, while feed was given twice a day i.e. in the morning and in the afternoon. The feedstuffs were locally available which mainly consisted of rice bran, dried rice, and fresh fish (*Leiognathidae* T. N. Gill, 1893) with proportion of 39.65 %, 25.11 % and 35.24 % respectively. The nutrient content were 26.38 % crude protein, 2.923 kcal kg^{-1} metabolic energy, 2.29 % Calcium and 0.78 % Phosphorus.

This study used Factorial method with treatments consisted of combinations between three doses of home–made (H) and commercial probiotics (C) with vitamin C 400 (C1) and 600 (C2) mg kg feed. There were four treatments units were replicated five times. Thus, totally were 20 flocks. Each flock had 50 female ducks; therefore this study involved 1 000 ducks. The treatments were applied each morning, mixed thoroughly in duck ration.

2.2.1 Data collected:

i. Heterophils to lymphocyte (H/L) ratio as a welfare status
   Blood samples were taken from two ducks per flock for determination of H/L ratio on 60th d of the study.
ii. Daily ambient air temperature and humidity as indicators of environmental condition were measured at 6 am, noon, and 3 pm.
iii. Laying percentage/duck day production (expressed as average number of eggs laid per day in relation to number of female ducks $flock^{-1}$), and average egg weight (g) $flock^{-1}$ as indicators of egg production.

2.2.2 Data analysis:

Data collected were analyzed using Analysis of Variance (ANOVA) and followed by Honestly Significant Different (HSD) test if any different effects.

3 Results and discussions

Blood H/L ratio, duck day production (DDP), and egg weight during this study were summarized in Table 1 below.

Table 1. H/L ratio, DDP, and egg weight during the study

Parameters	Treatments			
	HC1	HC2	CC1	CC2
H/L Ratio	0.56a	0.55a	0.75b	0.61b
DDP (%)	72.04a	68.75b	64.40b	66.39b
Egg weight (g)	67.28a	68.98a	69.35a	70.61a

Note: numbers at the same row with different alphabet indicate significant differences ($p < 0.05$).

3.2 Duck welfare

H/L ratio in this study ranges from 0.32. to 1.14 with an average of 0.62 ± 0.09. Statistical analyses indicated that H/L ratio in HC1 was significantly ($p < 0.05$) lower than that in other treatments. It indicated that combination of home–made probiotic and vitamin C supplementation has improved duck welfare. Figure 1 presents the average of H/L ratio of each group of treatment.

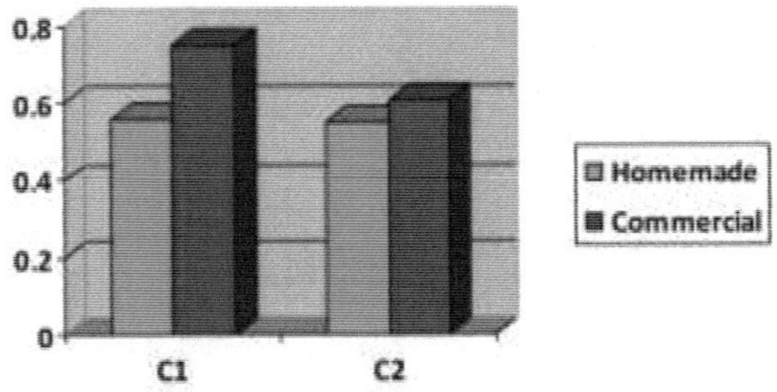

Fig. 1. The average of H/L ratio of each group of treatment

Results of this study indicate that ducks supplemented with combination of home–made probiotic and vitamin C had lower H/L ratio which means that the ducks were more welfare. [13] Stated that ducks in welfare condition has lower H/L ratio than those in stress condition.

3.2 Egg production

Average egg production was 66.99 % ± 3.29 % ranging from 47.86 % to 87.14 %. This finding confirms that previous results of [2, 14]. Egg production of control and treatment groups presented in Figure 2.

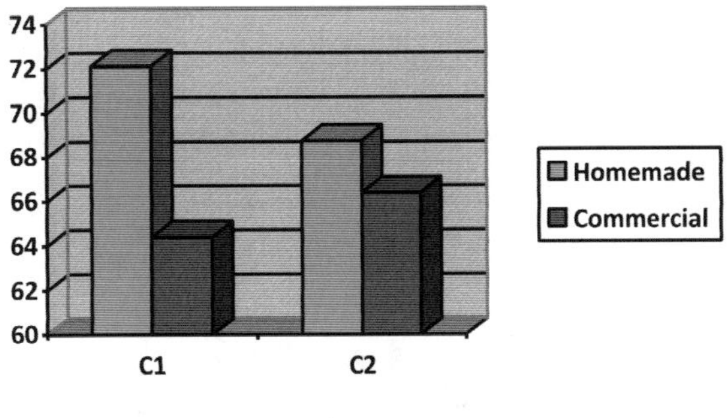

Fig. 2. Egg production of control and treatment groups

Statistical analysis showed that combination of home–made probiotic and vitamin C supplementation significantly ($p < 0.05$) increased duck day production. It seems that that the dose was the most effective quantity for supplementation. High requirement of vitamin occurs when poultry under stress condition since it has vital roles in metabolic reactions. One of the most crucial factor of vitamin C is in amino acid metabolism which affect egg production.

3.3 Egg weight

Egg weight ranges from 61.80 g to 75.20 g with an average of 69.05 g ± 1.37 g. This finding was confirmed the results of previous research which found that egg weight was 71.14 g ± 6.08 g [14]. The lowest egg weight was found at control ducks although it us not significant ($p > 0.05$). Figure 3 presents the average of egg weight during the study.

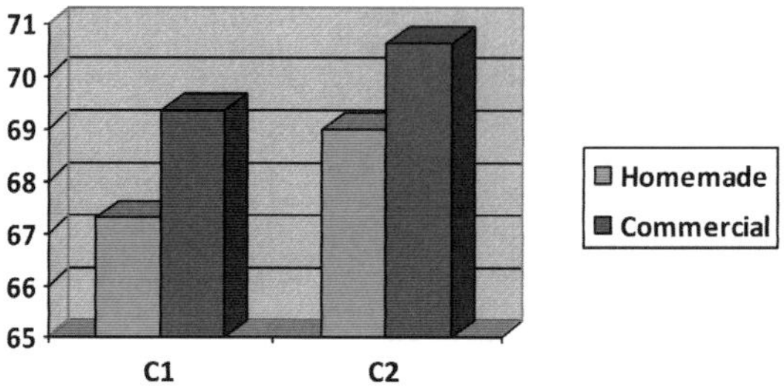

Fig. 3. The average of egg weight during the study

3.4 Environmental condition

Environmental condition at study site is presented in Table 2.

Table 2. Average ambient temperature and humidity at location study

Environmental condition	Morning	Noon	Afternoon	Average
Temperature (°C)	27.06 ± 0.64	31.91 ± 0.89	31.35 ± 0.85	29.35 ± 2.65
Humidity (%)	88.94 ± 3.81	79.43 ± 5.76	82.32 ± 3.50	83.56 ± 4.88

Table 2 indicates that environmental temperature of the study site was 29.35 °C ± 2.65 °C on average which was higher than the maximum temperature needed by poultry. Study site which was coastal region of Tegal City, Central Java, Indonesia is considerably suffer from climate change. At the same time relative humidity was also high with average of 83.56 % ± 4.88 %. It seems that environmental condition in the area was higher than the thermo–neutral zone required by ducks. Thermo–neutral zone for poultry is between 18 °C to 25 °C, and the most efficient temperature for ducks ranges between 23 °C to 25 °C [15]. Figure 4 presents the variation of average daily ambient temperature and humidity during the study.

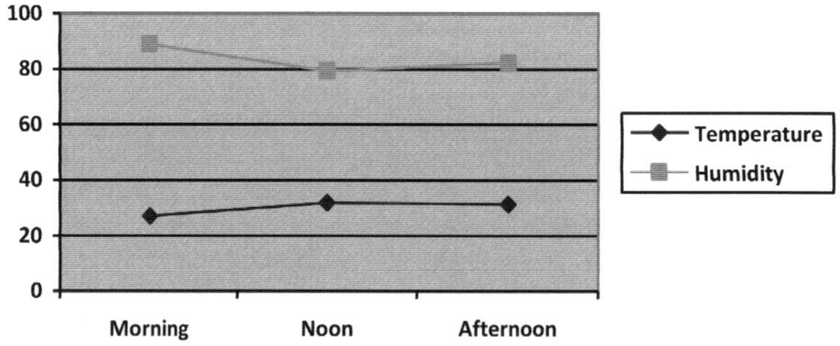

Fig. 4. Variation of average daily ambient temperature and humidity during the study

Controlling the ducks' environment, particularly temperature, humidity, litter moisture and ammonia is crucial to duck welfare [16]. If the ambient temperature is higher than the thermoneutral zone, panting will increase ten times which decreases productivity [17]. Qualitative observation along the study showed that ducks with treatments had more comfort condition indicated by less drinking and panting activities.

4 Conclusions

On commercial level, that administration of three doses of home–made probiotic and 600 mg kg^{-1} vitamin C has improved duck welfare and egg production, but did not increase the egg weight.

The authors would thank to the Directorate General of Higher Education and Rector of Jenderal Soedirman University for funding this research. The deepest thank also to duck farmers of '*Berkah Abadi*' group who made this study done.

References

1. LAPAN. *Perubahan iklim: Penyebab, dampak dan antisipasi [Climate change: Causes, impacts and anticipation]*. Bandung: Pusat Sains dan Teknologi Atmosfer (2017). p. 5–51. [in Bahasa Indonesia]. http://psta.lapan.go.id/files_uploads/buletin/Vol.2_No.2_Jul-Des2017.pdf
2. I. Suswoyo, Ismoyowati, I.H. Sulistyawan. Int. Jou of Poul. Sci. **13**,4: 214–217(2014). https://www.researchgate.net/publication/272964487_Benefit_of_Swimming_Access_to_Behaviour_Body_and_Plumage_Condition_and_Heat_Stress_Effect_of_Local_Ducks
3. P.F. Surai, V.I. Fisinin. Agricultural Biology. 4:3–13(2012). https://doi.org/10.15389/agrobiology.2012.4.3eng
4. S. Yadav, R Jha. J Animal Sci Biotechnol. **10**,2:1–11(2019). https://doi.org/10.1186/s40104-018-0310-9
5. H. Al–Khalaifa, A. Al–Nasser, T. Al–Surayee, S. Al–Kandari, N. Al–Enzi, T. Al–Sharrah, et al. Poultry Science. **98**,10:4465–4479(2019). https://doi.org/10.3382/ps/pez282
6. S.H. Rahimi, A. Khaksefidi. Iranian Journal of Veterinary Research. **7**,3: 23–28(2006). https://dx.doi.org/10.22099/ijvr.2006.2645
7. K. Deng, C.W. Wong, J.V. Nolan. Jou. of Anim. Phys. and Anim. Nutr. **90**, 1–2:81–86(2006). https://doi.org/10.1111/j.1439-0396.2005.00569.x
8. P.F. Surai. *Natural antioxidants in poultry nutrition: new developments*. Proceedings of the 16th European Symposium on Poultry Nutrition, (Scotlandia, UK 2007). Proceedings of the 16th European Symposium on Poultry Nutrition. 669–676(2007). http://www.feedfood.co.uk/download/1_Europ_Strasbourg_Symposium_2007.pdf
9. A.A. Fotina, V.I. Fisinin, P.F. Surai. Bulg J. Agric. Sci. **19**,5:889–896(2013). http://www.feedfood.co.uk/download/Bulgarian_Performax_2013.pdf
10. E. Kusnadi. JITV. **11**,4:249 253(2006). [in Bahasa Indonesia]. http://oaji.net/articles/2015/1610-1424247036.pdf
11. N. Bhatti, Z. Hussain, M. Mukhtar, A. Ali, M. Imran, A. Rafique, et al. Journal of Antivirals & Antiretrovirals. **8**,4:151–154(2016). https://doi.org/10.4172/jaa.1000152

12. P.P. Ketaren. Wartazoa. **17**,3:117–127(2007) (in Bahasa Indonesia). http://medpub.litbang.pertanian.go.id/index.php/wartazoa/article/view/877/886
13. I. Suswoyo, Rosidi. International Journal of Poultry Sci. **15**,6:235–239(2016). https://doi.org/10.3923/ijps.2016.235.239
14. I. Ismoyowati, D. Indrasanti, I.H. Sulistyawan. Buletin Peternakan. **42**,3: 197–202(2018). https://doi.org/10.21059/buletinpeternak.v42i3.34465
15. A.S.O. El–Badry, M.M. Hassanane, E.S. Ahmed, K.H. El–Kholy. Global J. Biotech. & Biochem. **4**,2:152–159(2009). http://citeseerx.ist.psu.edu/viewdoc/download?doi=10.1.1.482.7693&rep=rep1&type=pdf
16. T.A. Jones, M.S. Dawkins. British Poul. Sci., **51**,1:12–21(2010). https://doi.org/10.1080/00071660903421159
17. T. Ahmad, M. Sarwar. Feed Mix (The International Journal on Feed, Nutrition and Technology). **13**,4:15–17(2005). https://scholar.google.co.id/scholar?hl=id&as_sdt=0%2C5&q=Dietary+electrolytes+combat+heat+stress+in+broilers&btnG=

Entrapment Formulation for *env-Tm* Gene Based on Chitosan Low Molecular Weight as a Jembrana Disease Virus Vaccine Candidate

Indra Lesmana Rahayu[1,]* and *Asmarani* Kusumawati[1,2*]

[1]Department of Biotechnology, Postgraduate School, Universitas Gadjah Mada, Jl. Teknika Utara, Sleman, Yogyakarta 55281, Indonesia
[2]Department of Reproduction and Obstetrics, Faculty of Veterinary Medicine, Universitas Gadjah Mada, Jl. Fauna Karangmalang No.2, Yogyakarta 55281, Indonesia

Abstract. Jembrana disease (JD) caused by Jembrana Disease Virus (JDV) becomes an obstacle in Bali cattle (*Bos javanicus*). The development of JD vaccines has a critical meaning to prevent losses in the *B. javanicus* in Indonesia and is one of the models in the development of the Human Immunodeficiency Virus (HIV) vaccine. The development of vaccines for JDV has carried out DNA vaccines that are expected to provide better immunological effects. This study aimed to determine the low molecular weight chitosan (LC) entrapment towards pEGFP-C1-*env-Tm* in the formation of Chitosan Nanoparticles Low/pEGFP-C1-*env-Tm* complex. The *env-Tm* gene was inserted in pEGFP-C1 into the pEGFP-C1/*env-Tm* construct transformed on the *E. coli* DH5α host. The construct was formulated into LC/pEGFP-C1/*env-Tm* complex with a low molecular weight chitosan concentration of 0.06 % and the ratio of pEGFP-C1/*env-Tm*: LC (wt/wt) was 1:0.5-1:3. The complexes were then analyzed by gel retardation assay agarose 1 %. The results of this study indicated that the best entrapment results of low molecular weight chitosan to pEGFP-C1/*env-Tm* was in the mass ratio of pEGFP-C1/*env-Tm*: LC was 1:2. The best formulation entrapment for *env-Tm* by low molecular weight chitosan 0.06 % is in the rate 1:2.

Keywords: *Env-tm* gene, jembrana disease virus, low molecular

1 Introduction

The obstacle faced by Bali cattle is Jembrana Disease (JD), which is caused by bovine lentivirus, a member of the family *retroviridae*, subfamily *lentiviridae*. This disease can affect the fertility of Bali cattle, thereby reducing the production of Bali cattle, which can affect economic conditions in Indonesia [1–5]. Thus, prevention is needed to prevent the disease spread. DNA vaccines are proven effective in primates, stable in storage, easy to produce, safe (non-replicative), express only target antigens, cannot be returned pathogenic, and able to induce cellular and humoral immunity [6–8].

The delivery of pDNA using nonviral vectors results in gene expression, which encourages the development of new delivery systems that can protect pDNA, as well as localize its delivery, such as chitosan. Chitosan can protect therapeutic agents from nuclear

* Corresponding author : asmaranikusumawati2018@gmail.com

degradation both in vitro and in vivo, and can be formulated for the delivery of pDNA [9]. Therefore, this research was carried out on the use of low molecular weight chitosan as a protective molecule as well as carrying the target gene to cells

2 Materials and methods

2.1 Material

The gene encoding transmembrane portion of envelope protein from JDV, which had previously been isolated from infected cattle and cloned into the plasmid enhanced green fluorescent protein C terminus (pEGFP-C1) vector (Clontech, USA) to construct pEGFP-*env-Tm* JDV. Chitosan of low molecular weight (50 000 Da to 190 000 Da, based on viscosity: 20 cP to 300 cP) was purchased from Sigma (Singapore).

2.2 Methods

2.2.1 LC/pEGFP-C1-env-Tm complex preparation

Low molecular weight of chitosan as much as 1 g was dissolved in 100 mL of 1 % (v/v) acetic acid. The low molecular weight chitosan solution was then sterilized using a 0.2 μm syringe filter (Merck, Germany). The solution was then diluted using *nuclease-free water* (NFW) (TransGen Biotech, China) to a concentration of 0.06 %, and was used as a stock to make LC/pEGFP-C1/*env-Tm* complex. The preparation of the LC/pEGFP-C1/*env-Tm* complex was performed with the complex coacervation method with modifications [10]. A total amount of 1 μg plasmid and different amounts of low molecular weight chitosan were heated separately at 50 °C for 10 min. Then, plasmid construct and low molecular weight chitosan solution were mixed with a volume ratio by vortexing at 2 500 rpm (1 rpm = 1/60 Hz) for 30 s. The mass ratios of DNA and chitosan used were 1:0.5 to 1:3 (wt/wt).

2.2.1 Gel retardation essay analysis

The formation of LC/pEGFP-C1/*env-Tm* complex was examined using a gel retardation assay in agarose gel electrophoresis. The LC/pEGFP-C1/*env-Tm* complexes were electrophoresed on a 1 % agarose gel with Gel Stain (TransGen Biotech, China). Unbound pEGFP-C1-*env-Tm* and low molecular weight chitosan solution were used as negative controls.

3 Results

3.1 LC/pEGFP-C1/*env-Tm* complex

The formation of LC/pEGFP-C1/*env-Tm* complex was observed with a gel retardation assay on agarose gel 1 %. Several different DNA mass ratios and low molecular weight chitosan were used, and only in the 1:2 and 1:3 ratio were complexes formed, seen with entrapped LC/pEGFP-C1/*env-Tm* complexes in the well. Low molecular weight chitosan solution as a negative control cannot bind DNA, and it can show the absence of the band. Free plasmid DNA migrates into two DNA bands that show two different plasmid conformations. The bands of unbound pEGFP-C1/*env-Tm* could still be detected with a ratio of 1:0.5 and 1:1 but not detected in a higher LC/pEGFP-C1/*env-Tm* ratio complex

E3S Web of Conferences **226**, 00026 (2021) https://doi.org/10.1051/e3sconf/202122600026
ICoN BEAT 2019

(Figure 1). The results in this study indicate that the best entrapment results of low molecular weight chitosan to pEGFP-C1/*env-Tm* is in the mass ratio of pEGFP-C1/*env-Tm*: LC is 1:2 that entrapped the entire amount of plasmid DNA. This was the same as research conducted by Ishak (2019) regarding the use of medium molecular weight chitosan with pDNA (*env-Tm*) with a chitosan concentration of 0.02 %, which chitosan/pDNA complex is entrapped in well at a ratio of 1: 2 [11].

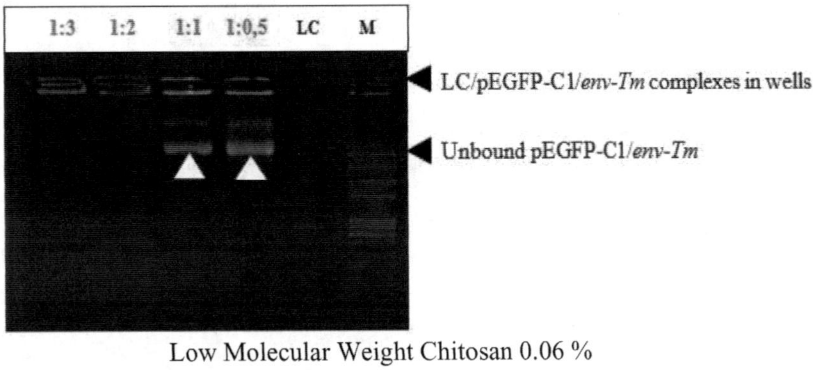

Fig. 1. Agarose gel retardation assay of LC/pEGFP/*env-Tm* complexes.
(M) = Marker, (LC)=Low Molecular Weight Chitosan

4 Discussion

Chitosan complex is widely used in the pharmaceutical world as a transport agent capable of releasing controlled molecules such as DNA, vaccines, peptides, or antibiotics [12]. This research focuses on the success of entrapped LC/pEGFP-C1/*env-Tm* complexes in well with 0.06 % chitosan concentration, 1:2 ratio (wt/wt).

Particle formation occurs at the N/P ratio, which is the ratio between nitrogen (N) chitosan per phosphate (P) DNA, which is essential for determining the charge on the surface of the particle. Chitosan has one primary amino group in repeated glucosidic residues that gives a positive charge. Chitosan has a unique cationic character. That is because chitosan has a major amino group that is responsible for providing strong electrostatic interactions with negative charges on DNA [13, 14].

The important thing in making a vaccine is that the complexes formed between the delivery agent and pDNA can enter the cell and express the target gene carried. Various factors that affect transfection efficiency are molecular weight (MW), degree of deacetylation (DD), chitosan-DNA molar ratio, DNA complex charge ratio (N: P ratio) and pH, and particle size. High load density, making chitosan with high MW, can't release the charge to the cytoplasm, but conversely, chitosan with low MW is much better at delivering genes. Additionally, in the formulation of making DNA vaccines, particle size and shape also depends on the conditions involved in the formulation process, including salt concentration, molecular weight complex, and chitosan concentration, and the mixing process [13, 15, 16].

5 Conclusion

The best formulation entrapment for *env-Tm* by low molecular weight chitosan 0.06 % is in the ratio 1:2.

References

1. I.W.M. Tenaya. Buletin Veteriner Udayana, **8**,2:187–202(2016).[in Bahasa Indonesia]. https://www.google.co.id/url?sa=t&rct=j&q=&esrc=s&source=web&cd=&ved=2ahUKEwiQ6o7Q68rrAhWL7HMBHaC_BzIQFjAAegQIARAB&url=https%3A%2F%2Fojs.unud.ac.id%2Findex.php%2Fbuletinvet%2Farticle%2Fdownload%2F23318%2F15323%2F&usg=AOvVaw1TSEHp-GG9ngaA-NpunpBQ
2. A. Kusumawati, I.A. Wulandari, A. Ratnawati, S. Hartati, T. Untari. Jurnal Veteriner, **16**,4:569–575(2015).[in Bahasa Indonesia]
3. https://ojs.unud.ac.id/index.php/jvet/article/view/17340/11393
4. F. Lebre, G. Borchard, H. Faneca, M.C.P.d. Lima, O. Borges. Mol. Pharm., **13**:472–82(2016). https://pubs.acs.org/doi/abs/10.1021/acs.molpharmaceut.5b00707
5. J.A. Williams. Vaccines, **1**,3:225–249(2013). https://www.ncbi.nlm.nih.gov/pmc/articles/PMC4494225/
6. I.W.M. Tenaya. Buletin Veteriner Udayana, **8**,2:187–202(2016).[in Bahasa Indonesia] https://www.google.co.id/url?sa=t&rct=j&q=&esrc=s&source=web&cd=&ved=2ahUKEwiQ6o7Q68rrAhWL7HMBHaC_BzIQFjAAegQIARAB&url=https%3A%2F%2Fojs.unud.ac.id%2Findex.php%2Fbuletinvet%2Farticle%2Fdownload%2F23318%2F15323%2F&usg=AOvVaw1TSEHp-GG9ngaA-NpunpBQ
7. G. Widera, M. Austin, D. Rabussay, C. Ulmer, S.W. Barnett, M. Chen, et al. J. Immunol., **164**,9:4635–4640(2018). https://pubmed.ncbi.nlm.nih.gov/10779767/
8. K. Zhao, Y. Zhang, X. Zhang, W. Li, C. Shi, C. Guo, et al. Int. J. Nanomedicine, **9**:389–402(2014). https://www.ncbi.nlm.nih.gov/pmc/articles/PMC3890423/
9. A.O. Abbas, M.D. Donovan, A.K. Salem. J. Pharm. Sci., **97**,7:2448–24641(2008). https://pubmed.ncbi.nlm.nih.gov/17918737/
10. K. Zhao, Y. Zhang, X. Zhang, C. Shi, X. Wang, X. Wang, et al. Int. J. Nanomedicine, **9**:4609–4619(2014) https://www.ncbi.nlm.nih.gov/pmc/articles/PMC4207079/
11. R. Martien, B. Loretz, M. Thaler, S. Majzoob, A. Bernkop-Schnurch. J. Biomed. Mater. Res. A. **82**,1:1–9(2007). https://www.researchgate.net/publication/6539973_Chitosan-thioglycolic_acid_conjugate_An_alternative_carrier_for_oral_nonviral_gene_delivery
12. J. Ishak, L. Unsunnidhal, R. Martien, A. Kusumawati. J. Vet. Res., **63**:7–16(2019). https://www.ncbi.nlm.nih.gov/pmc/articles/PMC6458558/
13. R. Martien, A. Adhyatmika, I.D.K. Irianto, V. Farida, D.P. Sari. J Pharm, **8**,1:133–139(2012).[in Bahasa Indonesia] https://jurnal.ugm.ac.id/majalahfarmaseutik/article/view/24067
14. L. Hu, Y. Suna, Y. Wu. Nanoscale, **5**:3103–3111(2013). https://pubs.rsc.org/en/content/articlelanding/2013/nr/c3nr00338h#!divAbstract
15. J. Nilsen-Nygaard, S.P. Strand, K.M. Vårum, K.I. Draget, C.T. Nordgård. Polymers, **7**,3:552–579(2015). https://www.mdpi.com/2073-4360/7/3/552
16. T. Kiang, J. Wen, H.W. Lim, K.W. Leong. Biomaterials, **25**,22:5293–5301(2004). https://pubmed.ncbi.nlm.nih.gov/15110480/
17. T. Delair. Eur. J. Pharm. Biopharm., **78**,1:10–18(2011). https://europepmc.org/article/med/21138767

Environment Identification of Plantation Plant Development for Competitive in Sukoharjo Districts, Indonesia

Irma Wardani* and *Tria* Rosana Dewi

Agribusiness Departement, Engineering, Sains and Agriculture Faculty, Islam Batik University Surakarta, Jl. Agus Salim No.10, Surakarta 57147, Indonesia

Abstract. Developing the competitiveness of an area needs to identify the environment to determine the potential of the region. This study aimed to determine the environment that affects the development of plantation crops in Sukoharjo. Data analysis techniques used were Analytical Hierarki Process and Strength Weakness Opportunities Threats. The results showed the direction of the development of plantation crops was productivity. The priority plantation crops developed in Sukoharjo Regency were Sugar Cane (0.2929), Coconut (0.2017), Java Tobacco (0.1650), Medicinal Plants (0.12834), Cashew (0.1259), and Kapok (0.0085). Identification on the development of estate crops as follows internal factors such as i) strengths: available land, farmer institutions, social culture, availability of human resources,ii) weaknesses: low farmer education, lack of farmer coaching, expensive labor. While the external factors were i) opportunities: availability of seeds from the government, mechanization assistance for land, market demand, land expansion, the presence of sugarcane processing factories, biotechnology ii) threats: land use change, sugarcane farmers competition, pest and disease attacks, farmers moving.

Keywords: Crops, farmers competition, productivity

1 Introduction

Development of identical plantations with large industrial components and sources of state revenue to make plantation products for export. The selection of priority commodities to be developed can contribute to regional development. This research aimed to determine priority plantation crops to be developed in Sukoharjo districts, Indonesia. The data analysis technique used is the AHP (Analytical Hierarki Process) method [1].

The results indicate the direction of the development of plantation crops is productivity. The priority plantation crops developed in Sukoharjo districts are Cane *(Saccharum officinarum* L.*)* (0.2929), Coconut (*Cocos nucifera* L.) (0.2017), Tobaco *(Nicotiana tabacum* L.*)* (0.1650), Herb and Spices (0.1283), Cashew *(Anacardium occidentale* L.*)* (0.1259), Cotton *(Ceiba pentandra* L.*)* (0.0085). Identification on the development of estate crops as follows internal factors including i) strengths: available land, farmer the association, social culture, availability of human resources, ii) weaknesses: low farmer education, lack of farmer coaching, expensive labor. The external factors are i)

* Corresponding author: wardaniirma6@gmail.com

 https://doi.org/10.1051/e3sconf/202122600027

opportunities: availability of seeds from the government, mechanization assistance for land, market demand, expansion of land, the presence of sugarcane processing factories ii) threats: land use change, sugarcane farmers competition, pest and disease attacks, farmers moving to other sectors.

2 Material and methods

2.1 Material

The study was conducted in Sukoharjo Regency, with primary data collection carried out by purposive sampling by interviews in the form of data and information or questionnaires.

2.2 Method

2.2.1 Analytical hierarki process analysis

The analysis used in this research was the AHP and SWOT analysis [2]. The procedure of the AHP was :

i. Arrange the hierarchy

Arrange the hierarchy of the problems faced, must first define the problem and determine the desired solution.

ii. Determine the priority

Table 1. Pairwise comparison matrix

G	A_1	A_2	A_3	A_N
A_1	A_{11}	A_{12}	A_{13}	A_{1n}
A_2	A_{21}	A_{22}	A_{23}	A_{2n}
....	....	....	....	....
A_n	A_{n1}	A_{n2}	A_{n3}	A_{nn}

G = Basic criteria of comparison

Ai,Aj = Element below level is load

Ij = 1,2,3, ..., n is the index of the element in the level which are the same and are jointly related to the criteria G.

Aij = the number given by comparing elements with the jth element, which is done with scale pairwise comparison.

iii. Synthesis

Add up the values of each column in the matrix and divide each value from the column by the total column in question, then add up the values of each row and divide by the number of elements to get the average value

Table 2. Criteria value matrix

G	A_1	A_2	A_n	Sum	Priority
A_1	A_{11}/a	A_{12}/b	A_{1n}/c	J	J /n
A_2	A_{21}/a	A_{22}/b	A_{2n}/c	K	K /n
A_n	A_{n1}/a	A_{n2}/b	A_{nn}/c	L	L /n

iv. Measure consistency

The making decisions, it is important to know how good consistency

v. Multiply each value in the column by the relative priority of the elemen

Table 4. Row addition matrix

G	A_1	A_2	A_n	Sum line
A_1	$(J/n)A_{11}$	$(J/n)A_{12}$	$(J/n)A_{1n}$	O
A_2	$(K/n)A_{21}$	$(K/n)A_{22}$	$(K/n)A_{2n}$	P
A_n	$(M/n)A_n$	$(M/n)A_{n2}$	$(L/n)A_{nn}$	Q

vi. Divide the results from the addition of rows with the relative elements concerned, then determine λ max

Table 5. Value determination λ maxs

G	Priority	Sum line	Sum	Λ max
A_1	J /n	O	O : J /n	
A_2	K /n	P	P : K /n	
A_3	L /n	Q	Q : L /n	
Σ			S	S/n

vii. Consistency index

Consistency index is described in Equation (1):

$$CI = \frac{\lambda_{max} - n}{n - 1} \quad (1)$$

n = Element

viii. Consistency ratio

Consistency ratio is described in Equation (2):

$$CR = \frac{CI}{RI} \quad (2)$$

CR = Consistency Ratio
IR = Index Random Consistency
CI = Consistency Index

The design of this study starts from observing data and information to agencies or institutions related to the research to determine the priority of estate crops commodities, then make a questionnaire, for AHP analysis. The results of AHP analysis in the form of priority commodities, then identified environmental factors that influence the development of plantation crops.

2.2.2 Environment analysis

Table 7. Environment identification

Environment identification			
1.	Strength :	3.	Opportunities:
2.	Weakness :	4.	Threat :

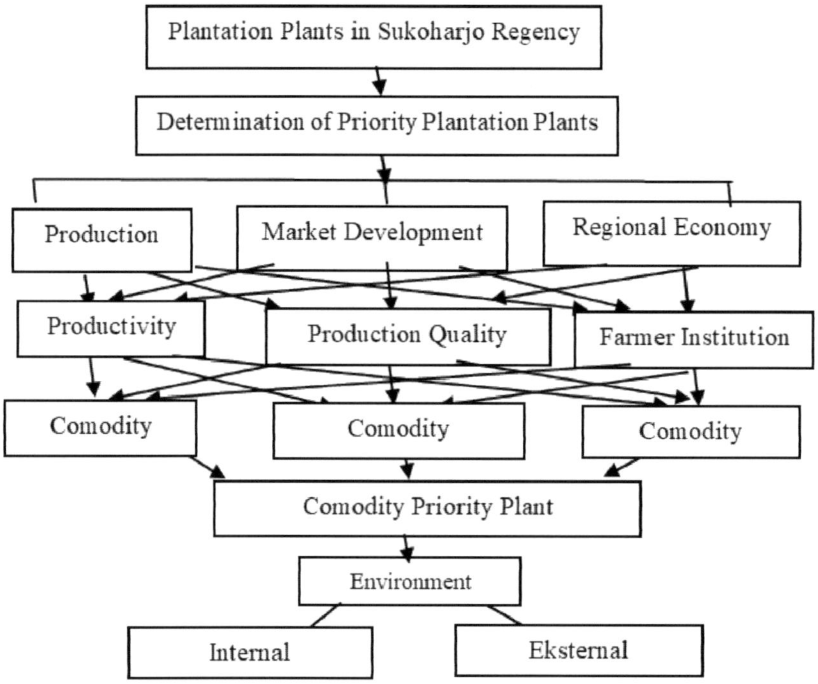

Fig 1. Research design

3 Results and discussions

The agricultural sector became one of the main saviors of the economy in Indonesia. It indirectly involves the role of the plantation subsector. The plantation sector holds an important role in the development of the economy in Indonesia [3].

The geographical, plantation district has an area of 466.66 km^2 which consists of 12 districts. Sukoharjo district has ten commodities in the plantation sector. Plantation as an integral part of the agricultural sector is a sub sector that has an important and strategic role in national development [4].

Table 8. The results of production smallholder plantation

No.	Commodities	Productivity (t)				
		2014	**2015**	**2016**	**2017**	**2018**
1.	Coconut (*Cocos nucifera* L.)	580.72	587.98	589.53	589.98	580.74
2.	Clove *(Syzygium aromaticum* L.*)*	0.90	0.98	1.03	1.01	1.01
3.	Kapok *(Ceiba pentandra* L.*)*	68.48	70.91	71.33	70.98	71.01
4.	Cashew *(Anacardium occidentale* L.*)*	58.73	63.06	68.29	68.29	69.14
5.	Cane *(Saccharum officinarum* L.*)*	4 212.23	4 413.62	3 852.06	2 084.58	3 582.28
6.	Candlenut *(Aleurites moluccanus* L.*)*	0.08	0.10	0.10	0.08	0.092
7.	Tobaco *(Nicotiana tabacum* L.*)*	2 067	257.50	196.27	307.95	150.3
8.	Herbs and spices	0	0	1 836.00	0	0
9.	Sesame *(Sesamum indicum* L.*)*	30.25	13.75	0	0	0
10.	Catton *(Ceiba pentandra* L.*)*	0	0	0	0	0

Source: Agriculture and Fisheries Agency Sukoharjo District, 2019

Plantation commodities in Sukoharjo District are much effort by local farmers scattered in 12 sub Districts in Sukoharjo District. The development of plantation commodities tended to increase every year, some commodities have not increased production for several years. Therefore, for the development of plantation commodities, need to determine priority commodities for Sukoharjo District, Indonesia.

The development of a priority commodity needs to determine the purpose so obvious for development of the commodity and to know the factors that influence in developing the priority commodity [5]. Analysis to determine the commodity using AHP then be identified internal and external factors. Determination of the purpose of the development of leading plantation commodities is the level of productivity, market development and improvement of the regional economy [6].

Table 9. Priority plantation

Commodity	Score priority	Number
Cane *(Saccharum)*	0.292 97	1
Kapok *(Ceiba pentandra)*	0.085 88	6
Cashew *(Anacardium occidentale)*	0.125 97	5
Coconut (*Cocos nucifera*)	0.20178	2
Tobaco *(Nicotiana tabacum)*	0.165 06	3
Herbs and spies	0.128 34	4

Source : Primary data , 2019

From Table 9, the sugarcane commodity was ranked 1, this shows the sugar cane commodity can be a superior commodity. Sugar cane is a plant that is widely cultivated to meet raw materials in more than 58 sugar factories in Indonesia. Cane plantation production has a high productivity compared to other plantation commodities. Cane is a plantation crop that has priority to be developed. Sugar cane plantation is located in Polokarto, Sukoharjo District, which has a land area of 680.23 ha.

Environmental conditions necessary for sugarcane cultivation include tropical or temperate climatic conditions with plentiful supply of water: approximately for more than 6 mo annually - either from rainfall or irrigation; with frost not favoring its growth and up to altitudes of 1 600 m. Sukoharjo Regency is located at an altitude of 125 m.a.s.l to 80 m.a.s.l. The highest place above sea level is the Subdistrict of Polokarto, which is 125 m.a.s.l., and the rainfall that occurs along. Geographical conditions support the development of sugarcane.

The first International Conference for Sustainable Plantation (The 1st ICSP), 101 presenters were presented in the first ICSP, discussing sustainable development of many estate crops. The major estate crops and plantations presented in the first ICSP include cocoa, coconut, coffee, nutmeg, oil palm, rubber, pepper, sago palm, sugar cane, sugar palm, tea, and other miscellaneous plantation crops and research areas related to the crops [7].

Plantation-based industries have the ability to be the leading sect of economic growth, employment, and also encourage increased income distribution. One of the downstream plantation industries is the sugar industry [8].

Other important roles can also be seen from the perspective of food security, investment absorption, and the extent of linkages in the downstream industry, such as the food industry, beverage industry, refined sugar industry, pharmaceutical industry, paper, MSG, particle board, and bio-energy.

Sugar industry is also highly related to local resources, thus high value commodities can be developed for the empowerment of the people's economy. Therefore, the presence of the sugar industry can be an economic asset and at the same time an important social asset [9]. The study prospect of developing sugarcane needs to identify internal and external factors [10]. The identification results shown in Table 10.

Table 10. Identification of internal and external factors of cane plants

Internal factors	**External factors**
Strength 1. Available land 2. Association farmers 3. Social Culture 4. Available people Weakness 1. Low education farmers 2. Lack of coaching for farmers 3. labor is expensive	Opportunity 1. Seed from the government 2. Subsidy mechanization for the land 3. Market demand 4. Expansion land 5. Manufacturing sugar cane 6. Biotechnology Threat 1. Transfer of land function 2. Competition of sugarcane farmers 3. Pests and diseases 4. Farmers move to other sectors

Source : Primary data , 2019

Polokarto, Sukoharjo District, is one of the suppliers of raw materials for the Tasikmadu sugar factory in Karanganyar District thus it has a significant opportunity to be developed and supported by an increase in the community consuming sugar. Sugarcane is managed by local farmers in Polokarto, but the amount of costs to pay farmers is a weakness, and it is feared that farmers will also switch professions. Biotechnology plays an important role for the improvement of Sugarcane varieties. Biotechnological approaches for sugarcane improvement in the following areas such as (i) Cell and tissue culture for rapid propagation genetic transformation and molecular breeding, (ii) Introduction of novel genes into commercial cultivars, (iii) Molecular detection of sugarcane pathogens,(iv) Development of genetic maps using molecular marker technology, (v) Understanding the molecular basis of sucrose accumulation, (vi) Molecular testing of plants for clonal fidelity, (vii) Variety identification, and (viii) Molecular characterization of various traits [11].

Sugarcane productivity can be disturbed by pests and diseases [12]. The sugar productivity often experiences obstacles, one of which is insect pests. Losses caused by pests and diseases are quite high, around 10 % decrease in sugar production. The results of the identification of internal and external factors are used as a reference in the strategy of developing sugarcane.

4 Conclusion

Sugarcane has the potential to be developed in Sukoharjo, Indonesia. Environmental factors are an important concern for developing sugarcane. The internal factors can also influence the development of sugarcane. Therefore, biotechnology as the application of technology is very important to be able to develop sugarcane varieties. Increasing the number of varieties, sugar cane plants can increase competitiveness in the market.

References

1. Brunelli, Matteo. *Introduction to the Analytic Hierarchy Process*. (2015).p.83 https://www.springer.com/gp/book/9783319125015
2. D.Oreski. TEM Journale **1**,4:283–291(2012). http://tem-journal.com/documents/vol1no4/pdf/Strategy%20development%20by%20using%20SWOT%20-%20AHP.pdf
3. W.S. Abir. Jurnal Ilmiah Universitas Batanghari Jambi. **16**,1:134–141(2016). [in Bahasa Indonesia]. http://ji.unbari.ac.id/index.php/ilmiah/article/view/93/88
4. H. Rakhmad. Social Economic of Agriculture. **2**,1:54–66(2013). https://jurnal.untan.ac.id/index.php/jsea/article/view/5119
5. F. Novitasari, R.V. Ayuningtyas, J. Regional and Rural Dev. Plan. **2**,3:218–227(2018). https://journal.ipb.ac.id/index.php/p2wd/issue/view/2285 / or DOI: https://doi.org/10.29244/jp2wd.2018.2.3.218-227
6. F.R. David. *Strategic Management: Concept and Case*. Prentice Hall. (2010).p.290 https://www.google.com/url?sa=t&rct=j&q=&esrc=s&source=web&cd=&cad=rja&uact=8&ved=2ahUKEwiO54K25N_qAhXSAnIKIIYvBABUQFjABcgQIARAB&url=http%3A%2F%2Fperpustakaan.kasn.go.id%2Findex.php%3Fp%3Dfstream-pdf%26fid%3D28%26bid%3D327&usg=AOvVaw0vXSkN9fXzvVyF3ciOLbGh
7. S. Sudarsono, A. Junaedi. IOP Conf. Series: Earth and Environmental Science. **418**:011001(2020).

https://www.google.com/url?sa=t&rct=j&q=&esrc=s&source=web&cd=&cad=rja&uact=8&ved=2ahUKEwjH-YP15N_qAhVVbn0KHZ1CC14QFjAAegQIBRAB&url=https%3A%2F%2Fiopscience.iop.org%2Farticle%2F10.1088%2F1755-1315%2F418%2F1%2F011001%2Fpdf&usg=AOvVaw0K1Xfncwd6OKMv3WHmSMNm or doi:10.1088/1755-1315/418/1/011001

8. Y.T.F. Marpaung, P. Hutagaol,W.H. Limbong, N. Kusnadi. Ind. J. Agric. Eco. **2**,1:1–14(2011). https://www.google.com/url?sa=t&rct=j&q=&esrc=s&source=web&cd=&cad=rja&uact=8&ved=2ahUKEwjtqMGi5d_qAhWEYysKHUbsDa0QFjAAegQIARAB&url=https%3A%2F%2Fid.linkedin.com%2Fin%2Fyanto-togi-ferdinand-marpaung-sp-mm-msc-808b4ba&usg=AOvVaw1jJPUwdclQq_SohxEuCe5I
9. S. Omwoma, M.N.T. Arowo, J. Lalah. Environ. Res. J. **8**,3:293–330(2015). DOI: 10.13140/2.1.2187.3448 Or https://www.google.com/url?sa=t&rct=j&q=&esrc=s&source=web&cd=&cad=rja&uact=8&ved=2ahUKEwiLrYKC5d_qAhVIU30KHZQxDSQQFjABegQIARAB&url=https%3A%2F%2Fwww.researchgate.net%2Fpublication%2F272182537_Environmental_Impacts_of_Sugarcane_Production_Processing_and_Management_A_chemist_Perspective&usg=AOvVaw1amJHvGrVfFIY2tAdqO4HQ
10. E. Gurel. J. Int. Soc. Res. **10**,51:994–1006(2017). http://sosyalarastirmalar.com/cilt10/sayi51_pdf/6iksisat_kamu_isletme/gurel_emet.pdf
11. A.K. Tiwari, Y.P. Bharti, S. Tripathi, N. Mishra, M. Lal, G.P. Rao, et al. Acta Phytopathol. Entomol. Hun. 45:235–249(2010). DOI: 10.1556/APhyt.45.2010.2.1 or https://www.google.com/url?sa=t&rct=j&q=&esrc=s&source=web&cd=&cad=rja&uact=8&ved=2ahUKEwiD-bWP5d_qAhXMAnIKHUwuBVwQFjAAegQIBBAB&url=https%3A%2F%2Fwww.researchgate.net%2Fpublication%2F235333195_Biotechnological_Approaches_for_Improving_Sugarcane_Crop_with_Special_Reference_to_Disease_Resistance&usg=AOvVaw3w1IuPPGaq6H0QGT6S1JoS
12. A.K. Srivastava, M.K. Rai. Biodiversitas Journal of Biological Diversity. **13**,4:214–227(2012). https://smujo.id/biodiv/article/view/144

Conceptual Design of Inventory Analysis Software to Support the Life Cycle Assessment in Palm Oil Production

Kiman Siregar[1,*], *Supriyanto* Supriyanto[2], *Devitra* Saka Rani[3], *Yanuar* Nurdiansyah[4], and *Feri* Wijayanto[5]

[1]Department of Agricultural Engineering of Syiah Kuala University, Jl.Tgk.Hasan Krueng Kalee No.3 Kopelma, Banda Aceh 23111, Indonesia
[2]Mechanical and Biosystem Engineering Department, Faculty of Agricultural Engineering and Technology, IPB University, PO BOX 220 Bogor, Indonesia
[3]Graduate School of Life and Environmental Sciences, University of Tsukuba, 1-1-1, Tennodai, Tsukuba, Ibaraki 305-8577, Japan
[4]Program Study of Information Technology, University of Jember, Jl. Kalimantan 37, Jember 68121, Indonesia
[5]Institute for Computing and Information Science, Radbound University Comenisulaan 4, 6525 HP Nijmegen, The Netherlands

Abstract. Life cycle assessment is a quantitative method to analyze the environmental impact that consists of four main activities: goal and scope definition, life cycle inventory, life cycle impact assessment, and interpretation. The application of the life cycle in palm oil industry are very important and already conducted by many researchers. However, the most difficult task in life cycle assessment are the life cycle inventory. In this research, this study proposed the software to support the life cycle inventory in palm oil production. The result of the study was the conceptual design of the life cycle inventory software.

Keywords: Environment friendly, life cycle inventory, software, sustainable palm oil

1 Introduction

Palm oil is a vital product in the world to produce cooking oil, salad oil, margarine, shortening, soap, fatty acid, biodiesel, and other valuable product. Indonesia, as the biggest producer of palm oil in the world [1]. One of the concerns in the palm oil industry is the environmental aspect of achieving the sustainability of the productions. The government and industry have an action by implementing the Roundtable Sustainable Palm Oil (RSPO) and Indonesian Sustainable Palm Oil (ISPO) as a standard to prevent the environmentally damaged during the productions [2, 3]. Another effort, by utilizing the Life Cycle Assessment (LCA) based on ISO 14040 included the series of the standard [4, 5].

LCA consists of several activities such as (i) scope and definition, (ii) life cycle inventory analysis, (iii) life cycle impact assessment, and (iv) interpretation. The range of

* Corresponding author: ksiregar.chairilcan@gmail.com

analysis has defined the boundary of the system included in the study. The scope included the definition of the boundary. Then, the inventory analysis is activities to collect the information on material flow, energy flow, system design, and information of the system will be analyzed.

Finally, the impact on the environment analyzed to get the result of the LCA. The users can interpret the result to get feedback on the system. In the palm oil industry, LCA implemented to analyze biodiesel production [6–10], land-use change analysis, and integrated utilizing the Palm Oil Mill Effluent (POME) for valuable product [11, 12]. In the previous study, the inventory analyzed conducted with specific cases and specific assumptions [13]. To extend the examined with another case, the tools of the life cycle inventory to assist evaluating the LCA [14, 15].

This research proposed the conceptual design of the life cycle inventory software for palm oil production. The conceptual framework consists of the system boundary of the system, actor design, and user interface design on the life cycle assessment application.

2 Materials and methods

The development of the software to support the inventory in the palm oil industry to conduct the LCA consists of the analysis, software design, and implementation.

2.1 Analysis

Palm oil production consists of land preparation, cultivation, transportation, palm oil mill, and conversation to the valuable product such as biodiesel and others. The boundary of the system in this research was defined (Figure 1). In palm oil plantation the input resources were needed are seed, fertilizer, herbicide, water, and diesel. In a life cycle assessment, inventory analysis was based on the boundary of the system. Based on the boundary, data was collected to evaluate the environmental impact of palm oil production. In this research, data will evaluate and collect from the Aceh Province, Indonesia.

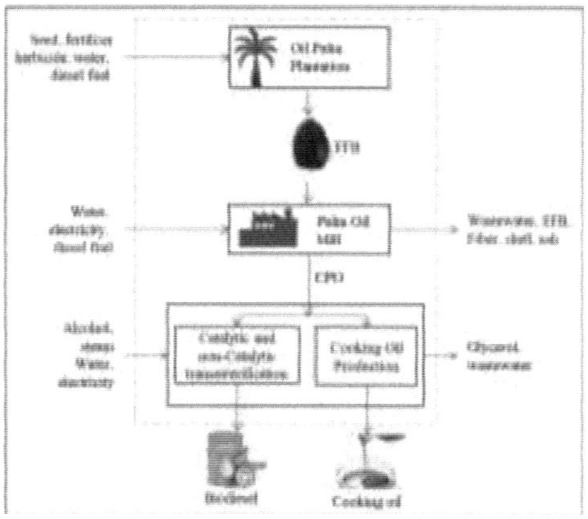

Fig. 1. Boundary condition of the system

The utilization of the resources mainly calculated in the area. Then, the Fresh Fruit Bunch (FFB) send to the palm oil mill (POM) to process the FFB to the Crude Palm Oil (CPO) [4, 15, 16]. The fiber and shell utilized to generate the energy to run the POM. The Empty Fruit Bunch (EFP) utilized as compos to use again as fertilizer. The CPO used to produce the biodiesel through transesterification process. An other products to provide the cooking oil. Finally, the boundary condition set regarding developing the life cycle inventory software.

2.2 Software design

This step, the design of Life Cycle Inventory (LCI) software was conducted by identifying the users related to the LCI Database. The data was collected from the researchers (expert) and practitioner by the software developer. Then, the data collector will collect the baseline data and input to the system. Data verification staff validates the data collected to the system. Finally, specific users will use the baseline data by information the new data with the case, edit the LCI database, and use (view) the database. Figure 2 shows the relation of the users in the database LCI.

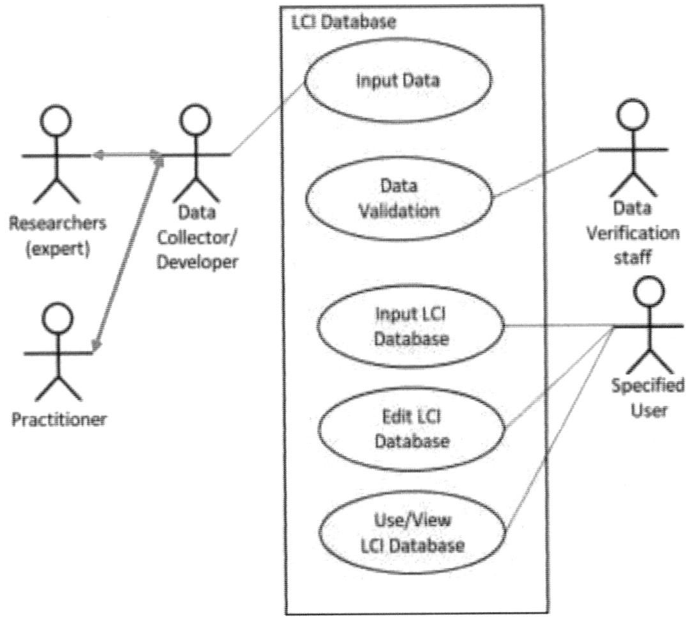

Fig. 2. Role of the users in the LCI database

2.3 Implementation

Software implementation was the process to transform the concept of software development. A database management system (DBMS) was implemented to store and manage the data from the inventory process. The MySQL DBMS was used to run the software. PHP and Javascript programming was used to develop the system interface. The user interface was developed to interact with the system.

3 Results and discussion

The result of this research was the conceptual design of the LCI software for the palm oil industry. Based on the analysis phased, the users with a specific role could access the system to input, edit, and make the new database from the cases. Then, the facility to input was designed in the user interfaces to make the system accessible to the users.

The user interface designed using the Bootstrap that consists of the Javascript, CSS, and HTML. At the beginning of the software, users should log in to the system with a specific username and password. The username set by the administrator to the specific user to conduct the inventory. In the main interface of the LCA, the user can access the menu to conduct the inventory analysis. The baseline dataset was input by the developers, and the user can edit the database based on specific industry conditions. On the left side, there are several menus to access the inventory analysis such as nursery, cultivation, plant maintenance, harvesting, transportation, POM, and process to biodiesel using the transesterification. The users can use the menu to access the inventory pages. The overall interface is shown in Figure 3.

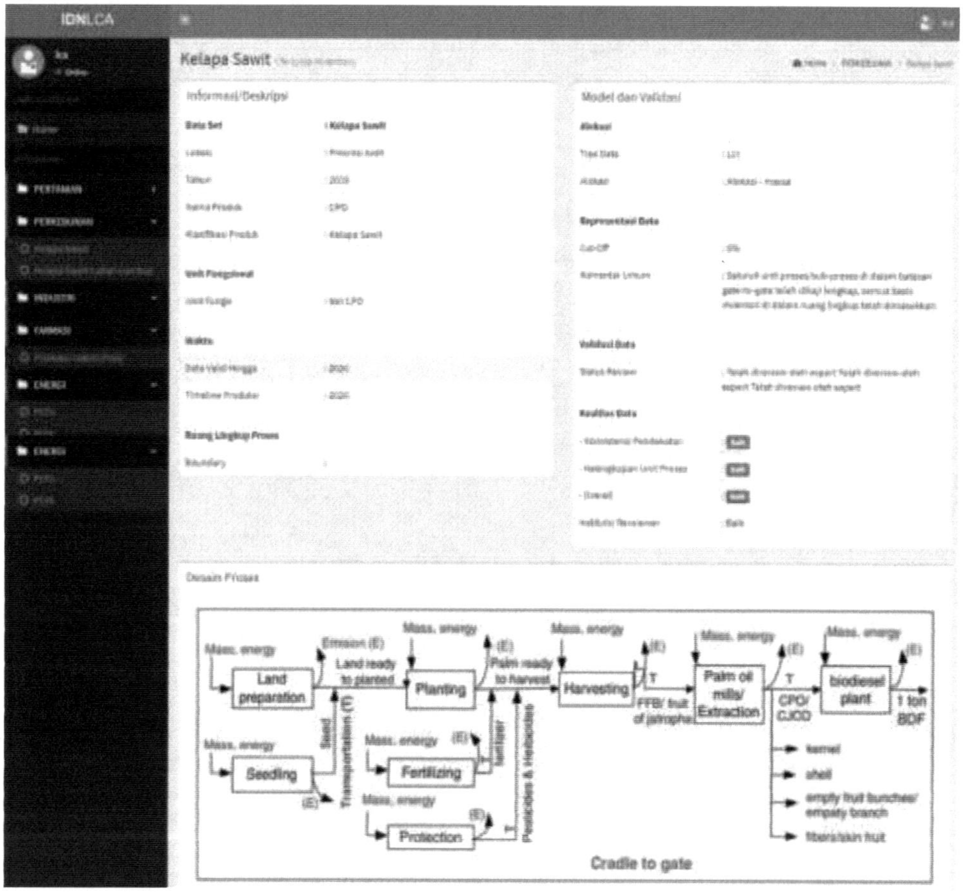

Fig. 3. The user interface of the Life-Cycle Inventory

4 Conclusion

The result of this study was the conceptual design of the Inventory analysis software to support the LCA of palm oil production. The software consists of the input facility, analysis, and visualizing the result of inventory. The inventory, analysis software can help to manage the data during the inventory process.

References

1. O. Pye, World Dev., **121**:218–228(2019). https://doi.org/10.1016/j.worlddev.2018.02.014
2. L.T. Gan, H. Cap, Planter, **92**:379–399(2016). http://www.isp.org.my/
3. S. Hutabarat, Agro Ekonomi. **28**:170–188(2017). https://doi.org/10.22146/jae.27789
4. K. Siregar, Ichwana, I.S. Nasution, Sholihati, I. Sofiah, T. Miharza,. IOP Conference Series: Earth Environmental Science. **542**:012046(2019). https://doi.org/10.1088/1755-1315/542/1/012046
5. M. Finkbeiner, A. Inaba, R. Tan, K. Christiansen, H.-J. Klüppel, Int. J. Life Cycle Assess, **11**:80–85(2006). https://doi.org/10.1065/lca2006.02.002
6. K. Siregar, A.H. Tambunan, A.K. Irwanto, S.S. Wirawan, T. Araki, *A Comparison of life cycle inventory of pre-harvest, production of crude oil, and biodiesel production on Jatropha curcas and palm oil as a feedstock for biodiesel in Indonesia*, (Yokohama, Japan, 2012), Proceeding of Ecobalance 2012 Conference, 21–24(2012). https://scholar.google.co.id/scholar?oi=bibs&cluster=16053431475488963685&btnI=1&hl=en
7. S.A. Archer, R.J. Murphy, R. Steinberger-Wilckens, Renew. Sustain. Energy Rev., **94**:694–704(2018). https://doi.org/10.1016/j.rser.2018.05.066
8. Y. Manik, A. Halog, Integr. Environ. Assess. Manag., **9**:134–141(2013). https://doi.org/10.1002/ieam.1362
9. Z. Sajid, F. Khan, Y. Zhang, Renew. Energy. **85**:945–952(2016). https://doi.org/10.1016/j.renene.2015.07.046
10. K. Siregar, S. Sholiati, I. Sofiah, T.Miharza, R.H. Setyobudi, O. Anne, et al. E3S Web of Conferences **190**:00021(2020). https://doi.org/10.1051/e3sconf/202019000021
11. T. Mumtaz, S. Abd-Aziz, N. Rahman, P.L. Yee, Y. Shirai, M. Hassan, African J. Biotechnol., **7**,21:3900–3905(2008). https://www.ajol.info/index.php/ajb/article/view/59466
12. S. Tabassum, Y. Zhang, Z. Zhang, J. Clean. Prod., **95**,148–155 (2015). https://doi.org/10.1016/j.jclepro.2015.02.056
13. R.H. Crawford, J. Environ. Manage., **88**:496–506(2008). https://doi.org/10.1016/j.jenvman.2007.03.024
14. J. Winkler, B. Bilitewski, J. Waste Manag., **27**:1021–1031(2007). https://doi.org/10.1016/j.wasman.2007.02.023
15. K. Siregar A.L. Machsun, S. Sholihati, R. Alamsyah, I. Ichwana, N. C. Siregar, et al., E3S Web of Conferences. **188**:00018(2020). https://doi.org/10.1051/e3sconf/202018800018
16. A. Ilyana, W. Hasrulnizzam, W. Mahmood, M.H. Fazli, M. Fauadi, M. Nizam et al., Pol. J. Environ. Stud., **24**,4:1463–1475(2015). http://www.pjoes.com/Author-Ilyana-Abdullah/8551

Fasciolosis Infection Level of Various Breed Cattle in Batu and Pujon District, East Java, Indonesia

Lili Zalizar[1,*], *Khusnul* Rahmawati[1], and *Abubakar* Yaro[2,3]

[1]Department of Animal Science, Faculty of Agriculture and Animal Science, University of Muhammadiyah Malang, Jl. Raya Tlogomas no 246, Malang 65145, East Java, Indonesia
[2]Africa Health Research Organization, 272 Bath Street, Glasgow G2 4JR, Scotland, UK
[3]Strathclyde Institute of Pharmacy & Biomedical Sciences (SIPBS), University of Strathclyde, SPBBS, 161 Cathedral St, Glasgow G4 0RE, Scotland, UK

Abstrack. Fasciolosis is a disease caused by *Fasciola gigantica* (Cobbold, 1855) or *Fasciola hepatica* (Linnaeus, 1758) that infected ruminants and caused liver damaged and loss of productivity. To control Fasciola disease in cattle, it is necessary to consider differences in the prevalence or number of cases in various cattle breed. This research was conducted in two slaughterhouses in Batu and Pujon District, Malang Regency, East Java, Indonesia involving 50 livestock in each slaughterhouse, either beef or dairy cattle. Observation was done for 2 mo by performing liver necropsy, observing the presence of fibrosis and calcification of the bile duct, counting the number of worms and weighing the removed liver tissue. Result of this research the number of fasciolosis in Ongole and Limousin cattle (30.95 % and 35.94 %) was fewer compared to Simmental and Frisian Holstein (58.33 %). Average number of worms head^{-1} in each cattle breed; Limousin, Simmental, Ongole, and FH were (105, 49, 27 and 129) worms. Meanwhile, fibrosis score in each group were 1.48, 1.89, 1.33 and 1.75 respectively. It is suspected that differences in cattle breed had correlation with susceptibility towards infection. It is indicated that Limousin cattle are more tolerant against liver worm infection.

Key words: Breed of cattle/cows, cow/cattle health, fasciola disease, fibrosis disease, manage disease infections

1 Introduction

Fasciolosis is a disease caused by *Fasciola gigantica* (Cobbold, 1855) or *Fasciola hepatica* (Linnaeus, 1758) that infected ruminants and caused liver damaged and loss of productivity. According to [1] the prevalence of these worms is negatively correlated with milk production. The total economic loss in Kenya during the 10 yr period as a result of condemnation of the infected livers was approximately USD 2 600 000. Fasciola infection can be transmitted from livestock to human and cause health problems [2].

Livestock breeds can affect the ability to manage disease infections. Genetic factors in these livestock can cause livestock in a resistance condition, livestock is able to fight infection [3]. Besides, even though the cattle were infected by pathogens, but only slightly affected, the condition of tolerance [4]. To control Fasciola disease in cattle, it is necessary

* Corresponding author: lilizalizar62@gmail.com

to consider differences in the prevalence or number of cases in various cattle breed. It needs to determine the most resistant or tolerant cattle breed. The purpose of this study was to determine the percentage of Fasciolosis cases in beef and dairy cattle that slaughtered at Batu and Pujon District slaughterhouse in Malang Regency, East Java, Indonesia and the impact of Fasciola infection in decreasing liver condition.

2 Materials and methods

This research was conducted in two slaughterhouses in Batu and Pujon District, Malang Regency, East Java, Indonesia involving 50 livestock in each slaughterhouse, either beef or dairy cattle. Observation was done for 2 mo (February 2019 to March 2019). All cattle older than 2 yr. The study was conducted by performing liver necropsy, observing the presence of fibrosis and calcification of the bile duct, counting the number of worms and weighing the weight of the removed liver tissue. The scoring of fibrosis was based on [4]. Data were analyzed descriptively.

3 Results

The results of liver necropsy from the slaughtered beef and dairy cattle in Batu and Pujon slaughterhouse for 2 mo showed that out of 50 cows slaughtered in Batu as many as 29 beef and dairy cattle (58 %) suffered Fasciolosis. While in Pujon, 10 out of 50 beef and dairy cattle (20 %) suffered Fasciolosis (Table 1). The prevalence of fasciolosis in the Pujon slaughterhouse is almost the same as presented by [5] in the Pontianak slaughterhouse that in beef cattle, the prevalence of *F. hepatica* has been found to reach 23.75 %. [5] Reported the prevalence of fasciolosis in Balinese cattles raised in four Prafi District villages, namely Udapi Hilir, Desay, Aimation, and Prafi Mulya Villages were 30.53 %, 30.61 %, 40.74 % and 38.24 % respectively. These results showed that there were no significant differences from the reseached conducted by [6], that Fasciola's infection rate in slaughtered houses in Punjab reached 22.60 %.

Table 1. Percentage of livestock infected by fasciolosis in Batu and Pujon slaughterhouses

No	Livestock type	District	Number of livestock (tail)	Positive fasciola infection (tail)	Percentage (%)
1.	Limousin	Batu	34	19	55.88
2.	Simmental		6	5	83.33
3.	PO*		7	2	28.57
4.	PFH**		3	3	100
Total			50	29	
1.	Limousin	Pujon	25	4	16.00
2.	Simmental		12	4	33.33
3.	PO		3	1	33.33
4.	PFH		6	1	16.67
5.	Brangus		4	0	0
Total			50	10	

* Ongole Cattle Breed
** Friesian Holstein Cattle Breed

The most common livestock slaughtered in the two slaughterhouses were beef cattle i.e. Limousin and Simmental. Fasciolosis were found in beef cattle i.e. Limousin, Simmental, Ongole and dairy cattle Friesian Holstein (FH) but none were found in Brangus cases. It was suspected that the number of Brangus cattle slaughtered during the study period was lesser than others, thus did not reflect the actual condition.

Table 2. Average percentage (%) of fasciolosis cases based on livestock type in Batu and Pujon slaughterhouse

No	Type of livestock	Percentage of fasciolosis cases		Average percentage of fasciolosis cases in both district
		Batu	Pujon	
1.	Limousin	55.88	16.00	35.94
2.	Simmental	83.33	33.33	58.33
3.	Ongole	28.57	33.33	30.95
4.	FH	100	16.67	58.33

The average percentage of Fasciolosis in both slaughterhouses that infected Limousin, Simmental, Ongol and FH were 35.94 %, 58.33 %, 30.95 % and 58.33 % respectively. Fasciolosis infection needs to be controlled since it zoonoses that endangers human life. According to [7], control of *F. hepatica* infection may have an influence on shedding of *E. coli* 0157 in beef cattle aimed at the human food chain.

Previous studies stated that breed factors need pay attention in an effort to find the types of livestock that are most resistant to infections including the *F. gigantica* [8] reported that the level of susceptibility cattle is higher than buffalo. According to [4], Indonesian thin–tailed sheep are more resistant to artificial infections of *F. gigantica* compared to Merino sheep in terms of the smaller number of worms found in the liver and differences in immune response.

Therefore, research that observed cases of Fasciolosis in various types of livestock such as this research needs to be continued in order to obtain information to develop types of livestock that are more resistant to fight against Fasciola infection. In this study it was observed that fewer cases of Fasciola in Ongole and Limousin cattle compared to Simmental and FH.

Table 3. Average number of worms per livestock and fibrosis score on each type of livestock in both slaughterhouse

No	Type of livestock	Average number of worms/head	Average fibrosis score*
1.	Limousin	105	1.48
2.	Simmental	49	1.89
3.	PO	27	1.33
4.	PFH	129	1.75

Note: Mazeri et al. (4)

In Table 3, the average number of worms head^{-1} in each cattle breed; Limousin, Simmental, Ongole and FH were (105, 49, 27, and 129) worms respectively. While the fibrosis score of each breed was reached 1.48, 1.89, 1.33 and 1.75 respectively. According to [4], fibrosis scores are classified in scores from 0 to 3. Where a score of zero indicates no fibrosis; score 1 indicates the presence of local fibrosis; score 2 shows a severe local fibrosis or mild general fibrosis and score 3 indicates the presence of severe local fibrosis accompanied by calcification of the bile duct or severe general fibrosis. Thus on average in all cattle breed in this study had local fibrosis. Liver fibrogenesis and increased activity of stellate cells are associated with *F. hepatica* infection [9]. Formation of fibrosis occured because the liver cells are damaged due to the presence of juvenile worms that enter the liver tissue. Fibrosis can also be said as an attempt by the body to localize the infection [10].

It is suspected that different breed of cattle has a role in developing immunity against the parasitic infection. In Limousin and FH cattle the number of worms is higher, but the fibrosis score is lower than Simmental (Table 3).

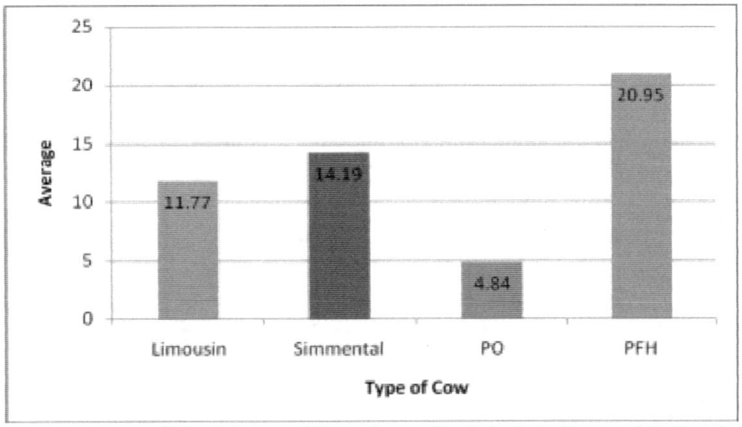

Fig. 1. Average percentage (%) of damaged liver due to fasciolosis in Batu and Pujon slaughterhouse according to cattle breed

Figure. 1 showed that the percentage of liver damaged in Limousin, Simmental, Ongole, and FH were 11.77 %, 14.19 %, 4.84 % and 20.95 % respectively. When comparing Table 2 with Figure 1 it can be concluded that in Limousin, although the number of worms head^{-1} was higher than Simmental, the fibrosis score and the percentage of damaged liver tissue were lower than Simmental. It is possible that Limousin had tolerance to liver worm infection.

The average weight of the damaged liver by fasciolosis from all cattle breed was 10.98 kg on average or 1.42 kg head^{-1}. If in a month 20 cattle suffered Fasciolosis in Batu, East Java, Indonesia the number of liver removed is around 28.40 kg mo^{-1} or 340.80 kg yr^{-1}. The average price of cattle liver kg^{-1} in Batu was IDR 60 000 kg^{-1} so that the estimated economic loss per year reaches IDR 20 448 000.

4 Conclusions

Both beef cattle (Limousin, Simmental, Ongole) and dairy cattle (FH) can be infected by Fasciola sp. The number of Fasciolosis in Ongole and Limousin cattle (30.95 % and 35.94 %) was fewer compared to Simmental and FH (58.33 %). Average number of worms head^{-1} in each cattle breed; Limousin, Simmental, Ongole, and FH were (105, 49, 27 and 129) worms. Meanwhile, fibrosis score in each group were 1.48, 1.89, 1.33 and 1.75 respectively. It is suspected that differences in cattle breed had correlation with susceptibility towards infection. It is indicated that Limousin cattle are more tolerant against liver worm infection.

The authors would say thanks you to the officials and employees of the Batu City abattoir and Pujon, District Pujon for helping this research.

References

1. F. Raunelli, S. Gonzalez. Int. J. App. Vet. Med. **7**,4:145–152(2009). http://www.jarvm.com/articles/Vol7Iss4/Gonzalez.pdf
2. K. Cwiklinski, S.M. O'Neill, S. Donnelly, J.P. Dalton. Parasite Immunology. **38**:558–568(2016). https://doi.org/10.1111/pim.12343
3. S.C. Bishop, C.A. Morris. Small Ruminant Research. **70**,1:48–59(2007). https://doi.org/10.1016/j.smallrumres.2007.01.006
4. S. Mazeri, G. Rydevik, I. Handel, M. B.M.deC. Bronsvoort, N. Sargison. Scientific Reports. **7**,1:1–15(2017). https://doi.org/10.1038/s41598-017-07396-1
5. Purwaningsih, Noviyanti, and Rizki Pratama Putra. Acta Veterinaria Indonesiana. **5**,2:120–126(2017). [in Bahasa Indonesia]. https://doi.org/10.29244/avi.5.2.120-126
6. U.J. Khan, A. Maqbool. Pakistan J. Zool. **44**:1193–1196(2012). http://www.zsp.com.pk/pdf44/1193-1196%20_1_%20PJZ-694-11%20Pakistan%20J.pdf
7. A.K. Howell, S.C. Tongue, C. Currie, J. Evans, D.J.L. Williams, T.N. McNeilly. Preventive veterinary medicine. **150**:70–76(2018). https://doi.org/10.1016/j.prevetmed.2017.12.007
8. M. Hambal, A. Sayuti, A. Dermawan. J. Med. Vet. **7**,1:49–53(2013). ISSN : 0853–1943 49. [in Bahasa Indonesia]. http://www.jurnal.unsyiah.ac.id/JMV/article/view/2921
9. L.A. Marcos, Angelica Terashima, P. Yi, R. Andradet, F.J. Cuberot, E. Albanist, et al. J. Parasitol. **97**,1:82–87(2011). https://doi.org/10.1645/GE-2420.1
10. J.M. Kithuka, N. Maingi, F.M. Njeruh, J.N. Ombui. Onderstepoort Journal of Veterinary Research. **69**,4:255–262(2002). https://repository.up.ac.za/bitstream/handle/2263/18393/32kithuka2002.pdf?sequence=1&isAllowed=y

Factors that Influence Farmer's Behavior Towards Risk

Muhamad Nurdin Yusuf*, *Agus* Yuniawan Isyanto, and *Sudradjat* Sudradjat

Department of Agribusiness, Faculty of Agriculture, Galuh University, Jl. R. E. Martadinata No.150, Ciamis 46274, Indonesia

Abstract. The research was carried out with the aim to find out the behavior of farmers towards risk and the factors that influence it. The research sample was 100 paddy farmers in flood-prone area paddy fields in Pangandaran District, West Java Province, Indonesia. Farmer's behavior towards risk was analyzed using quadratic utility functions, while the factors that influence farmer's behavior towards risk were analyzed using logistic regression. The results showed farmers 87 was risk neutral, while 13 farmer risk takers were farmers. Education, familys size and income significantly influence farmer's behavior towards risk; while age, experience, land area, production risk, price risk, income risk and group did not significantly influence farmer's behavior towards risk.

Keywords: Land area, logistic regression, risk taker, quadratic utility.

1 Introduction

The impact of global climate change has an effect on paddy production in Indonesia in the form of flooding which results in crop failure or a decrease in paddy production [1]. Climate change has occurred in Indonesia with indications of an increase in temperature and changes in rainfall patterns. Agriculture is very vulnerable to the effects of climate change with an indication of the high level of danger in decreasing paddy production as a result of increasing temperatures and changes in rainfall patterns. The average decline in paddy production is 1.37 % yr^{-1} which has the potential to cause a decline in national food production [2].

The threat of flooding in paddy fields can lead to reduced harvest area and paddy production [3]. Production is related to the nature of farming which is always dependent on nature supported by risk factors [4]. The risk of failure in farming comes from the use of new technology, prices of agricultural production, capital, government policies and individual behavior of farmers in dealing with outsiders [5], as well as climate change and weather that are not in accordance with crop needs [6]. The main risks farming include flooding [7].

Farmer's behavior towards risk consists of risk averter, risk neutral, and risk taker. Farmer behavior is the basis of farmers' decision making in carrying out their farming [1]. Farmer's behavior towards risk has an important role in influencing the productivity of agricultural products which has an impact on production efficiency [8]. Factors that

* Coresponding author: muhamadnurdinyusuf@gmail.com

influence farmer behavior towards risk are the area of planting, age, education, experience, family size, income and productivity risk [8].

2 Research methodology

Amount ten villages in Padaherang and Kalipucang Subdistricts in Pangandaran District were taken purposively as a sample area with consideration of being areas that have flood-prone rice fields. From each village a sample of ten farmers was taken, so that the total number of samples was 100 farmers.
Farmer's behavior towards farming risk was analyzed using the quadratic utility Equation (1):

$$U = \tau_1 + \tau_2 M + \tau_3 M^2 \quad (1)$$

Where:
U : utility for expected income (in util)
τ_1 : intercept
M : expected income at the balance point (rupiah value from certainty equivalent (CE)
τ_2 : indifference income coefficient (CE)
τ_3 : farmer risk coefficient

The risk preference coefficient shows the farmer's attitude to risk, namely:
$\tau_3 = 0$: Risk neutral
$\tau_3 < 0$: Risk averse
$\tau_3 > 0$: Risk taker

The factors that influence farmers' behavior towards risk are analyzed using ordinal logit regression with the following Equation (2):

$$P_i = F(Y_i)= F(\alpha + \beta_1X_1 + \beta_2X_2 + \beta_3X_3 + \beta_4X_4 + \beta_5X_5 + \beta_6X_6 + \beta_8D_8+ \beta 9D9 + \beta_{10}D) \quad (2)$$

Y_i values are calculated using the following Equation (3):

$$Y_i = Log\ [P_i/(1-P_i)] = (\alpha + \beta_1X_1 + \beta_2X_2 + \beta_3X_3 + \beta_4X_4 + \beta_5X_5 + \beta_6X_6 + \beta_7X_7 + \beta_8X_8 +\beta_9X_9+ \beta_{10}D + e) \quad (3)$$

Where:

Y_i = Opportunities for farmers to make decisions, where:
Y_1 = 1 for the farmer who are risk averse
Y_2 = 2 for the farmer who are risk neutral
Y_3 = 3 for the farmer who are risk taker
α = Intercept
β_i = Parameter regression coefficient (i = 1, 2, 3,….10)
X_1 = Age (year)
X_2 = Education (year)
X_3 = Family size (person)
X_4 = Experience (year)
X_5 = Land area (ha)
X_6 = Farm income (IDR)
X_7 = Production risk
X_8 = Price risk
X_9 = Income risk
D = Membership in groups (1 if being a member of a group, 0 if not)
e = error term

3 Results and discussion

3.1 Farmer's behavior towards risk

Farmer's behavior towards risk in paddy farming in flood-prone paddy fields can be seen in Table 1.

Table 1. Farmer's behavior towards risk

Farmer's behavior towards risk	Number of people	Percentage
Risk Neutral	87	87.00
Risk Taker	13	13.00
Total	100	100.00

Table 1 shows that most farmers (87 %) are risk neutral, while the rest (13 %) are risk takers. There are no risk averse farmers. According to [9], the absence of risk avers farmers show that there are no farmers who are willing to sacrifice their income or potential income to reduce opportunities for loss or low income. According to [1], farmers will try to avoid failure and not get big profits by taking risks. Such behavior is called safety first, which is characteristic of most farmers.

Farmers who are risk neutral are farmers who have a rational attitude in facing risks [9]. Risk neutral farmers tend to cultivate based on hereditary habits. They only seek income that can meet their family's needs [10]. If there is additional capital for risk neutral farmers, then they might add input to get higher income [11]. The availability of capital for farmers is a risk factor that is considered to affect farmers to the farming they are doing [12].

Risk taker farmers are farmers who are willing to allocate and use their production factors to the maximum, even though there are risks that must be faced with the aim of obtaining optimal results. [8] states that conceptually farmers are able to reduce production risk and price risk by improving their productivity, the use of diversification, the use of appropriate cropping patterns, strengthening farmer institutions, and bargaining position of farmers can increase farmers' production and income [7].

3.2 Factors that influence farmer's behavior toward risk

The results of the analysis of factors that influence farmer behavior on risk in paddy farming in flood-prone paddy fields can be seen in Table 2.

Table 2. Factors that influence farmer's behavior towards risk

Variabel	B	Wald	Exp(B)
Age	0.788	0.020	2.199
Education	7.139	3.452**	1.260E3
Family size	4.983	4.303*	145.958
Experience	-0.149	0.003	0.861
Land area	0.687	0.279	1.988
Farm income	2.896	4.313*	18.103

(Continued on next page)

Table 2. Continued

Variabel	**B**	**Wald**	**Exp(B)**
Production risk	1.029	0.407	2.799
Price risk	-0.834	1.016	0.434
Income risk	-0.134	0.014	0.875
Group	1.795	0.946	6.020
Constant	-69.854	3.703**	0.000
Model Summary			
-2 Log likelihood	Cox & Snell R Square	Nagelkerke R Square	Chi-square
19.158	0.441	0.819	18.307*

*,** = significant at 5 %, 10 %

Table 2 shows that age is not significant influence farmers' behavior towards risk. This shows that the difference in age of farmers does not affect the behavior of farmers in dealing with risk. Education significantly influences farmer's behavior towards risk. The results of this study are in accordance with the results of research from [13]. The higher the level of education of farmers, the more courageous farmers face risks [5, 7]. According to [13], the level of education of farmers who are still low makes the main cause of the majority of farmers who choose safety first (zero risk) behavior in developing their farming.

Family size significantly influences farmer's behavior towards risk. The results of this study are in accordance with the results of research by [14]. According to [9], family size affects the outpouring of time that can be allocated for farming. According to [14], the more number of members of the farmer's family, the higher the capacity of the workforce owned by the farmer in facing the risk.

Experience does not significantly influence farmers' behavior towards risk. The longer the farmer's experience, the more careful it will be in carrying out farming so that it tends to be more neutral to risk. The results of this study are in accordance with the results of research from [7].

Land area does not significantly affect farmer's behavior to risk. The results of this study are consistent with the results of a study from [5] which shows that the addition or reduction of land area will not reduce risk aversion or that farmers are neutral towards risk.

Revenue significantly influences farmer's behavior towards risk. The greater the income the farmer receives from the farming that is carried out, the farmer will be more courageous in accepting the risk. The results of this study are in accordance with the results of research from [8, 14, 5]. According to [8], the higher the farm income received by farmers, the farmers will be more willing to accept the risk.

Production risk does not significantly influence farmer's behavior towards risk. According to [7], the magnitude of the risk of production faced by farmers due to the uncertainty of results as a result of natural factors and income as a result of price fluctuations, causes farmers to tend to reject the possibility of accepting the risks and uncertainties of the business. According to [15], efforts to handle production risk can be done by implementing a diversification and agricultural insurance program.

Price risk does not significantly influence farmer's behavior towards risk. According to [16], farmers' household attitudes in carrying out production activities can be seen from variations in prices as a measure of price risk. According to [17], the behavior of farmers who accept price risk is caused by the expectation of expected price reductions compared to the actual prices that make farmers continue to carry out farming as long as they provide benefits. According to [18], the behavior of farmers does not dare to risk due to fluctuations in production and selling prices which will have an impact on farmers' income.

Income risk does not significantly influence farmer behavior towards risk. The higher the income risk faced by farmers, the more farmers will behave neutrally against risk. Membership in groups does not significantly influence farmers' behavior towards risk. According to [19], group membership is one of the efforts of farmers to reduce the risk of farming by cooperating with each other, exchanging information in managing their farming with fellow farmers.

4 Conclusion

Amount 87 farmers are risk neutral and 13 farmers are risk takers. Education, family size and income significantly influence farmer's behavior towards risk; while age, experience, land area, production risk, price risk, income risk and group did not significantly influence farmer's behavior towards risk.

References

1. C. Gardebroek, M.D. Chavez, A.O. Lansink, J. Agric. Econ. **61**,1:60–75(2010). https://doi.org/10.1111/j.1477-9552.2009.00222.x
2. Ruminta, Handoko, T. Nurmala. Jurnal Agro. **5**,1:48–60(2018). [in Bahasa Indonesia]. https://journal.uinsgd.ac.id/index.php/ja/article/download/1607/pdf_16
3. Shahab. E. Saqib, Mokbul. M. Ahmad, Sanaullah. P, Irfan. A Rana, Int. J. Disaster Risk Reduc. **18:**107–114(2016). https://www.researchgate.net/profile/Shahab_E_Saqib/publication/304305997_An_Empirical_Assessment_of_Farmers%27_Risk_Attitudes_in_Flood-Prone_Areas_of_Pakistan/links/5bf55cd2299bf1124fe32898/An-Empirical-Assessment-of-Farmers-Risk-Attitudes-in-Flood-Prone-Areas-of-Pakistan.pdf?origin=publication_detail
4. C. Gardebroek, Eur. Rev. Agric. Econ. **33**,4:485–510(2006). https://doi.org/10.1093/erae/jbl029
5. M.P. Lucas, I.M. Pabuayon., Asian J. Agric. Dev. **8**,2:61–77(2011). https://tind-customer-agecon.s3.amazonaws.com/9fdd6f1e-7e97-4413-a5f2-ffb38a8fea0c?response-content-disposition=attachment%3B%20filename%2A%3DUTF-8%27%27AJAD_2011_8_2_5Lucas.pdf&response-content-type=application%2Fpdf&X-Amz-Algorithm=AWS4-HMAC-SHA256&X-Amz-Expires=86400&X-Amz-Credential=AKIAXL7W7Q3XHXDVDQYS%2F20200908%2Feu-west-1%2Fs3%2Faws4_request&X-Amz-SignedHeaders=host&X-Amz-Date=20200908T165030Z&X-Amz-Signature=20ca172c04b02c087149aa1ba32cc86309bf574faede65f1910b98ade0e76e34
6. T. Tiedemann, U.L. Lohman, J. Agric. Econ. **64**,1:73–96(2012). https://doi.org/10.1111/j.1477-9552.2012.00364.x
7. J.N. Nmadu, G.P. Eze, A.J. Jirgi. British J. Econ. Manag. Trade, **2**,2:92–108(2012). https://www.journaljemt.com/index.php/JEMT/article/download/30022/56339
8. R. Darma, T.N.A. Rahmadanih, R. Amandaria. Int. J. Agric. Syst. **2**,1:77–89(2014). http://pasca.unhas.ac.id/ojs/index.php/ijas/article/download/24/23
9. E.D. Oruonye, IJFSE. **3**,4:113–117(2013). http://www.ijfse.com/uploadedfiles/IJFSEArchive/IJFSE2013/3(4)/01.pdf
10. M.C. Plaxedes, P. Mafongoya, Clim. Risk Manag. **16**:145–163(2017). https://doi.org/10.1016/j.crm.2017.02.001

11. Y. Ferrianta, M.F. Makki, Suprijanto, Rifiana. Int. J. Agric. Manag. Dev. **5**,2:133–139(2014). https://www.researchgate.net/publication/272667859_Risk_Analysis_and_Strategy_of_Rice_Farmers_in_Swampland_in_the_Face_of_Climate_Change_Impact_Case_in_South_Kalimantan_Province-Indonesia/fulltext/57a79def08aee07544c1d003/Risk-Analysis-and-Strategy-of-Rice-Farmers-in-Swampland-in-the-Face-of-Climate-Change-Impact-Case-in-South-Kalimantan-Province-Indonesia.pdf?origin=publication_detail
12. A.D. Brauw, P. Eozenou. J. Dev. Econ. **111:**61–74(2014). https://www.researchgate.net/profile/Patrick_Eozenou/publication/266617635_Measuring_risk_attitudes_among_Mozambican_farmers/links/5435ad990cf2643ab9867e0d/Measuring-risk-attitudes-among-Mozambican-farmers.pdf?origin=publication_detail
13. C.A. Kalu, J.A. Mbanasor. NAPReJ, **1**,1:1–10(2016). https://econpapers.repec.org/scripts/redir.pf?u=https%3A%2F%2Fdoi.org%2F10.22004%252Fag.econ.292052;h=repec:ags:naprej:292052
14. M.M. Haque, S. Bremer, S. Aziz, J.P. Sluijs, Clim. Risk Manag. **16:**43–58(2017). https://dx.doi.org/10.1016/j.crm.2016.12.002
15. S. Dadzie, N. Kwesi, H. de Graft. A, Int. J. Agric. For. **2**,2:29–37(2012). https://www.researchgate.net/profile/Samuel_Dadzie/publication/241699647_Attitudes_Toward_Risk_and_Coping_Responses_The_Case_of_Food_Crop_Farmers_at_Agona_Duakwa_in_Agona_East_District_of_Ghana/links/0046351caa436af152000000/Attitudes-Toward-Risk-and-Coping-Responses-The-Case-of-Food-Crop-Farmers-at-Agona-Duakwa-in-Agona-East-District-of-Ghana.pdf?origin=publication_detail
16. S.R. Fajri, E. Fauziyah. J. Hort. Indonesia, **9**,3:188–196(2018). [in Bahasa Indonesia] https://journal.ipb.ac.id/index.php/jhi/article/download/26496/17128/
17. C. Mulwa, P. Marenya, D.B. Rahut, M. Kassie, Clim. Risk Manag. **16**:208–221(2017). https://reader.elsevier.com/reader/sd/pii/S2212096317300062?token=8C6D3F91A4BEF28A0A8335A476A7B6B75EB4D84ED7742D9C358F0465D68675FD002D8C4A371BC054A01FAF3CD18289A8
18. A. Ali, O. Erenstein, Clim. Risk Manag. **16**:1–12(2017). https://www.researchgate.net/profile/Olaf_Erenstein/publication/311879490_Assessing_farmer_use_of_climate_change_adaptation_practices_and_impacts_on_food_security_and_poverty_in_Pakistan/links/58b71affaca27261e51a4134/Assessing-farmer-use-of-climate-change-adaptation-practices-and-impacts-on-food-security-and-poverty-in-Pakistan.pdf?origin=publication_detail
19. A. Tripathi, A.K. Mishra. Clim. Risk Manag. **16**:1–37(2017). https://www.researchgate.net/publication/310815962_Knowledge_And_Passive_Adaptation_to_Climate_Change_an_Example_From_Indian_Farmers/fulltext/583a4d6308ae3a74b49ea7e4/Knowledge-And-Passive-Adaptation-to-Climate-Change-an-Example-From-Indian-Farmers.pdf?origin=publication_detail

Effectiveness of Mycorrhiza, Plant Growth Promoting Rhizobacteria and Inorganic Fertilizer on Chlorophyll Content in *Glycine max* (L.) cv. Detam-4 Prida

Muhammad Muhammad[1]**, *Umi* Isnatin[1], *Peeyush* Soni[2], and *Praptiningsih* Gamawati Adinurani[3]

[1]Department of Agrotechnology, University of Darussalam Gontor, Jl. Raya Siman, Km 5 Siman Ponorogo 63471, Indonesia
[2]Department of Agricultural and Food Engineering Indian Institute of Technology Kharagpur, 21302 Kharagpur, West Bengal, India
[3]Department of Agrotechnology, Merdeka University of Madiun, Jl. Serayu No.79, Madiun 63133, East Java, Indonesia

Abstract. This study aimed to find an effective combination of mycorrhiza, PGPR (Plant Growth Promoting Rhizobacteria), and inorganic fertilizers. Whereas the specific purpose was observed in effective mycorrhiza to increased chlorophyll content. This study used a completely randomized design (CRD) with three factors: the 1st factor is mycorrhiza application, the 2nd factor is PGPR and the 3rd factor is the application of nitrogen and phosphorus fertilizers. The data were analyzed with Analysis of Variance to determine the effect of the treatment being tried. Continued with the Least Significance Different test at a 95 % confidence level. The results indicated that the most effective application in increasing total chlorophyll content was (i) the "Commercial Mycorrhiza" without being combined with PGPR and fertilizer (TP.TR.MP:80 g mL^{-1}). (ii) "Brawijaya Mycorrhiza" is combined with PGPR without fertilizer (TP.R.MB: 83 g mL^{-1}). (iii) "Unida Mycorrhiza" without combined with PGPR and without fertilizer (TP.TR.MU: 80 g mL^{-1}).

Keywords: Environmental friendly, increase soybean productivity, microbial fertilizer,

1 Introduction

Photosynthesis is the process by which plants harvest sunlight to produce sugars from carbon dioxide and water [1]. Photosynthesis is a metabolic process in plants to form carbohydrates that use CO_2 from free air and water from the soil with the help of sunlight and chlorophyll. The process of photosynthesis will occur if there is light and an intermediate pigment, chlorophyll. Chlorophyll is a leafy or pigment green substance that has a major role in photosynthesis. The main function of photosynthesis is the formation of fat and protein energy sources from glucose. The results of the process of photosynthesis in

* Corresponding author: muhammad07@unida.gontor.ac.id

the form of glucose. The chlorophyll content is a benchmark of plant growth related to crop production [2]. The higher the chlorophyll content of a plant, the higher the yield potential [3].

The main problem in increasing the production of black soybean [*Glycine max* (L.) Merr.] lies in the absorption of the element nitrogen, phosphorus and potassium. Not yet the maximum absorption of nutrients by black soybean plants will cause low yield production. Therefore, one alternative that can be done to increase production is to look for environmentally friendly technologies, namely using mycorrhiza and PGPR to optimize the absorption of nutrients, especially phosphorus and nitrogen [4, 5]. Mycorrhizal fungi can help plants absorb phosphorus and water absorption, while PGPR can provide nitrogen for black soybean plants. PGPR application can support the formation of mycorrhiza and can improve various functions adequately, such as Arturson's research inoculation with strains of the bacterium *Paenibacillus brasilensis* has been shown to increase the level of root colonization by mycorrhizal fungi and can absorb phosphorus elements so as to increase plant growth [6].

The general objective of this study is to find an effective combination of mycorrhizae, PGPR and inorganic fertilizers. While the specific purpose of this research is to find mycorrhizae that are effective in nutrient absorption to increase chlorophyll content.

2 Methods

Research has been carried out at the greenhouse of the Department of Agrotechnology, University of Darussalam, Gontor, Ponorogo, Indonesia. The design used was a completely randomized design (CRD) with three factors: the first factor was mycorrhizal application, the second factor was PGPR and the third factor was the application of nitrogen, phosphorus and potassium fertilizers (P). Mycorrhiza used in this study were collections from the Darussalam University – UNIDA (MU) Agrotechnology Laboratory, mycorrhiza produced by Brawijaya University (MB) and mycorrhiza produced by the manufacturer (commercial mycorrhiza) (MP). Whereas PGPR uses the types of *Pseodomonas* sp. (Migula, 1894) and *Bacillus Polymixa* (Prazmowski, 1880, Macé 1889) from the Agrotechnology Laboratory, Darussalam University. While the black soybean [*Glycine max* (L.) Merr.] cv. Detam-4 Prida was obtained from BALITKABI (*Balai Penelitian Tanaman Aneka Kacang dan Umbi* – Indonesian Legumes and Tuber Crops Research Institute).

Stages the chlorophyll content test has been carried out as follows; leaf sampling on the 40 d after the application of treatment by taking samples of the leaves of the bottom, middle and top of the plant, then weighed 1 g. Each sample was crushed with cold mortal and 1 mL of boric acid was added, the results of the grinding were transferred to the ependof and 4 mL of distilled water was added. Furthermore centrifuged at 6 000 rad s^{-1} for 25 mL. The resulting chlorophyll (pellet) extract was transferred to the reaction tube and 96 mL of 5% ethanol was added. Chlorophyll extracts were incubated in a dark room for 30 min. The chlorophyll content of absorbance was measured using a spectrophotometer at wavelengths of 649 nm and 665 nm.

The amount of chlorophyll-a was calculated using the Formula (1), and the amount of chlorophyll-b was calculated using the Formula (2)

$$(13.7 \times A665) - (5.76 \times A649) \quad (1)$$

$$(25.8 \times A649) - (7.60 \times A665) \quad (2)$$

while the amount of chlorophyll the total is calculated using the formula chlorophyll-a plus chlorophyll-b.

The observational data were analyzed with Analysis of Variance (ANOVA) to determine the effect of the treatment being tried. If the results of the analysis have a significant effect then proceed with the LSD test (Least Significance Different) at a 95 % confidence level [7].

3 Results and discussions

3.1 Level of group dynamics

Chlorophyll is a leafy or pigment green substance that has a major role in photosynthesis. The main function of photosynthesis is the formation of fat and protein energy sources from glucose. The results of the process of photosynthesis in the form of glucose. The process of chlorophyll synthesis through photoreduction of protochlorophyllide to chlorophyllide a and subsequently by esterification of phytol to form chlorophyll a is catalyzed by the chlorophyllase enzyme [8]. Chlorophyll content is a benchmark of plant growth related to crop production. The higher the chlorophyll content of a plant, the higher the yield potential [3].

Chlorophyll-a is the main pigment in photosynthesis which functions to electron donors in the photosynthetic electron transport chain. Chlorophyll-a most effectively absorbs the spectrum with wavelengths of 429 nm and 659 nm. This chlorophyll reflects the blue-green color. Factors that influence the chlorophyll content are nutrients N, Mg, Fe and Light. All green plants contain chlorophyll-a and chlorophyll-b, chlorophyll-a making up 75 % of the total chlorophyll [8]. The results showed that there were very significant differences between mycorrhizal, PGPR and fertilizer treatments according to Analysis of Variance (ANOVA) analysis.

Table 1. Clorophyll a content

No.	Treatment	Average	Notation
1	TP TR TM	27	bc
2	TP TR MU	29	d
3	TP TR MB	28	c
4	TP TR MP	27	bc
5	TP R TM	29	d
6	TP R MU	28	c
7	TP R MB	26	a
8	TP R MP	28	c
9	P TR TM	29	d
10	P TR MU	29	d
11	P TR MB	32	e
12	P TR MP	26	a
13	P R TM	29	d
14	P R MU	26	a
15	P R MB	28	c
16	P R MP	28	c

Notes: Figures in the same column and followed by the same letter are not significantly different based on the LSD 5 % further test.

The results of the analysis of chlorophyll-a absorbance value of mycorrhizal treatment without fertilizer and without PGPR were the highest "Unida Mycorrhiza" (TP.TR.MU:g mL^{-1}), while the lowest treatment of mycorrhizal treatment (TP.TR.MP: 27 g mL^{-1}). The highest mycorrhizal treatment plus PGPR without chlorophyll-a fertilizer is

"Unida Mycorrhiza" and "Commercial Mycorrhiza" (TP.R.MU: 28 g mL^{-1}), while the lowest is "Brawijaya Mycorrhiza" (TP.R.MB: 26 g mL^{-1}). The treatment of mycorrhizae plus fertilizer without PGRP the highest value of chlorophyll a content is "Brawijaya Mycorrhiza" (P.TR.MB: 32 g mL^{-1}), while the lowest is the "Commercial Mycorrhiza" (P.TR.MP: 26 g mL^{-1}). The highest treatment of mycorrhizae plus fertilizer and PGPR chlorophyll-a content is control (P.R.TM: 29 g mL^{-1}), while the lowest is "Unida Mycorrhiza" (P.R.MU: 26 g mL^{-1}).

In this study, it was found that there was a different response between mycorrhizae with other mycorrhizae towards fertilizer and PGPR. The interaction of fertilizers between plants and microorganisms in accumulating and translocating metabolites between stems and roots is an indicator of the symbiotic status between plants and microorganisms. In Juge's research on soybean plants showed that inoculation of three microorganisms when applied together had a negative impact on the shoot biomass of soybeans. The association of three microorganisms will reduce the root depth for nodulation during formation and reduce the number of nodules at the end of the experiment which can show competition when the three microorganisms are applied together [9]. "Unida Mycorrhiza" in increasing chlorophyll-a content is more effectively applied alone without a mixture of fertilizer and PGPR (TP.TR.MU: 29 g mL^{-1}) in Table 1. Application of "Brawijaya Mycorrhiza" more effectively combined with fertilizers without PGPR in increasing chlorophyll-a content compared to with other combinations, and get the highest value (P.TR.MB: 32 g mL^{-1}). For the application of "Commercial Mycorrhiza" is more effectively combined with PGPR without fertilizer compared to other combinations and can increase the chlorophyll-a content by 28 g mL^{-1} (TP.R.MP).

Table 2. Clorophyll b content

No	Treatment	Average	Notation
1	TP TR TM	52	cdefg
2	TP TR MU	46	bcde
3	TP TR MB	46	bcd
4	TP TR MP	53	cdefg
5	TP R TM	45	bc
6	TP R MU	49	bcdef
7	TP R MB	58	g
8	TP R MP	51	bcdefg
9	P TR TM	45	bc
10	P TR MU	42	b
11	P TR MB	26	a
12	P TR MP	55	defg
13	P R TM	44	bc
14	P R MU	55	defg
15	P R MB	47	bcde
16	P R MP	47	bcde

Notes : Figures in the same column and followed by the same letter are not significantly different based on the LSD 5 % further test.

Chlorophyll-b is an accessory pigment that is also responsible for the photosynthesis process. Chlorophyll has a function to collect light energy and enter chlorophyll-a during the process of photosynthesis. Chlorophyll-b is the most effective absorbing spectrum with wavelengths of 455 nm and 642 nm. Chlorophyll-b reflects the yellow-green color.

The results of research on the chlorophyll-b content in Table 2 show that mycorrhizal application without fertilizer and without PGPR is ineffective. This study found a decrease in the content of chlorophyll-b when "Unida Mycorrhiza" and "Brawijaya Mycorrhiza", except the "Commercial Mycorrhiza". Applications of mycorrhizae show no significant difference from the treatment without mycorrhiza. Mycorrhizal treatment combined with PGPR without fertilizer shows that the effective application is "Brawijaya Mycorrhiza" (TP.R.MB: 58 g mL^{-1}) while the lowest is the application without mycorrhiza (TP.R.TM: 45 g mL^{-1}). This study found that the application of mycorrhizae combined with PGPR without fertilizer is still able to increase the chlorophyll-b content of both "Unida Mycorrhiza" or "Commercial Mycorrhiza". Mycorrhizal application treatment combined with fertilizer without PGPR shows that there is a decrease in chlorophyll-b content except the treatment of "Commercial Mycorrhiza" (P.TR.MP: 55 g mL^{-1}). The most significant reduction was the treatment of "Brawijaya Mycorrhiza" (P.TR.MB: 26 g mL^{-1}). Mycorrhizal application treatment Fertilizer and PGPR showed that there was an increase in chlorophyll-b content when compared with those without mycorrhizal application, and the highest value was obtained by "Unida Mycorrhiza" treatment (P.R.MU: 55 g mL^{-1}).

Table 3. Total clorophyll content

No	Treatment	Average	Notation
1	TP TR TM	79	bcde
2	TP TR MU	80	cde
3	TP TR MB	74	bcd
4	TP TR MP	80	cde
5	TP R TM	73	bc
6	TP R MU	78	bcde
7	TP R MB	83	e
8	TP R MP	79	bcde
9	P TR TM	74	bc
10	P TR MU	71	b
11	P TR MB	58	a
12	P TR MP	81	cde
13	P R TM	73	bc
14	P R MU	81	cde
15	P R MB	75	bcde
16	P R MP	75	bcd

Notes : Figures in the same column and followed by the same letter are not significantly different based on the LSD 5 % further test.

Chlorophyll content is a benchmark of plant growth related to crop production [1]. The results of the study of total chlorophyll content of mycorrhizal applications without fertilizer and without PGPR showed that all mycorrhizal applications can increase the total chlorophyll content of both "Unida Mycorrhiza", "Brawijaya Mycorhiza" and "Commercial Mycorrhiza". This has also been explained by Hassani in his research inoculation of mycorrhizal fungi and genus *Bacillus* can increase chlorophyll content and P uptake at the initiation of tuber potato plants. Interaction between mycorrhizae and plants can increase P uptake and chlorophyll content by 1 % compared to without mycorrhizal application [10]. The process of taking phosphate nutrients in the soil by mycorrhiza involves hyphae. Phosphorus nutrient flow in hyphae follows cytoplasmic flow, while nutrient transfer from fungi to host plants through subtle branches in host cells [11]. Increased phosphorus due to mycorrhizal inoculation in plants will affect photosynthetic activity, mycorrhizal infection in plant roots can increase nutrient translocation to the top of the plant resulting in an increase in the rate of photosynthesis and the use of assimilates in the canopy and an increase in photosynthate supply from leaves to roots. Further, phosphorus is found as phosphate in some plant minerals and is a key element of protoplasm. Phosphorus is present in water as orthophosphate [12]. Phosphate exists in three forms namely H_2PO4, HPO_4^2, and PO_4^{3-}. Phosphate is generally absorbed by plants in the form of primary orthophosphate ions $H_2PO_4^-$ or secondary orthophosphate HPO_4^{2-} whereas PO_4^{3-} is more difficult to be absorbed by plants.

The highest treatment of mycorrhizal application is unida mycorrhiza (80 g mL^{-1}), while the lowest treatment value is the treatment without mycorrhiza. The treatment of mycorrhizal application combined with PGPR and without fertilizer shows that the brawijaya mycorrhiza gets the highest value (TP.R.MB: 83 g mL^{-1}), while the lowest PGPR application without Mycorrhizae and without fertilizer (TP.R.TM: 73 g mL^{-1}). Vafadar's research explains that using mycorrhizae and PGPR can significantly increase leaf dry mass production and stevioside concentration. The application of double inoculation of mycorrhizal fungi and N-binding bacteria is most effective for increasing plant growth, plant nutrient uptake, total chlorophyll content and stevioside content [13]. Mycorrhizal application treatment combined with fertilizer without PGPR shows that mycorrhizal application has decreased total chlorophyll content except the application of mycorrhizal (P.TR.MP: 81 g mL^{-1}), while the lowest mycorrhizal treatment is "Brawijaya Mycorrhiza" (P.TR.MB: 58 g mL^{-1}).

Mycorrhizal application treatment combined with PGPR and fertilizer shows that all mycorrhizal applications experience an increase in total chlorophyll content, the highest mycorrhizal application is "Unida Mycorrhiza" (PRMU: 81 g mL^{-1}), while the lowest mycorrhizal application is PGPR application and fertilizer without mycorrhizae (PRTM: 73 g mL^{-1}). This study found that the most effective application of mycorrhizae in increasing total chlorophyll content was "Commercial Mycorrhiza" without combined with PGPR and fertilizer (TP.TR.MP: 80 g mL^{-1}) because if combined with PGPR and fertilizer it would decrease the total chlorophyll content (PRMP: 75 g mL^{-1}). The most effective application of "Brawijaya Mycorrhiza" in increasing total chlorophyll content is "Brawijaya Mycorrhiza" combined with PGPR without fertilizer (TP.R.MB: 83 g mL^{-1}), if the application of "Brawijaya Mycorrhizae" combined with fertilizer will dramatically decrease the total chlorophyll content (P .TR.MB: 58 g mL^{-1}). The most effective application of "Unida Mycorrhiza" in increasing total chlorophyll content is "Unida mycorrhiza" without combined with PGPR and without fertilizer (TP.TR.MU: 80 g mL^{-1}), if unida mycorrhiza is combined with PGPR and fertilizer can still increase the total chlorophyll content (PRMU: 81 g mL^{-1}) and if "Unida Mycorrhiza" is combined with fertilizer without PGPR, it will decrease the total chlorophyll content (P.TR.MU: 71 g mL^{-1}).

4 Conclusion

The most effective application in increasing total chlorophyll content was (i) the "commercial mycorrhiza" without being combined with PGPR and fertilizer (TP.TR.MP:

80 g mL^{-1}). (ii) "Brawijaya mycorrhiza" is combined with PGPR without fertilizer (TP.R.MB: 83 g mL^{-1}). (iii) "Unida mycorrhiza" without combined with PGPR and without fertilizer (TP.TR.MU: 80 g mL^{-1}).

Gratitude is devoted to the Ministry of Research and Technology and Higher Education of Republic of Indonesia, for funding this research by Beginner Lecturer Research Program with decision letter No. 113/SP2H/LT/DRPM/2019. Gratitude is also directed to Directorate of Research and Community Service, University of Darussalam Gontor, for the opportunity to carry out this research.

References

1. L. Guanter, Y. Zhang, M. Jung, J. Joiner, M. Voigt, J.A. Berry, et al. Proceedings of the National Academy of Sciences, **111,**14:E1327–E1333(2014). https://www.pnas.org/content/pnas/111/14/E1327.full.pdf
2. E. Proklamasiningsih, I.D. Priyambodo, D. Rachmawati, R.P. Sancayaningsih, Bionatura Jurnal Ilmu-ilmu Hayati dan Fisik, **14**,2:107–114(2012). [in Bahasa Indonesia] https://repository.ugm.ac.id/101262/
3. E.D. Purbajanti, F. Kusmiyati, W. Slamet, P.G. Adinurani,. AIP Conference Proceedings. **1755,**130013:1–4 (2016); https://doi.org/10.1063/1.4958557
4. P.G. Adinurani, S. Rahayu, L. S. Budi, A. Nindita, P. Soni, M. Mel,. MATEC Web Conference. **164,** 01035: 1–5(2018). https://doi.org/10.1051/matecconf/201816401035
5. P.G. Adinurani, S. Rahayu, L. S. Budi, S. Pambudi, P. Soni, IOP Conference Series: Earth and Environmental Science **293,**012032:1–7(2019). doi:10.1088/1755-1315/293/1/012032
6. S.M. Nadeem, M. Ahmad, Z.A. Zahir, A. Javaid, M. Ashraf, Biotechnol. Adv. **32,**2:429–448(2014). https://www.sciencedirect.com/science/article/abs/pii/S073497501300222X
7. L.J. Williams, H. Abdi, Encyclopedia of research design. **218**:840–853(2010). http://dx.doi.org/10.4135/9781412961288.n154
8. A.J. Pratama, *Analisis Kandungan Klorofil Gandasuli (Hedychium gardnerianum Shephard ex Ker-Gawl) pada Tiga Daerah Perkembangan Daun yang Berbeda* [Gandasuli Chlorophyll Content (*Hedychium gardnerianum* Shephard ex Ker-Gawl) in Three Different Leaf Development Areas]. Seminar Nasional Konservasi dan Pemanfaatan Sumber Daya Alam 2015. Sebelas Maret University, Solo.(2015).p.216–219. [in Bahasa Indonesia]. https://core.ac.uk/download/pdf/289792435.pdf
9. C. Juge, D. Prévost, A. Bertrand, M. Bipfubusa, F.P. Chalifour, Appl. Soil Ecol. **61,**147–157(2012). https://www.researchgate.net/profile/Christine_Juge2/publication/257382118_Growth_and_biochemical_responses_of_soybean_to_double_and_triple_microbial_associations_with_Bradyrhizobium_Azospirillum_and_arbuscular_mycorrhizae/links/59e4f3a5458515250246ec50/Growth-and-biochemical-responses-of-soybean-to-double-and-triple-microbial-associations-with-Bradyrhizobium-Azospirillum-and-arbuscular-mycorrhizae.pdf
10. F. Hassani, M. Ardakani, A. Asgharzade, F. Paknezhad, A. Hamidi, Int. J. Biosci. **4,**1:244–251(2014). http://citeseerx.ist.psu.edu/viewdoc/download?doi=10.1.1.651.8551&rep=rep1&type=pdf
11. M. Muhammad, H. Setyaningrum, Jurnal Agroqua: Media Informasi Agronomi dan Budidaya Perairan, **15,**2:1–12(2017). https://journals.unihaz.ac.id/index.php/agroqua/article/view/94

12. M. Muhammad, U. Wasit, Kajian Kesuburan Tanah di Kabupaten Madiun [Soil Fertility Study in Madiun Regency], Seminar Nasional Dalam Rangka Dies Natalis UNS Ke 43 Tahun 2019. (2019).p. 42–53. [in Bahasa Indonesia] https://core.ac.uk/reader/295746928
13. F. Vafadar, R. Amooaghaie, M. Otroshy, J. Plant Interac. **9**,1:128–136(2014). https://www.tandfonline.com/doi/pdf/10.1080/17429145.2013.779035

Microalgae Microbial Fuel Cell (MMFC) using *Chlorella vulgaris* and "Batik" Wastewater as Bioelectricity

Nadiyah Faizi Polontalo[1], *Falvocha* Alifsmara Joelyna[1], *Abdullah* Malik Islam Filardli[1], *Hadiyanto* Hadiyanto[1,2*] and *Zainul* Akmar Zakaria [1,3]

[1]Chemical Engineering Department, Engineering Faculty, Diponegoro University, Indonesia
Jl. Prof. Soedarto No. 13, Tembalang, Semarang 50275, Indonesia
[2]Centre of Biomass and Renewable Energy (C-Biore), Diponegoro University, Indonesia
Jl. Prof. Soedarto No. 13, Tembalang, Semarang 50275, Indonesia
[3]Bioprocess and Polymer Engineering Department, Faculty of Chemical and Energy Engineering, University Teknologi Malaysia Jl. Iman, 81310 Skudai, Johor, Malaysia

Abstract. Nowadays, Indonesia is faced with an increase in human growth, and followed by increasing electricity demand. One of the environmental friendly alternative energy that can solve this problem is microbial fuel cell, which utilizes organic matter as a substrate of bacteria in carrying out its metabolic activities to produce electricity. In this study, investigated the electrical energy produced by Microalgae Microbial Fuel Cell (MMFC) using *Chlorella vulgaris* and "Batik" wastewater. This study aims to assess the performance of the MMFC system based on the influence of yeast (8 g L^{-1} and 4 g L^{-1}), "Batik wastewater" concentration (50 % and 100 %), and graphite electrodes (1:1 and 2:2). The MMFC system was carried out by filling anode chamber with "Batik" wastewater and the cathode with *C. vulgaris*. MMFC simulation was operated for 7 d. Concentration of 100 % "Batik" wastewater and 2:2 number of electrodes gave the best result in MMFC with voltage 0.115 Volt, algae absorbance 0.666. The COD decreased from 824 mg L^{-1} to 752 mg L^{-1} after the MMFC. The addition of 8 g L^{-1} yeast gave the optimum of bioelectricity production reached 0.322 Volt and the microalgae grew until the absorbance reached 1.031.

Keywords: Wastewater, bioelectricity, *Chlorella vulgaris*, microbial fuel cell, waste to energy

1 Introduction

Nowadays, a crucial problem with the increase of human growth in Indonesia is the increasing electricity demand [1]. It is estimated to continue growing by 4.6 % yr^{-1} and less effort to increase electricity production, this condition will triple by 2030. This condition made the availability of fossil energy, especially crude oil, increasingly scarce lead to Indonesia become a net importer of crude oil and its derivative products [2]. Amount one environmentally friendly alternative energy is a microbial fuel cell (MFC) by utilizing organic material (substrate) as an energy source of bacteria that its metabolic activities produce electricity. The system utilized wastewater as a substrate, could be used as an ideal tool for processing [3]. By utilizing *Chlorella vulgaris* [Beijerinck, 1890] and "batik" waste

[1] Corresponding author: hady.hadiyanto@gmail.com

that convert biochemical compounds directly into electrical energy, yeast is used as a biocatalyst in improving the performance of MFC in the MMFC process [4].

C. vulgaris is a photosynthetic organism that has great CO_2 fixation ability. This ability indirectly has a positive impact on reducing the effects of global warming [5]. *C. vulgaris* is also able to be used as a substrate in the system of microbial fuel cells which is an alternative source of electricity through hydrolysis and fermentation in a unit process [6].

Microbial fuel cell is a technology that utilizes microbes to produce energy in the form of electricity. Microbes convert various kinds of organic compounds into CO_2, water, and energy [10]. Recently, one of MFC technologies developed is MMFC (Microalgae Microbial Fuel Cell) technology. MMFC is an alternative that be used as a source of electricity generator through hydrolysis and fermentation of microalgae in one process unit. MMFC consists of an anode and cathode that connected through a load (usually a resistor). Anode chamber is containing microorganism cultures that catalyses the composition of organic materials into electrons and protons. Power will be produced by the reduction of oxygen or other compounds at the cathode chamber [6]. On the operation of MFC, electrogene which will produce CO_2. Microalgae in the cathode chamber use CO_2 as a carbon source for growth [7].

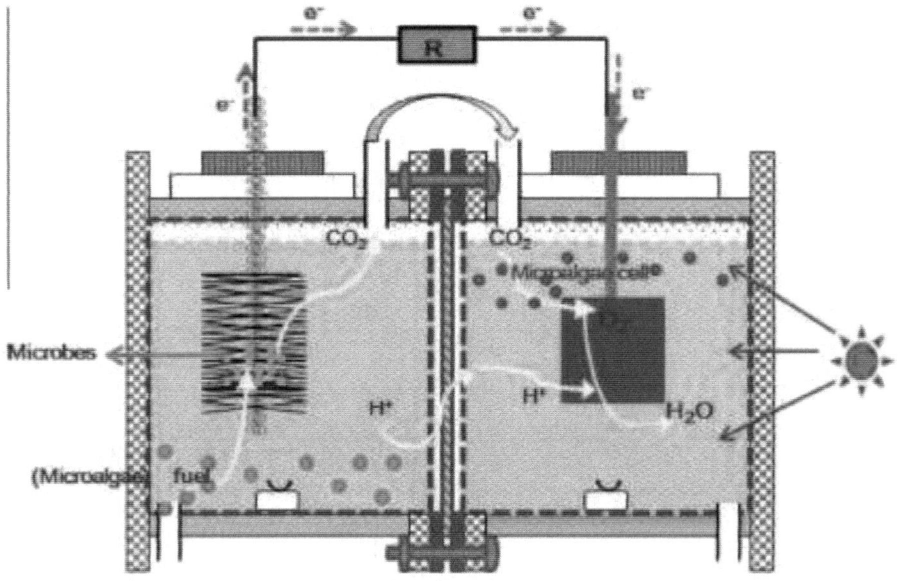

Fig. 1. MMFC operating system [7].

The application of *Chlorella* sp. in liquid waste, reduce levels of pollutants such as BOD, COD with a range 89 % to 96 % in the logarithmic phase and 80 % to 95 % in the stationary phase [8]. In this study, research was conducted on the electrical energy produced by MMFC using *C. vulgaris* and "batik" waste (from Figa Batik Collection) in Semarang, Indonesia. The results of this study are also expected to be one step forward to obtain renewable energy sources so that the use of MFC systems with wastewater substrates can reduce fossil energy consumption [9].

2 Materials and methods

2.1 Materials

The materials used in this study are *C. vulgaris* as catholyte which obtained from the Centre of Biomass and Renewable Energy (C-Biore) Integrated Laboratory Diponegoro University, "batik" liquid waste as anolyte with 100 % and 50 % concentration which obtained from *Griya Batik* Semarang (Figa Batik Collection Rejomulyo Semarang). Graphite electrode with a length of 10 cm and a diameter of 10 mm with a ratio 1:1 and 2:2 electrode obtained from an online store in Jakarta, yeast with mass of 4 g L^{-1} and 8 g L^{-1}, $NaHCO_3$ 0.32 g L^{-1}, Urea 0.05 g L^{-1}, and TSP 0.05 g L^{-1}. Besides, the utilization used in this study were glass box, digital multimeter, connecting wire, CO_2 wire, aeration wire, aerator, graphite electrode, lamp, ballast, and lux meter.

2.2 Methods

2.2.1 Cultivation of chlorella vulgaris

Amounted 1 L of *C. vulgaris* was diluted to 2 L, then add the nutrients of 0.64 g $NaHCO_3$, 0.1 g urea, and 0.1 g TSP, then stirred. During cultivation, lighting and oxygen were given through the aerator, then cover them with clear plastic. The absorbance or Optical Density (OD) check daily. If the absorbance approaches 0.6, then *C. vulgaris* was ready to use.

2.2.2 MMFC making process

"Batik" wastewater was diluted to concentration according to treatments 100 % and 50 %, then check the OD absorbance using a spectrophotometer. The anode chamber (anolyte) is filled with 1 L of waste. Meanwhile, at the cathode chamber (catholyte) is filled with 1 L of *C. vulgaris*. Followed, putting in the graphite electrodes with a number of 1:1 and 2:2 bar, and 4 g and 8 g of yeast added. Then, turn on the lamp, 20 cm far from the reactor.

2.2.3 Simulation of MMFC

The MMFC system was operated for 7 d, then every 24 h, the voltage and current are measured as well as the COD levels in the reactor. Data in the form of current strength (I) and voltage (V) would be calculated into power density, which was the power per unit area of the electrode (A), by using the Equation (1):

$$P = \frac{V \times I}{A} \quad (1)$$

Note:

P = Power density (Watt)
V = Voltage (mV)
I = Current (mA)
A = Area of electrode (mm^2)

2.2.4 Measurement of chemical oxygen demand level

Chemical oxygen demand (COD) was defined as the amount of oxygen equivalents consumed in the organic compounds of samples by oxidizing agents. The higher COD, the higher the amount of pollutant in the water sample. "Batik" wastewater measured its COD level before and after the MMFC process with a COD photometer with an operating condition by 150 °C for 2 h in the COD reactor.

3 Results and discussions

3.1 The effect of the concentration of "batik" wastewater on MMFC performance

3.1.1 Bioelectricity production

Bioelectricity refers to electrical potentials and currents occurring within or produced by living organisms. It results from the conversion of chemical energy into electrical energy. The influence of wastewater concentration on *C. vulgaris* growth and bioelectricity production presented in Figure 2 and Figure 3.

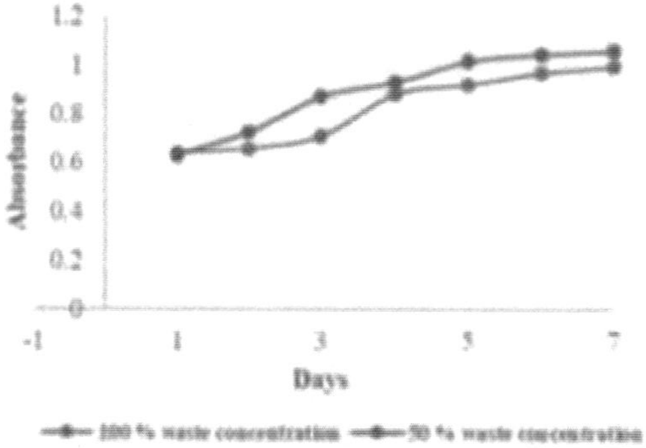

Fig. 2. Effect of wastewater concentration on *C. vulgaris* growth

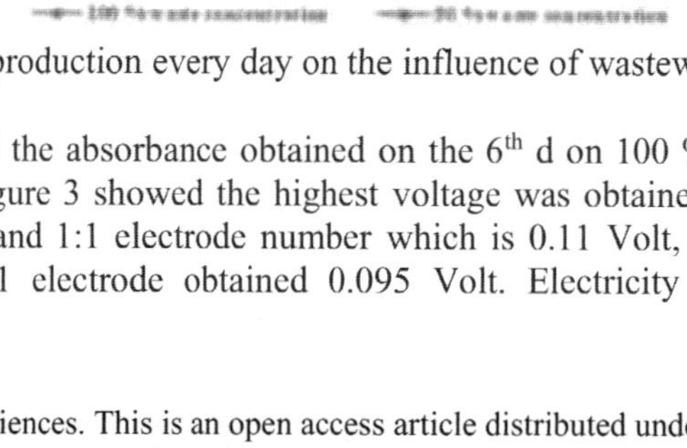

Fig. 3. Bioelectricity production every day on the influence of wastewater concentration

Figure 2 defined the absorbance obtained on the 6th d on 100 % waste concentration was 1.045. While, Figure 3 showed the highest voltage was obtained on the 6th d. 100 % waste concentration, and 1:1 electrode number which is 0.11 Volt, while on 50 % waste concentration and 1:1 electrode obtained 0.095 Volt. Electricity production has been decreased on the last day of the MMFC process, because microalgae could produce biofilms which is have the effect of increasing the resistance in the anode and caused a decrease in the electric voltage and current generated.

3.1.2 Chemical oxygen demand analyzation

The highest COD was obtained in 100 % waste concentration. The value before the MMFC implementation process was 460 mg L^{-1} and after the process, the COD was decreased by 266 mg L^{-1}. The reactor generated power density amounted to 27.07 mW m^{-2}. The result found that the MMFC process has made the value of the COD was decreased, and the wastewater would be safe to the environment.

3.2 The effect of the number of electrode on MMFC performance

3.2.1 Bioelectricity production

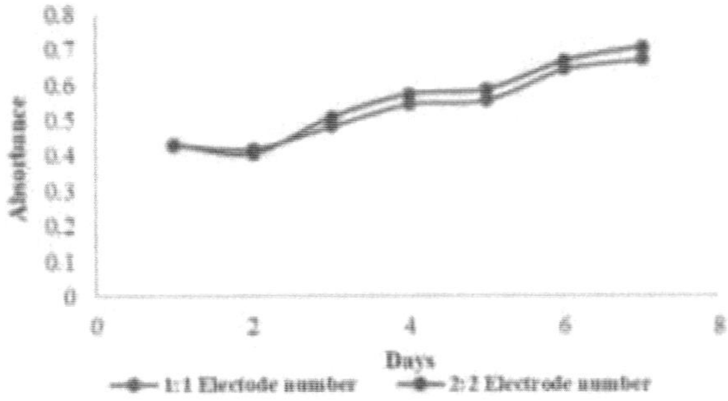

Fig. 4. Effect of number of electrodes on *C. vulgaris* growth

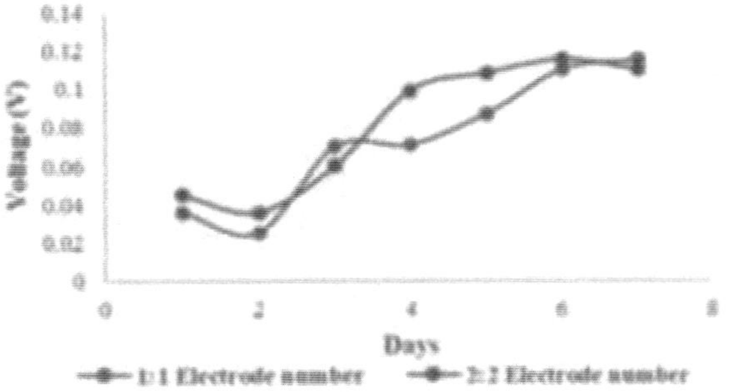

Fig. 5. Bioelectricity production every day on the influence of the number of electrodes

Growth of C. *vulgaris* showed same patterns with the electrodes number (Fig. 4). From 1 d to 5 d, the voltage tends to increase daily, but there is decreasing on the 1 d and 2 d for both variables with 1:1 and 2:2 number of electrodes, became 0.025 Volt and 0.036 Volt. In line with algae growth that still in stage of adaptation to the environment on the first 2 d. The optimum result was obtained on variable 2:2 electrodes on the 6th d, a voltage of 0.115 Volt and algae absorbance of 0.666.

3.2.2 Chemical oxygen demand analyzation

The results of the COD analysis on the wastewater showed that the COD value decreased from 824 mg L^{-1} to 752 mg L^{-1} after the MMFC process was carried out. This indicated that the dissolved oxygen amount in the wastewater decreased approach the maximum threshold, which is 150 mg L^{-1} [11]. Based on the research, it has been proven in the MMFC process, and there was a synergetic relationship between the degradation of components in wastewater with the cultivation of microalgae used to produce environmental friendly bioelectricity [7].

3.3 The effect of the addition of yeast on MMFC performance

3.3.1 Bioelectricity production

Effect of yeast addition on C. *vulgaris* growth and bioelectricity production presented on Figure 5 and Figure 6.

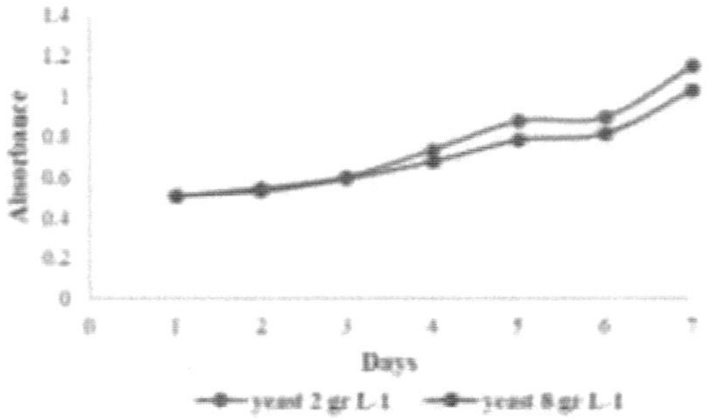

Fig. 5. Effect of yeast addition on *C. vulgaris* growth

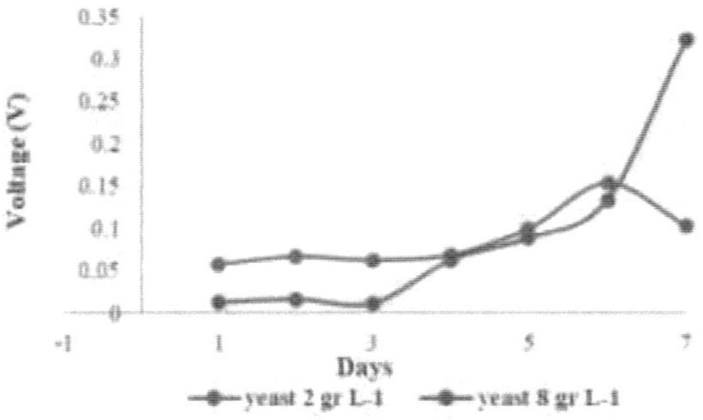

Fig. 6. Bioelectricity production every day on the influence of the addition of the yeast

The growth of *C. vulgaris* slightly higher on yeast addition of 2 g L^{-1} (Figure 5). The voltage increased daily, except for the addition of yeast variable 2 g L^{-1}, the voltage decreases on 7 d. The optimum value from the addition of 2 g L^{-1} yeast is 0.103 Volt, while the addition of 8 g L^{-1} yeast is 0.322 Volt. The decreasing voltage is caused by the metabolism of bacteria producing hydrogen, that accumulated over time and cover the electrodes, thus reducing electron transfer and decreased electricity production [4]. Meanwhile, the optimum bioelectricity is obtained by adding yeast 8 g L^{-1}, the voltage

 https://doi.org/10.1051/e3sconf/202122600032

increased daily and the highest value reaches 0.322 Volt. Increasing number of yeasts, improve the chance of oxidizing waste and electron exchange.

3.3.2 Chemical oxygen demand analyzation

The results of COD measurements carried out on "Batik" wastewater with the addition of 2 g L^{-1} yeast showed that the COD value decreased from 957 mg L^{-1} to 568 mg L^{-1} after the MMFC process, and become 493 mg L^{-1} in 8 g L^{-1} yeast. The addition of yeast to the waste increased MMFC's ability to degrade components of the waste. Yeast as a catalyst in oxidizing waste made better electron transfer.

4 Conclusion

Concentration of 100 % "Batik" wastewater and 2:2 number of electrodes gave the best result in MMFC process with the highest voltage found to be 0.115 Volt with the algae absorbance of 0.666. The COD value also decreased from 824 mg L^{-1} to 752 mg L^{-1} after the MMFC process. Yeast addition to "Batik" wastewater has a significant effect on the results of the MMFC process. The addition of 8 g L^{-1} yeast gave the optimum of bioelectricity production, which is reached 0.322 Volt and the microalgae also grew until the absorbance reached 1.031.

References

1. I. Kholiq. J. IPTEK. **19**,2:75–91(2015). [in Bahasa Indonesia] https://doi.org/10.31284/j.iptek.2015.v19i2.12
2. P. Y. So. J. Grad. Unpar. **1**,1:1–13(2014). [in Bahasa Indonesia]. http://journal.unpar.ac.id/index.php/unpargraduate/article/view/837
3. B. Ibrahim, P. Supitjah, Z. N. Adjani. JPHPI. **20**,2:296–304(2017). [in Bahasa Indonesia]. http://jurnal.ipb.ac.id/index.php/jphpi/article/view/19813/13679
4. L. Iradati, G. Samudro, S. Sumiyati. Jurnal Teknik Lingkungan. **4**,2:200–206(2015). [in Bahasa Indonesia]. https://www.neliti.com/publications/192345/pengaruh-konsentrasi-chemical-biological-demand-cod-dan-ragi-terhadap-kinerja-du
5. Dianursanti, R. Nuzulliany, A. Wijanarko, M. Nasikin. Jurnal Teknik Kimia Indonesia. **8**,3:87–93(2009). [in Bahasa Indonesia]. https://doi.org/10.5614/jtki.2009.8.3.3
6. S. B. Velasquez-Orta, T. P. Curtis, B. E. Logan. Biotech and Bioeng. **103**,6:1068–1076(2009). https://doi.org/10.1002/bit.22346
7. Y. Cui, N. Rashid, N. Hu, M. S. U. Rehman, J. Han. Erg. Conv. and Mgt. **79**:674–680(2014). https://doi.org/10.1016/j.enconman.2013.12.032
8. F. Hartini, F. Restuhadi, T. Dahril. JOM FAPERTA. **4**,1:1–13(2017). https://www.neliti.com/publications/202320/pemanfaatan-mikroalga-chlorella-sp-dalam-menurunkan-baku-mutu-polutan-limbah-cai
9. L. He, P. Du, Y. Chen, H. Lu, X. Cheng, B. Chang, Z. Wang. Ren. and Sustain. Erg. Rev. **71**:388–403(2017). https://doi.org/10.1016/j.rser.2016.12.069

10. E. Kristin. *Produksi energi listrik melalui microbial fuel cell menggunakan limbah industri tempe*. [Electricity production through microbial fuel cell using wastewater from tempeh industry] [Thesis] Teknologi Bioproses, Universitas Indonesia (2012). p.5. [in Bahasa Indonesia]. http://lib.ui.ac.id/file?file=digital/20311042-S43242-Produksi%20energi.pdf
11. R. H. Olivera, A. D. Gonza, U. Y. Pari, P. Vega, M. R. Urgarte, J. Tapia, L. Molina, A. L. Rivera, D. G. P. Salazar, M. Esparza. Elc. J. of Biotech. **31**:34–43(2018). https://doi.org/10.1016/j.ejbt.2017.10.013
12. L. E. Wall, Sharon B, Velasquez-Orta, E. R. Frasca, P. Leary, I. Yanez, Noguez, M. T. O. Ledesma. Biochem. Eng. J. **151**:107319(2019). https://doi.org/10.1016/j.bej.2019.107319
13. C. Safi, B. Zebib, O. Merah, P. Y. Pontailer, C. V. Garcia. Renew. Sust. Energ. Rev. **35**: 265–278(2014). https://doi.org/10.1016/j.rser.2014.04.007
14. D. P. Aghni. *Pengolahan Air Limbah Batik dengan Reaktor yang Berisi Tanaman Eceng Gondok* [Batik Wastewater Treatment with Reactors Containing Water Hyacinth Plants] [Thesis] Teknik Sipil, Universitas Sebelas Maret (2017). [in Bahasa Indonesia].p.13. https://eprints.uns.ac.id/36460/

Addition of *Moringa oleifera* Lam. Leaves Flour for Increasing the Nutritional Value of Modified Cassava Flour–Based Breakfast Cereal

Novian Wely Asmoro[1,*], *Agustina* Intan Niken Tari[1], *Retno* Widyastuti[1] and *Chandra* Kurnia Setiawan[2, 3]

[1]Department of Agricultural Product Technology, Faculty of Agriculture, Universitas Bangun Nusantara, Sukoharjo, Jl. Letjend Sujono Humardani No.1, Bendosari, Kabupaten Sukoharjo 57521, Central Java, Indonesia
[2]Department of Agrotechnology, Faculty of Agriculture, Universitas Muhammadiyah Yogyakarta, Jl. Brawijaya, Kasihan, Bantul 55183, Special Region of Yogyakarta, Indonesia
[3]Department of Horticulture, Faculty of Agriculture, Kasetsart University at Kamphaeng Saen Campus, 1 M.6 Kamphaeng Saen District, Nakhon Pathom 73140, Thailand

Abstract. Many Indonesian children have a habit of not eating breakfast according to balanced nutritional needs. Breakfast provides a third of daily nutritional needs. This study aims to evaluate the potential and determine the formulation of breakfast cereals using Mocaf Cassava Flour (MOCAF) enriched with Moringa Leaves Flour (MLF) to add nutritional value to the product. The experimental design was Completely Randomized Design (CRD) with one treatment, addition of dry Moringa leaf flour as much as 0 % as a control, and 5 % w w^{-1}, 10 % w w^{-1}, and 15 % w w^{-1} as the treatment. The results showed that the addition of MLF in MOCAF flour had the potential to increase the nutritional value of breakfast cereal products. The addition of MLF with a percentage of 5 %, 10 %, and 15 % had a significant effect ($p < 0.05$) on water content, ash content, and fat content. The addition of MLF up to 15 % increases ash content by 2.79 % and fat content by 8.2 %.

Key words: *Cassia fistula*, drumstick tree, fungsional food, increase nutrition, kelor

1 Introduction

More than 50 % of Indonesian children have a habit of not eating breakfast according to balanced nutritional needs. Breakfast provides daily nutritional needs. It is increasing concentration when receiving lessons at school. The schools in Pacitan, East Java, Indonesia about 74 % had no breakfast habits which resulted in the emergence of malnutrition and obesity [1]. Lifestyle, business, and the not practical preparing breakfast are the causes of not eating breakfast. There needs an alternative practical breakfast [2]. Comercial breakfast foods must meet the nutrients needed by the body. Developing healthy

* Corresponding author: novianwelyasmoro@gmail.com

breakfast foods is done by making composite products by mixing various ingredients to produce the expected nutritional content. High–quality breakfast food made by mixing of ingredients.

Breakfast foods containing carbohydrates as the main energy source come from tubers, a source of protein obtained from mixing nuts, and a source of vitamins and fiber from fruit [3]. Foods that provide health benefits can reduce the risk of disease was called functional foods. Functional food components can naturally come directly from the foodstuff but can also be obtained from the fortification process by adding functional components from other foodstuffs [4]. One of the plants that can be used to increase the functional value of food is *Moringa oleifera* Lam. leaves. Previous studies mention that *M. oleifera* leaves have high protein, vitamin C, and calcium values of 23.78 %, 56 mg, and 3 467 mg 100 g^{-1} of dried leaves respectively. Other studies show moringa leaves have seven times the vitamin C content of oranges, four times the calcium content of milk, four times the vitamin A content of carrots, three times the potassium content of bananas, and two times the protein content of yogurt [5].

Breakfast cereal formulations can use a mixture of various ingredients, but basically must meet at least 300 Kcal nutritional needs of breakfast and meet the organoleptic aspects preferred by consumers, including texture/crispness, taste, and aroma. Research related to the addition of Moringa Leaves Flour (MLF) conducted by [6] showed that the acceptance of wet noodles by panelists on the addition of MLF to 2 % in terms of color, taste, and texture, although the aroma aspect was less favored. As well as increasing levels of wet noodle protein with the addition of 2 % MLF by 14.84 %.

This study aims to evaluate the potential and determine the formulation of breakfast cereals using Modified Cassava Flour (MOCAF) enriched with MLF to add nutritional value to the product.

2 Methods

Mocaf flour obtained from Wonogiri, Indonesia. Cereal was made from mocaf flour, wheat flour, and MLF. The process of making cereal begins with the preparation of MLF as a composite of cereal raw materials. Mix all raw materials into dough and form into sheets with a thickness of 1 mm to 2 mm, then cut into squares. Pieces of dough are steamed and then roasted using a temperature of 120 °C, subsequently tested the chemical and nutritional content of the cereal product.

Completely Randomized Design (CRD) was used in this study with treatment 0 % of MLF as a control (K), 5 % (K1), 10 % (K2), and 15 % (K3) (w w^{-1}). All treatment replicated four times. Each unit was tested with three samples.

3 Results and discussions

3.1 Cereal

Cereal products produced from variations in the addition of MLF can be seen in Figure 1. In Figure 1A, 1B, 1C, and 1D, respectively are cereal products with the addition of MLF as much as 0 %, 5 %, 10 %, and 15 %. The green color in the leaves of Moringa powder caused discoloration of cereals from yellowish–white to green. The higher concentration of moringa leaves, the greener cereal product.

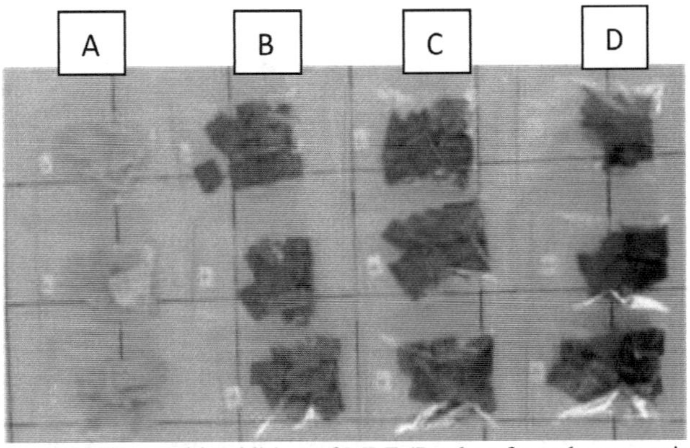

Fig. 1. Breakfast cereal products with addition of MLF (Product formula successively A, B, C and D with the addition of 0 %, 5 %, 10 % and 15 % MLF)

3.2 Water content

The result showed that water content was different among treatments ($p < 0.05$). Figure 2 showed that the addition of MLF value of water content decreased at 0 %, 5 %, 10 % and 15 % by 8.45 %, 7.56 %, 6.22 % and 5.85 % respectively. Water content, in general, is an important component in food because water can affect the appearance, texture, and taste of food [7]. Water content is influenced by the main ingredients of breakfast cereal products, i.e wheat flour, mocaf flour, and MLF. The addition of MLF caused the percentage of other ingredients (mocaf flour and flour) in cereal product formulations decrease. Meanwhile, mocaf flour is one of the ingredients that can bind water to the dough. It caused water content decreased in line with the addition of MLF.

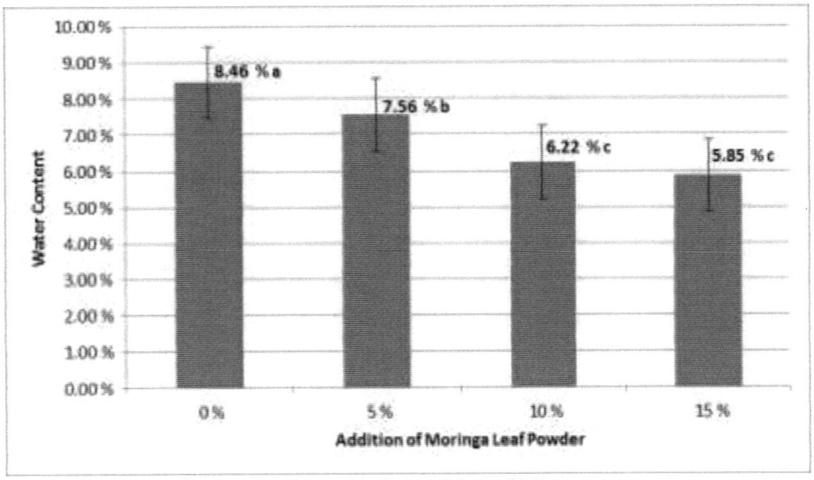

Fig. 2. Moisture content of cereal product with the addition of MLF

3.3 Ash content

Figure 3 showed that the addition of MLF successively increases the value of ash content, where the addition of MLF by 0 %, 5 %, 10 %, and 15 % increases the value of ash content

1.03 %, 1.80 %, 2.24 % and 2.79 % respectively. Statistical analysis of ash content showed that there were significant differences among breakfast cereal products ($p < 0.05$). Ash content generally indicated the mineral content of the product. The addition of MLF caused an increase in mineral content of breakfast cereal products. That was because the mineral content found in Moringa leaves [8]. Moringa leaves had calcium of 440 mg g^{-1} and potassium of 259 mg g^{-1} [9]. It also had magnesium, phosphorus, and sodium of 8.79 g kg^{-1}; 3.70 g kg^{-1} and 0.6 g kg^{-1}, respectively [10].

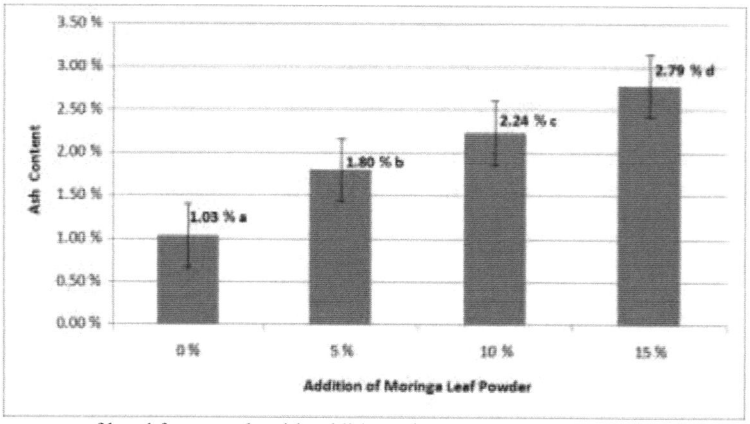

Fig. 3. Ash content of breakfast cereals with addition of MLF

3.4 Fat content

The addition of MLF had a positive influence on increasing fat content of breakfast cereal products, from 6.4 % to 8.2 % (figure 4). Compared to control, the addition of MLF caused an increase in fat content. The results of the statistical analysis of fat showed no significant difference in the concentration of 5 % to 15 %. Increased fat content in cereal products was due to the presence of fat in MLF by 3.2 g 100 g^{-1} ingredients [6].

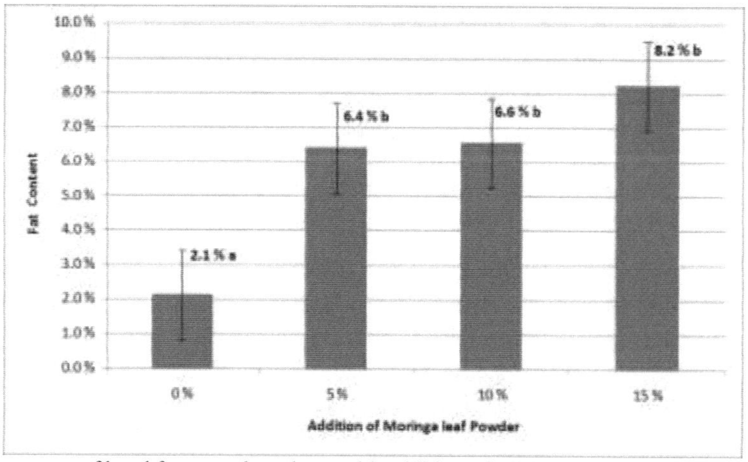

Fig. 4. Fat content of breakfast cereal products with addition of MLF

4 Conclusions

The addition of MLF in MOCAF flour had the potential to increase the nutritional value of breakfast cereal products. The addition of MLF with a percentage of 5 %, 10 %, and 15 % had a significant effect ($p < 0.05$) on water content, ash content, and fat content. The addition of MLF up to 15 % increases ash content by 2.79 % and fat content by 8.2 %.

Thank you to the Institute for Research and Community Services (LPPM) Bangun Nusantara Sukoharjo Veterans University for funding support for research activities conducted through the internal funding scheme of the Competitive University of Science 2019.

References

1. N.W. Asmoro, A.I.N. Tari, A. Afriyanti, Pro Food Journal. **3**,2:230–234(2017). https://doi.org/10.29303/profood.v3i2.54
2. E. Chukwuebuka. Int J Food Sci Nutr. **4**,6: 624–630(2015). http://www.sciencepublishinggroup.com/journal/paperinfo?journalid=153&doi=10.11648/j.ijnfs.20150406.14
3. A.A. Kiswandono. Jurnal Sains Natural Universitas Nusa Bangsa. **1**,1:45–51(2011). [in Bahasa Indonesia] https://doi.org/10.31938/jsn.v1i1.13
4. A.I. Kusumanegara, Jamhari, Y. Erwanto. Animal Husbandry Bulletin. **36**,1: 19–24(2012). [in Bahasa Indonesia]. https://doi.org/10.21059/buletinpeternak.v36i1.1272
5. K.T. Mahmood, T. Mugal, I.U. Haq. J. Pharm. Sci. Res. **2**,11:775–781(2010). http://citeseerx.ist.psu.edu/viewdoc/download?doi=10.1.1.193.7882&rep=rep1&type=pdf
6. S.P. Mishra, S. Pankaj, S. Sanjay. Bull Env. Pharmacol. Life Sci. **2**,1:28–31(2012). http://asset-0.soupcdn.com/asset/6710/0253_0c3c.pdf
7. F. Perdana, Hardinsyah. Journal of Nutrition and Food. **8**,1:39–46(2013). https://doi.org/10.25182/jgp.2013.8.1.39-46
8. S. Sahay, U. Yadav, S. Srinivasamurthy. Int J Food Sci Nutr . **2**,5:31–37(2017) https://www.researchgate.net/profile/Upasana_Yadav2/publication/330872655_Potential_of_Moringa_oleifera_as_a_functional_food_ingredient_A_review/links/5de600e9299bf10bc33a8b6b/Potential-of-Moringa-oleifera-as-a-functional-food-ingredient-A-review.pdf
9. E. Sukasih, Setyadjit. J. Postharvest. **9**,2:70–76(2012) http://repository.pertanian.go.id/bitstream/handle/123456789/1501/Formulasi%20Pembuatan%20Flake%20Berbasis%20Talas%20Untuk%20Makanan%20Sarapan%20%28Breakfast%20Meal%29%20Energi%20Tinggi%20Dengan%20Metode%20Oven.pdf?sequence=1&isAllowed=y
10. N Zakaria, A Tamrin Media Gizi Pangan. **21**,1:73–78(2016) [in Bahasa Indonesia]. http://mediagizipangan.org/wp-content/uploads/2017/03/12.-zakaria.pdf

Adoption Level of Integrated Farming System Based on Rice–Cattle and Its Determinants Related to Sustainable Agriculture

Novitri Kurniati[1,*], *Ketut* Sukiyono[2], *Purmini* Purmini[3], and *Mutyarsih* Oryza Sativa[4]

[1]Faculty of Agriculture, Department Agribussiness, University of Muhammadiyah Bengkulu, Jl. Bali, Bengkulu 38119, Indonesia
[2]Faculty of Agriculture, Department of Agricultural Social Economics, University of Bengkulu, Jl. Kandang Limun, Bengkulu 38119, Indonesia
[3]Faculty of Economics and Business, Department of Development Economics, University of Bengkulu, Jl. Kandang Limun, Bengkulu 38119, Indonesia
[4]Faculty of Agriculture, Department of Animal Science, Khonkaen University, 123 Moo 16 Mittraphap Road, Khon kaen 40002, Thailand

Abstract. A way to restore the fertility of rice fields is through improving soil structure and microbes by using organic fertilizer derived from livestock waste. Therefore it is necessary to implement an Integrated Farming System (IFS) particularly rice and cattle. The study aimed to analyze the determinants of farmers' decisions in adopting IFS of rice in Bengkulu Province, Indonesia. This research applied a survey method, which was conducted in Seluma and Rejang Lebong Regencies, Bengkulu Province, Indonesia. The data included primary and secondary and analyzed with the use of Multinomial Logistic Regression. The results indicated that simultaneously all the predictor variables had a significant effect on the response variable, while the income, land area, number of cattle and farmers' perceptions had a very significant effect on the adoption of the integration system while the variable costs of production, farming experience and labor did not have a significant effect on the decision to adopt a rice and cattle integration system.

Keyword: Farmer perception, organic fertilizer, plant crop system

1 Introduction

The role of the agriculture sector in Bengkulu Province is also adequately strategic where nearly 60 % of the productive workforce works in the agricultural sector and this sector redounds 39.84 % in the Bengkulu Province Gross Domestic Product, Indonesia [1]. Rice is the backbone of the development of the food crops sub-sector and enact an important role in achieving food security.

The rate of increase in rice productivity in Bengkulu tends to be stationer, indicated by an alleviation in rice production of around 2.28 % [1]. The rice intensification system that has been implemented but it has no longer increased production. This situation is due to a less integrated way of managing land and violating the rules of land conservation and the

* Corresponding author : noviewin@gmail.com

environment. Intensive and continuous exploitation of rice fields affect declined soil fertility and physical properties due to lack of organic materials.

A way to remediate soil fertility is through improving soil structure and microbes by using organic fertilizer derived from livestock waste. To overcome the problems faced by most rice farmers, it is necessary to implement the IFS which combines rice and cattle.

Cattle still becomes a farmer income source in rural areas, where cows as a superior cattle commodity by most farmers in Bengkulu province. The high interest in raising cattle due to suitable natural conditions that has been a priority to support the development of the livestock sector.

According to Al Mamun, et al. [2] the problem faced by farmers in generally practice subsistence farming are high degree of risk because of seasonal, environmental and decreasing productivity. Integrated Farming System (IFS) can eradicate these constraints. Meanwhile, in accordance with Wibawa and Silviani [3], a new crop integration system was adopted by farmers in Bengkulu. Practically, the implementation of the IFS has not yet attained the target due to the limited assistance and technology dissemination, as well as farmers' views or perceptions about the IFS which is still low. Therefore, a study is needed of what factors affect the decision of farmers to adopt rice's IFS in Bengkulu Province. This study aimed to analyze the determinants of farmers' decisions in adopting IFS based on rice-cattle in Bengkulu Province.

2 Methods

This study employed the Survey method that takes samples from a population and used questionnaires as a primary data collection tool [4].

The study was conducted from January to July 2019. The study located in the Province of Bengkulu included the lowland area in Seluma Regency and the highland areas in Rejang Lebong Regency, Indonesia. The data included primary data from farmers and secondary data from relevant agencies.

The sampling method applied the Simple Random Sampling technique with a sample size as much as ten times of the independent variables number in each study area [4]. In this study, the number of samples taken was 200 rice farmers who raised cattle.

Data analyzing techniques employed Multinomial Logistics Regression [5], with the following Equation (1):

$$\text{INT} = \log \frac{p(xi) \quad p(xi)}{1-p(xi)1-p(xi)} \text{bo} + \text{b}1Pi + \text{b}2BP + \text{b}3LL + \text{b4Put} + \text{b}5JTk + \text{b}6P + \mu \quad (1)$$

Explanation :

INT = Integration of cow rice

P(Xi) = farmer decision opportunities to adopt integration systems

1 = not integration

2 = low integration

3 = medium integration

4 = high integration

Pi = integrated farming income (IDR mo^{-1})

BP = integration production costs (IDR mo^{-1})

LL = land area (ha)

Put = experience of farming (yr)

JTk = number of family workers (person)

P = Farmer's perception (score)

bo = constanta

b1...b6 = coefficient (guess parameter) and μ = error

Parameter testing was enforced simultaneously and partially. The test statistic employed was to use *Odds Ratio* and *Likelihood Ratio*.

3 Results and discussions

3.1 Factors that predispose farmers' decisions to adopt a cattle rice integration system

The model that used to analyze the factors swaying the decision of farmers to adopt rice's IFS is the Multinomial Logistic Regression Analysis Model. The dependent variable (Y) used is the farmer's decision to adopt IFS in the form of nominal data with four categories, namely non, low, medium, and high integration respectively. While the independent variable (X) used are seven variables, videlicet farm income, production costs, land area, number of cows, farming experience, the number of workers in the family and farmers' perceptions.

3.2 Model conformity test

Model testing implemented the *goodness-of-fit* test to like the compatibility of different models, or there was no difference between the observations with the model of outcome prediction. Table of *Goodness-of-fit* for testing the suitability model are provided in Table 1.

Table 1. Model conformance test output from goodness-of-fit table values

	Chi-Square	**Sig**
Pearson	374.154	1.000
Deviance	326.400	1.000

Source: Data processed by SPSS 16 (2019)

Based on Table 1, it found that the Sig value of Pearson and the Deviance Sig was similar as 1.000, implied that the model is fit to apply to outcome prediction because p value > α; (1.000 > 0.01). In addition, the value of Deviance (326.400) < Chi-Square table (497.199) then H0 was accepted, indicating that the model is suitable even though no significant difference between the observations and the possible predictions of the model.

3.3 Model determination coefficient

The model suitability test is carried out by looking at the value of the coefficient of determination listed on the Pseudo R-Square. From Table 2, the Nagelkerke R2 value was 0.650 which identified the variable variability of income, production costs, land area, number of cattle, farming experience, labor and perceptions able to explain the variability of integration system adoption decisions by 65 %, while the remaining 35 % is explained by other variables outside the model.

Table 2. Model suitability test from the pseudo r-square table value

Nagelkerke	0.650

Source: Data Processed by SPSS 16 (2019)

3.4 Simultaneous parameter estimation testing

Simultaneous testing is performed using the *likelihood ratio test* as presented in Table 3.

Table 3. Simultaneous parameter estimation test

	Likelihood Ratio test			
Model	-2 log likehood	Chi-square	Sig.	Decision
Intercept	508.944			Reject Ho
Final	326.400	182.543	0.000**	Accept H1

Source: Data Processed by SPSS 16 (2019).
Information : $\alpha = 0.01 = 42.98$
Df = 24
** = significant at 99 % confidence level

Based on Table 3, it can be cross checked that the Final Sig. value is 0.000. Test criteria are if the Sig. (0.000) < α (0.01) or X2 count > X2 table then its decision rejected H0 and accepted H1. It means income variable (X1), farming production costs (X2), land area (X3), number of cattle (X4), farming experience (X5), labor (X6) and farmers' perception (X7) simultaneously very significant effect on ($p < 0.01$) the decision of farmers to adopt the integration system (Y).

3.5 Partial variable estimation testing

Partial test results shown in Table 4 that it recognized partially four predictor variables i.e. income (X1), land area (X3), number of Cows (X4) and farmers' perceptions (X7), that it significantly affected on adoption decision variables system integration (Y), while three other variables namely production costs (X2), experience (X5), and the number of labor (X6) have no significant effect on the adoption of the integration system decision.

Table 4. Partial variable estimation test

	Likelihood Ratio Test			
Variabel	-2 likelihood	Chi-square	Sig.	Decision
Intercept	3.264 E2	0.0 00		
Income (X1)	345.4 51	19.0 51	0.0 00**	Reject Ho Accept H1
Production cost (X2)	333.1 27	6.7 27	0.0 81	Reject Ho Accept H1
Land area (X3)	342.0 98	15.6 98	0.0 01**	Reject Ho Accept H1
Number of Cows (X4)	343.6 12	17.2 11	0.0 01**	Reject Ho Accept H1
Experience (X5)	332.8 65	6.4 64	0.0 91	Reject Ho Accept H1
Labor (X6)	328.6 28	2.2 28	0.5 27	Reject Ho Accept H1
Perception (X6)	382.4 41	56.0 40	0.0 00**	Reject Ho Accept H1

Source: Data processed by SPSS 16 (2019).

The effect of each predictor variable on the response variable can be explained as follows:

i. Income (X1)

In the income variable, it is gained value X2 count > X2 table or 19.051 > 11.34 and also Sig. < α or 0.000 < 0.01, then, the decision taken is to reject Ho and Accept H1 meaning that farm income partially has a very significant effect on the decision to adopt a system of integration of rice and cattle.

One of the motives of farmers in adopting new technologies is viewed from the income acquired from these new technologies. If the income obtained is greater, then, it will encourage farmers to adopt something new [6].

Several studies have indicated that the integration of livestock crops are able to increase farmers' incomes. The benefit of productivity improvement by 30 % to 50 % depending upon the number and kind of livestock products [7]. Besides that, Lindawati, et al [8] research's demonstrates that integrated farming income has a significant effect on the decisions of farmers' households in adopting integration system. This is in line with the research of Soni et al. [9] and Jaishankar, et al, [10] who show that the analysis of costs and benefits and farm income affect the decision to adopt a livestock crop integration system.

ii. Production Cost (X2)

Production costs is the cost spent on rice farming and cattle farming. In the production cost variable, it was obtained value X2-count < X2-table or 6.727 < 7.815 and Sig. > α or 0.081> 0.01, then the decision taken is H0 is accepted and H1 is rejected. This means that the production cost has a partially insignificant effect on the decision to adopt a rice and cattle integration system.

iii. Land area size (X3)

In the land area variable, it gained the value of X2-count > X2-table or 15.689 > 11.34 and Sig. < α or 0.001 < 0.01, then the decision taken is Ho is rejected and H1 is accepted. This means that the land area partially has a very significant effect on the decision to adopt an integrated rice and cattle system. This is because the more extensive rice fields cultivated, the more straw can be used to feed cattle. Farmers in the study area averagely cultivate 0.7 ha of rice land per farmer household.

Menozzi's research [11] suggests that the decision to adopt innovations by farmers is affected by many factors including farmer's attitude. The perception and adoption of the integration system is affected by, among others, socioeconomic factors including respondent's age, educational background, benefits, farm size and perception [12,13]. Livestock-crop integration is an agricultural system that is characterized by close links between crop components and livestock in farming or in an area. The synergistic relationship produced allows livestock to provide high profits per unit area of land used besides providing fertilizer for plants.

iv. Number of Cows (X4)

In the number of cows variable, it is obtained the value of X2 count > X2 table or 17.211 > 11.34 and Sig. < α or 0.001 < 0.01, then the decision taken is Ho is rejected and H1 is accepted. This means that the number of cows partially has a very significant effect on the decision to adopt the integration system of rice and cattle. This is because livestock waste can be used in rice plants as compost made manure, and as biourine that is processed from cow urine, and straw can be used directly as cattle feed.

Initially, the motivation of the rice farmers to establish cattle was to get income from the cattle fattening business and obtain cattle tillers. With the increasing number of cattle being farmed, it encouraged farmers to process fecal and urine into compost and biourine used in rice plants. Currently in the study area, on average the farmers keep 4.8 cattle per farmer household.

According to Kariyasa and Pasandaran [14], rice farming combined with livestock or using manure is able to produce around 6.9 % to 8.8 % higher than partially managed rice farming without using manure. One hectare of rice fields capable of producing about 10 t to 20 t of straw (fresh weight at harvest and depending on season) and used for feeding one to two adult cattle throughout the year. With rice straw production of 5 t ha^{-1} per harvest to 8 t ha^{-1} per harvest can be given two to three adult cattles throughout the year. If the rice cropping pattern is two times a year, it means that it can fulfill the requirements for feeding four cattles throughout the year to six cattles throughout the year.

v. Farming Experience (X5)

Farming experience is the length of time a farmer has done farming activities. In the farming experience variable, it is obtained the value of X^2 count < X^2 table or 6.464 < 7.815 and Sig. > α or 0.091 > 0.05, then the decision taken is H0 is accepted and H1 is rejected. This means that the farming experience partially has insignificant effect on the decision to adopt an integrated system of rice and cattle.

vi. Labor (X6)

Labor is the number of workers in families owned by farm households that consist of wives, children or relatives who live in one house and are the responsibility of the farmers. In the labor variable, it is obtained value X2 count < X2 table or 2.222 < 7.815 and Sig. > α or 0.527 > 0.05, then the decision taken is H0 is accepted and H1 is rejected. This means that labor partially has insignificant effect on the decision to adopt an integrated system of rice and cattle. This is because at the location of the study the number of family workers was relatively small at an average of 3.56 people per household. Meanwhile, implementing an integrated system requires relatively more labor than non-integrated systems.

vii. Perception (X7)

Perception is the farmers' views of the integrated system of rice and cattle. The perception is assessed with a score of one for low perception, two for medium perception, and three for high perception. In the number of cattle variables, it is obtained the value X2 count > X2 table or 56.040 > 16.81 and Sig. < α or 0.000 < 0.01, then the decision taken is Ho is rejected and H1 is accepted. This means that farmers' perceptions partially have a very significant effect on the decision to adopt a rice and cattle integration system.

Measurement of farmers' perceptions is carried out to know the effectiveness of an innovation that has been applied by farmers. Perception arises after the farmers implement farming integration systems and further can affect the adoption process of integration. Farmers' perceptions of the rice and cattle integration are viewed from the aspects of i) relative superiority, ii) suitability, iii) complexity, iv) can be tried, and v) can be observed [15].

The results showed that 34 farmers had low perception, 122 farmers had medium perception and 44 farmers had high perception of the integration system. On average farmers' perceptions at the research location are at a medium level.

3.6 Model interpretation

Model interpretation is carried out by considering the value of Exp (B) which is the value of the odds ratio in the SPSS output. In this study, four categories were used for the dependent variable, namely the category of no integration, low integration, medium integration, and high integration. One of the categories must be used as a reference variable. In this research, the category of no integration is used as a reference variable.

Table 5. Model interpretation results Parameter Estimation

Decision	**Variabel**	**B**	**SE**	**Wald**	**Sig**	**Exp (B)**
Low Integration	Interception	3.5 59	1.0 59	11.2 1	0.0 01	
	Income	0.0 00	0.0 00	0.3 49	0.5 55	1.0 00
	Prod. cost	0.0 00	0.0 00	1.0 46	0.3 06	1.0 00
	Land area	0.4 15	0.4 86	0.7 28	0.3 94	0.6 61
	Number of cows	0.3 89	0.1 43	7.4 15	0.0 06*	0.6 78
	Farming experience	-0.0 05	0.0 23	0.0 52	0.8 19	0.9 95

(Continued on next page)

Tabel 5. Continued

Decision	Variabel	B	SE	Wald	Sig	Exp (B)
Low Integration	Labor	-0.1 78	0.1 69	1.1 15	0.2 91	0.8 37
	Perception = low	3.3 32	0.9 97	11.1 69	0.0 01*	0.0 36
	Perception = medium	1.8 15	0.8 41	4.6 57	0.0 31*	0.1 63
	Perception = high	10^b				
Medium integration	Interception	2.7 77	1.1 14	6.2 17	0.13	
	Income	0.0 00	0.0 00	4.8 90	0.0 27*	1.0 00
	Prod. cost	0.0 00	0.0 00	4.9 56	0.0 26*	1.0 00
	Land area	1.7 01	0.5 55	9.4 06	0.0 02*	0.1 82
	Number of cows	0.5 76	0.1 85	9.7 42	0.0 02*	0.5 62
	Farming experience	-0.0 45	0.0 30	2.2 37	0.1 35	0.9 56
	Labor	-0.0 04	0.1 91	0.0 00	0.9 85	0.9 96
	Perception = low	3.9 10	1.3 48	8.4 15	0.0 04*	0.0 20
	Perception = medium	2.4 16	0.8 78	7.5 75	0.0 06*	0.0 89
	Perception = high	0^b				
High integration	Interception	-4.9 42	4.3 81	1.2 72	0.2 59	
	Income	0.0 00	0.0 00	8.1 01	0.0 04*	1.0 00
	Prod. cost	0.0 00	0.0 00	3.5 17	0.0 61	1.0 00
	Land area	3.7 18	1.4 21	6.8 50	0.0 09*	0.0 24
	Number of cows	1.2 29	0.5 60	4.8 15	0.0 28*	0.2 93
	Farming experience	0.1 09	0.0 81	1.8 17	0.1 78	1.1 15
	Labor	0.5 22	0.7 42	0.4 96	0.4 81	1.6 86
	Perception = low	-22.2 82	0.0 00	-	-	2.1×10^{-10}
	Perception = medium	-24.7 42	7 658.4	0.0 00	0.9 97	1.8×10^{-11}
	Perception = high	0^b				

Reference category: No Integration
Source: Data processed, 2019.

From the table of parameter estimation results above, there are three multinomial logit models formed. All three models have independent variables that have a statistically significant effect.

i. Multinomial logit 1 equation (Comparison between Low Integration category (Y = 2) and No Integration (Y = 1) Equation (2):

$$\text{Ln (P2/P1)} = 3.5\,59 + 0.0\,00X1 + 0.0\,00X2 + 0.4\,15X3 + 0.3\,89X4 - 0.0\,05X5 - 0.178X6 + 0.332X7R + 1.815X7S \quad (2)$$

The coefficient value of the cow numbers *(X4)* is significant at α 0.05. The coefficient value is 0.389 and Exp (B) is 0.678 which means that more cows have 0.678 times chance to adopt a low level integration system than not integration.

The low and medium coefficient value of farmer's perception (*X7R* and *X7S*) is significant at α 0.05. The coefficient value is 3.332 and Exp (B) is 0.036, and the coefficient value is 1.815 and Exp (B) is 0.163 which means low and medium perception have the opportunity of 0.036 and 0.163 times to adopt a low level integration system compared to no integration system.

ii. Multinomial logit two equation (Comparison between the Medium Integration category (Y = 3) and No Integration (Y = 1) Equation (3)

$$\text{Ln (P3/P1)} = 2.777 + 0.000X1 + 0.000X2 + 1.701X3 + 0.576X4 - 0.045X5 - 0.004X6 + 3.910X7R + 2.416X7S \quad (3)$$

The coefficient of income (*X1*) is statistically significant at α 0.05. The coefficient value is 0.000 and Exp (B) is 1.000, which means the higher farm income, the opportunity to do a medium level integration system is 1.000 times compared to no integration.

The coefficient of production cost (*X2*) is statistically significant at α 0.05. The coefficient value is 0.000 and Exp (B) is 1.000, which means that the higher the production cost, the opportunity to do a medium level integration system is 1.000 times compared to no integration.

The land area size coefficient value (*X3*) is statistically significant at α 0.05. The coefficient value is 1.701 and Exp (B) is 0.182, which means that the larger land area size has a 0.661 times opportunity to adopt a medium level integration system than no integration.

The low and medium coefficient value of farmers' perception (*X7R* and *X7S*) is significant at α 0.05. The coefficient value is 3.910 and Exp (B) is 0.020 and the coefficient value is 2.416 and Exp (B) is 0.089 which means low and medium perception have an opportunity of 0.020 times and 0.089 times to adopt a medium level integration system compared to no integration system.

iii. Multinomial logit three equation (Comparison between the High Integration category (Y = 4) and No Integration (Y = 1) Equation (4)

$$\text{Ln}\,(P4/P1) = -4.9\,42 + 0.0\,00X1 + 0.0\,00X2 + 3.7\,18X3 + 1.2\,29X4 + 0.1\,09X5 + 0.5\,22X6 - 22.2\,82X7R - 24.7\,42X7S \quad (4)$$

The coefficient of income (*X1*) is statistically significant at α 0.05. The coefficient value is 0.000 and Exp (B) is 1.000, which means the higher farming income, the opportunity to do a high level integration system is 1.000 times compared to no integration.

The land area size coefficient value (*X3*) is statistically significant at α 0.05. The coefficient value is 3.718 and Exp (B) is 0.024, which means the larger land area size has 0.024 times the opportunity to adopt a high level integration system rather than no integration.

The coefficient value of the cow numbers (*X4*) is statistically significant at α 0.05. The coefficient value is 1.292 and Exp (B) of 0.293 which means that the more cow numbers has 0.293 times the opportunity to adopt a higher level of integration system than not integration.

4 Conclusion

In conclusion, this study indicated that simultaneously all the predictor variables had a significant effect on the response variable, while the income, land area, number of cattle and farmers' perceptions had a very significant effect on the adoption of the integration system while the variable costs of production, farming experience and labor did not have a significant effect on the decision to adopt a rice and cattle integration system.

References

1. Biro Pusat Statistik. Bengkulu Dalam Angka. Bengkulu (2017). https://bengkulu.bps.go.id/publication/2017/08/11/e3f44cfdc2b9cf36c2891704/provinsi-bengkulu-dalam-angka-2017.html
2. S. A. Mamun, N. Fouzia, R.D. Momota. J. Environ. Sci Natural Res., **4**,2:127–136 (2011). https://www.banglajol.info/index.php/JESNR/article/view/10161
3. Wibawa, Wahyu dan Silviyani. Sistem Pertanian Bioindustri yang Berkelanjutan Berbasis Integrasi Padi – Sapi. BPTP. Bengkulu. (2015).p.121. [in Bahasa Indonesia]
4. K. Sukiyono. Penelitian Survai dan Teknik Sampling. Badan Penerbitan Fakultas Pertanian UNIB. Bengkulu. (2018).p.233. [in Bahasa Indonesia]
5. A.M. El-Habil. Pakistan J. Stat. Oper. Res. **8**,2:271–291(2012) https://www.researchgate.net/publication/274394043_An_Application_on_Multinomial_Logistic_Regression_Model

6. B. Situmorang, S. Edwina, E. Maharani. Jurnal Online Mahasiswa Fakultas Pertanian Universitas Riau. **2**,1:1–12(2015). [in Bahasa Indonesia]. https://www.neliti.com/id/publications/200956/adopsi-inovasi-petani-kelapa-sawit-terhadap-sistem-integrasi-sapi-kelapa-sawit-s
7. S. B. Manjunatha, D. Shivmurthy, A. S. Sunil, M. V. Nagaraj, K. N. Basavesha,. J. Agr. Allied Sci. **3**,4:30–38(2014) https://www.researchgate.net/publication/272158685_Research_and_Reviews_Journal_of_Agriculture_and_Allied_Sciences_Integrated_Farming_System_An_Holistic_Approach_A_Review
8. Lindawati. *Analisis Faktor yang Mempengaruhi Perilaku Ekonomi dan Kesejahteraan Rumahtangga Petani Usahatani Jawa Barat* [Analysis of Factors Affecting the Economic Behavior and Welfare of Farm Farmers in West Java]. [Thesis Phd]. Institut Pertanian Bogor (2017).p.140. [in Bahasa Indonesia]. https://repository.ipb.ac.id/jspui/bitstream/123456789/85350/1/2017lin.pdf
9. R.P. Soni, K. Mittu, L. Rajesh, IOSR J. Agricul Veterinary Sci. **7**,10:36–42(2014) https://pdfs.semanticscholar.org/3a42/2fb51a31a8c1dbe60e34b5db68e16294b149.pdf
10. N. Jaishankar, B.S. Janagoudar, K. Basavaraj, V. P. Naik, S. Siddayya. *Integrated Farming for Sustainable Agriculture and Livelihood Security to Rural Poor*. International Conference on Chemical, Biological and Environmental Science. May 12–13. Kuala Lumpur. Malaysia (2014).p22–24 https://www.researchgate.net/publication/282148497_Integrated_Farming_for_Sustainable_Agriculture_and_Livelihood_Security_to_Rural_Poor
11. D. Menozzi, M. Fioravanzi, M. Donati. BAE . **4**,2:125–147(2015). https://www.researchgate.net/publication/282133518_Farmer's_motivation_to_adopt_sustainable_agricultural_practices
12. S. Patidar dan H. Patidar. Int. J. Appli. Innov Eng. Manag. **4**,3:269–277(2015) https://www.ijaiem.org/Volume4Issue3/IJAIEM-2015-03-31-91.pdf
13. S.D. Argade, G. Sankhala, S. Wadkar. Int. J. Agr. Ext. **03**,1:25–30(2015). https://www.researchgate.net/publication/277773800_International_Journal_of_Agricultural_Extension_FARMER'S_PERCEPTION_TOWARDS_INTEGRATED_FARMING_SYSTEMS_IN_MAHARASHTRA_A_METHODOLOGICAL_APPROACH
14. K. Kariyasa, Y. A. Dewi. J. Int. Food Agric. Econ. **1**,2:29-38(2013) https://www.researchgate.net/publication/283715033_Analysis_of_factors_affecting_adoption_of_integrated_crop_management_farmer_field_school_Icm-Ffs_in_swampy_areas
15. E.M. Rogers. Diffusion of Innovation. The Free Press. New York (2004).p.576. https://books.google.co.id/books/about/Diffusion_of_Innovations_5th_Edition.html?hl=id&id=9U1K5LjUOwEC&redir_esc=y

Eco-agriculture and Farming in the Anthropocene Epoch: A Philosophical Review

Rangga Kala Mahaswa[1,*], *Agung* Widhianto[2], and *Nurul* Hasanah[3]

[1]Faculty of Philosophy, Universitas Gadjah Mada, Jl. Olahraga, Caturtunggal, Depok, Sleman, Yogyakarta 55281, Indonesia
[2]Department of Political Science, Umeå University, SE-901 8 7, Umeå, Sweden
[3]The School of Materials, The University of Manchester, Manchester M13 9PL, United Kingdom

Abstract. This article examines the complex interactions between agriculture, farming, and the Anthropocene environment. It discusses the challenges facing modern agriculture as a significant contributor to land degradation and climate change related to the planetary boundaries scale. Criticism of current agriculture is the effort to approach a philosophical view in considering eco-agriculture as part of environmental ethics. The holistic resolution that aligns the sustainability orientation for future agriculture is necessarily needed by social and political transformative movement. Therefore, the result finds the human moral value of land farming responsibility that agriculture is an ethical act requiring reflection at all planetary aspects, including food resilience, socio-economic changes, climate change adaptation, and natural preservation.

Keywords: Environmental movement, planetary boundaries, sustainabilization, utilitarianism.

1 Introduction

The future challenge of the 21st century concerns current global agriculture and food management, which are following an ideal consensus to produce food good quality in terms of a sustainable goal. Geologically, modern agriculture has impacted the beginning of the 'Anthropocene,' a new geological epoch, represented by the fact that human intervention changes the Earth's surface structurally [1]. It recognizes that agriculture as a counterpart of the Anthropocene is at the point of anthropogenic activities in natural cycles (phosphorus, nitrogen, carbon, water, etc.) [2]. It means that the Anthropocene narrative draws the turning point of the agriculture revolution, from hunter-gathered to modern agriculture.

In the last Holocene epoch transition, the 'stable conditions' have allowed humans to develop their basic civilization communities [3]. This stability is possible to evolve agriculture innovation. However, this frame has drawn attention related to asynchronous interconnected phenomenon in terms of global ecological change during the last century based on the anthropogenic footprints such as the rate of extinction and climate changes [4, 5]. Several significant human impacts on the environment are used to the indicator or potential point of the Anthropocene beginning; one of them is farming impact. The origins of agriculture have estimated in the Neolithic revolution of the Neolithic Demographic

*Corresponding author: rangga.mahaswa@gmail.com

Transition (NDT) as an evolutionary concept of agriculture about 8 000 yr ago [6]. Agriculture significantly increased as the way for a transition from hunter-gather society to agriculturalist, as the vision to widespread the process of land usage and changes. On the other hand, the notion of ultrasocial revolution can lead to an understanding of the relation of agriculture and the Anthropocene [7].

Modern agriculture is one of the most significant contributors to greenhouse gas emissions. The issues of climate change from agriculture also supply a negative impact on land use. Farmers are aware of the climate change impact, and they deal directly with capricious seasonal patterns such as floods, drought, soil degradation, waste, and damaging pollutants. In most critical cases, anthropogenic behavior intertwines with and impact cyclical activity on Earth [8]. Thus, human actions collectively at the planetary level align with the natural cycle's interaction. The role of agriculture also correlates to deforestation in driving the carbon cycle under the various issues by region found around the world.

Given the nature of Anthropocene discourse being the concerning issues of this article, especially pertaining to the question of the human-nature relationship, agriculture on the planetary context, then it is sound to state that this issue needs a philosophical reflection. Precisely, the specific inquiry on how a philosophical approach examines a prudent and safe practice of agriculture to sustain humanity, without compromising issues pertinent to ecological damage and the consequence of climate change? However, despite its philosophical nature that perhaps often regarded to enshrine a problem-solving tendency, this paper is purely hypothetical and thus, potentially inspiring a set of real practical importance that eventually produce an appropriate praxis of agriculture. Orienting this 'hard-science' notion such as in this case, agriculture, in a philosophical manner has become an accepted norm within the scientific community to introduce alternative perspective or reinforcing the importance of praxis that this so-called 'science' is simply a mere idea without human intervention to be considered along the way. In brief, this paper aims to examine the philosophical concept of subjectivity and solidarity to understand how humans deal with the ecological risk and planetary boundaries of the Anthropocene based on the concept of eco-agriculture, especially in the evolving moral status of the decision making in the agriculture sustainability and food resilience.

2 Literature review

2.1 Eco-agriculture and environmental ethics

Eco-Agriculture emerged in the 1970s, a fundamentally a self-aware movement, and alternative to agriculture in response to the dominance of 'conventional agriculture' due to its accepted practice of dependence on chemical biocides and energy intensiveness [9]. From a philosophical point of view, this movement is fundamentally a holistic system, meaning that it complies with the natural order law as an ecological creativity movement rather than the merely mechanistic industrial end. Farmers' roles, as nature's partner, is under the responsibility and obligation to be a respectful and caring steward of life [9]. Agriculture, therefore, becomes an art of farming that needs the dynamic interplay of science and technology to adapt to the environment. Soil is a separate entity with its complexity that is organic and fragile and relies on external protection and care.

As a movement, eco-agriculture is a radically individualistic action. Each farm is its separate realm and heavily personalized. Hence, they have their unique methods and materials in order to meet the specific need and the natural condition of the land. This means that every single farm has its own 'philosophical' finesse, yet the real problem is situated in the development of a self-awareness movement that is based on previous learning and experience. Eco-agriculture is a fundamental concept that can be traced back to the history of human civilization itself, stretching as far as the prehistoric (the soil community and spiritualistic farming), ancient (basic methods of irrigation, crop rotations, cultivation), and modern (mainstream agriculture—controversial 'organic' vs. chemical) [4, 10]. On the positive side, however, archeological evidence shows that ancient farmers often preserve topsoil and rebuild damaged land [10, 11]. Furthermore, contemporary agriculture is possible to conserve the land in order to anticipate the loss of biodiversity and climate change. Nevertheless, this movement subsequently generates a significant deal cost for transforming global understanding of future agriculture ecologically.

Specifically, environmental ethics inspire Eco-Agriculture perspective [12, 13]. This paper will bring the context of geological time—the Anthropocene to specify the spatial and temporal scale in framing the right philosophical approach. According to the Ruddiman hypothesis, greenhouse-gas has driven climate change since the Neolithic revolution, then initiated the Agriculture revolution as a part of the Anthropocene thesis [14, 15]. The two dominating moral paradigms that were implemented in the discourse of environmental ethics: deontology and utilitarianism, began to emerge in the 1990s [16]. In the present day, there are dissolved into holistic and affective ethics [12, 17].

Agriculture within the discourse of Anthropocene is the pinnacle of anthropogenic activity that is radically altering the natural relationships of the biome. In responding to the Anthropocene relations, Miles and Craddock propose the orientation of internal value called Biome ethics to anticipate a catastrophic global disequilibrium [18]. Some are optimistic and simultaneously pessimistic about human progress and the apex of society where prudent decision-making becomes the norm that can potentially affect the biosphere, particularly the possible collapse of global biome in the future. Moreover, in relation to prudent decision-making and ethical outlook in dealing with the environment, three ethical notions should be considered, such as i) conservationism (preservation natural spaces and wild species aesthetically), ii) environmentalism (against chemical industrial and preservation of diversity), iii) apocalypticism (global human catastrophe) [18]. The Anthropocene epoch presents a set of detrimental effects of anthropogenic activity on the life-sustaining capability. Therefore, it needs to extend beyond a focus on prudentialism or responsibility of this situation. The Anthropocene is holistic and primarily concerns with the natural cycle, thus the inclusion of past human activity in the narrative, as well as the product of colonization and capitalism that did not merely result in the production of the high carbon oxide. Still, the idea of various exploitation and oppression become the core of this issue that affected living- and non-living being. Economic factors disrupted local farming by opting for the most fertile land for industrial agriculture industries without ethical reasoning and its impending consequence. Such considerations may pertinent to the political-ecological connectivity, human-nonhuman continuum, and shared suffering to portend the possibility of life rather than apocalyptical peril.

2.1 Planetary boundaries in the Anthropocene view

The new realities of the Anthropocene face the issues of limited global natural resources. Johan Rockström (the Stockholm Resilience Centre) proposes the central concept of Planetary boundaries to define the notion of safe operating space for humanity [19]. It evaluates the human impacts on the biosphere, indicating a planetary scale that anthropogenic activities are substantially altering four out of nine crucial Earth system indicators in ways that affect the stability conditions of Earth's life [20, 21]. Planetary boundaries include climate change, biodiversity loss, biogeochemical flows (nitrogen and phosphorous cycle), stratospheric ozone depletion, ocean acidification, freshwater use, land-system use, atmospheric aerosol loading, and the introduction of novel entities. These four of the earth-system processes that have been transgressed to its critical condition are i) climate change, ii) biodiversity loss, iii) land-system use, N-P cycle, and iv) change in land use. Based on Table 1. and Figure 1., the current value due to transgressions of planetary boundaries are particularly alarming for two reasons: first, it is causing the severe adverse effect of future of global resource and human well-being, and second, the boundary transgression will result in altering the other systems. Therefore, the sense of preserving a safe operating space for ecological resilience is the highest priority to limit the consequence of transition.

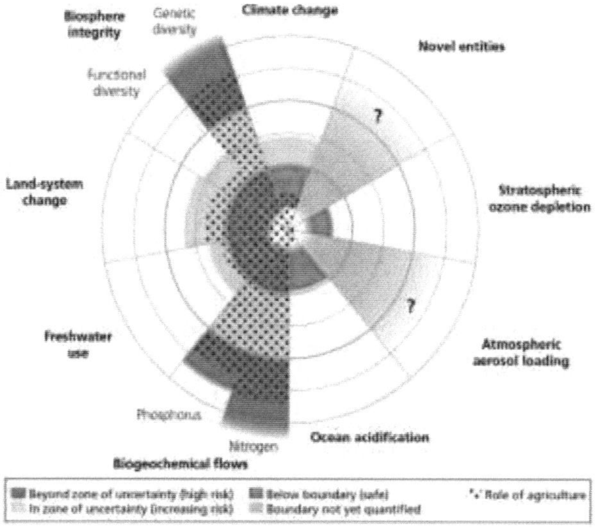

Fig. 1. The role of agriculture in planetary boundaries [2].

Table 1. The current value of planetary boundaries [22]

Earth-system process	Control variable	Planetary boundary	Current value
Climate change	Atmospheric concentration of carbon dioxide change in radiative forcing	≤ 350 ppm ≤1 W m^{-2}	396.5 ppm 2.3 W m^{-2}
Biodiversity loss	Global extinction rate in E/MSY or one per million species per year	≤10 E MSY^{-1}	100 E MSY^{-1} to 1 000 E MSY^{-1}
Nitrogen and phosphorous cycle	Reactive nitrogen removed from the atmosphere Phosphorus flowing into oceans	≤62 Tg N yr^{-1} ≤11 Tg P yr^{-1}	150 Tg N yr^{-1} 22 Tg P yr^{-1}
Stratospheric ozone depletion	Stratospheric concentration of ozone measured in Dobson units (DU)	≤ 5 % below preindustrial levels (290 DU)	~200 DU over Antarctica in Austral spring
Ocean acidification	Mean saturation state to aragonite in the oceans	≥ 80 % of the preindustrial level	84 % of the preindustrial level
Freshwater use	Freshwater consumption	≤ 4 000 km^3	~ 2 600 km^3
Change in land use	Area of forested land as a percentage of original forest cover	≥ 75 %	62 %
Novel entities	NA	NA	NA
Atmospheric aerosol loading	Aerosol optical depth	Regional limit of ≤ 0.25	0.3 AOD over South Asian region

The role of agriculture in altering the stability at the Earth scale has fully transgressed two planetary boundaries at high risk from biosphere integrity and biogeochemical flows [23]. Agriculture is the prime factor at increasing risk level for an uncertain situation in land-system change and freshwater use, and even climate change. Several interventions will be needed in relation to agriculture and the food system in more comprehensive aspects. Biogeochemical use (nitrogen and phosphorous) as part of modern agriculture into terrestrial and aquatic ecosystem have affected the land-system and biodiversity loss. Whereas in the case of the zone of uncertainty, the N-P boundary related to agriculture is more than 200 % transgressed at high-level risk, including food processing-consumption [24].

Further, the concept of planetary boundaries is inspired by the Holocene environment, which relatively stable for the last 12 000 yr [25]. Still, it is now already been replaced by the Anthropocene due to the glorious human domination. Since the start of the Anthropocene epoch, the persistent utilization of chemical products on a global scale has contributed massively to the agriculture industries since the 1950s [26]. A recent Anthropocene research finds a novel and relevant stratigraphically distinct sediments contained persistent chemicals and radionuclides. Thus, there has to be an agriculture reform within the framework of public policy on a global scale to collectively assume responsibility, especially in reducing land degradation and global ecological change.

3 Results and discussion

3.1 Agricultural sustainability as the Anthropocene context

Visualizing the condition of the Anthropocene becomes a way to contemplate the position of humans today. The focus of the Anthropocene discussion is often directed at geological time scale issues rather than considering aspects of the scale of resilience and sustainability of natural resources for humans. This idea may sound strange and seems to be anthropocentric. Yet, the development of sustainability in the agricultural system cannot be separated from the development of technology and practices that have a direct impact on the environment as a space for farming, as well as influencing improvisation in food production for farmers. However, the extraordinary progress in modern agricultural productivity has led to increased use of fertilizers, irrigation systems, agricultural machinery and equipment, pesticides, and land expansion [27]. It is one of the most fundamental considerations of how relations between farmers, communities, and the environment can guarantee stability conditions in the future. Therefore, to answer these challenges, a philosophical and ethical approach that integrates biological and ecological processes in food production is needed to reduce the use of non-renewable resources, environmental damage, and public health.

Archaeological evidence shows that agricultural development is attached to the community's perspective in determining survival choices. For example, marginal zone theory claims that humans tend to choose agriculture when the most optimal areas for hunting and gathering are no longer sufficient to meet the needs of the community [11]. Whereas a 'feasting model' further explains the development of technology that can dominate the origin of agriculture. The development of agriculture became very intense after humans began to develop the division of labor and knowledge in managing seeds and soil so that the issue of farming has become an interesting perspective in the analysis of origins of farming and agricultural intensification. Based on the social anthropological view of farming, three things underlie the agricultural perspective in an area, which are related to i) power and capital relations, ii) social relations dependency with values, race, and gender, and iii) physical environment [11]. However, anthropologically speaking, farming undergoes different approaches related to the severe planetary change conditions due to the dominant influence on the environment and climate, all of which stem from anthropogenic activities, capital, and plantation agriculture. This condition directs the political agenda in overcoming the ecological crisis based on the logic of environmental modernization, homogeneity, and control, which are developed on the history of plantations. Haraway argues that the emerging Plantationocene as a critique of the dark side of the system of human control over nature and racial policy in terms of ecological justice [28].

The interpretation of agricultural sustainability as an ideology is to fulfill the continual strategies of land stewardship-integrated with agriculture. In 1996, Hansen criticized the concept of agricultural sustainability as a property of agriculture and motivation of changes about threats to agriculture itself [29]. For sustainability agriculture, its characterization should be following i) literal (consistency), ii) system-oriented (an objective property reform), iii) quantitative (quality of human health and earth-wellbeing), iv) predictive (future orientation), v) stochastic (variability recognition), and vi) diagnostic (an integrated measure of sustainability and its weakness and threats). Furthermore, Table 2 shows a comparison between conventional agriculture and sustainable agriculture.

Table 2. The characterization of conventional and sustainable agriculture approaches

Conventional	Sustainable
Symptoms	Causes, prevention
Reductionist notion	Holistic view
Eliminate 'enemies' and narrow focus (neglects side-effects)	Respond to indicators (planetary crisis) and Broad focus
Short-term / Instant	Long-term / Future responsibility
Competitive and centralized (value secondary)	Co-operative and decentralized (human scale with higher values)
Dependent on chemical	Self-maintaining for bio-ecological
Homogeneity	Heterogeneity
Oriented towards imported management and products	Oriented towards locality management for work and service and processes

Organic farming orientation trends have been quite popular among the people today due to its practicality and its credulity of ideal values that aim to protect the environment. As a concept and ideal, organic agriculture movement began in the 1960s, at the height of the counterculture with its idealist and agriculturist readily prepared to resolve the perennial problem of agriculture, such as low quality of food and livestock feed, erosion, soil-degradation, and rural poverty. Inexhaustibly, the reason people trust in organic farming because of two enduring questions, for example, i) is organic food healthier? and ii) do organic crops resist pests? [30]. So, it needs a soil management strategy called humus farming by integrating cultural and biological adaptation, advanced green technology, promoting ecological balance, and biodiversity conservation. The organic farming and products must meet a check and balance in terms of the quality standard; that is, organic certification system plans to record the land integrity management.

Thompson [31] clarifies the hidden philosophical assumption of agricultural sustainability in two main substantive approaches: resource sufficiency and functional integrity as sophisticated alternatives for conceptualizing the nature of human responsibility. It is; therefore, the concept of sustainability has associated with democracy and social justice. These principles also help to develop an agricultural system for improving natural capital rather than social capital. Both modern biological approaches and agronomic management address the operation of resilience energy flows and nutrient cycling for food productivity. Primary challenges, however, persist in terms of the relationship between Anthropocene and the sustainability concept that can be followed these two possible case scenarios [4]. First, society should live with a new perspective in human-nature relations within the planetary boundaries of the physical Earth, including water resources, biogeochemical cycles, biosphere, and land use. Second, sustainability has its anthropogenic limitation. In the future, agriculture will always depend on the physical earth condition and in the dynamic interplay of the socio-political systems. Therefore, the sustainability goal could be the measurement of the natural resource's management like social well-being, eco-green urban movement, environmental ethics, carbon footprint, water management, food consumption, waste, renewable energy, and recycling in terms of improving agricultural and farming in the Anthropocene context.

3.2 Future agriculture: A philosophical reflection

In this section, a thoughtful review will be applied to analyze the future of agriculture. This paper follows two studies: in Indonesia, it is mainly concerning the nature of the agricultural production system in Syuaib [32] and coffee farming in the Anthropocene research by Naponen et al. [33]. First, Syuaib argues that the challenge of agricultural sustainability in Indonesia is because of the shifting agriculture paradigm in the last four decades, as well as the growing population, industrialization, and urbanization. On the other hand, the problem of modern farming technology in developing countries is due to a high dependency on chemicals in several intensive farming systems. Second, Naponen et al. solve the general problem of chemical farming by changing the perspective of coffee farming in facing global issues. If Syuaib focuses on the local problem, so Naponen et al., primarily demonstrate that the reality of climate change is the underlying fundamental problem of future agriculture, farming, and agro-forestry. Sustainable coffee production, for example, has been a subject of the agriculture research network. It means that this target area needs extraordinary efforts to support farmers, especially in terms of maintaining soil fertility, water supply, and control of pests and disease in traditional farm-management challenges.

According to climate-smart agriculture (CSA) [33], sustainable agriculture should integrate the three dimensions of economic, social, and climate challenge, and extended to environmental protection, and sustainably increasing productivity and resilience. Its concept is assisting farmers to adapt the global climate change and secure sustainable well-being, economic, and social justice. For example, CSA projects have been applied in Oaxaca, Mexico, with smallholder coffee farmers to develop community-based CSA. In response to the market challenges affecting the coffee supply chain, both in short- or long-term adaptation strategies for increasing productivity and resilience of sustainability agricultural. Recent research in flowering plants as the political agenda brings agriculture as geopolitical issues [34]. It is because human civilization depends on flowering plants in a retrospective view of evolution, even crucial for providing biomass for social needs. Conversely, human appropriation of biomass is altering planetary boundaries, therefore, agriculture plan in the future involves long-term resilience rather than short-term performance.

The discussion of ethics in agriculture is the reason for the ethical life action that needs the attention of agriculturalists by being aware of the reality of the embedded value in the agricultural practice. There are several ethical approaches that can be a model for the best agriculture practice as a policy or new science. In this paper, the implicit utilitarianism of agriculture will be examined as the common ethical justification for action, to improve and develop the greatest good for the greatest number of people [35]. It will consider utilitarian ethics as retaining essential notions to solving agriculture moral dilemmas. The Multiple Strategies Utilitarianism suggest that the focus of ethics recognize two critical matters: maximize the goal of general welfare and the future-oriented rather than favoring the established status quo. Zimdahl [36] suggests that agriculture should judge the value of farming technology and practice, and all agricultural action is maximizing the general welfare. Therefore, Table 3 shows 11 criteria for approaching the agriculture ethical decision, but all value may not be applicable in a particular case.

Table 3. The Value Criteria Suggested for Agriculture

No.	Value Criteria	Goal
1	Equity	Questioning fairness of all species
2	Food security	Not only for human needs, but also security includes nutritional adequacy, availability, and access
3	Environmental soundness	Evaluating agricultural technology and practice as a cause of environmental degradation
4	Profitability	Investigating the real cost of non-technological user
5	Safety and risk	Measuring risk is an empirical scientific activity, but judging safety is a normative political activity
6	Quality of life and human dignity	Minimizing the conflict of interest
7	Aesthetics	Agriculture should appear better aesthetically
8	Human and animal health	Providing eco-friendly technology
9	Consent	Developing agricultural democratic
10	Sustainability	Protecting human well-being, species, biodiversity, and environmental as highly valued
11	Institutional roles	Evaluating the decision making and role of agriculture institution

Lastly, this paper improves the efforts in applying the value criteria above based on the logic in reality (LIR), which defines the basis of the logic of moral responsibility known as sustainabilization. Hofkirchner in [37] develops this concept as a process of suppression and rejection of anthropogenic breakdown and alternative safeguard development of the threshold of endangering biodiversity and the maintenance of society. Thus, by understanding the logical relationships between the human and non-human as the domain of nature, it will be fundamental strategies for evolving the socio-epistemology of society in real transformative action. Alternatively, the agrobiodiversity knowledge framework [38], for example, is essential to guide the transformative planetary for sustainable development goals with socio-economic changes, biodiversity conservation, social justice, and food security. Nevertheless, there needs to sharpen the kind of independent awareness related to the shift from subjectivity desires to solidarity collectively [39–43]. All movements must begin immediately because this condition is not only regarding human existence but also for owning responsibility for the uncertain fate. As a said wise-man, "you are what you eat." Eco-agriculture is an art that emphasizes moral responsibility, thus guarantee an appropriate and adequate human food consumption and also respecting non-human position as relevant to human development itself.

4 Conclusions

This article concludes that a philosophical review for eco-agriculture and farming in the Anthropocene epoch is based on the planetary boundaries concepts and environmental ethics to solve the challenge of global climate change. It is because the issues of the Anthropocene are not only challenging the geological finding but also the gradual emergence of the essential contribution to modern agriculture that is distressing for nature; such as those that caused land degradation, biodiversity loss, biogeochemical flow, and climate change. For environmental ethics, this research finds that future agriculture needs the fundamental value criteria to achieve the sustainability goal related to the logic of ecological reality. Finally, the future of agriculture depends on our action where humanity must be free from dominating egoism towards nature and replace capital-oriented decision to transformational policy in terms of agricultural resilience.

References

1. U. Svedin, *Global conditions for the future of agriculture in the Anthropocene*, [Online] from http://regardssurlaterre.com/en/global-conditions-future-agriculture-anthropocene (2012), [Accessed on 28 July 2020]
2. B.M. Campbell, D.J. Beare, E.M. Bennett, J.M. Hall-Spencer, J.S.I. Ingram, F. Jaramillo, et al., Ecology and Society, **22**,4:8(2017). https://doi.org/10.5751/ES-09595-220408
3. R.K. Mahaswa, A. Widhianto, SHS Web of Conferences, **76**:01040(2020). https://www.shs-conferences.org/articles/shsconf/abs/2020/04/shsconf_icsh2020_01040/shsconf_icsh2020_01040.html
4. Roka. Anthropocene and climate change. In: *Climate Action*. Encyclopedia of the UN Sustainable Development Goals. W. L. Filho, A. Azul, L. Brandli, P. Özuyar, T. Wall (Eds). Cham: Springer (2019), p. 1–13. https://doi.org/10.1007/978-3-319-71063-1_26-1
5. L. Head, Anthropocene Review, **1**,2:113–125(2014). https://doi.org/10.1177/2053019614529745
6. J. Gowdy, L. Krall, J. Bioecon., **16**:179–202(2013). http://dx.doi.org/10.1007/s10818-013-9156-6
7. J. Gowdy, L. Krall, Ecological Economics, **95**:137–147(2013). https://doi.org/10.1016/j.ecolecon.2013.08.006
8. N. Castree, Anthropocene and planetary boundaries, In: *International Encyclopedia of Geography*, D. Richardson, N. Castree, M.F. Goodchild, A. Kobayashi, W. Liu, R.A. Marston (Eds). US: John Wiley & Sons (2018). https://doi.org/10.1002/9781118786352.wbieg0027.pub2
9. M.C. Merrill, Biological Agriculture and Horticulture, **1**,3:181–210(1983). https://doi.org/10.1080/01448765.1983.9754395
10. J.A. Sandor, J.A. Homburg, Journal of Ethnobiology, **37**,2:196–217(2017). https://bioone.org/journals/Journal-of-Ethnobiology/volume-37/issue-2/0278-0771-37.2.196/Anthropogenic-Soil-Change-in-Ancient-and-Traditional-Agricultural-Fields-in/10.2993/0278-0771-37.2.196.short
11. A. Ofstehage, *Farming*. [Online] from Cambridge Encyclopedia of Anthropology. p. 1–21. https://www.anthroencyclopedia.com/entry/farming (2020). [Accessed on 28 July 2020]
12. J.B. Callicott, Environmental ethics, In: *Encyclopedia of the Anthropocene*, D.A. DellaSala, M.I. Goldstein (Eds.). Oxford, Waltham MA: Elsevier (2018), p. 1–10. https://www.sciencedirect.com/science/article/pii/B9780128096659103039
13. M. Sagoff, Ecomodernism and the Anthropocene, In: *Encyclopedia of the Anthropocene*, D.A. DellaSala, M.I. Goldstein (Eds.). Oxford, Waltham MA: Elsevier (2018), p. 61–66. https://www.sciencedirect.com/science/article/pii/B9780128096659103027
14. W. Ruddiman, S. Vavrus, J. Kutzbach, F. He, Anthropocene Review, **1**:147–153(2014). https://journals.sagepub.com/doi/10.1177/2053019614529263
15. W.F. Ruddiman, Rev. Geophys., **45**,4:1–37(2007). https://doi.org/10.1029/2006RG000207

16. S.M. Gardiner, S. Caney, D. Jamieson, H. Shue, *Climate ethics: Essential readings*. New York: Oxford University Press (2010). https://global.oup.com/academic/product/climate-ethics-9780195399622
17. J.W. Walters, L.F. Greer, Evolving moral status, In: *Encyclopedia of the Anthropocene*, D.A. DellaSala, M.I. Goldstein (Eds.). Oxford, Waltham MA: Elsevier (2018), p. 53–60. https://www.sciencedirect.com/science/article/pii/B9780128096659104422
18. S.H. Miles, S. Craddock, Ethics for the Anthropocene epoch, In: *Encyclopedia of the Anthropocene*, D.A. DellaSala, M.I. Goldstein (Eds.). Oxford, Waltham MA: Elsevier (2018), p. 21–27. https://www.sciencedirect.com/science/article/pii/B9780128096659104446
19. J. Rockström, W. Steffen, K. Noone, Å. Persson, F.S. Chapin III, E. Lambin, et al., Ecology and society, **14**,2:1–33(2009). https://www.jstor.org/stable/26268316
20. W. Steffen, K. Richardson, J. Rockström, S.E. Cornell, I. Fetzer, E.M. Bennett, et al. Science, **347**,6223:1259855(2015). https://science.sciencemag.org/content/347/6223/1259855.abstract
21. W. Steffen, J. Rockström, K. Richardson, T.M. Lenton, C. Folke, D. Liverman, et al., Proceedings of the National Academy of Sciences, **115**,33:8252–8259(2018). https://www.pnas.org/content/115/33/8252?mod=article_inline
22. K. Meyer, P. Newman, *Planetary Accounting*. Singapore: Springer (2020). https://doi.org/10.1007/978-981-15-1443-2_3
23. H. Washington, Overshoot, In: *Encyclopedia of the Anthropocene*, D.A. DellaSala, M.I. Goldstein (Eds.). Oxford, Waltham MA: Elsevier (2018), p. 239–246. https://www.sciencedirect.com/science/article/pii/B9780128096659095173
24. T. Meier, *Planetary boundaries of agriculture and nutrition–an Anthropocene approach*. In: Proceedings of the symposium on communicating and designing the future of food in the Anthropocene. (Bachmann Verlag, Berlin, 2017). p. 67–76. http://www.nutrition-impacts.org/media/2017_TMeier_planetary_boundaries_agriculture_nutrition.pdf
25. L.H. Berry, Common resource governance. In: *Encyclopedia of the Anthropocene*, D.A. DellaSala, M.I. Goldstein (Eds.). Oxford, Waltham MA: Elsevier (2018), p. 239–246. https://www.sciencedirect.com/science/article/pii/B9780128096659104847
26. R. Kanianska, Agriculture and its impact on land-use, environment, and ecosystem services. In: *Landscape ecology-The influences of land use and anthropogenic impacts of landscape creation*, A. Almusaed (Ed), Croatia: InTech(2016), p. 1–26. https://www.intechopen.com/books/landscape-ecology-the-influences-of-land-use-and-anthropogenic-impacts-of-landscape-creation/agriculture-and-its-impact-on-land-use-environment-and-ecosystem-services
27. J. Pretty, Philosophical Transactions of the Royal Society B: Biological Sciences, **363**,1491:447–465(2008). https://doi.org/10.1098/rstb.2007.2163
28. D. Haraway, Environmental Humanities, 6,1:159–165(2015). https://read.dukeupress.edu/environmental-humanities/article-abstract/6/1/159/8110
29. J.W. Hansen, Agricultural Systems, **50**,2:117–143(1996). https://doi.org/10.1016/0308-521X(95)00011-S
30. G. Kuupeer, *A Brief Overview of the History and Philosophy of Organic Agriculture*, USA: Kerr Center for Sustainable Agriculture (2010). https://kerrcenter.com/wp-content/uploads/2014/08/organic-philosophy-report.pdf
31. P.B. Thompson, International Journal of Agricultural Sustainability, **5**,1:5–16(2007). https://doi.org/10.1080/14735903.2007.9684809
32. M.F. Syuaib, Agricultural Engineering International: CIGR Journal, **18**,2:170–184(2016). http://cigrjournal.org/index.php/Ejounral/article/view/3747

33. M.R.A. Noponen, C. Góngora, P. Benavides, A. Gaitán, J. Hayward, C. Marsh, et al., Environmental Sustainability-Farming in the Anthropocene. In: The Craft and Science of Coffee, B. Folmer (Ed), USA: Academic Press(2017). p. 81–107.
34. I. Negrutiu, M.W. Frohlich, O. Hamant, Trends in Plant Science, **25**,4:349–368(2020). 10.1016@j.tplants.2019.12.008
35. M.J. Mariola, Agriculture and Human Values, **22**,2:209–223(2005). https://doi.org/10.1007/s10460-004-8281-1
36. R.L. Zimdahl, *Agriculture's Ethical Horizon*, USA: Elsevier (2012). p. 53–80. https://doi.org/10.1016/b978-0-12-416043-9.00004-0
37. J. Brenner, Philosophies, **3**,2:1–16(2018). https://doi.org/10.3390/philosophies3020016
38. K.S. Zimmerer, S. de Haan, A.D. Jones, H. Creed-Kanashiro, M. Tello, M. Carrasco, et al., Anthropocene, **25**:100192(2019). https://www.sciencedirect.com/science/article/pii/S2213305419300037
39. J. Fanzo, A. Hood, C. Davis, Physiology and Behavior, **222**:1–4(2020). https://doi.org/10.1016/j.physbeh.2020.112929
40. M. Powers, Sustainability and resilience, In: *Encyclopedia of the Anthropocene*, D.A. DellaSala, M.I. Goldstein (Eds.). Oxford, Waltham MA: Elsevier (2018), p. 29-37. https://www.sciencedirect.com/science/article/pii/B9780128096659104914
41. R. ter Meulen, Solidarity In: *Encyclopedia of the Anthropocene*, D.A. DellaSala, M.I. Goldstein (Eds.). Oxford, Waltham MA: Elsevier (2018), p. 111–117. https://www.sciencedirect.com/science/article/pii/B9780128096659104525
42. M.B. Morrissey suffering in the Anthropocene era: Contributions of phenomenology to understanding the world-constituting role of subjectivity, In: *Encyclopedia of the Anthropocene*, D.A. DellaSala, M.I. Goldstein (Eds.). Oxford, Waltham MA: Elsevier (2018), p. 147–150. https://www.sciencedirect.com/science/article/pii/B9780128096659103088
43. I. Douglas, Ecosystems and human well-being, In: *Encyclopedia of the Anthropocene*, D.A. DellaSala, M.I. Goldstein (Eds.). Oxford, Waltham MA: Elsevier (2018), p. 185–197. https://www.sciencedirect.com/science/article/pii/B9780128096659092065

The Strategy of Salt Business Development: A Case Study in Sumenep, Indonesia

Rika Diananing[1], *Amilia* Destryana[1,*], *Ribut* Santosa[1], *Noor* Illi Mohamad Puad[2], and *Agustine* Christela Melviana[3]

[1]Faculty of Agriculture, Jl. Raya Sumenep–Pamekasan, Wiraraja University, Sumenep 69451, Indonesia
[2]Kulliyyah of Engineering, International Islamic University Malaysia P.O. Box 10, 50728 Kuala Lumpur, Malaysia
[3]School of Life Sciences and Technology, Bandung Institute of Technology, Jl. Ganesha No. 10, Bandung 40132, West Java, Indonesia

Abstract. Sumenep is one of the salt producers in Indonesia. The problem experienced by farmers is the production of salt using evaporation by solar energy that depends on the weather and the low price, caused by worse business management. Salt is a potential commodity, because its market is still wide open. This research aims to develop the salt production method and development strategy of salt business in Sumenep Regency by using SWOT analysis. The result of the analysis concludes that the priority of salt development business strategy in Sumenep Regency are: i) Geo-isolator technology application strategy to produce good quality of salt; ii) cooperation strategy in group mechanism to build power and increasing the bargaining value of the farmers; iii) capital strengthening strategy through partner cooperatives; iv) broader marketing management management strategies to industrial salt user sectors.

Keywords: Environmental friendly, gap of technology in farmer salt, Geo-isolator, increase salt productivity, increase salt quality

1 Introduction

Coastal and marine areas have a very high potential to be developed, therefore in the management of marine resources and fisheries coastal and marine areas need to be carefully planned and in accordance with the characteristics of the region. Based on data from the Department of Marine Affairs and Fisheries of Sumenep Regency, the largest production of the last 5 yr in 2012 amounted to 232 409.27 t.

The large number of production that contribute greatly to the national production has not been able to improve the welfare of farmers [1]. The process of making salt is generally done by the method of evaporation of seawater with solar energy, so that the minerals in it settles as desired. Another way is by the method of evaporation of seawater / brine / brine in fuel, electrodialysis and salt mining from salt rock. Traditional folk salt is generally made by drawing sea water, then incorporating evaporation fields to produce crystals. There are gaps in technology so resulting salt is still crunchy, contains low NaCl, a lot of

* Corresponding author: amiliadestryana@gmail.com or amiliadestryana@wiraraja.ac.id

soil/low impurities, dull color, high water content. Meanwhile, in terms of marketing system. One concern related to seawater desalination addresses the disposal of the concentrated saline, resulting from seawater desalination, which could cause serious environmental problems in the marine ecosystem [2–8].

There are gap of technology so resulting salt is still crunchy, contain low NaCl, a lot of soil /low impurities, dull color, high water content. Meanwhile, in terms of marketing system, salt is weak and vulnerable position because it does not have direct market access. The integration of seawater desalination and sea salt production has been reported in the world [9, 10].

Most salt ponds around the Indonesia are being either abandoned or converted for aquaculture, mainly due to decreases in the profitability of salt production in recent decades [11]. The new method is needed to increase the profitability of salt production, such as desalination to obtain the frewshwater and salt production [12], ion-exchange membrane electrodialysis [13], osmosis and nanofiltration [14]. The research objectives are i) to know the method of salt production in Sumenep Regency and ii) to determine the strategy of salt people business development in Sumenep regency.

2 Methods

This research was conducted in the salt center of Sumenep Regency, Indonesia with primary data collection technique by giving questionnaire and interview and secondary data with article or data.

2.1 Population and sampling

In this study, the population is the stakeholders, especially the salt farmers in accordance with the extent of land they are working on. Based on data from the Department of Marine and Fisheries, 2015, salt ponds spread in Sumenep Regency are presented in the table below.

Table 1. Land area of public salt production

No	Districts	Labors	Land Area Production (ha)	Land Area (%)
1	Gapura	445	302.7	15.57
2	Dungkek	239	143.5	7.38
3	Pragaan	390	257.96	13.25
4	Saronggi	576	338.75	17.42
5	Kalianget	787	495.05	25.46
6	Kota	8	4.72	0.24
7	Giligenting	506	285.54	13.14
8	Ra'as	109	65.84	3.27
9	Sapeken	100	50.76	2.61
10	Arjasa	49	24.5	1.26
11	Talango	10	7.59	0.39
	Total	3 222	1 977.21	100

Source: *Dinas Kelautan dan Perikanan* [15]

2.2 Equations and mathematics

Sampling conducted in this study using purposive sampling method is the selection of respondents conducted deliberately in the District Kalianget, District Gapura and District Saronggi. In this study, to determine the number of samples representing the population used the Slovin Umar formula [3] in Equation (1):

$$n = \frac{(1)\mathrm{N}}{1 + Ne^2} \tag{1}$$

Explanation:
N = sample size
N = population size
E = critical value

In this study the number of customer populations with the desired error limit is 10 %. Based on the above formula, the samples taken in this study as in Table 2.

The method of analysis used to know the strategy of salt business development using SWOT analysis model which is one method to describe the condition and evaluate a problem, project or business concept based on internal factor and external factor that is Strengths, Weakness, Opportunities and Threats. This method is most often used in business evaluation methods to find the strategy to be performed. SWOT analysis only describes the situation that occurs not as a problem solver.

Table 2. Data of population and sample

No	Districts/Area	Population		Sample	
		Geoisolator	Traditional	Geoisolator	Traditional
1	Gapura/Gersik Putih	12	16	7	8
2	Saronggi/ Kebundadap Barat	9	18	6	9
3	Saronggi/ Kebundadap Timur	2	2	1	1
4	Saronggi/Saroka	2	1	1	0
5	Kalianget/ Karanganyar	15	14	9	4
6	Kalianget/ Marengan Laok	3	15	1	8
7	Kalianget/ Pinggirpapas	4	14	1	7
8	Kalianget/Kertasada	8	6	2	3
	Total	55	86	28	40

Source: Processed Data, 2016

3 Results and Discussion

3.1 Potential salt pond in Sumenep Regency

Potential salt pond in Sumenep Regency spread in several areas, among others: Pragaan Sub-district, Giligenting Sub-district, Saronggi Sub-District, Kalianget Sub-District, Gapura Sub-District, Dungkek Sub-district, Talango Sub-District, Raas Sub-District, Arjasa District and Kangayan Sub-District. Salt pond in Kab. Sumenep by ownership status can be distinguished into private/public property and belongs to PT. Garam Persero.

Table 3. Data of land area of salt regency of Sumenep

No	Land ownership	Land area (ha)
1	Public	2 068
2	PT. Garam Persero	3 300
	Total	5 368

Source : *Dinas Kelautan dan Perikanan*, [15]

Productivity salt from public land area in the last 5 yr the samples taken in this study as in Table 4

Table 4. Production of salt in Sumenep regency 2011 to 2015

No	Production (t)				
	2011	2012	2013	2014	2015
1	160 683.82	232 409.27	185 385.54	292 051.54	236 117.97

Source: *Dinas Kelautan dan Perikanan* [15]

3.2 Analysis of study

3.2.1 Identification analysis of internal and external factors (S, W, O, T)

In the identification of internal factors consists of two parts of strengths and weaknesses (weaknesses) that exist in the internal development of salt public business in Sumenep Regency with the description as follows:

1. Strengths
   a. Experienced workers
   b. There is no salt substitution yet
   c. Simple production equipment
   d. Raw material production abundant
   e. Potential land
2. Weakness
   a. Lack of facilities and infrastructure
   b. The bargaining position of salt farmers is weak
   c. Lack of capital
   d. Lack of technology
   e. Limited financial services

The identification of external factors consists of two parts, Opportunity and Threat that exist in the development of salt public business in Sumenep Regency, Indonesia with the following description:

3. Opportunity
   a. High and unfulfilled market demand
   b. Relatively high selling prices in certain seasons
   c. Government policy support
   d. Technological innovation iodine salt development
   e. Accompaniment
4. Threads
   a. Free market
   b. Higher quality salt production is circulating within the same consumer region
   c. Business security
   d. Weather conditions
   e. Wholesaler

3.2.2 *Matrix of quality IFAS and EFAS*

In performing the SWOT analysis, it is necessary to consider the internal weighting matrix of Factor Analysis Strategy (IFAS) and External Factor Analysis (EFAS). Observations and interviews with stakeholders of IFAS matrix identification can be seen in Table 5 and identification of EFAS matrix presented in Table 6

Table 5. Matrix of IFAS of salt public business in Sumenep

No	Strategic Internal Factor	Quality (a)	Rating (b)	Quality vs Rating (axb)
	Strengths			
1	Experienced workers	0.197	4	0.788
2	There is no salt substitution yet	0.196	4	0.784
3	Simple production equipment	0.198	4	0.72
4	Raw material production abundant	0.213	4	0.852
5	Potential land	0.197	4	0.788
	Total	1.000	19	
	Weakness			
1	Lack of facilities and infrastructure	0.182	3	0.546
2	The bargaining position of salt farmers is weak	0.213	4	0.852
3	Lack of capital	0.214	4	0.856
4	Lack of technology	0.198	3	0.594
5	Limited financial services	0.194	3	0.582
	Total	1.000	17	

Source*:* Processed data, 2016

Table 6. Matrix of EFAS of salt public business in Sumenep

No	Strategic Internal Factor	Quality (a)	Rating (b)	Quality vs Rating (axb)
Opportunity				
1	High and unfulfilled market demand	0.215	3	0.645
2	Relatively high selling prices in certain seasons	0.190	3	0.57
3	Government policy support	0.197	3	0.591
4	Technological innovation iodine salt development	0.207	3	0.621
5	Accompaniment	0.191	3	0.573
	Total	1.000	15	
Thread (s)				
1	Free market	0.188	2	0.376
2	Higher quality salt production is circulating within the same consumer region	0.206	2	0.412
3	Business security	0.178	2	0.356
4	Weather conditions	0.221	3	0.663
5	Wholesaler	0.206	2	0.412
	Total	1.000	11	

Source: Processed data, 2016

3.2.3 Strategy alternative formula

The alternative strategy is based on the interaction of internal factors and external factors obtained based on the results of the analysis. Formulation of alternative salt business strategy in Sumenep Regency presented in Table 7.

From the table of matrix in Table 7, with strengths, weaknesses, opportunities and threats can be described as follows,

I. S-O Strategy

a. Encourage increased production through extensification to produce standard quality salts according to market needs. This strategy is carried out by utilizing the newly available labor force and potential untapped land. At the same time the development of new processing and innovation technologies for long-time laborers and self-sustaining production land to direct iodized salt production by utilizing financial / banking services. Implementation of this strategy can be done by extensification of land that produces salt without iodine according to quality standards so far. With the strength of the land area and the availability of sufficient labor then the implementation of this strategy is very possible to be applied so that the opportunity to reach the market and increase production volume in the short term is very potential can be achieved.

b. Intensification of capacity building activities in marketing management. This strategy is one effort that aims to optimize the utilization of opportunities for development of professional salting technology in order to enter the era of tight market competition in the future. Training programs, internships, pilots and comparative studies are essential to improve business-oriented business management skills. Amount one of the key opportunities associated with seasonal variability is the price at which certain seasons of high salt prices can be exploited by developing marketing management capabilities.

II. W-O Strategy

Maximize the assistance given to the farmers to improve accessibility of some of the production and potential areas and to improve the quality of human resources, especially in marketing management so that the selling price of salt production as much as possible can be achieved. The maximization of government assistance in salt development programs should be well targeted and not only to form the physical aid areas in the form of facilities and facilities but should include improving the quality of human resources related to marketing. Improved quality allows to exploit potential opportunities such as temporal fluctuations in prices through marketing management and production rhythms with the highest selling price.

III. S-T Strategy

Improving access to market information and to the short-term context the quantity of salt production using traditional technology is maximized by utilizing the remaining potential land area to dominate the market, especially unmet demand. This strategy is strived as soon as possible in order to excel in the competition so that the threat of free competition can be avoided. Production of salt with sufficient quantity and quality allows for market expansion in areas where competitors have not reached it so that the projection and expansion of market information is still running and superior in the speed of controlling the market or consumer.

IV. W-T Strategy

Encourage accelerated improvement of human resource capacity and other resource utilization including financial institution services to compete in the local and regional salt industry. This strategy should be synergistic and simultaneous so that salt development efforts can avoid threats with all the weaknesses they have. Weaknesses in the form of weak human resources must be addressed through intensification of training activities and comparative studies so that the insights and skills of the farmers and salt entrepreneurs are able to compete in a tight competition especially with the implementation of free market in the future.

Table 7. Matrix of SWOT of development strategy salt business in Sumenep.

IFAS / EFAS	Strengths a. Experienced workers b. There is no salt substitution yet c. Simple production equipment d. Raw material production abundant e. Potential land	Weakness a. Lack of facilities and infrastructure b. The bargaining position of salt farmers is weak c. Lack of capital d. Lack of technology e. Limited financial services
Opportunity a. High and unfulfilled market demand b. Relatively high selling prices in certain seasons c. Government policy support d. Technological innovation iodine salt development e. Accompaniment	S-O STRATEGY a. Encourage increased production through extensification to produce standard quality salts according to market needs b. Intensification of capacity building activities in marketing management.	W-O STRATEGY Maximize the assistance given to the farmers to improve accessibility of some of the production and potential areas and to improve the quality of human resources
a.		

(Continued on next page)

Table 7. Continued

Threads	S-T STRATEGY	W-T STRATEGY
b. Free market c. Higher quality salt production is circulating within the same consumer region d. Business security e. Weather conditions f. Wholesaler	Improving access to market information and to the short-term context the quantity of salt production using traditional technology is maximized by utilizing the remaining potential land area to dominate the market, especially unmet demand	Encourage accelerated improvement of human resource capacity and other resource utilization including financial institution services to compete in local and regional salt industry.

4 Conclusion

Modern methods based on geo-isolator can accelerate the production of local salt because the evaporation process is much more perfect by using a pool container using plastic, the salt produced is much more. The salt productivity by using a 100 % Geoisolator is much greater than in the traditional way. Comparison of salt production per hectare using traditional methods can only produce 60 t to 80 t once harvested. But with Geoisolator harvest salt per hectare can reach 120 t ha^{-1} to 140 t ha^{-1}. Community salt development strategy in Sumenep Regency using aggressive strategy that is by utilizing power to exploit opportunities by increasing productivity and technology to increase profit and expand marketing network and innovate in processing and production of salt.

References

1. I. Ihsannudin, S. Pinujib, S. Subejo, B.S. Bangko. Econ. Dev. Analysis J. **5**,4:395–409(2016). [in Bahasa Indonesia] https://journal.unnes.ac.id/sju/index.php/edaj/article/view/22177
2. H. Wang H, X. Xu, G Zhu. Sustainability. **7**,8:10078–10097(2015). https://www.mdpi.com/2071-1050/7/8/10078
3. D.A. Roberts,E.L. Johnston, N.A. Knott. Water Res.**44**,18:5117–5128(2010). https://www.sciencedirect.com/science/article/pii/S0043135410002903
4. S. Lattemann S, T. Höpner. Desalination. **220**,1–3:1–15(2008). https://doi.org/10.1016/j.desal.2007.03.009
5. J.J. Sadhwani, J.M. Veza, C. Santana. Desalination. **185**,1–3:1–8(2005). https://www.sciencedirect.com/science/article/pii/S0011916405006041
6. G.L.M.V. Medeazza. **185**,1–3:57–70(2005). https://www.sciencedirect.com/science/article/pii/S0011916405006089
7. R. Einav, F. Lokiec. Desalination. **156**,1:79–85(2003). https://www.academia.edu/download/49616404/Environmental_aspects_of_a_desalination_20161015-6177-txj6wp.pdf
8. A. Antonites. J. Anthropol Archaeol. **44**:31–42(2016). https://www.sciencedirect.com/science/article/pii/S0278416516300976
9. A. Ravizky, N. Nadav. Desalination. **205**,1–3:374–9(2007). http://atominfo.ru/files/brine/Brine%20from%20desalination%20plants_06.pdf
10. Y. Tanaka, R. Ehara, S. Itoi, T. Goto. J. Memb. Sci. **222**,1–2:71–86(2003). https://www.sciencedirect.com/science/article/pii/S0376738803002175
11. M.P. Dias. Waterbirds. **32**,4:531–537(2009). https://bioone.org/journals/waterbirds/volume-32/issue-4/063.032.0406/Use-of-Salt-Ponds-by-Wintering-Shorebirds-throughout-the-Tidal/10.1675/063.032.0406.short

12. I.G. Wenten, D. Ariono, M. Purwasasmita, Khoirudin. *Integrated processes for desalination and salt production: A mini-review*. AIP Conference Proceedings 1818. AIP Publishing LLC. (2017).p.20065. https://aip.scitation.org/doi/abs/10.1063/1.4976929
13. Y. Tanaka, M. Reig, S. Casas, C. Aladjem, J.L. Cortina. Desalination. **367**:76–89(2015). https://www.sciencedirect.com/science/article/pii/S0011916415001897
14. Chauhan VM, Alnouri SY, Linke P, Abdel-Wahab A. Desalination. **400,**25–37(2016). https://www.sciencedirect.com/science/article/pii/S0011916416306683
15. Dinas Kelautan dan Perikanan. Profil Pengembangan Usaha Garam Rakyat [Profile of People's Salt Business Development]. Sumenep (2015)p.2 [in Bahasa Indonesia]

Analysis of Taking Decision of Farmers in Choosing Rice Cultivars: Case of Pakel District, Tulungagung, Indonesia

Rima Dewi Oryza Sativa[1,*] , *Jabal* Tarik Ibrahim[2], and *Sutawi* Sutawi[3]

[1]Master's Degree, Department of Agriculture Science, Postgraduate Program, University of Muhammadiyah Malang, Jl. Raya Tlogomas No. 246, Malang 65145, East Java, Indonesia
[2]Department of Agribusiness, Faculty of Agriculture and Animal Sciences, University of Muhammadiyah Malang
[3]Department of Animal Sciences, Faculty of Agriculture and Animal Sciences, University of Muhammadiyah Malang

Abstract. Rice, one of the main crops in Indonesia, is the largest staple food. To increase national rice production, the government improved productivity and expanded rice fields including the attitudes and preferences of farmers to choose and use the right superior seeds. The study was conducted in Pakel Subdistrict, Tulungagung Regency, East Java Province in February 2019 to April 2019.The results showed that the farmers' decision-making process for seeds was strongly influenced by price, yield, and resistance to pests and diseases. In addition to the promotion factor, the availability of seeds at the trader and product quality determines the stage of purchasing the seeds to be planted.

Keyword: Increase rice productivity, preference, price, promotion, seed

1 Introduction

Rice is one of the main crops in Indonesia which produces the largest staple food. Indonesia is the country with the third largest rice consumption after China and India. This is supported by more than 90 % of Indonesian people consuming rice every day thus instability in food handling issues, especially rice will have an impact on various aspects of life such as social conditions, economic stability, employment, and others. The government is always trying to improve food security through rice self-sufficiency. Therefore, rice production needs to be increased thus it can be met from domestic production (self-sufficiency). Many modern rice varieties (MVs) have been released but only a few have been widely adopted by farmers [1].

Efforts to increase national rice production are by increasing productivity and expanding the region. The technology used is the use of superior seeds. The use of quality seeds is the first key to success in rice farming. Varieties or cultivars are important technological components that have a large contribution in increasing agricultural

* Corresponding author: rimadewioryza@gmail.com

production and income [2]. This technological component is very instrumental in changing the rice farming system, from subsystem to commercial rice farming [3].

Rice superior cultivars themselves have developed in Indonesia since before 1970. In general the names of these superior cultivars use river names in Indonesia including: *'Bengawan'*, *'Brantas'*, *'Citarum'*, and others. Food self-sufficiency, especially rice, from 2007 to 2016 the Agricultural Research and Development Agency through the Indonesian Rice Research Center (ICRR) has released various New Superior Varieties (VUB) for agroecosystem rice cultivation. Starting in 2008, the naming of New Superior Varieties (VUB) no longer uses the name of the river in Indonesia but uses a new name, called INPA for inbred rice and HIPA for hybrid rice.

Farmer decision making also influences the development of high yielding varieties. This is related to the nature possessed by superior cultivars [5]. Farmers generally want high-yielding varieties, good taste (specifically for the region), early maturity, proper height of plants, and resistant to pests and major diseases such as brown plant hopper (*Nilaparvata lugens* Stål, 1854), tungro (rice tungro bacilliform virus and rice tungro spherical virus).

In an effort to increase national rice productivity, the role of innovation in superior variety technology is very large. Around 90 % of the superior cultivars used by farmers in Indonesia today are cultivars produced by the ICRR. The use of improved cultivars by farmers in addition to increasing national rice production, also has a multiplier effect on the national economy, including the development of bioindustrial farming systems, the seed industry, improving farmers' welfare, opening jobs and others, estimated at IDR 1.5×10^{12} yr^{-1} [5].

Farmers are the main actors in the agricultural business. Therefore the attitudes and characteristics of rice farmers towards seed varieties are very important, because farmers have a dual role, as a producer of rice and consumers of seed products [6]. As producers, farmers must be able to meet market demand thus farmers' preferences for seed varieties cannot be separated from meeting the interrelated market desires. Although many new varieties have been discovered and introduced to farmers, if they do not meet the criteria of farmers as users, farmers will return to using old varieties [7].

Research aims to describe the characteristics of farmers based on the selected rice cultivars to be planted in their paddy fields and the decision-making process of farmers in choosing rice cultivars, analyzing the factors that influence decision making in selecting rice cultivars for farmers in Pakel District, Tulungagung Regency, Indonesia.

This research provides information about the decision making process and the factors that influenced in choosing rice cultivars. Research information can be used as a reference in policy making by the government in developing rice cultivars, for seed companies it can be used as a reference in marketing, and can be used as literature for other researchers.

2 Methods

The method used descriptive tabulation, Customer Satisfaction Index (CSI) and Multiple Linear Analysis. Descriptive analysis aimed to describe the various data collected without intending to make general conclusions or generalizations. Descriptive analysis according to data is in the form of tables and grouped based on the same answer Tanty et al. [8]. The results obtained were presented based on the number of respondents, and the largest percentage is the dominant factor of the analyzed variables. The Customer Satisfaction Index is used to measure satisfaction with product performance and desired attributes. Multiple linear regression is one of the statistical data analysis techniques that is often used to study the relationship between several variables and predict a variable [9].

3 Results

3.1 Characteristics of farmers

The characteristics of farmers in this study were 50 yr to 60 yr, most of the respondents' education was primary school. Farming was the main occupation of the respondent farmer. The results ranged from 4 t ha^{-1} to 7 t ha^{-1}. The farmers' decision-making process for seeds was strongly influenced by price, yield, and resistance to pests and diseases [10]. In addition to the promotion factor, the availability of seeds at the trader and product quality determined the stage of purchasing the seeds to be planted.

3.2 Customer satisfaction index

The level of importance and the level of performance of each attribute by calculating the Customer Satisfaction Index (CSI) [11]. In Table 1, the CSI value was 76.42 % (0.764 2). This value ranged from 0.60 to 0.80 which means farmers are satisfied with the performance attributes found in superior variety seeds. In subsequent studies the satisfaction index must be increased to close to 100 %.

3.3 Analysis of testing requirements

3.3.1 The 'Ciherang', 'Logawa', and 'Inpari' normality test

i. Classic assumption test

In the Kolmogorof Smirnov table, cultivars produce asymptotic significance ≥ 0.05. The normality test was on a normal graph, spreading points around a diagonal line and following a diagonal line. The histogram pattern showed that the distribution pattern was close to normal, so it can be concluded that the data is normally distributed and the regression model meets normality.

ii. Heteroskedasticity test

On the plot graph the prediction value of the variable was bound to the residual, random distribution of points above or below the zero point of the Y axis, then it is concluded that there is no heteroscedasticity.

iii. Multicollinearity test

The results of the Variance Inflation Factor (VIF) value using SPSS, the VIF value < 10 and tolerance value > 0.10. It can be concluded that there is no multicollinearity between the independent variables.

3.3.2 The 'Ciherang' multiple linear model

A simple linear regression model is used to test the effect of variables X1, X2, X3 on Y

Table 1. The '*Ciherang'* multiple linear analysis

Model	Unstandardized coefficients		Standardized coefficients	t	Sig.	Collinearity statistics
	B	Std. error	Beta			Tolerance
Constant	28.022	5.154		5.437	0.000	
Price	–0.272	0.523	–0.078	–0.521	0.605	0.819
Where to get seeds	0.894	0.370	0.369	2.417	0.020	0.785
Promotion	1.056	0.685	0.214	1.543	0.130	0.954

Based on the results of data processing using SPSS, a constant coefficient value of 28.022, grain price coefficient of –0.272, place coefficient of getting grain 0.894, promotion coefficient of 1.056. Then the regression equation can be Equation (1):

$$Y = 28.022 - 0.272X1 + 0.894\ X2 + 1.056X3 \quad (1)$$

i. The '*Ciherang*' determination test

The coefficient of determination (R2) from the results of multiple regression showed that the dependent variable (product) was influenced by the independent variable (price, place to get seeds, promotion).

Table 2. The '*Ciherang'* determination test

Model	R	R square	Adjusted R square	Std. error of the estimate
	.441[a]	.195	0.140	3.206 23

The test resulted in a coefficient of determination (R2) of 0.195 or 19.5 %. This showed that 19.5 % of products were influenced by price, place to get seeds, promotion. While 80.5 % is influenced by other variables not included in this research model.

ii. The '*Ciherang'* t test

t test is done by comparing the t-value obtained, with a significance level of 0.05 with a level of freedom (df) = nk = 48-4 = 44, with this provision the table 2015 is obtained.
The conclusion:
A. Price (X1)
H0 is accepted and Ha is rejected, so it can be concluded that the price variable partially has no effect on the product variable. As for the combined variables, price affects the product.
B. Place (X2)
H0 is accepted and Ha is accepted, so it can be concluded that the place variable partially influences the product variable. Whereas the shared place variable affects the product.
C. Promotion (X3)
H0 is accepted and Ha is rejected, so it can be concluded that the promotion variable partially has no effect on the product variable. Whereas the joint promotion variable affects the product.

iii. The '*Ciherang*' F test

ANOVA test using SPSS version 19, the number of F was 3.546. Based on table F with a significance level (α) = 5 %, it is known that table T with dfl = k-1 = 4-1 = 3 and df2 = nk = 48-4 = 44, then F table (dfl) (df2) = 2.816. Based on the calculation results obtained F count = 3.546 while F table = 2.816, then F count > F table. If seen from the significance value in table 0.022 < 0.05 this showed that the price, place to get seeds, and promotions affect the product.

Table 3. The '*Ciherang*' F test

Model	Sum of squares	df	Mean square	F	Sig.
Regression	109.352	3	36.451	3.5 46	.022[b]
Residual	452.315	44	10.280		
Total	561.667	47			

3.3.3 The 'Logawa' multiple linear model

Table 4. The '*Logawa*' multiple linear analysis

Model	Unstandardized coefficients		Standardized coefficients	t	Sig.	Collinearity statistics
	B	Std. Error	Beta			Tolerance
Constant	20.353	5.134		3.964	0.000	
Price	0.517	0.634	0.1 13	0.816	0.419	0.810
Location	0.618	0.309	0.2 75	1.997	0.052	0.812
Promotion	2.250	0.583	0.4 81	3.858	0.000	0.993

Based on the results of data processing using SPSS, a constant coefficient value of 20.353, grain price coefficient of 0.517, place coefficient of getting grain 0.618, promotion coefficient of 2.250. Then the regression equation can be Equation (2):

$$Y = 20.353 + 0.517X1 + 0.618\ X2 + 2.250\ X3 \quad (2)$$

i. The '*Logawa*' determination test

The coefficient of determination (R2) from the results of multiple regression showed the dependent variable (product) is influenced by the independent variable (price, place to get seeds, promotion).

Table 5. The '*Logawa*' determination test

Model	R	R square	Adjusted R square	Std. error of the estimate
	0.567[a]	0.321	0.275	3.350 64

The results of the coefficient of determination (R2) were 0.321 or 32.1 %. This showed that 32.1 % of the products were influenced by price, place to get seeds, promotion, while 67.9 % was influenced by other variables not included in this study.

ii. The '*Logawa*' t test

Degrees of freedom (df) = n-k = 48-4 = 44, with this provision a Table of 2.015 was obtained.

The conclusion:

i. Price (X1)

H0 is accepted and Ha is rejected, so it can be concluded that the price variable partially has no effect on the product variable. Whereas for the joint variable Price affects the product.

ii. Place (X2)

H0 is accepted and Ha is rejected, so it can be concluded that the place variable partially has no effect on the product variable. Whereas for shared place variables affect the product.

iii. Promotion (X3)

H0 is accepted and Ha is accepted, so it can be concluded that the promotion variable partially influences the product variable. Whereas for joint promotion variables affect the product.

iii. The '*Logawa*' F test

According to ANOVA test using SPSS version 20, the F count was 6.946. Based on table F with a significance level (α) = 5 %, it is known that T table with dfl = k-1 = 4-1 = 3 and df2 = n-k = 48-4 = 44, then F table (dfl) (df2) = 2.816. Based on the calculation results obtained F count = 6.946 while F table = 2.816, then F count > F table. If seen from the significance value in the table 0.0 01 < 0.05 this showed that the price, place to get seeds, and promotions affected the product.

Table 6. The '*Logawa*' F test

Model	Sum of squares	df	Mean square	F	Sig.
Regression	233.938	3	77.979	6.946	0.001[b]
Residual	493.979	44	11.227		
Total	727.917	47			

3.3.4 The 'Inpari' multiple linear models

Table 7. The '*Inpari*' multiple linear analysis

Model	Unstandardized coefficients		Standardized coefficients	t	Sig.	Collinearity statistics
	B	Std. Error	Beta			Tolerance
Constant	26.136	4.876		5.360	0.000	
Price	0.449	0.588	0.106	0.763	0.450	0.886
Location	0.960	0.317	0.434	3.026	0.004	0.836
Promotion	0.228	0.586	0.053	0.390	0.699	0.935

Based on the results of data processing using SPSS, a constant coefficient value of 26.136, grain price coefficient 0.449, grain coefficient obtained 0.960 grain, promotion coefficient 0.228. Then the regression equation can be Equation (3):

Y = 26.136+ 0.449X1 + 0.960 X2 + 0.228X3 (3)

i. The '*Inpari*' determination test

The coefficient of determination (R2) from the results of multiple regression showed the dependent variable (product) was influenced by the independent variable (price, place to get seeds, promotion).

Table 8. The '*Inpari*' determination test

Model	R	R square	Adjusted R square	Std. error of the estimate
	0.493[a]	0.243	0.192	3.397 83

Test results for the coefficient of determination (R2) of 0.243 or 24.3 %. This showed that 24.3 % of products were influenced by price, place to get seeds, promotion, while 75.7 were influenced by other variables outside this study.

i. The '*Inpari*' t test

The conclusion:
i. Price (X1)
H0 is accepted and Ha is rejected, so it can be concluded that the price variable partially has no effect on the product variable. Whereas for the joint variable Price affects the product.
ii. Place (X2)
H0 is accepted and Ha is accepted, so it can be concluded that the place variable partially influences the product variable. Whereas for shared place variables affect the product.
iii. Promotion (X3)
H0 is accepted and Ha is rejected, so it can be concluded that the promotion variable partially has no effect on the product variable. Whereas for joint promotion variables affect the product.

ii. The '*Inpari*' F Test

ANOVA test using SPSS version 20, the F count was 4.720. Based on table F with a significance level (α) = 5 %, it is known that T table with dfl = k-1 = 4-1 = 3 and df2 = n-k = 48-4 = 44, then F table (dfl) (df2) = 2.816. Based on the calculation results obtained F count = 4.720 while F table = 2.816, then F count > F table. If seen from the significance value in the table 0.006 < 0.05 this showed that the price, place to get seeds, and promotions affect the product.

Table 9. The '*Inpari*' F Test

Model	Sum of squares	df	Mean square	F	Sig.
Regression	163.488	3	54.496	4.720	0.006[b]
Residual	507.992	44	11.545		
Total	671.479	47			

4 Discussion

The results of the F test showed that price, place, promotion jointly affected the product. Result t test indicated the price did not affect the product. This explained that the price set for a particular product did not have a positive effect on the buying interest of the respondent farmers because rice seed is a prestige product that more reasonably the price the buying interest of the respondent farmers. While the current promotion of rice seeds has not yet fully attracted the interest of farmers, the reason being that promotion is limited to banners, advertisements, without any direct promotion to farmers [5]. Moreover, the places influenced on products because respondents tend to look for products that are easily accessible through kiosks or farmer groups. This showed that the equitable distribution of products to farmers is needed.

5 Conclusion

The farmers' decision-making process for seeds was strongly influenced by price, yield, and resistance to pests and diseases. In addition to promotion factors, the availability of seeds at traders and product quality determine the stage of purchasing seeds to be planted. Based on CSI calculations, farmers were satisfied with the performance of superior seed attributes.

References

1. A.G. Laborte, N.C. Paguirigan, P.F. Moya, A. Nelson, A.H. Sparks, G.B. Gregorio. Plos One **10**,8: e0136562(2015). https://doi.org/10.1371/journal.pone.0136562
2. M. Mwangi, S. Kariuki. J. Econ. Sustain. Dev. **6**,5:208–216(2015). https://www.iiste.org/Journals/index.php/JEDS/article/viewFile/20710/21632
3. M.A. Altieri, C.I. Nicholls, A. Henao, M.A. Lana. Agron. Sustain. Dev. **35**:869–90(2015). doi: 10.1007/s13593-015-0285-2. or https://link.springer.com/article/10.1007/s13593-015-0285-2
4. W. Naivit, L.C. Page, G. Trebuil, N. Gajaseni. Environ. Model. Softw. **25**,11:1345–1358(2010). https://www.sciencedirect.com/science/article/pii/S1364815210000381
5. S. Rahayu, Y.P. Wanita, M. Kobarsih. Agrin. **15**,1:36–44(2011). [in Bahasa Indonesia]. http://garuda.ristekbrin.go.id/documents/detail/747295
6. M. Mehar, A. Panda, T. Yamano. Asian J. Agric. Dev. **14**,1:17–36(2017). https://ageconsearch.umn.edu/record/259241/
7. S.K. Killenga, P. Tongoona, J. Derera, Z. Kanyeka. Int. J. Dev. Sust. **3**,6:1257–1271(2014). https://isdsnet.com/ijds-v3n6-2.pdf
8. H. Tanty, R.D. Bekti, A. Rahayu. Jurnal Mat. Stat., **13**, 2:97–104(2013). [in Bahasa Indonesia]. https://www.google.com/url?sa=t&rct=j&q=&esrc=s&source=web&cd=&cad=rja&uact=8&ved=2ahUKEwi96snGpOPrAhVIVH0KHTOnCmkQFjAAegQIBRAB&url=http%3A%2F%2Flibrary.binus.ac.id%2FeColls%2FeJournal%2F04_Heruna%2C%2520Rokhana%2C%2520Anita.pdf&usg=AOvVaw02kVGxmPJ8NPzBV84zBdOw
9. K.A. Garrett, G.A. Forbes, S. Savary, P. Skelsey, A.H. Sparks, C. Valdivia C, et al. Plant. Path. **60**,1:15–30(2011). https://bsppjournals.onlinelibrary.wiley.com/doi/abs/10.1111/j.1365-3059.2010.02409.x

10. K.A. Garrett, G.A. Forbes, S. Savary, P. Skelsey, A.H. Sparks, C. Valdivia, et al. Plant Disease. **95,**10:1204–1216(2011). https://apsjournals.apsnet.org/doi/pdf/10.1094/PDIS-04-11-0316
11. S. Nikou, H. Selamat, R.C.M. Yusoff, M.M. Khiabani. Int. J. Adv Sci. Technical Res. **6**,6:29–46(2016). https://www.researchgate.net/publication/311607113_Service_Quality_Customer_Satisfaction_and_Customer_Loyalty_A_Comprehensive_Literature_Review_1993-2016

Pyrolysis of Water Hyacinth [*Eichhornia crassipes* (Mart.) Solms] for Liquid Smoke Production

Rita Dwi Ratnani[1,*], *Hadiyanto* Hadiyanto[2], *Widiyanto* Widiyanto[3], and *Maizirwan* Mel[4]

[1]Department of Chemical Engineering, Wahid Hasyim University, Jl. Menoreh Tengah X/22 Sampangan, Semarang 50236, Indonesia
[2]Department of Chemical Engineering, Diponegoro University, Jl. Prof. Sudarto No.13, Tembalang, Kec. Tembalang, Semarang 50275, Indonesia
[3]Department of Animal Sains, Diponegoro University, Jl. Prof. Sudarto No.13, Tembalang, Kec. Tembalang, Semarang 50275, Indonesia
[4]Department of Biotechnology Engineering, Kulliyyah of Engineering, International Islamic University Malaysia, P.O. Box 10, 50728 Kuala Lumpur, Malaysia

Abstract. This research aimed to determine the chemical composition of Rawapening, Semarang District, Indonesia water hyacinth and to study the effect of pyrolysis time on the volume of liquid smoke products. The water hyacinth was cut into pieces with a size of about 2 cm to 3 cm, and placed in pyrolisator. A total of 700 g of water hyacinth sample was fed into the pyrolysis. The product was observed for its content of hemicellulose, cellulose and lignin. The pyrolysis was carried out at a temperature of 117 °C, 400 °C and 683 °C. The results showed that water hyacinth contains hemicellulose 36 % d.b, cellulose 21 % d.b and lignin 7 % d.b which it potential to be a raw material for liquid smoke production. The effect of temperature was significant on liquid smoke products. It was proven that the higher the pyrolysis temperature, the greater the liquid smoke produced. The liquid smoke was obtained for 0.2 mL per 100 g, 12.3 mL per 100 g and 16.3 mL per 100 g dried biomass water hyacinth at a temperature of 117 °C, 400 °C and 683 °C, respectively.

Keywords: Biomass, lignocellulose, renewable energy, waste to energy

1 Introduction

Environmental sustainability around Rawa Pening, Semarang District, Indonesia has changed due to the density of water hyacinth weeds [Eichhornia crassipes (Mart.) Solms]. The growth of water hyacinth is rapid due to eutrophication. Eutrophication is the presence of excess nutrients in the form of nitrate and phosphate which are sourced from agricultural activities, fisheries, settlements, and livestock from the land around the lake. This is the factor made water hyacinth grew quickly and short time to multiply [1]. To control the growth water hyacinth is by utilizing in another product i.e. for liquid smoke as it contains high cellulose.

* Corresponding author: ritadwiratnani@unwahas.ac.id

The pyrolysis is used for converting this biomass to liquid smoke. Biomass pyrolysis is usually carried out at temperatures above 150 °C. Pyrolysis is divided into two based on the level of the process, i.e. primary and secondary. Primary pyrolysis takes place at temperatures less than 600 °C. The main result is carbon or charcoal. Primary pyrolysis is further divided into two namely fast and slow primary. Slow primary pyrolysis occurs at 150 °C to 300 °C heating. The results slow primary pyrolysis are carbon, water, carbon monoxide and carbon dioxide. Rapid primary pyrolysis that occurs at temperatures over 300 °C will produce gas, carbon and steam [2].
Subsequent pyrolysis is pyrolysis that occurs at temperatures above 600 °C and at this temperature, the result is carbon monoxide, hydrogen, hydrocarbons and tar around 1 % to 6 % [3, 4].

Decomposition reaction of biomass undergoing pyrolysis will produce as much as 50 g to 70 g of steam. The general reaction is as Formula (1):

$$\underset{(100\ \text{g})}{\text{Biomass}} \longrightarrow \underset{(50\ \text{g to } 70\ \text{g})}{\text{Steam}} + \underset{(4\ \text{g to } 10\ \text{g})}{\text{Gas}} + \underset{(10\ \text{g to } 20\ \text{g})}{\text{Charcoal}} + \underset{(13\ \text{g to } 25\ \text{g})}{\text{Water}} \quad (1)$$

The purpose of this study was to determine the chemical composition of Rawapening's water hyacinth and to study the effect of pyrolysis time on the volume of liquid smoke products.

2 Materials and methods

2.1 Material

The materials used water hyacinth, distilled water, Phenolphthalein (PP) indicator, 0.1 N NaOH, 2 % $Na_2S_2O_3$, follicular ciocalteu, 2,4-dinitrophenyl hydrazine, concentrated HCl, KOH 1 N.

2.2 Equipment

The experiment was carried in a pyrolysis equipment as shown in Figure 1.

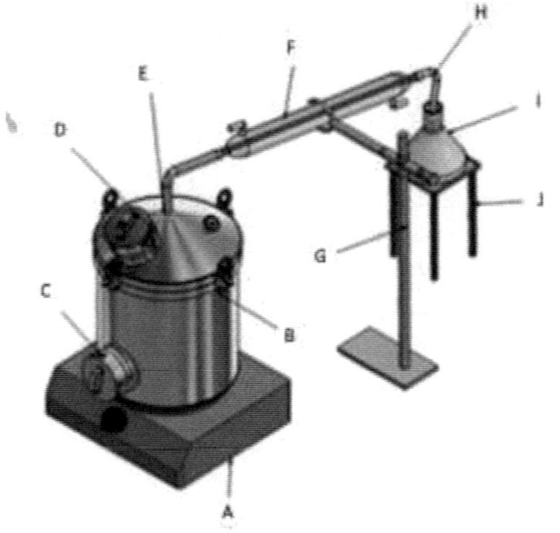

Fig. 1. Pyrolysis processing Stove A), Pyrolysis outlet hole B), Reactor C), Pyrolysis inlet hole D), Steam outlet E), Condenser F), Statip G), Adapter H), Erlenmeyer I), and Leg J).

2.2.1 Procedure

The research was conducted through two stages of processing. The first step was the preparation of water hyacinth for chemical composition analysis in hemicellulose, cellulose and lignin. Chemical composition analysis used the Chasson Data Method. The second stage was the pyrolysis to determine the effect of temperature on the volume of liquid smoke. For the pyrolysis stage, water hyacinth was cut into pieces with a size of about 2 cm to 3 cm and then put the pyrolysator. Pour as much as 700 g of the dried water hyacinth into the 316 stainless steel pyrolysis reactor with a length of 360 mm and an internal diameter of 240 mm under atmospheric vacuum. The reactor was heated using a gas stove. Temperature variations were designed in 117 °C, 400 °C and 683 °C. Furthermore, liquid smoke generated in the Erlenmeyer after passing through the condenser is accommodated as a result.

3 Results and discussion

The chemical composition of the water hyacinth Rawa pening, Semarang District, Indonesia shown in Figure 2 was hemicellulose, cellulose and lignin content as 36 %, 21 % and 7 % dry weight, respectively. Figure 2 also depicts other materials, i.e. hardwood, softwood, coconut shell and cassava stems. Lignocellulose as biomass in the pyrolysis turns into charcoal, liquid smoke and fuel [5]. The chemical composition of water hyacinth has the highest hemicellulose content, while the smallest lignin when compared to the others such as softwood, hardwood, coconut shell, and cassava stems, This is inversely proportional to the coconut shell, where lignin is greater than hemicellulose. Hemicellulose pyrolysis containing pentosan and hexosan will produce furan and its derivatives and produce carboxylic acid as well as acetic acid. Pyrolysis of cellulose will produce acetic acid and homologous, water, furan and phenol. Lignin pyrolysis is an essential factor in providing smoke odor for smoked products. The resulting compounds are phenols and phenol esters [6, 7]. These compounds in liquid smoke potentially a preservative.

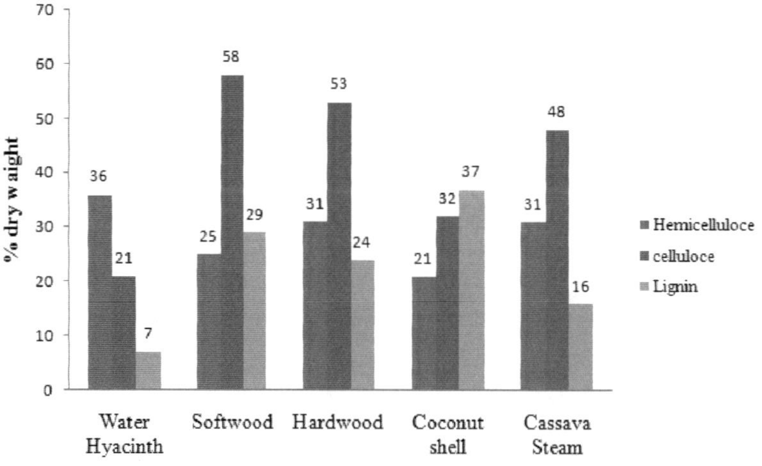

Fig. 2. Chemical Composition of water hyacinth and several other biomass

The temperature influence on the volume of liquid smoke, that the higher the temperature, the more liquid smoke produced. In Figure 3 the amount of liquid smoke produced is 0.2 mL per 100 g, 12.3 mL per 100 g, 6.3 mL per 100 g dried biomass for temperature 117 ° C, 400 °C, and 683 °C respectively. Hemicellulose pyrolysis occurs at temperatures of 200 °C to 250 °C, 280 °C to 320 °C of cellulose pyrolysis and 400 °C for lignin pyrolysis [8]. So it can be underlined that the pyrolysis of water hyacinth at a temperature of 117 °C is expected to occur decomposition of hemicellulose compounds. However, the decomposition of hemicellulose that starting at 200 °C has not been reached thus resulted less volume. But at 400 °C cellulose and lignin decomposed, resulting more volume of liquid smoke. According to Utomo et al. at 683 °C pyrolysis produced carbon monoxide, hydrogen, hydrocarbons and tar [8–10]. Therefore it need further research to find out the chemical compound content of the liquid smoke produced.

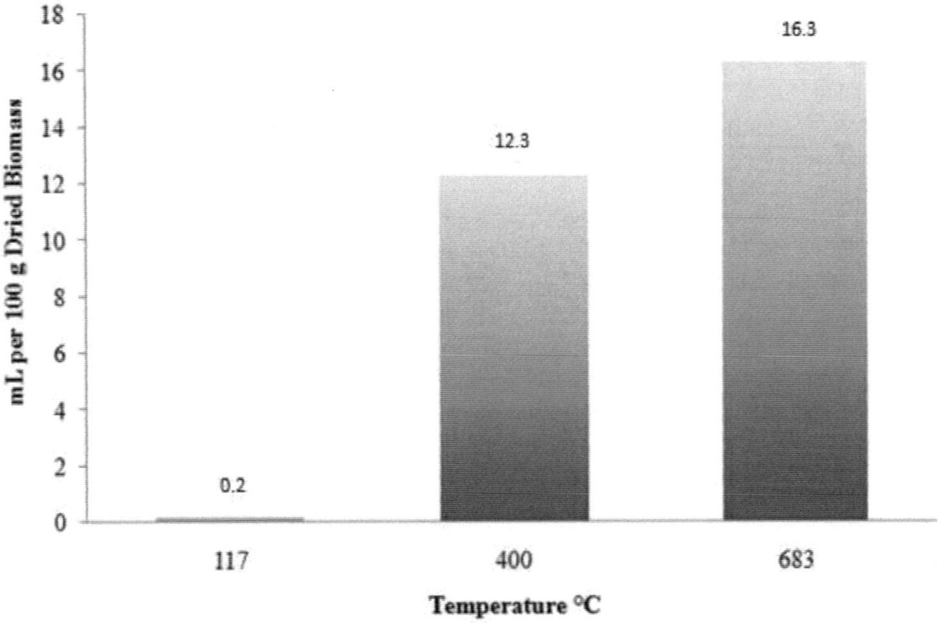

Fig. 3. Effect of Temperature on the volume of liquid smoke produced

4 Conclusion

The conclusion of this research is the effect of temperature was significant on liquid smoke products. It was proven that the higher the pyrolysis temperature, the greater the liquid smoke produced. The liquid smoke was obtained for 0.2 mL per 100 g, 12.3 mL per 100 g and 16.3 mL per 100 g dried biomass water hyacinth at a temperature of 117 °C, 400 °C and 683 °C, respectively

References

1. D.A. Wulandari, D. Kurniani, S. Edhisono, F. Ardianto, D. Dahlan. MATEC Web of Conferences, **270**, 04018(2019). https://doi.org/10.1051/matecconf/201927004018
2. S. Sadaka, A.A. Boateng. *Pyrolysis and Bio Oil*. University of Arkansas Agriculture and Natural Resources. (2009).p.5. https://scholar.google.co.id/scholar?hl=id&as_sdt=0%2C5&q=2.%09Sadaka+S%2C+Boateng+AA.+Pyrolysis+and+Bio+Oil.+Agric+Nat+Resour.+2017&btnG=

3. G. Ozbay, N. Ayrilmis. *Effect of Pyrolysis Temperature On Bio-Oil Production From vacuum Pyrolysis of Waste From Wood Industry*. Proceedings of 117th The IIER International Conference, Helsinki, Finland. 56–58(2017) https://www.google.com/url?sa=t&rct=j&q=&esrc=s&source=web&cd=&cad=rja&uact=8&ved=2ahUKEwi86NLDv-jqAhXSbn0KHSWYBfkQFjAAegQIBBAB&url=http%3A%2F%2Fwww.worldresearchlibrary.org%2Fup_proc%2Fpdf%2F1015-150521687756-58.pdf&usg=AOvVaw3N3zaTEmrLxovQ4MQ8fYCD
4. T. Kan, V. Strezov, T.J. Evans, Renewable Sustainable Energy Reviews. **57**:1126-1140(2016). doi:10.1016/j.rser.2015.12.185 or https://www.sciencedirect.com/science/article/abs/pii/S1364032115015683
5. T. Dickerson, J. Soria, Energies. **6,**1:514–538(2013). doi:10.3390/en6010514 or https://www.mdpi.com/1996-1073/6/1/514
6. R. Hadanu, D.A.N. Apituley, Makara Journal of Science. **20,**3:95–100(2016). https://doi.org/10.7454/mss.v20i3.6239
7. G. Ozbay, A. Ozcifci, E.S. Kokten, Tirkish Journal og Agriculture and Forestry. **40**:705–714(2016). https://journals.tubitak.gov.tr/agriculture/abstract.htm?id=19386
8. M. Faisal, A.R. Yelviasunarti, H. Desvita, Rasayan Journal Chemistry. **11,**2:871–876(2018). DOI: 10.31788/RJC.2018.1123035
9. H. Ma, T. Li, S. Wu, X. Zhang, Bioresource Technology, **309**,123351(2020). https://doi.org/10.1016/j.biortech.2020.123351
10. F. Swastawati, E. Susanto, B. Cahyono, W. Aji, APCBEE Procedia, **2**:1–6(2012). https://doi.org/10.1016/j.apcbee.2012.06.001

Effect and Effectivity of Granular Organic Fertilizer on Growth and Yield of Lowland Rice

Rohmad Budiono[1,*], *Fuad* Nur Aziz[1], *Endang* Dwi Purbajanti[2], *Tsitsino* Turkadze[3], and *Praptiningsih* Gamawati Adinurani[4]

[1]Assesment Institute for Agricultural Technology, Jl. Raya Karangploso Km. 4, Karang Ploso, Malang 65152, East Java, Indonesia
[2]Department of Agriculture, Diponegoro University, Jl. Prof. Sudarto No.13, Tembalang, Semarang 50275, Central Java, Indonesia
[3]Department of Chemical and Environmental Technologies, Akaki Tsereteli State University, 59 Tamar mepe str. Kutaisi 4600, Georgia, UAE
[4]Department of Agrotechnology, Merdeka University of Madiun, Jl. Serayu No.79, Madiun 63133, East Java, Indonesia

Abstract. The tendency to use inorganic fertilizers in very high doses has causesvarious negative impacts, namely damage to ecosystems, loss of soil fertility to health problems. Therefore, the use of organic fertilizers is encouraged to overcome these problems. The purpose of this study was to determine the effectiveness of Granules Organic Fertilizer (GOF) on the growth and yield of rice plants. The design used was a Randomized Block Design (RBD) with three replications and eight treatments. The GOF treatment was arranged in six levels: 1 000 kg ha^{-1} to 6 000 kg ha^{-1}, combined with Urea 300 kg ha^{-1} and Phonska 300 kg ha^{-1}, one standard fertilization treatment (Petroganic 2 000 kg ha^{-1} + Urea 300 kg ha^{-1} and Phonska 300 kg ha^{-1}) and one treatment without fertilization as a control. The results showed that GOF had a significant effect on the growth and yield of rice plants. The use of GOF 3 000 kg + 300 kg Urea + 300 kg Phonska gave the highest grain yield of 7.21 t ha^{-1} with RAE value of 116 % or an increase of 7.45 %, and with an R/C ratio of 1.4.

Key words: Balance fertilizer, fertilizer efficiency, food security, slow realese fertilizer, soil fertility

1 Introduction

Nowadays, rice farmers use production factors is not efficient [1] among others, rely on inorganic fertilizers with very high doses. That tendency causesvarious negative impact of using inorganic fertilizers, ranging from damage to the ecosystem, loss of soil fertility to health problems [2, 3]. Organic fertilizer refers to materials used as fertilizer that occur regularly in nature, usually as a by product or end product of a naturally occurring process. Organic fertilizers are sustained sources of nutrients due to slow release during decomposition [4]. Therefore, the use of organic fertilizer is encouraged to overcome these problems [5]

Organic fertilizer is fertilizer derived from dead plants, animal dung or animal parts, and other organic waste. It has been through an engineering process and can be enriched with

*Correspondingauthor: rohbudi68@yahoo.com

minerals and microbes. Organic fertilizer is useful for increasing nutrient content and soil organic matter, and improving physical, chemical, and biological soil characteristics [2]. Organic fertilizers produced through the process of drying, cooking, composting [6], chopping, grinding and fermenting [7], or other method.

Nutrient composition in organic fertilizer is very dependent on the source of origin of the basic ingredients. The sources of organic fertilizer were from agricultural and non–agricultural activities. The source from agriculture activities can be in the form of crop residues and livestock manure, while from non–agriculture can come from municipal organic waste, industrial waste, and so on.

Organic fertilizer technology is growing rapidly. The use of Granules Organic Fertilizer (GOF) in rice plants is expected to improve land conditions, technically easy to apply, and improve the growth and yield of rice plants. Organic fertilizer senhance the natural soil processes, which have long–termeffects on soil fertility [4].

The purpose of this study was to determine the effectiveness of GOF on the growth and yield of rice plants.

2 Research methods

The study was conducted in the Mojosari research station, Mojokerto, East Java, Indonesia from February 2017 to August 2017. The materials used in this study included: Inpari 30 variety of rice seeds, GOF, comparative organic fertilizer (Petroganik), Urea fertilizer, and Phonska fertilizer. The equipment used were a tillage machine (hand tractor), hoe, and equipment for observation and harvest.

A previous study showed that soil at the Mojosari research station had a Neutral Acidity (pH). Organic matter content and macronutrients were low, but P and Ca were high.

The experimental design used was a Randomized Block Design (RBD). The treatment of organic fertilizer was arranged in six levels of organic fertilizer (1 000 kg ha^{-1}, 2 000 kg ha^{-1}, 3 000 kg ha^{-1}, 4 000 kg ha^{-1}, 5 000 kg ha^{-1}, and 6 000 kg ha^{-1}). Each of the level was combined with 300 kg of Urea and 300 kg of Phonska. As a comparison, it used one standard fertilizing treatment (Petroganik 2 000 kg ha^{-1} + Urea 300 kg ha^{-1} and Phonska 300 kg ha^{-1}) and one treatment without fertilization as a control. All treatments were repeated three times. Data were analyzed using ANOVA. Further tests were carried out using the Least Significant Difference Test (LSD) at the 5 % level.

RAE (Relative Agronomic Effectiveness) was used to compare the effectiveness of each organic fertilizer against standard fertilizer. RAE is the ratio between the increase in yield due to the use of a fertilizer with the increase in yield with the use of a standard fertilizer multiplied by 100 [2], with the Formula (1):

$$\text{RAE} = \frac{\text{Yield of tested fertilizer} - \text{results on control}}{\text{Standard fertilizer yields} - \text{results on control}} \times 100\ \% \quad (1)$$

To find out the feasibility level of rice farming, the simplest financial analysis approach using R/C was the ratio between revenue and cost. If R/C > 1, meansthe business is feasible to be continued, and if R/C < 1 then the business is not feasible/inefficient to be continued and R/C = 1 means the farming carried outreaches breakeven. Comparison between the amount of revenue and cost using the Formula (2) [8], as follows:

$$\text{R/C ratio} = \text{TR/TC} \quad (2)$$

TR = Total Revenue
TC = Total Cost

3 Results and discussions

Table 1 showed that at the age of 21 dap (day after plantation), the highest average plant height was shown in the treatment of 3 000 kg ha^{-1} GOF, but not significantly different with the treatment of 2 000 kg ha^{-1}, 4 000 kg ha^{-1} and 6 000 kg ha^{-1} GOF. At 70 dap, the highest average was shown in the treatment of 3 000 kg ha^{-1} GOF with average value of 89.4 but not significantly different from the standard treatment, 2 000 kg ha^{-1}, 4 000 kg ha^{-1}, 5 000 kg ha^{-1}, and 6 000 kg ha^{-1} GOF. At the age of 35 and 56 dap the highest average was shown in the treatment of 4 000 kg ha^{-1} GOF. At 56 dap, the addition of GOF organic fertilizer to 6 000 kg ha^{-1} did not show any significant difference. This indicated that the nutrients contained in the treated organic fertilizer did not affect the height of the rice plant because the GOF cannot be directly utilized by the plant and also the amount of nutrients contained in the GOF was low. According to [9], granules fertilizer or pellets can prevents the occurrence of segretion, prevents an overdose of plants against shedding sudden nutrition. The results of this study are different from [10] that plant height were significantly increased in all the treatments over control.

Table 1. Effect of GOF on plant height

Treatment (kg ha^{-1})	Plant height (cm)			
	21 dap	35 dap	56 dap	70 dap
Control (without fertilizer)	46.9 c	54.9 c	55.9 b	77.2 c
Standart (2 000 Petroganik + 300 Urea + 300 Phonska)	51.5 b	57.2 bc	69.7 a	85.3 ab
(1 000 GOF + 300 Urea + 300 Phonska)	52.2 b	56.9 bc	68.5 a	80.7 bc
(2 000 GOF + 300 Urea + 300 Phonska)	53.5 ab	58.5 bc	70.5 a	88.0 a
(3 000 GOF + 300 Urea + 300 Phonska)	57.7 a	61.3 ab	69.0 a	89.4 a
(4 000 GOF + 300 Urea + 300 Phonska)	54.7 ab	64.5 a	71.7 a	88.0 a
(5 000 GOF + 300 Urea + 300 Phonska)	52.9 b	59.9 ab	69.6 a	87.7 a
(6 000 GOF + 300 Urea + 300 Phonska)	55.1 ab	59.7 abc	70.5 a	88.0 a
Anova	*	*	*	*
CV (%)	4.62	4.78	5.22	3.15
LSD 5 %	4.29	4.95	6.23	4.72

Note: GOF = Granular Organic Fertilizer, dap = Day After Planting
The number followed by the same letter in one column was not significantly different based on the LSD test of 5 %
* = Significantly different at 5 %

Table 2 showed that the highest average number of tillers at the age of 21 dap was shown in the dose of 4 000 kg ha^{-1} GOF with an average value of 19.6. At the age of 35 dap, the highest average number of tillers was at the dose 5 000 kg ha^{-1} GOF, but not significantly different with the treatment of 3 000 kg ha^{-1} GOF. At the age of 56 and 70 dap, the highest number of tillers was obtained in the treatment of 5 000 kg ha^{-1} GOF. At the age of 70 dap, the number of tillers in the treatment was not significantly different with the treatment of 4 000 kg ha^{-1} GOF, 6 000 kg ha^{-1} GOF.

Table 2. Effect of GOF on number of tillers

Treatment (kg ha^{-1})	Number of tillers			
	21 dap	35 dap	56 dap	70 dap
Control (without fertilizer)	11.3 e	13.3 e	13.7 e	13.9 e
Standart (2 000 Petroganik + 300 Urea + 300 Phonska)	15 cd	15.0 d	17.8 d	17.9 d
(1 000 GOF + 300 Urea + 300 Phonska)	14.1 d	16.2 c	19.5 cd	19.5 cd
(2 000 GOF + 300 Urea + 300 Phonska)	17.3 b	19.6 b	21.0 bc	21.4 bc
(3 000 GOF + 300 Urea + 300 Phonska)	17.0 b	20.7 ab	21.0 bc	23.1 ab
(4 000 GOF + 300 Urea + 300 Phonska)	19.6 a	19.6 b	21.4 b	23.5 a
(5 000 GOF + 300 Urea + 300 Phonska)	16.3 bc	21.0 a	24.2 a	24.2 a
(6 000 GOF + 300 Urea + 300 Phonska)	15.5 bc	20.9 a	22.0 b	23.2 ab
Anova	**	**	**	**
CV (%)	7.12	3.68	5.12	5.84
LSD 5 %	1.96	3.11	1.80	2.13

Note: GOF = Granular Organic Fertilizer, dap = Day After Planting
The number followed by the same letter in one column was not significantly different based on the LSD test of 5 %
* = Significantly different at 5 %

The number of panicles per clump ranged from 7.9 to 14.6 panicles. The lowest number of panicles was found in the treatment without fertilization while the highest was achieved in the treatment of 3 000 kg ha^{-1} GOF, but not significantly different with the treatments of 2 000 kg ha^{-1} GOF, 4 000 kg ha^{-1} GOF, 5 000 kg ha^{-1} GOF, and 6 000 kg ha^{-1} GOF (Table 3).

The number of filled grains per panicle ranged from 86.6 to 172.0. Filled Grain $panicle^{-1}$ is lowest in the treatment without fertilization, while the highest is in the treatment of 5 000 kg ha^{-1} GOF, which is not significantly different from the number of filled grain in the treatment of 3 000 kg ha^{-1} GOF, 4 000 kg ha^{-1} GOF and 6 000 kg ha^{-1} GOF (Table 3).

The number of empty grains $panicle^{-1}$ ranges from 9.2 to 22.0. The highest number of empty grain is found in the highest fertilizer treatment (6 000 kg ha^{-1} GOF). While the lowest number of empty grains was found in the treatment of 2 000 kg ha^{-1} GOF, which was not significantly different from the treatment of 1 000 kg ha^{-1} GOF (Table 3).

Table 3. Effect of GOF on number of panicles per clump, number of filled and empty grain $panicle^{-1}$

Treatment (kg ha^{-1})	Number of panicles $clump^{-1}$	Number of grain per panicles	
		Filled	Empty
Control (without fertilizer)	7.9 d	86.6 c	11.7 d
Standart (2 000 Petroganik + 300 Urea + 300 Phonska)	12.5 bc	124.8 b	11.2 d
(1 000 GOF + 300 Urea + 300 Phonska)	11.2 c	121.7 bc	11.0 de
(2 000 GOF + 300 Urea + 300 Phonska)	13.3 ab	125.3 b	9.2 e
(3 000 GOF + 300 Urea + 300 Phonska)	14.6 a	152.9 ab	12.3 d
(4 000 GOF + 300 Urea + 300 Phonska)	14.4 a	148.8 ab	14.8 c
(5 000 GOF + 300 Urea + 300 Phonska)	14.5 a	172.0 a	18.1 b
(6 000 GOF + 300 Urea + 300 Phonska)	13.3 ab	154.8 ab	22.0 a
Anova	**	**	**
CV (%)	7.20	14.44	7.34
LSD 5 %	1.61	34.39	1.77

Note: GOF = Granular Organic Fertilizer
The number followed by the same letter in one column is not significantly different based on the LSD test of 5 %
** = Significantly different at 1 %

Table 4 showed that the weight of 1 000 seeds no significant difference among treatments. It showed that the GOF and petroganik did not affect the weight of seeds. For variable harvested driedgrain yield, the treatments showed very significant differences. Harvested dried grain yield ranged from 3.57 t ha^{-1} to 7.21 t ha^{-1}.

The lowest harvested dried grain yield was found in the treatment without fertilization, while the highest was achieved in the treatment of 3 000 GOF but no significant difference with the 4 000 GOF and 5 000 GOF (Table 4).

This research showed that the increase in the amount of GOF was not followed by the rate of increase in production. The reasons were, i) GOF utilization had the same effect as petroganik fertilizer, ii) In production parameters, GOF and petroganik gave the same advantages effect. Besides nutrients as a determinant of productivity, rice varieties also determined the yields.

Organic fertilizer is useful for improving the texture and structure of the soil, there by accelerating the process of absorption of nutrients by plants. Improved soil structure, a season–long supply of nutrients, and an increased water–holding capacity are some of the benefits of using organic fertilizers [11, 6]. The absorption of nutrients following the needs of plants havea positive impact on plant growth and yield. The generative phase begins with the emergence of rice panicles, followed by the process of fruit set and seed set. Environmental conditions and the availability of nutrients is very influential in the filling phase of rice grains. The different environmental conditions results in the difference in temperature and radiation which ultimately affected grain yield formation in rice [12, 13]. Therefore, providing the right type of nutrition and the right amount can increase rice yield and productivity.

Table 4. Effect of GOF on 1 000 seed weights and harvested dried grain yield

Treatment (kg ha^{-1})	1 000 seed weights (g)	Harvested dried grain yield (t ha^{-1})
Control (without fertilizer)	26.1 ab	3.57 e
Standart (2 000 Petroganik + 300 Urea + 300 Phonska)	25.9 ab	6.71 cd
(1 000 GOF + 300 Urea + 300 Phonska)	25.9 ab	6.65 d
(2 000 GOF + 300 Urea + 300 Phonska)	26.8 a	6.71 cd
(3 000 GOF + 300 Urea + 300 Phonska)	25.9 ab	7.21 a
(4 000 GOF + 300 Urea + 300 Phonska)	25.7 b	7.11 ab
(5 000 GOF + 300 Urea + 300 Phonska)	26.17 ab	7.15 a
(6 000 GOF + 300 Urea + 300 Phonska)	25.6 b	6.89 bc
Anova	ns	**
CV (%)	2.16	2.05
LSD 5%	0.98	0.23

Note: GOF = Granular Organic Fertilizer
The number followed by the same letter in one column is not significantly different based on the LSD test of 5 %
ns = Not significantly different
** = Significantly different at 1 %

The results of RAE analysis showed that the treatment GOF 1 000 kg had not been able to match the treatment Standart, because it had the RAE value of 98 %, which means that treatment GOF 1 000 kg was not feasible to apply. While the treatment GOF 3 000 kg was effective based on agronomic with the RAE value was 116 %. This means that the dose of the fertilizer can increase yield by 1.16 times (Table 5).

Table 5. RAE value of GOF treatments on rice plants

Treatments (kg ha^{-1})	RAE (%)
Control (without fertilizer)	–
Standart (2 000 Petroganik + 300 Urea + 300 Phonska)	–
(1 000 GOF + 300 Urea + 300 Phonska)	98
(2 000 GOF + 300 Urea + 300 Phonska)	100
(3 000 GOF + 300 Urea + 300 Phonska)	116
(4 000 GOF + 300 Urea + 300 Phonska)	113
(5 000 GOF + 300 Urea + 300 Phonska)	114
(6 000 GOF + 300 Urea + 300 Phonska)	106

Table 6 showed the results of the economic feasibility analysis of each treatment. The revenue rate is the result of multiplying the selling price of grain with grain yield. Based on the selling price of grain at the time of the study was IDR 3 500 kg^{-1}, then the average income of farmers ranged from ID R12 495 000 to IDR 25 235 000. The lowest income was shown in treatment without fertilizer while the highest was in treatment GOF 3 000 kg.

Table 6. Recapitulation of simple economicanalysis of GOF treatment on rice

Treatments (kg ha^{-1})	Income (x IDR 1 000)	Cost (x IDR 1 000)	Benefit (x IDR 1 000)	R/C
Control (without fertilizer)	12 495	13 480	(985)	0.93
Standart (2 000 Petroganik + 300 Urea + 300 Phonska)	23 485	15 740	7 745	1.49
(1 000 GOF + 300 Urea + 300 Phonska)	23 275	16 640	6 635	1.40
(2 000 GOF + 300 Urea + 300 Phonska)	23 485	18 540	4 945	1.27
(3 000 GOF + 300 Urea + 300 Phonska)	25 235	20 440	4 795	1.23
(4 000 GOF + 300 Urea + 300 Phonska)	24 885	22 340	2 545	1.11
(5 000 GOF + 300 Urea + 300 Phonska)	24 885	24 240	785	1.03
(6 000 GOF + 300 Urea + 300 Phonska)	24 115	26 140	(2 025)	0.92

Description of price per kg :
Grain yield = IDR 3 500, Urea = IDR 1 900, Phonska = IDR 2 300, Petroganik = IDR 500, GOF = IDR 1 900.

Farming costs consist of land rent, labor costs, seeds, organic fertilizer (Petroganik), GOF fertilizer, Phonska, Urea, fungicide, and insecticide in one planting season. The lowest total cost was shown in treatment without fertilizer of IDR 13 480 000 and the highest cost for treatment GOF 6 000 kg of IDR 26 140 000 (Table 6).

The benefit is the difference between revenue and costs incurred. Table 6 showed that the GOF treatment reached the highest profit was in treatment GOF 1 000 kg ha^{-1} of IDR 6 635 000 with an R/C ratio of 1.40 but still below the standard treatment (Table 6). Increased use of GOF 5 000 kg ha^{-1} resulted in reduced profits. On the use of GOF 6 000 kg ha^{-1} GOF suffered a loss of IDR 2 025 000 with an R/C ratio of 0.92. This showed that the increase in the use of GOF is not proportional to the rate of increase in production and results in reduced profits.

4 Conclusions

Granule Organic Fertilizer (GOF) significantly affected the growth and yield of rice plants. The use of GOF 3 000 kg + 300 kg Urea + 300 kg Phonska ha^{-1} had the highest grain yield

(7.21 t ha^{-1}) with an RAE value of 116 % or an increase of 0.5 t (7.45 %) compared to grain yield at the use of standard fertilizers (2 000 kg ha^{-1} Petroganic + 300 kg ha^{-1} Urea + 300 kg ha^{-1} Phonska). The highest profit rate for using GOF 1 000 kg ha^{-1} + 300 kg ha^{-1} Urea + 300 kg ha^{-1} Phonska was IDR 6 635 000 with an R/C ratio of 1.4.

References

1. R. Budiono, P.G. Adinurani. Efficiency analysis of production factors utilization in upland rice farming. KnE Life Sciences: 180–187(2017). https://doi.org/10.18502/kls.v3i5.991
2. S. Savci. APCBEE Procedia. **1**:287–292(2012). https://doi.org/10.1016/j.apcbee.2012.03.047
3. R Budiono, P.G. Adinurani, P. Soni. IOP Conf. Series: Earth and Environmental Science 293 (2019) 012034. p.10. https://doi.org/10.1088/1755-1315/293/1/012034
4. R. Singh. *Organic fertilizers types, production and environmental impact*. Nova Science Publisher, New York (2012). p. 1–261. https://www.researchgate.net/publication/274896698_Organic_Fertilizers_Types_Production_and_Environmental_Impact
5. E.D. Purbajanti, F. Kusmiyati, W. Slamet, P.G. Adinurani. AIP Conference Proceedings, **1755**,130013:1–4(2016). https://doi.org/10.1063/1.4958557
6. D. Dadi, G. Daba, A. Beyene, P. Luis, B.V. der Bruggen. Int J Recycl Org Waste Agric. **8**:263–277(2019). https://doi.org/10.1007/s40093-019-0256-8
7. M.A.H. Chontal, C.J.L. Collado, N.R. Orozco, J.V. Velasco, A.L. Gabriel, G.L. Romero. Int J Recycl Org Waste Agric. **8**:309–315(2019). https://doi.org/10.1007/s40093-019-0248-8
8. D.K.S. Swastika. Jurnal Pengkajian dan Pengembangan Teknologi Pertanian. **7**,1: 90–103(2004). [in Bahasa Indonesia]. http://103.213.119.214/index.php/jpengkajian/article/download/1503/1282
9. N.W.A. Utari, T. Tamrin, S. Triyono. Jurnal Teknik Pertanian Lampung. **3**,3: 267–274(2015). [in Bahasa Indonesia]. https://docs.google.com/viewerng/viewer?url=http://jurnal.fp.unila.ac.id/index.php/JTP/article/viewFile/666/608
10. M. Siavoshi, A. Nasiri, S.L. Laware. Journal of Agricultural Science. **3**,3: 217–224(2011). https://doi.org/10.5539/jas.v3n3p217
11. G. Silva. Michigan State University Extension. (2018). https://scholar.google.co.id/scholar?hl=id&as_sdt=0%2C5&q=What+organic+fertilizers+mean+to+plants+and+soil&btnG= or https://www.canr.msu.edu/news/what_organic_fertilizers_mean_to_plants_and_soil
12. Z. Mo, W. Li, S. Pan, T.L. Fitzgerald, F. Xiao, Y. Tang, et al. Rice. **8**,1:9(2015). https://doi.org/10.1186/s12284-015-0040-y
13. S. Li, H. Jiang, J. Wang, Y. Wang, S. Pan, H. Tian, et. al. Scientific Reports. **9**:10618(2019). https://doi.org/10.1038/s41598-019-47031-9

Techno Economic Analysis of Biomass to Methanol Plant Based on Gasification of Palm Empty Fruit Bunch

Rudy Heryadi[1,*] and *Syukri* Muhammad Nur[2]

[1]Department of Mechanical Engineering, Bina Tunggal Technology College, Jl. H. Wahab Affan I, Bekasi 17134, West Java, Indonesia
[2]Graduate School of Renewable Energy, Darma Persada University, Jl. Taman Malaka Selatan, Pondok Kelapa, Jakarta 13450, Indonesia

Abstract. Biomass from palm empty fruit bunch (EFB) is considered as potential renewable energy sources to be developed in Indonesia. Palm EFB can be efficiently converted into valuable and useful methanol products through gasification. Research about methanol synthesis from various biomass has been done. However, Little research about methanol production based on palm empty fruit bunch gasification has been found, and no research yet on technoeconomic analysis of methanol plant construction based on palm empty fruit bunch gasification. The purpose of this research is to find early-stage feasibility of small-scale methanol plant model based on gasification of palm EFB. Method methanol plant design, investment cost estimation, operating cost estimation, and calculation to estimate profitability in terms of economic feasibility parameters. Economic analysis calculated that the total investment to build a methanol plant in this study was USD 87 508 479 with manufacturing cost per annum USD 6 127 014. Net Present Value (NPV) calculated was USD -51 721 064 and IRR at 1.1 %. Based on NPV and Internal Rate of Return (IRR) value, the construction of methanol plant is not feasible. Feasibility predicted to be obtained when the capacity of the plant increased fivefold.

Keywords: Dimethyl ether, feasibility methanol plant, renewable energy, syngas, waste to energy

1 Introduction

The utilization of biomass to be used as one of the energy sources can be done through a gasification process. Gasification is a thermochemical process that converts biomass into a gas called producer gas or synthetic gas (syngas). Syngas is a fuel-based mixture consisting mainly of Hydrogen (H_2), Carbon Monoxide (CO), and Methane (CH_4) [1]. Syngas is a raw material for most other chemical products, such as methanol, ammonia, and dimethyl ether (DME). Biomass that has big potential to be utilized is palm EFB waste. In 2015, in Indonesia, the total EFB waste generated was 30.6×10^6 t yr^{-1} [2]. Based on Inayat et al. [3], Indonesia is currently the world's largest producer of palm oil resulted in very large EFB. Every kg of Crude Palm Oil (CPO) produced, it will produce 4 kg of dry biomass.

* Corresponding author : rudy.heryadi@live.com

About 33 % of biomass is EFB and the rest are palm leaves and stems [4]. Abundant EFB resources have the potential to be used as raw material for further chemical synthesis through the gasification route.

This study, EFB is gasified and further processed to obtain methanol. Methanol has chemical formula CH_3OH and often abbreviated MeOH. Methanol is toxic, light, colorless, volatile, and flammable liquid with a distinctive odor that like ethanol. Major methanol uses as a fuel. It can be used as a gasoline blendstock, bunker fuel for ships, reactant of bio-diesel making reaction, fuel for power generation, and feed for further chemical processes [5].

Research related to the utilization of EFB as a renewable energy source of biomass has been done. Some of these are studies related to the manufacture of biomass pellets to be utilized as feeds of gasifier and boiler reactors by Salomon et al. [6] research related to the utilization of EFB into bio-oil products is conducted by Sukiran et al. [7] which discusses the conversion of EFB to bio-oil through pyrolysis pathway, conversion of EFB into bio-syngas by using fluidized bed reactors performed by Mohammed et al. [8], and research by combining pyrolysis catalytic cracking method to produce bio-syngas with gasification by Pradana and Budiman [9]. Some of the research related with the techno-economic analysis of chemical and fuel products based on gasification of EFB was conducted by Do and Lim [10], that was discussed about techno-economic analysis of EFB conversion into bio-ethanol, and jet-fuel, conversion into power, and fast pyrolysis and bio-oil upgrading, and also research by Do et al. [11] analyze the techno-economic of bio-oil production.

Studies on simulation of gasification process based on biomass gasification to methanol that was discussing methanol production based on gasification of woody biomass in a fluidized bed two stages gasifier, had been done by Clausen [12], and study about techno-economic analysis of methanol production in existing pulp and paper mill were conducted by Andersson et al. [13]. Other research for methanol synthesis based on gasification were study about simulation of methanol production based on biomass gasification in interconnected beds by Zhang et al. [14], study about modeling cost of methanol production by gasification based on geographical location by Leduc et al. [15], and study about economic and technical analysis of methanol production from residues of palm oil biomass and research done by Gomez et al. [16]. The specific research developed for methanol based on EFB gasification has not yet been found. Therefore, this study arises to provide an economic analysis of methanol production plant based on EFB gasification.

2 Methodology

2.1 Modelling steps undertaken

The following steps were undertaken to perform the modeling.

i. Collecting information on the existing model of methanol plant that was available.
ii. Performing the simulation by utilizing process engineering simulation tools. In this study, Aspen Hysys was utilized.
iii. Preparing material stream and energy stream to find methanol yield, process efficiency, and energy efficiency.
iv. Equipment sizing through Aspen Hysys Economic Analyzer and manual calculation sizing.
v. Costing and Economic parameters calculation.

Simplified Block Flow Diagram (BFD) of methanol synthesis was developed and shown in Figure 1. The technology and process configuration used for simulation purposes on the model developed was adopting on some research [12–14, 17]. Table 1 summarized the methanol plant design step and showed the step of designing methanol plants from the preparation of simulation tools, biomass feedstock, and steam generation, and power generation.

Table 1. The steps of Methanol plant design

Simulation tools	The simulation was done by using Aspen Hysys V 8.8. The input of EFB as a hypothetical solid component was based on the ultimate analysis of the EFB measured feedstock (daf) base. From weight percentage of components C, H, O, N, S.
Biomass feedstock	The composition of the EFB was based on a dry ash-free base (daf), and raw composition taken from [18]. The chemical formula for Aspen hysys hypothetical component was $C_{4.306}$ $H_{6.176}$ $O_{2.587}$ $N_{0.048}$ $S_{0.003}$.
Biomass pre-treatment to pellet	The energy required for EFB pellet making is 210 kWh/dry long fiber [5]. It was assumed that dried EFB is the same with dry long fiber (DLF).
Gasification	The process of gasification occurs in Circulating Fluidized Bed (CFB). Gasification temperature at 900 °C and pressure at 10 bar. Detailed of the gasification simulation was based on earlier work on the simulation of gasification in Circulating Fluidized Bed gasifier [19].
Water gas shift	WGS reactor adjusted molar ratio $H_2/CO = 2$ [12]. Temperature = 400 °C and pressure = 1 000 kPa based on [20]. The type of WGS used is High-Temperature WGS because it favors a greater reaction rate [20]. WGS reactor volume was obtained by using multiplying volume of reactor with residence time which is the ideal reactor formula [21], The residence time value taken was 1.8 s [20]. Catalyst initial fill is 60 % volume of WGS reactor, and the density of catalyst is 1 600 kg m^{-3} [20].
Water removal and syngas cleaning	Water removal was modeled using separator as a dryer. and acid gas was removed by the Rectisol process [22]. After water removed, syngas entering Acid removal modeled as a component splitter. CO_2 and H_2S assumed removed at 90 % and 100 %, respectively [20].
Methanol synthesis	Sweet syngas is compressed into 5 000 kPa (5 MPa) [23]. The reactor used is Gibbs reactor, and reactions involved in equilibrium [24]. Then, 79 % of unconverted is recycled and 21 % is sent to off-gas for electricity generation [12]. Catalyst's initial filling quantity was based on Almeland et al. [25]. To determine catalysts total required in kg, the density of the catalyst is 1 140 kg m^{-3} [26]. The liquid outlet of Gas-Liquid separator flows to the purification section.
Steam generation and power generation	Steam generation for the process and power generation was from the waste heat. Steam generated from waste heat of CFB gasifier and combination of waste heat from WGS reactor, Methanol reactor, and a cooler of syngas outlet gasifier. Power was generated by using one steam turbine system, one Genset driven by Internal Combustion Engine (ICE) that was fueled by off-gas, and one genset driven by ICE, fueled by methanol fuel/product.

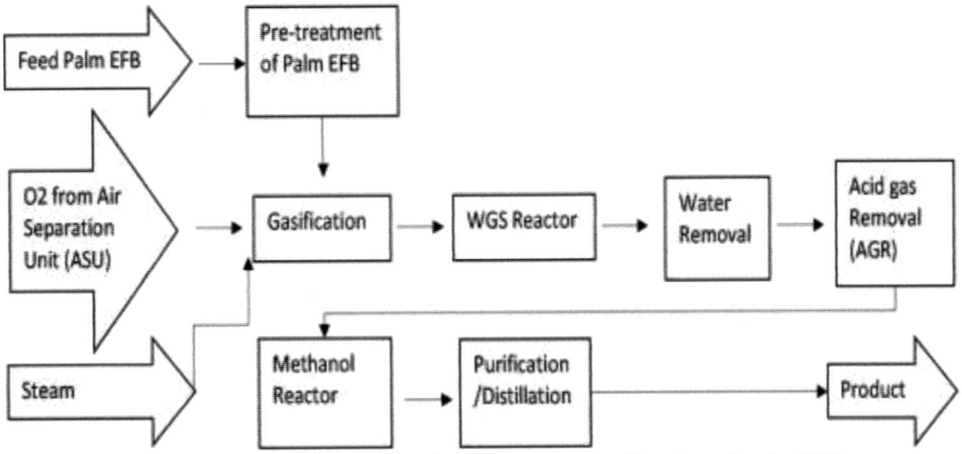

Fig. 1. Simplified BFD of Methanol synthesis based on gasification of palm EFB

2.2 Economic analysis

2.2.1 Market aspect

It was assumed that the market for methanol is already available, and the location is in one of the palm oil mills that was near to the main road. Location can be selected elsewhere in Indonesia. The price of methanol will follow the amount of Methanex FOB price of Asia Pacific at USD 295 t^{-1} in October 2019 [27].

2.2.2 Chemical engineering plant cost index

The investment was a basis to find the chemical engineering plant cost index (CEPCI) that can be calculated by comparing the cost index on the present year to index of reference year [28]. Reference year for main equipment was the year 2014 as a cost basis of Aspen Hysys version 8.8 [29], gasifier and high-pressure feeding system was based on the year 2011 [30], EFB biomass pre-treatment was based on the year 2015 [31], air separation unit was based on year 2007 [32], acid gas removal (Rectisol) was based on year 2007 [32], and the rest of the equipment/section was based on year 2014 [33].

2.2.3 Rule of six tenth

Cost for equipment such as gasifier, high pressure (HP) feeding system, air separation unit, and acid gas removal was obtained by using the rule of six-tenths. The basic equation expresses the rule of six-tenths shown by Formula (1) that was obtained from [34].

$$C_B = C_A \left(\frac{S_B}{S_A}\right)^{0.6} \quad (1)$$

Where, C_B is the approximate cost of equipment having to size S_B (could be any unit) and C_A is the known cost of equipment having size S_A (same unit with S_B).

2.2.4 Estimation of fix capital cost

The basic components in the methanol manufacturing plant are the cost of equipment, overhead costs, administrative costs of doing business, and profits. From the list of existing equipment and acceptable raw specifications, estimates can be made. The most important thing is to make a base of the process equipment used from the equipment list. Equipment installation costs can be searched as a percentage of the total equipment cost [28. In this study, the method used to estimate fix capital cost was the percentage method of total plant installation cost [28].

2.2.5 Estimation of manufacturing cost

Estimated manufacturing costs covered are as follows: raw material, utility costs, including steam; compressed air, refrigeration, electricity; water treatment or water treatment costs; labor costs; maintenance cost; insurance and taxes; factory overhead costs, all fees on process facilities that cannot be charged into other costs; expenses related to sales and marketing; research and development costs; and administrative costs. This estimation is done by Anderson [35].

2.2.6 Profit measurement

Profit is a clear goal, and it is something that can be quantified in an economic evaluation. The measurement by considering the time value of money are:

i. NPV, the value can be obtained from the NPV function of Microsoft excel.
ii. IRR, the value can be obtained from the IRR function of Microsoft Excel.

While for the profitability measurement that is not considering the time value of money is PBP, where PBP can be found by using Formula (2) based on Seider et al. [28].

$$PBP = \frac{\text{Total Capital Investment}}{\text{Annual Cash Flow Average}} \tag{2}$$

3 Results and discussions

3.1 Process simulation result

The process simulation result is presented in Table 2. The energy efficiency calculation was calculating heat integration on the process to ensure optimal energy utilization.

Table 2. Summary of process simulation result

Parameter	Value
Methanol production rate (t d^{-1})	126
Overall thermal efficiency, LHV basis (%)	85
Total energy consumption (kW)	3 417
Air Separation Unit (kW)	1.103
- Syngas Compressor (kW)	996
- Pre-treatment EFB (kW)	1.294
- Acid Gas Cleaning (kW)	11.65
- CO_2 Compressor (kW)	12.35

3.2 Economic analysis

The investment for a methanol plant based on EFB gasification was estimated based on the components listed in Table 3. The price of the equipment was obtained from other works by using the six-tenths rule and CEPCI adjustment. It was seen that the gasification equipment purchase was very cost-intensive compares to others. To further analyze the total plant investment, an approach used for the total plant cost was based on delivered-equipment cost for solid-fluid processing plant [28]. The base for production capacity was based on 90 % capacity [28].

Table 3. Summary of plant equipment cost

Equipment/Section	Ref. Cost (× 10^6 USD)	Ref. Size	Ref.CEPCI	Cost year 2017 CEPCI =572.8 × 10^6 (USD)
Main Equipment	2 143 (Hysys, cost basis 2014 [29])		576.1	2.155
Genset Methanol	0.196			0.196
Genset Off-gas	0.5			0.52
Steam Turbine	0.338 (Hysys, cost basis 2014 [29])		576.1	0.336
WGS Reactor	0.064 (Hysys, cost basis 2014 [29])		576.1	
WGS catalyst	0.027 calculated based on Florence and Bour [20]		556.8	0.028
Methanol Reactor	0.0214 (Hysys, cost basis 2014 [29])		576.1	0.021 3
Methanol catalyst	0.790 calculated based on Almeland et al. [25]		576.1	0.078 5
CFB Gasifier	28.171 [30]	41.7	585.7	8.750
Air Separation Unit	82.700 [32]	2 965	525.4	16.210
AGR (rectisol)	28.800 [32]	8 928	525.4	5.733
HP Biomass Feed	24.550 [30]	42	585.7	0.759

The lifetime of the plant was assumed at 30 yr [36]. Purchase equipment cost was marked up at 10 % value to compensate for the delivery since the price quoted usually in the Free on Board (FOB) basis [28]. Total Plant Cost (TPC) as shown in Table 4, based on the delivered-equipment cost of the solid-liquid plant [28] was USD 87 508 479. From that total cost, equipment cost was USD 77 495 647, and the working capital was USD 10 012 832.

Table 4. Methanol plant capital cost

Component	Estimation	Cost in USD
Main Equipment Cost Plus delivery	110 % E	13 350 442
Installation	39 % E	5 206 673
Piping	31 % E	4 138 637
Electricity	10 % E	1 335 044
Instrumentation	26 % E	3 471 115
Yard improvement	12 % E	1 602 053

(Continued on next pages)

Table 4. Continued

Component	Estimation	Cost in USD
Service facilities	55 % E	7 342 743
Building	29 % E	3 871 628
Total Direct Cost (DC)	**USD 40 318 336**	
Engineering and Supervision	32 % E	4 272 142
contruction expenses	34 % E	4 539 150
Legal expenses	4 % E	534 018
Contractors Fee	19 % E	2 536 584
Contingency	37 % E	4 939 664
Total Indirect Cost (IC)	**USD 16 821 558**	
Total main Plant Cost (TPC)	USD 57 139 894	
Total Additional Plant equipment (AGR, ASU, Pre-treatment)	USD 20 355 753	
Fix Capital Investment	USD 77 495 647	
Working capital	11 % TPC	USD 10 012 832
Total Plant Cost (TPC)	**USD 87 508 479**	

Manufacturing cost was estimated based on Anderson [35]. The cost of raw material was assumed 0 because EFB used is a waste and obtained from palm oil mill since the location of methanol plant was assumed integrated with palm oil mill. The total operating cost per year as shown in Table 5 is USD 6 127 014.

Table 5. Methanol plant manufacturing cost

Cost Component	Cost (USD)	Remark(s)
Operating Labor cost	158 571	Estimation based on average labor cost in Jakarta, Indonesia.
Non Operating Labor cost	95 143	60 % of Operating labor cost
Supplies	47 571	30 % of Operating labor cost
Administration and overhead	142 714	90 % of operating and labor cost
Maintenance	1 549 913	2 % of fix capital investment
Raw Material	0	EFB waste
Utilities (Water, refrigeration, water treatment)	774 956	1 % of fix capital investment
Miscelleneous	774 956	1 % of fix capital investment
Depreciation	2 583 188	Straight line depreciation of Total Direct Cost for 30 yr
Total Manufacturing Cost	**USD 6 127 014**	

A feasibility analysis was conducted to know the profitability of this methanol plant project. The discount rate used assumed the same as the corporate loan interest rate in Indonesia at 10.25 % [37]. Parameters to analyze the profitability of this Methanol plant were calculated and shown in Table 6.

From Table 6, it was shown for the capacity at 6.16 t h^{-1} EFB feed, the project was not feasible. Small scale DME plant based on EFB gasification attached to main palm oil mill is not feasible by the current interest rate at 10.25 %. To make a project feasible, the option for higher production capacity should be considered for further feasibility study.

The initial prediction of the feasible capacity of DME plant at 10.25 % discount rate can be done by using sixth-tenths rule for investment cost, and percentage of investment cost for operating cost. At the current study annual operating cost was 5.52 % of total plant cost, this value was assumed constant for 30 yr plant lifetime, and revenue from sales assumed increased linearly according to production capacity. Annual operating cost in percentage was used for scale-up feasibility calculation.

Table 6. Feasibility parameters

Feasibility Parameters	Unit	Value
NPV	USD	- 51 721 064
IRR	%	1.1 %
PBP	Years	> 30

Based on Figure 2, feasibility can be achieved at 30.8 t h^{-1} EFB biomass input or fivefold from 6.16 t h^{-1} palm EFB biomass input. This feasibility needs to be analyzed and studied further, including sensitivity analysis to see the impact of changes that can affect the feasibility of the project.

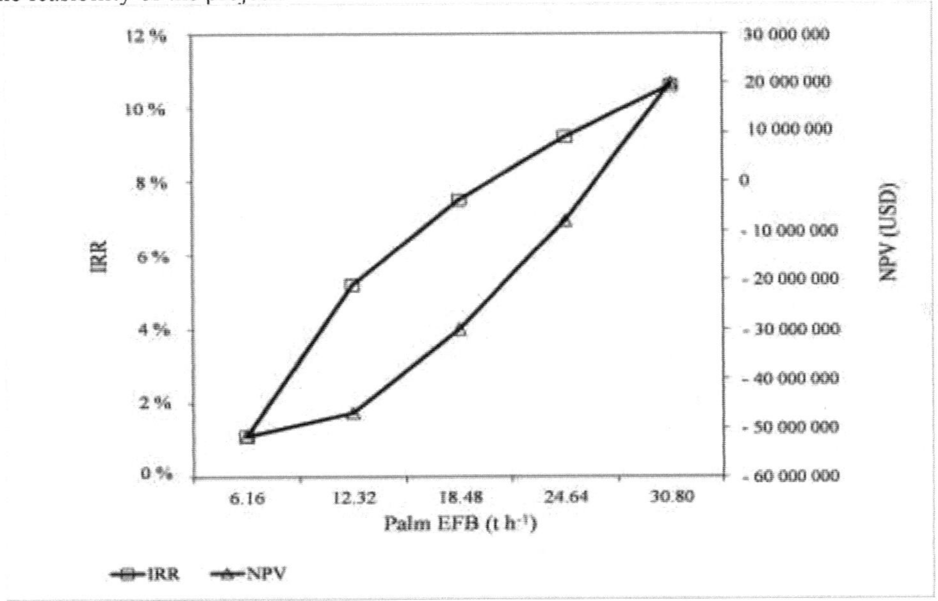

Fig. 2. Prediction of methanol plant feasibility for increased capacity

4 Conclusion

This paper exhibits the mass and energy balance and economic aspect of the proposed methanol plant based on gasification of EFB and the model of a methanol plant was designed and simulated by using Aspen Hysys. Total energy consumption was 3 417 kW and the performance of the methanol plant shown by energy efficiency at 85 %.

Based on economic analysis, methanol plant at 6.16 t h^{-1} EFB Feed was not feasible to be built at 10.25 % discount rate, to make project feasible discount rate must be below 1.1 %, option for higher production capacity should be considered for a further feasibility study. Initial predictions for feasibility based on production capacity have been made and it was found that a fivefold increase in capacity makes the project feasible.

References

1. P. Basu, *Biomass gasification and pyrolysis: practical design and theory*, Academic press, 2010. https://www.elsevier.com/books/biomass-gasification-and-pyrolysis/basu/978-0-12-374988-8
2. E. Hambali, M. Rivai, IOP Conference Series: Earth and Environmental Science 65, 012050(2017) http://iopscience.iop.org/article/10.1088/1755-1315/65/1/012050
3. A. Inayat, C. Ghenai, M. Naqvi, M. Ammar, M. Ayoub, M. Hussin, Energy Procedia, **105**:1242–1249(2017). https://www.sciencedirect.com/science/article/pii/S1876610217304721
4. M. Maitah, P. Prochazka, A. Pachmann, K. Šrédl, H. Řezbová, International Journal of Energy Economics and Policy. **6**:35–38(2016). https://www.econjournals.com/index.php/ijeep/article/view/1501
5. R. Heryadi, A. Uyun, E. Yandri, S. Nur, K. Abdullah, O. Anne, IOP Conference Series: Earth and Environmental Science. **293**, 012036(2019). https://iopscience.iop.org/issue/1755-1315/293/1
6. M. Salomon, M. F. Gómez, C. Erlich, A. Martin, Biomass Convers. Biorefin., **3**:213–229(2013). https://link.springer.com/article/10.1007/s13399-013-0075-5
7. M. A. Sukiran, N. K. A. Bakar, C. M. Chin, J. Oil Palm Res., **21**:653–658(2009). http://palmoilis.mpob.gov.my/publications/joprv21dec09-azri.pdf
8. M. Mohammed, A. Salmiaton, W. W. Azlina, M. M. Amran, Bioresour. Technol, **110**:628–636(2012). https://www.sciencedirect.com/science/article/pii/S0960852412000624
9. Y. S. Pradana, A. Budiman, Eng. Sci. Technol. an Int. J., **10**:1–8(2015). https://repository.ugm.ac.id/id/eprint/138936/contents
10. T. X. Do, Y.-i. Lim, Renewable Energy, **90**:307–318(2016). https://www.sciencedirect.com/science/article/abs/pii/S0960148116300301
11. T. X. Do, Y.-i. Lim, H. Yeo, Energ. Conver. Manage., **80**:525–534 (2014). https://www.sciencedirect.com/science/article/pii/S0196890414000685
12. L. R. Clausen, *Design of novel DME/methanol synthesis plants based on gasification of biomass*,[Thesis Phd], Technical University of Denmark, (2011).p.295 http://orbit.dtu.dk/files/6321613/Phd%20afhandling%20-%20Lasse%20R%20Clausen.pdf
13. J. Andersson, J. Lundgren, M. Marklund, Biomass Bioenergy, **64**:256–268 (2014). https://doi.org/10.1016/j.biombioe.2014.03.063
14. Y. Zhang, J. Xiao, L. Shen, Ind. Eng. Chem. Res. **48**:5351–5359(2009). https://doi.org/10.1021/ie801983z
15. S. Leduc, E. Schmid, M. Obersteiner, K. Riahi, Biomass Bioenergy, **33**:745–751(2009). https://doi.org/10.1016/j.biombioe.2008.12.008
16. R. Gomez, L. Corredor, A. Avila, J. Mendoza, A. Bula, Proceedings Paper: Energy Sustainability. 1175–1182(2011) http://proceedings.asmedigitalcollection.asme.org/proceeding.aspx?articleid=1636684

17. P.-C. Chen, H.-M. Chiu, Y.-P. Chyou, C.-S. Yu, World Acad. Sci. Eng. Technol., **41**:989–996 (2010). http://citeseerx.ist.psu.edu/viewdoc/download?doi=10.1.1.294.1465&rep=rep1&type=pdf
18. A. Wijono, *PLTU Biomasa tandan kosong kelapa sawit studi kelayakan dan dampak lingkungan* [PLTU oil palm empty bunch biomass feasibility studies and environmental influences]. Simposium Nasional RAPI XIII 2014, Surakarta.FT UMS.111–18(2014). [in Bahasa Indonesia] https://www.google.com/url?sa=t&rct=j&q=&esrc=s&source=web&cd=&cad=rja&uact=8&ved=2ahUKEwjD6OWi18rrAhX7lEsFHTodDBUQFjAAegQIBBAB&url=https%3A%2F%2Fpublikasiilmiah.ums.ac.id%2Fbitstream%2Fhandle%2F11617%2F5492%2F17.Agung%2520Wijono.pdf%3Bsequence%3D1&usg=AOvVaw0yyzeGFzAGXHshMgiJylvB
19. R. Heryadi, A. S. Uyun, E. Yandri, S. M. Nur, K. Abdullah, E3S Web of Conferences. **67**,02043:1–10(2018) https://www.e3s-conferences.org/articles/e3sconf/pdf/2018/42/e3sconf_i-trec2018_02043.pdf
20. H. Florence, A. Bour, Norway: NTNU, (2016). https://daim.idi.ntnu.no/masteroppgaver/015/15871/masteroppgave.pdf
21. Ideal Reactor Design Equations and Formulas Calculator [Online] Available from https://www.ajdesigner.com/phpreactor/reactor_equations_mean_residence_time.php. [Accessed on 27 July 2019]
22. Lurgi, Air Liquide, 1–12(2011). https://www.engineering-airliquide.com/sites/activity_eandc/files/2017/11/22/air-liquide-ec-technology-handbook-2017.pdf
23. J. Ott, V. Gronemann, F. Pontzen, E. Fiedler, G. Grossmann, D. B. Kersebohm, G. Weiss and C. Witte, *Ullmann's Encyclopedia of Industrial Chemistry 6th edition*, (2011). https://onlinelibrary.wiley.com/doi/10.1002/14356007.a16_465.pub3
24. K. S. Ng, J. Sadhukhan, Biomass Bioenergy, **35**, 115–169, (2011). https://doi.org/10.1016/j.biombioe.2010.12.003
25. S. K. Almeland, K. Meland, D. G. Edvardsen, S. Skogestad, M. Panahi, *Norway: Norwegian University of Science and Technology, Department of Chemical Engineering*, (2009) http://folk.ntnu.no/skoge/diplom/prosjekt09/methanol-project/FINAL%20DOCUMENT.doc
26. F. Samimi, M. R. Rahimpour, A. Shariati, Catalysts, **7,11**:332(2017). https://doi.org/10.3390/catal7110332
27. Methanex Monthly Average Regional Posted Contract Price History [Online] Available From https://www.methanex.com/our-business/pricing. [Accessed on 1 October 2019]
28. W. G. Seider , J. D. Seader, D. R. Lewin, S. Widagdo, *Product and Process Design Principles : Synthesis, Analysis, and Evaluation 3rd Edition*, John Wiley & Sons Inc., 2009 https://www.academia.edu/16568227/208468464-Product-and-Process-Design-Principles-Synthesis-Analysis-and-Design-Third-Edition_1_
29. O. J. Symister, Florida: Florida Institute of Technology (2016). https://pdfs.semanticscholar.org/9e18/9b6e79fb68f4ee8b3414973010 51ada0ea36.pdf
30. M. Worley, J. Yale, *Biomass gasification technology assessment: Consolidated report*, National Renewable Energy Lab.(NREL), Golden, CO (United States), 2012

https://www.nrel.gov/docs/fy13osti/57085.pdf
31. GEMCO Pellet Plant in Malaysia / Malaysian Oil Palm Biomass Study [Online] Available from http://www.biomass-energy.org/blog/efb-pellet-plant-in-malaysia.html. [Accessed on 25 July 2019]
32. K. M. Holmgren, *IVL Swedish Envrionmental Research Institute*, (2015). https://www.ivl.se/download/18.7e136029152c7d48c202a1d/1465298345076/B2221.pdf
33. Matches' Process Equipment Cost Estimates [Online] Available from http://matche.com/equipcost/Default.html. [Accessed on 25 July 2019]
34. R. Smith, Chemical Process: Design and Integration, John Wiley & Sons, 2005 https://imtk.ui.ac.id/wp-content/uploads/2014/02/Chemical-Process-Design-and-Integration.pdf
35. J. Anderson, Cep, 27-31, (2009) http://www.janecki.pr.radom.pl/394/CostEstn.pdf
36. E. C. Tan, M. Talmadge, A. Dutta, J. Hensley, J. Schaidle, M. Biddy, D. Humbird, et al., *Process Design and Economics for the Conversion of Lignocellulosic Biomass to Hydrocarbons via Indirect Liquefaction. Thermochemical Research Pathway to High-Octane Gasoline Blendstock Through Methanol/Dimethyl Ether Intermediates*, National Renewable Energy Lab.(NREL), Golden, CO (United States), 2015 https://www.nrel.gov/docs/fy15osti/62402.pdf
37. Danamon Prime Lending Rate [Online] Available from https://www.danamon.co.id/en/Tentang-Danamon/SBDK. [Accessed on 31 July 2020]

E3S Web of Conferences **226**, 00041 (2021) https://doi.org/10.1051/e3sconf/202122600041
ICoN BEAT 2019

Risk Analysis of Partnership Broiler Farm in Blitar District, East Java, Indonesia

Septi Nur Wulan Mulatmi[1,*], *Apriliana* Devi Anggraini[1], *Febrianti* Shaywont Pratami[1], and *Martina* Motalova[2]

[1]Department of Animal Science, Faculty of Agriculture and Animal Sciences, University of Muhammadiyah Malang, Jl. Raya Tlogomas no 246 Malang 65145, East Java, Indonesia
[2]Department of Infectious Diseases and Microbiology, Faculty of Veterinary Medicine, University of Veterinary and Pharmaceutical Sciences Brno, Palackého tř. 1946/1, 61242 Brno–Královo Pole, Czech Republic

Abstract. Contract farming is a form of business cooperation between two or more parties in a certain period of time to help farmers provide production facilities. The superiority of the partnership is as a partner to share the risk burden of production between core companies and small farmers. The risk of production in broiler chicken farming includes DOC quality, feed, and disease. Therefore, production risk management must be able to manage these factors well so they cannot cause losses. The material in this research is thirty partnership farmers in Blitar, East Java, Indonesia. The method used is a survey by observation and interview. The sources of data used are primary and secondary data sources from the results of interviews and observations. The analysis used is production risk analysis to determine the level of production risk in Blitar, East Java, Indonesia and descriptive analysis to find out the production risk management used by partnership farmers in Blitar, East Java, Indonesia. The results of this study can be concluded that the risk of production in Blitar, East Java, Indonesia is categorized as high risk caused by death. Meanwhile, production risk management carried out by farmers in Blitar, to reduce mortality is spraying disinfectants.

Key words: Contract farming, risk management, risk production

1 Introduction

Contract farming (CF) has attracted considerable at–tention over the past decades. Several studies show that CF increases farm productivity, profitability, farmers' income, and food security [1–3]. The partnership is a form of business cooperation between two or more parties (the core company and the breeder) for a certain period of time to assist the farmers in providing production facilities in the form of DOC, feed, medicines, vaccines, vitamins, and marketing of broiler products for mutual benefit. Given the large cost of broiler chicken production, the partnership can be a solution for small farmers to build a broiler chicken farming business [4]. The advantage of the partnership is as a partner for sharing the burden of production and marketing risk between the core company and small farmers [5]. Agriculture with a contract farming system have a negative impact on small farmers, including contracting, manipulation of inputs, contracts that do not benefit farmers (unprofitable contracts), manipulation of weighing (under weighing of poultry), giving an incorrect index or ranking, and grading problems [6]. The high volatility in production cost and the fluctuation price for broiler also become another cases. It was evidence that

* Corresponding author: septi.mulatmi@gmail.com

majority (91.2 %) of poultry farmers practice with intensive farming system still low adopted risk management strategy.

The distribution of the risk burden on the partnership pattern between the company and farmers makes the small broiler farmers prefer the partnership, although the risks that occur during production cannot be avoided. One of the highest sources of risk is the source of production risk. Production risks in the business of broiler chickens include DOC quality, feed, and disease. Therefore, various production risks that can be caused by these production factors must be managed properly so that they cannot cause losses [7].

The role of broiler partnership scheme in mitigating market risks occurred in the sixth harvest time through financial performance achievement on the basis of high profit (IDR 399/Kg live body weight), the low category of enterprise and economic rentabilities (REnt=4.07 % and REco = 1.68 %), and a limited payment capability toward total liabilities as the solvency level in the caution category (DTAR = 55.87 %) [8].

Broiler chicken farmers decide to partner to minimize risk, especially the risk of production during the maintenance period which is marked by the fluctuation of chicken mortality rates each period. Disease outbreaks often afflict broiler chickens and have a direct effect as a trigger for production risks. The disease is difficult to detect, can occur suddenly, and can cause high rates of mortality. If the temperature becomes too low, this causes the temperature of the air inside the cage to be high, causing high humidity in the cage. During the rainy season, the temperature in the cage becomes cold, and the air in the cage becomes humid. Conversely in the dry season, the temperature of the air inside the cage gets hot, carbon dioxide levels increase and the air in the cage feels more stuffy. Such conditions are difficult to avoid and result in death with a fairly high mortality rate. This can increase the breeding of germs in broiler chickens [9].

The economic feasibility of the production system increases with the level of welfare improvements for a sufficiently high price level for broiler meat and low volatility in producer prices. If this is not the case, however, risk attitudes of farmers become important as well as the use of potential risk management instruments. [10]

2 Materials and methods

This research was conducted in February 2019 to March 2019 in Blitar, East Java, Indonesia. Determination of this location was based on certain considerations of purposive sampling with the consideration that Blitar, East Java, Indonesia has a partnership system farm with a population of 5 000 to 10 000 broilers per plasma farmer.

The research method used in data collection is the survey method. The survey method was a technique of collecting data or information from a sample through interviews or filling out research questionnaires.

Some measures that could be used to measure deviations include variance, standard deviation, coefficient of variation, and lower–income limits [5]. Descriptive analysis was a method of examining the status of a group of people, an object, a condition, a system of thought in the present [6]. The descriptive statistical analysis aims to make a systematic, actual and accurate description, picture or painting of the facts, nature, and relationships between the phenomena being observed that were usually presented in the form of diagrams, graphs, etc. from the data that had been obtained from respondents using a questionnaire tool. The results of the descriptive statistical analysis obtained were used to determine the production risk management that was widely applied by partnership broiler farmers in Blitar, East Java, Indonesia.

3 Result and discussions

3.1 Characteristics of respondents

The selection of respondent characteristics was one of the things that will affect the validity of the data to be used. Characteristics of respondents in the research conducted could be divided into four characteristics, i.e. characteristics based on age, education, and business scale.

The result showed that breeders in Blitar, East Java, Indonesia were still productive both physically and in developing their livestock business. Productive age had a higher level of productivity compared to workers who are old [11, 12].

The majority of farmers in Blitar, East Java, Indonesia have a high school/equivalent level of education. The higher the farmer's education level, the better the maintenance management because the farmer could adopt innovation and change the way of thinking and problem solving more mature [13].

The range of business scale owned by farmers in Blitar, East Java, Indonesia is on a business scale of 5 000 to 7 000 individuals. This includes a small and medium business scale.

Table 1. Characteristic of partnership broiler farmers in Blitar, East Java, Indonesia

Parameter	Number of partnership broiler farmer (%)
Ages	
21 to 30	23.07
31 to 40	46.15
41 to 50	30.76
Formal Education	
Junior High School	26.90
Senior High School	57.69
Bachelor	15.38
Business Scale	
5 000 to 7 000	88.46
8 000 to 10 000	11.53

3.2 Risk analysis

Broiler chicken farmers decide to partner for risks minimizing, especially the risk of production during the maintenance period which is marked by the fluctuation of chicken mortality rates each period [14].

The value of variance shows that the greater the value of the variance, the greater the value of the deviation, the greater the risk faced in business activities [15]. The variation value obtained by partnership broiler farmers in Blitar, East Java, Indonesia is very large. These results indicate the level of risk of partnership broiler farmer production in Blitar East Java, Indonesia was very large.

The greater the standard deviation, the greater the risk faced [16]. The standard deviation value indicates the risk faced by partnership broiler farmers in Blitar, East Java, Indonesia is very large. The production risk was caused by high mortality rates caused by disease and restrictions on the use of Antibiotics Growth Promoters (AGP).

Table 2. Variance calculation of partnership broiler farm

Parameter	Factors affecting production risk				
	Average gain	**FCR**	**Mortality**	**Age of harvest**	**Benefit**
Variance (s^2)	169.26	113.83	0.14	51 412.60	IDR 2 616 917 654 478 460.00
Standard deviation (s)	13.01	10.67	0.37	226.74	IDR 51 155 817.00
Expected return $E=\frac{\sum_{i=1}^{n} Ei}{n}$					IDR 9 233 333.00
Variance coefficient $CV=\frac{s}{E}$					IDR 5.54
Lower limit $L=E-2s$					IDR 93 078 301.00

The expected return was obtained from the average net income throughout the observation period. The expected results depend on the income earned by the farmer each period. The current net income obtained by farmers is valued at IDR 1 000 head^{-1}.

The coefficient of variation was a number that shows the comparison between the risk that might be borne with the cash income to be obtained or the coefficient of variance was used to compare the risks faced with the income received. The coefficient of variation showed that the risk borne by farmers is 554 % of the return value obtained by farmers. This means that every IDR 1 income received will result in a risk of IDR 5.54. The coefficient value of variation greater than 0.5 indicated that the partnership broiler chicken farming business in Blitar, East Java, Indonesia had a large production risk in each period.

The lower–income limit showed the lowest nominal net income that a farmer might receive. If the lower–income limit was lower than zero, the farmer will suffer a loss. The lower boundary value showed that the broiler chicken farmers in Blitar, East Java, Indonesia suffer losses. The amount of the lower–income limit was caused by several cases of livestock death that could not be predicted every period. The difficulty of preventing mortality was a major factor in the high production risks experienced by partnership broiler farmers in Blitar, East Java, Indonesia.

3.3 Risk management

Efforts to reduce the high risk of production in broilers were to implement risk management. Corporate risk management was a way of dealing with all risks in the company without choosing certain risks. Risk management could be considered as a function of management. Several management functions were well known, namely Planning, Organizing, Directing, and Controlling (POAC) [17].

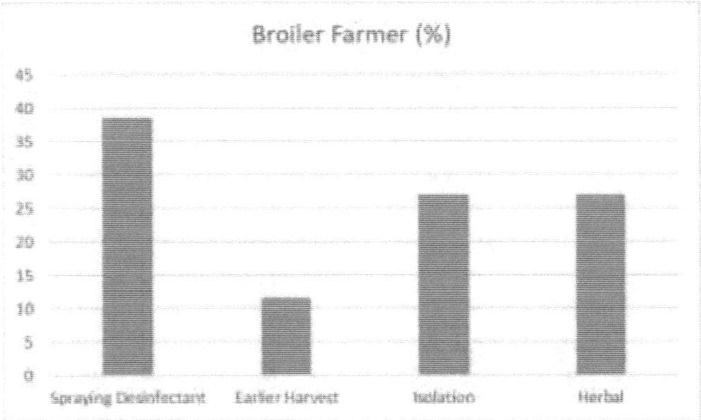

Fig 1. Farmers applied risk management

The results of the study stated that the factor that caused the high risk of partnership broiler production in Blitar, East Java, Indonesia was the death caused by the disease. Diseases that often attack chickens were *E. coli*, Newcastle Disease (ND), and Gumboro.

Risk management applied to farms includes spraying disinfectants, earlier harvest, isolation, and the use of herbal medicines. As many as 38.46 % of farmers choose to spray disinfectants to avoid the wider spread of the disease. The next risk management undertaken by farmers was to conduct an earlier harvest. This effort was carried out by farmers whose chickens were attacked by diseases and caused high mortality of around 80 % of the population. Early harvesting was the last choice by farmers, especially if the age of chickens was still too small. As many as 11.53 % of farmers make early harvest efforts when their chickens in the maintenance period yesterday were stricken with the disease.

Isolation or separation of farm locations was risk management of production that was mostly done by farmers. 26.92 % of farmers choose to isolate their animals to minimize contracting disease. The isolation or separation of breeding sites aimed to create a broiler breeding environment that was protected from carriers transmitted by humans, fomites, wild animals, infected birds, air, water, etc [18].

Production contracts with longer terms, lower upfront deposit requirements and higher cost sharing with enterprises for technology adoption may make farmers more likely to adopt technology, to adopt technology early and to invest more [19]

Some farmers choose to use herbal medicine to reduce the risk of disease. Farmers in Blitar, East Java, Indonesia were accustomed to providing additional herbal medicine to increase the immune power of the chicken body. Other farmers work to reduce risk by cleaning the surrounding yard, removing wet litter, or sowing husks on wet litter.

4 Conclusions

Based on the results of the study it could be concluded that the level of risk of partnership broiler farm production in Blitar, East Java, Indonesia was high. This was caused by the high mortality rate caused by the disease.

The risk management of broiler farm production implemented by farmers in Blitar, East Java, Indonesia focuses on preventing the spread of disease. The applied production risk management was the addition of disinfecting spraying in cages, separating sick animals in isolation cages, providing herbal medicine, cleaning around the cage, and conducting early harvests.

References

1. C.B. Barrett, M.E. Bachke, M.F. Bellemare, H.C. Michelson, S. Narayanan, T.F. Walker. World Dev. **40**,4:715–730(2012). https://doi.org/10.1016/j.worlddev.2011.09.006
2. M.F. Bellemare, L. Novak. Am. J. Agric. Econ. **99**,2:357–378(2017). https://doi.org/10.1093/ajae/aaw053
3. H.H. Wang, Y. Wang, M.S. Delgado. Am. J. Agric. Econ. **96**,5:1257–1271(2014). https://doi.org/10.1093/ajae/aau036
4. R. Ratnasari, W. Sarengat, A. Setiadi. Animal Agriculture Journal. **4**,1: 47–53(2015). [in Bahasa Indonesia] https://scholar.google.co.id/scholar?hl=id&as_sdt=0%2C5&q=ANALISIS+PENDAPATAN+PETERNAK+AYAM+BROILER+PADA+SISTEM+KEMITRAAN+DI+KECAMATAN+GUNUNG+PATI+KOTA+SEMARANG&btnG=
5. A.H Hoddi, M.B. Rombe, Fahrul. Jurnal Agribisnis. **10**,3:25–32(2011). [in Bahasa Indonesia]. https://scholar.google.co.id/scholar?hl=id&as_sdt=0%2C5&q=ANALISIS+PENDAPATAN+PETERNAKAN+SAPI+POTONG+DI+KECAMATAN+TANETE+RILAU%2C+KABUPATEN+BARRU+%28Revenue+Analysis+Cattle+Ranch+In+Sub+Tanete+Rilau+Barru&btnG=
6. M. Murthy, S.B. Madhuri. Asia Pacific J. Mark. Manage. Review. **2**,5: 58–68(2013). http://indianresearchjournals.com/pdf/APJMMR/2013/May/7.pdf
7. O.R. Iheke, C.A. Igbelina. NJAFE. **12**,1:67–74(2016). https://scholar.google.co.id/scholar?hl=id&as_sdt=0%2C5&q=RISKS+MANAGEMENT+IN+POULTRY+PRODUCTION+IN+IKEDURU+LOCAL+GOVERNMENT+AREA+OF+IMO+STATE%2C+NIGERIA&btnG=
8. Sutawi. *Kemitraan sebagai strategi manajemen risiko* [Partnership as a risk management strategy]. Poultry Indonesia. (2009). [in Bahasa Indonesia]. https://www.semanticscholar.org/paper/Kemitraan-Sebagai-Strategi-Manajemen-Risiko-Sutawi/9847cbdc4c540926f09b6eeb83594ed297d8b111#paper-header
9. H.D. Utami, B.A. Nugroho, U.W. Ningsih, L.E. Radiati, H. Nugroho. J. Adv. Agric. Technol. **5**,3:222–226(2018). https://scholar.google.co.id/scholar?hl=id&as_sdt=0%2C5&q=Broiler+Partnerships+Scheme+Advantages+toward+Minimising+Its+Production+and+Marketing+Risks+at+Mojokerto+Indonesia&btnG=
10. D. Crownia, A. Fariyanti. *Analisis faktor–faktor yang memengaruhi risiko produksi ayam broiler (Studi kasus: Peternak mitra CV Anjawani Purwakarta).* [Analysis of factors affecting the risk of broiler chicken production (Case study: breeder CV CV Anjawani Purwakarta)] [Skripsi] Ekonomi dan Manajemen, Institut Pertanian Bogor. (2017) [in Bahasa Indonesia]. p. 1–61. http://repository.ipb.ac.id/handle/123456789/87344
11. É. Gocsik, A.G.J.M.O Lansink, H.W. Saatkamp. Poult. Sci. **92**,12: 3314–3329(2013). https://doi.org/10.3382/ps.2013-03221
12. A. Fariyanti, N.F.N. Kuntjoro, S. Hartoyo, A. Daryanto. *Perilaku ekonomi rumah tangga petani sayuran dalam menghadapi risiko produksi dan harga produk di Kecamatan Pangalengan Kabupaten Bandung*. [Economic behavior of vegetable farmer households in facing production risk and product prices in Pangalengan District, Bandung Regency]. [Thesis Phd]. Institut Pertanian Bogor, Jawa Barat, Indonesia (2008). [in Bahasa Indonesia]. p. 178–206. http://dx.doi.org/10.21082/jae.v25n2.2007.178-206

13. R. Hafni. *Mitigasi risiko dalam pembiayaan murabahah: Studi di PT. BPRS Bhakti Sumekar Kantor Pusat Sumenep*. [Risk mitigation in murabahah financing: Study at PT. SRB Bhakti Sumekar Sumenep Head Office] [Thesis] Ekonomi Syariah, UIN Sunan Ampel Surabaya. (2016) [in Bahasa Indonesia]. p. 1–121. https://scholar.google.co.id/scholar?hl=id&as_sdt=0%2C5&q=Mitigasi+Risiko+dalam+Pembiayaan+Murabahah+%3A+Studi+di+PT.+BPRS+Bhakti+Sumekar+Kantor+Pusat+Sumenep&btnG=
14. D. Tanto, S.M. Dewi, S.P. Budio. Rekayasa Sipil. **6**,1:69–82(2012). [in Bahasa Indonesia]. https://scholar.google.co.id/scholar?hl=id&as_sdt=0%2C5&q=Faktor-Faktor+yang+Mempengaruhi+Produktivitas+Pekerja+pada+Pengerjaan+Atap+Baja+Ringan+di+Perumahan+Green+Hills+Malang&btnG=
15. A.D. Mahendra, N. Woyanti. *Analisis pengaruh pendidikan, upah, jenis kelamin, usia, dan pengalaman kerja terhadap produktivitas tenaga kerja (Studi di industri kecil tempe di Kota Semarang)*. [Analysis of the effects of education, wages, gender, age, and work experience on labor productivity (Study in tempe small industries in Semarang City)]. [Thesis Phd]. Universitas Diponegoro, Semarang, Indonesia (2014). [in Bahasa Indonesia]. p. 1–70. https://scholar.google.co.id/scholar?hl=id&as_sdt=0%2C5&q=Analisis+pengaruh+pendidikan%2C+upah%2C+jenis+kelamin%2C+usia%2C+dan+pengalaman+kerja+terhadap+produktivitas+tenaga+kerja+%28Studi+di+industri+kecil+tempe+di+Kota+Semarang%29&btnG=
16. Murtiyeni, D. Priyanto, D. Yulistiani, *Karakteristik peternak domba atau kambing dengan pemeliharaan digembala atau angon dan hubungannya dengan tingkat adopsi inovasi teknologi* [Characteristics of sheep or goat breeders with herding or angon rearing and their relationship with the level of adoption of technology innovation]. Seminar Nasional Teknologi Peternakan dan Veteriner, (Bogor, Indonesia 2005). [in Bahasa Indonesia]. https://scholar.google.co.id/scholar?hl=id&as_sdt=0%2C5&q=karakteristik+Peternak+Domba+atau+Kambing+dengan+Pemeliharaan+Digembala+atau+Angon+dan+Hubungannya+dengan+Tingkat+Adopsi+Inovasi+Teknologi&btnG=
17. N.U. Handayani, M.A. Wibowo, D.P. Sari, Y. Satria, A.R. Gifari. Jurnal Ilmiah Bidang Ilmu Kerekayasaan. **39**,2:78–85(2018). [in Bahasa Indonesia]. https://doi.org/10.14710/teknik.v39i2.15918
18. I.B.K. Ardana. Buletin Veteriner Udayana. **3**,1:51–59(2011). [in Bahasa Indonesia]. https://scholar.google.co.id/scholar?hl=id&as_sdt=0%2C5&q=STRATEGI+PENCEGAHAN+PENYAKIT+INEFEKSIUS+PADA+PETERNAKAN+BROILER+BERBASIS+LABORATORIUM&btnG=
19. H. Mao, L. Zhou, J. Ifft, R. Ying. China Econ. Rev. **54**:147–159(2019). https://doi.org/10.1016/j.chieco.2018.10.014

Hydrolysis of Cellulose from Oil Palm Empty Fruit Bunch using *Aspergillus niger*

Sri Sugiwati[1,*], *Suaidah* Suaidah[1*], *Eka* Triwahyuni[1], *Muryanto* Muryanto[1], *Yosie* Andriani[2] and *Haznan* Abimanyu[1]

[1]Research Center for Chemistry, Indonesian Institute of Sciences, Kawasan Puspiptek Serpong, Tangerang 15314, Indonesia
[2]Institute of Marine Biotechnology, University Malaysia Terengganu, Mengabang Telipot 21030, Kuala Nerus, Terengganu, Malaysia

Abstract. Oil palm empty fruit bunch (OPEFB) constitutes a great source of lignocellulosic biomass, mainly comprising of 66.97 % of holocellulose (cellulose and hemicellulose) and 24.45 % of lignin. This present work aimed to hydrolyze cellulose present in OPEFB to form glucose with the aid of *Aspergillus niger*. *A. niger* is a type of filamentous fungi able to produce cellulase, a multi-enzyme complex consisting of an endoglucanase, exoglucanase, and β-glucosidase, able to converting cellulose into glucose. The glucose produced is then fermented to produce bioethanol. The present study compared hydrolytic activity of cellulose between OPEFB with pretreatment using NaOH 10 % and OPEFB without pretreatment, concerning temperature, pH, and hydrolysis time. The concentration of reducing sugar derived from cellulosic hydrolysis was determined by using a glucose assay of 3.5-dinitrosalicylic acid. The results showed that the optimum temperature for hydrolysis of cellulose OPEFB (pretreated and untreated) was at 40 °C and the optimum pH was 5.0 for OPEFB-untreated and 5.5 for OPEFB-pretreated. Hydrolysis of cellulose at 40 °C and 3 d yielded reducing sugar 13.01 mg mL^{-1} and 1.16 mg mL^{-1} for OPEFB-untreated and OPEFB-pretreated, respectively.

Keywords: 2nd bioethanol generation, biomass, cellulose hydrolysis, lignin, renewable energy

1 Introduction

Bioethanol constitutes an advanced source of renewable energy which may potentially replace the use of fossil fuel. Ethanol for transportation fuel can contribute to the alleviation of the greenhouse effect in the atmosphere [1]. Agricultural and forestry biomass can be an imperative source since they are rich in cheap lignocellulose which is potentially used as a substrate for fermentation in producing bioethanol [2].

Oil Palm Empty Fruit Bunches (OPEFB) is considered as a residual problem derived from palm oil production, and it is present in a large quantity. OPEFB contains a significant portion of cellulose and hemicellulose (66.97 %), and lignin (24.45 %) [3]. Cellulose and hemicellulose are possible to convert into fermentable sugars.

* Corresponding author: sri_sugiwati@yahoo.co.id

Hydrolysis process becomes of a foremost factor in bioconversion of lignocellulose into bioethanol. Commonly, the stage is performed using two methods: acid and enzymatic hydrolysis using cellulases [4]. Acid hydrolysis commonly used sulphuric and hydrochloric acids as catalysts for hydrolysis of lignocellulosic biomass [5]. Acid hydrolysis often uses either concentrated acids or diluted acids. When concentrated acid is used, it induces corrosion in the instruments, while diluted acid may need a higher temperature to decompose hemicellulose which leads to the formation of toxic compounds such as furfural and 5-hydroxymethyl furfural (HMF). Presence of these noxious components is harmful to the growth of yeasts and then retards fermentation, enabling to alleviate the production of ethanol [6]. On the other hand, an enzymatic process for decomposing cellulose can be more satisfying in comparison with inorganic acids, considering that enzyme catalytic reaction works at the specific substrate and can be performed in mild process conditions [4].

Enzymatic hydrolysis of cellulosic biomass is carried out using cellulase which is a multienzyme complex consisting of Carboxymethyl cellulase (CMCase) or endo-β-glucanase (EC 3.2.1.4), exo-β-glucanase (EC 3.2.1.91) and β-glucosidase (EC 3.2.1.21) [7]. These enzymes worked synergistically to hydrolyze cellulose into glucose [8].

Filamentous fungi, particularly *Aspergillus niger* [van Tieghem, 1867]and *Trichoderma reesei* [Simmons, 1977] have attracted most attention due to their high cellulase productivity and safe use in industry [9]. In this work, enzyme-induced hydrolysis of OPEFB involved local isolate *A. niger* InaCC F98, obtained from Indonesian Culture Collection (InaCC), Research Center for Biology, Indonesian Institute of Sciences. *A. niger* has been reported capable of hydrolyzing cellulose present in OPEFB since this species enabled to produce cellulase. Prior to hydrolysis, OPEFB used as substrate was pre-treated using NaOH 10 %, while the result was then compared with the untreated substrate. Hydrolytic product was evaluated according to the content of reducing sugar using the 3,5-dinitrosalisilic acid method [10].

2 Materials and methods

2.1 Materials

Oil palm empty fruit bunches (OPEFB) were collected from a palm oil plantation in Palembang, Indonesia. A local fungal isolate of *A. niger* InaCC F98 was obtained from the Indonesian Culture Collection (InaCC), Research Center for Biology Indonesian Institute of Sciences, Cibinong, Bogor. Potato Dextrose Agar (PDA), sodium citrate buffer, dinitrosalicylic acid, sodium potassium tartrate. The experimental parameters included temperature, pH, and hydrolysis time.

2.2 Alkaline Pretreatment of OPEFB

The alkaline pretreatment of OPEFB was conducted using a bench-scale reactor at the Research Center for Chemistry, Indonesian Institute of Sciences (LIPI). 500 g of OPEFB was heated with 2 500 mL of 10 % NaOH solution at 150 °C for 30 min. The pressure was controlled by 4 bar at early heating. After pretreatment, the substrate was pressed in order to separate solidity from liquid fraction. Furthermore, solidity fraction was neutralized until the pH of 7.0 and dried in the oven at 50 °C to 60 °C overnight [11].

2.3 Regeneration of *Aspergillus niger*

A. niger was grown on Potato Dextrose Agar (PDA) slants for 5 d at 28 °C. Furthermore, the spores were harvested by scraping from PDA slants with 10 mL of distilled water to give spore suspension (1×10^7 spores) for the inoculum of the hydrolysis process.

2.4 Hydrolysis of oil palm empty fruit bunches

Sample of 2.5 g each from OPEFB-untreated and alkali pretreated OPEFB was placed in 50 mL of 0.05 M citrate buffer in 100 mL Erlenmeyer flask. Prior to the addition of the spore suspension of *A. niger* to the OPEFB-untreated and alkali pretreated OPEFB, the substrate and buffer mixture was autoclaved for 20 min at 121 °C to sterilize the medium and prevent contamination. After cooling, into each of these Erlenmeyer flasks was inoculated with 1 mL spore suspension (10^7 spores) of *A. niger* InaCC F98 wich 5 d old. Furthermore, these Erlenmeyer flasks containing sterile substrate medium were incubated at 150 rad s^{-1} at defined condition were used to be optimized for 72 h. Samples from each set were taken after 24 h, 48 h, and 72 h and centrifuged for 10 min at 4 °C. The reducing sugars content in the supernatant was measured using dinitrosalicylic acid (DNS) method [10].

To determine the optimal temperature, the hydrolysis process was performed at different temperatures (30 °C, 35 °C, 40 °C, and 45 °C) and an incubation time of 72 h. The hydrolysis optimum pH was determined by adjusting the initial medium pH in range 4.0 to 6.5 using 1 mmol L^{-1} HCl or 1 mmol L^{-1} NaOH. The hydrolysis process was performed at an incubation temperature of 40 °C for 72 h.

3 Results and discussions

In the present study, enzymatic hydrolysis of OPEFB with or without pre-treatment using NaOH 10 % was compared. For hydrolysis, isolate *A. niger* InaCC F98 was applied for its role as producer of cellulase. To achieve optimum results, the hydrolytic process was performed at a variety of temperature of incubation, medium pH, and time of hydrolysis.

3.1 Effect of temperature on hydrolysis

In this part, untreated and treated OPEFB was used at 5 % in the citric buffer (0.05 M; pH 4.5). The optimum temperature for hydrolysis of OPEFB-untreated and alkali-treated OPEFB was determined at different temperatures of 30 °C to 45 °C for 24 h, 48 h and 72 h of incubation. The content of reducing sugars obtained during the hydrolysis of OPEFB at different temperatures of incubation can be seen in Fig. 1 and Fig. 2. The results showed that the hydrolysis rate was increased with the increasing of temperature and incubation time of hydrolysis. Reducing sugars production at 30 °C and 35 °C ranged from 0.68 mg mL^{-1} to 0.93 mg mL^{-1} for OPEFB-untreated and 0.13 – 0.19 mg mL^{-1} for alkali-treated OPEFB after 72 h of incubation. The optimum yield was achieved at 40 °C for 72 h of incubation, resulting in reducing sugar of 13.01 mg mL^{-1} (OPEFB-untreated) and 1.16 mg mL^{-1} (alkali-treated OPEFB).

The usage of temperature (above or below optimum level) significantly alleviated hydrolysis rate. The hydrolysis process of lignocelluloses proceeds at low temperature it would be desirable for industrial application because of less input of energy [12, 13]. Optimum temperature in this study is lower than that found by Sridevi et al. [12], i.e., 50 °C, in which they investigated the usage of the crude enzyme (cellulase) from *A. niger* for depolymerizing sawdust (native and treated). Production of reducing sugar in native and pretreated sawdust after 72 h incubation at 50 °C reached 5.4 % and 14 %, respectively.

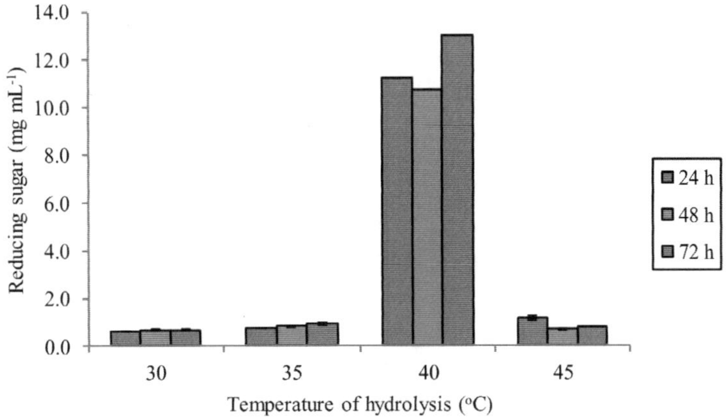

Fig. 1. Effect of temperature on hydrolysis of OPEFB-untreated using *A. niger*.

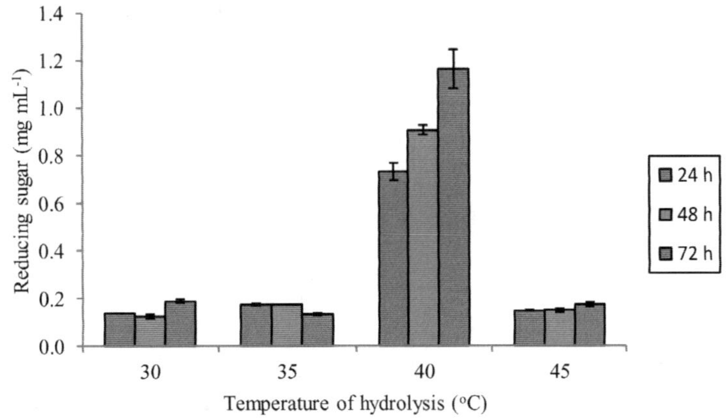

Fig. 2. Effect of temperature on hydrolysis of alkali-treated OPEFB using *A. niger*.

3.2 Effect of pH on hydrolysis

One of the imperative factors in the hydrolysis process is the pH of the medium since it could affect the growth of microorganisms. These inhibitions of microorganism's growth, therefore, affects the production of cellulase and reducing sugars during hydrolysis. The hydrolysis of OPEFB-untreated and alkali-treated OPEFB was carried out in the medium pH range 4.0 to 6.0 to obtain the optimum pH of hydrolysis. The effect of pH on the production of reducing sugars during the hydrolysis of OPEFB can be seen in Fig. 3 and Fig. 4. The results showed that reducing sugars content at the medium pH 4.0 to pH 6.0 after 72 h of incubation were a range of value of 0.74 mg mL^{-1} to 1.58 mg mL^{-1} for

OPEFB-untreated and 0.17 mg mL^{-1} to 0.23 mg mL^{-1} for alkali-pretreated OPEFB, respectively.

The optimum pH value was observed at pH 5.0 (OPEFB-untreated) and pH 5.5 (alkali-treated OPEFB). Similarly, the level of pH was augmented by the former report of Sridevi et al. [12] for hydrolyzing alkali-treated sawdust using crude extract of cellulase from *A. niger*. In this regard, the pH of nutrient medium and substrate sigficantly contributed to the performance of fungi *Aspergillus* growth [14]. This is noteworthy that pH of the medium can properly promote changes in microbial morphology and secretion of cellulase [15]. The enzyme secretion is further important for the production of reducing sugar during hydrolysis.

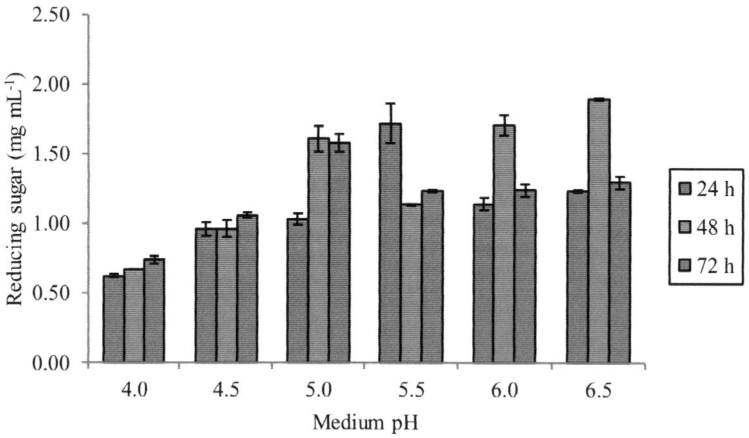

Fig. 3. Effect of medium pH on hydrolysis of OPEFB-untreated using *A. niger*.

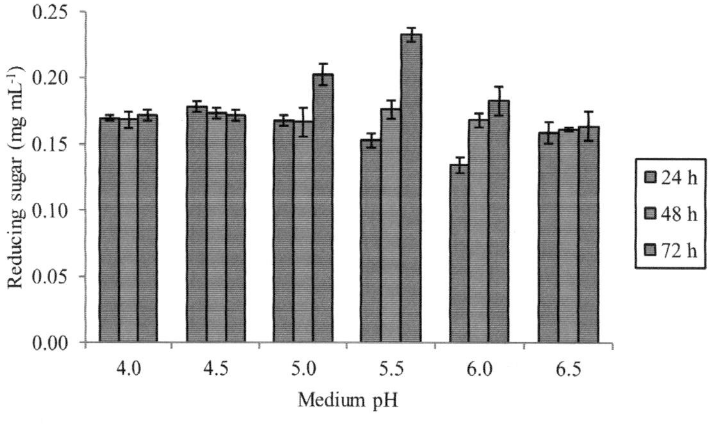

Fig. 4. Effect of medium pH on hydrolysis of alkali-treated OPEFB using *A. niger*.

4 Conclusion

Enzymatic hydrolysis for depolymerization of both untreated and pre-treated OPEFB performed, with the aid of cellulase generated by isolate *A. niger* InaCC F98. The optimum process was achieved at 40 °C, while the pH of medium differed between substrates, namely pH 5.0 for untreated OPEFB and pH 5.5 for pretreated OPEFB. This condition could produce reducing sugar at 13.01 mg mL^{-1} for untreated OPEFB and 1.16 mg mL^{-1} for pre-treated OPEFB after 72 h of incubation.

The authors would like to thank the Korea Institute of Science and Technology (KIST) for providing research funding in the form of KIST school partnership project 2019 (Number: B-797/IPT.2/OT.01/II/2019).

References

1. N. Sarkar, S. K. Ghosh, S. Bannerjee, K. Aikat, Renew. Energ. **37**,1:19–27, (2012). https://doi.org/10.1016/j.renene.2011.06.045
2. A. Mulyaningtyas and W. B. Sediawan, Biocatal. Agric. Biotechnol. **21**:101292(2019). https://doi.org/10.1016/j.bcab.2019.101292
3. A. Rodríguez, L. Serrano, A. Moral, A. Pérez, L. Jiménez, Bioresour. Technol. **99**,6:1743–1749(2008). https://doi.org/10.1016/j.biortech.2007.03.050
4. A. Verardi, I. De Bari, E. Ricca, V. Calabrò, Bioethanol. In Tech:. (2012).p.290. https://www.intechopen.com/books/bioethanol/hydrolysis-of-lignocellulosic-biomass-current-status-of-processes-and-technologies-and-future-perspe
5. P. Lenihan, A. Orozco, E. O'neill, M. Ahmad, D. Rooney, G. Walker, Chem. Eng. J. **156**,2:395–403(2010). https://doi.org/10.1016/j.cej.2009.10.061
6. A. M. J. Kootstra, H. H. Beeftink, E. L. Scott, J. P. Sanders, Biochem. Eng. J. **46**,2:126–131(2009). https://doi.org/10.1016/j.bej.2009.04.020
7. J. Kaur, B. S. Chadha, B. A. Kumar, H. S. Saini, Bioresour. Technol. **98**,1:74–81(2007). https://doi.org/10.1016/j.biortech.2005.11.019
8. H.-Q. Liu, Y. Feng, D.-Q. Zhao and J.-X. Jiang, Biodegradation, **23**:465–472, (2012). https://doi.org/10.1007/s10532-011-9525-6
9. J.J. Abdullah, D. Greethan, N. Pensupa, G.A. Tucker, C. Du, J.Fundam. Renew. Energy Appl. **6**,3:1–10(2016). https://www.longdom.org/archive/jfra-volume-6-issue-3-year-2016.html
10. J. Itelima, A. Ogbonna, S. Pandukur, J. Egbere, A. Salami, (2013). http://hdl.handle.net/123456789/871
11. Y. Sudiyani, J. Waluyo, E. Triwahyuni, D. Burhani, Muryanto, P. Primandaru, et al., *Optimization pretreatment condition of sweet sorghum bagasse for production of second generation bioethanol*. AIP Conference Proceedings. **1803**,1:020015(2017). https://doi.org/10.1063/1.4973142
12. A. Sridevi, G. Narasimha, G. Ramanjaneyulu, K. Dileepkumar, B. R. Reddy, P. S. Devi, 3 Biotech. **5**,883–892, (2015). https://doi.org/10.1007/s13205-015-0284-7
13. P. Pathak, N. K. Bhardwaj, A. K. Singh, Appl. Biochem. Biotechnol. **172**:3776–3797(2014). https://doi.org/10.1007/s12010-014-0758-9
14. K.M. Grigoryan, L.L. Hakobyan, *Proceedings of The Yerevan State University*, **3**, 23–28, (2015). http://eph.am/science/en/1449232721

15. M. Remaz, A. Abd Elrsoul, A. Shami Elhaj, Jordan J. Biol. Sci. **11**:31–36(2017). http://jjbs.hu.edu.jo/files/v11n1/Paper Number 5m.pdf

Lignin and Cellulose Content of Fermented Rice Straw with *Aspergillus niger* (van Tieghem) and *Trichoderma mutan* AA1

Sri Sukaryani[1,*], *Engkus* Ainul Yakin[2], *Yos* Wahyu Harinta[2], *Zane* Vincēviča–Gaile[3], and *Endang* Dwi Purbajanti[4].

[1]Department of Animal Husbandry, Faculty of Agriculture, Veteran Bangun Nusantara University, Jl. Letjen Sujono Humardani No. 1, Sukoharjo 57521, Central Java, Indonesia
[2]Department of Agribusiness, Faculty of Agriculture, Veteran Bangun Nusantara University, Jl. Letjen Sujono Humardani No. 1, Sukoharjo 57521, Central Java, Indonesia
[3]Department of Environmental Science, University of Latvia, Jelgavas Street 1, Room 302, Riga LV-1004, Latvia
[4]Department of Agriculture, Faculty of Animal Science and Agriculture, Diponegoro University, Jl. Prof. Sudarto No.13, Tembalang, Semarang 50275, Central Java, Indonesia

Abstract. The rice straw has potential to be used as an alternative ruminant feed. However, it has limiting factors i.e low crude protein, high crude fiber, lignin, cellulose, and silica content. To overcome the limiting factors, immersion in a solution of alkaline (lime) or fermentation by using inoculum microbial cellulolytic and lignocellulolytic (*Trichoderma mutan AA1* and *Aspergillus niger*.). The research method was experimental, with four treatments and repeated five times. Completely randomized design was used and if there are differences among treatments a further test with DMRT was carried out (level 1 % and 5 %). These treatments were T0: The rice straw without t fermentation; T1: Fermented with *A. niger*; T2: Fermented with *T. mutan* AA1; T3: Fermented with a combination *A. niger* and *T. mutan* AA1. The results showed that the rice straw fermented with *A. niger* and *T. mutants* AA1 very significantly increased the cellulose and decreased lignin content. The highest cellulose reached on T3 (20.297 %) followed by T2 (18.191 %), T1 (17.712 %) and T0 (16.747 %), respectively. While the lowest content of lignin reached on T3 (14.793 %), followed T2 (26.063 %), T1 (26.421 %) and T0 (38.164 %).

Keywords: Fermentation, *fungus*, increase feed quality, inoculum, waste to feed

1 Introduction

The rice production in Indonesia is abundant, namely in 2017, the production of dry rice around 81 382 451 × 10^6 t. Based on the calculations, each of ton dry grain will produce the rice straw with equivalent weight. Therefore, rice straw availability enormous potential in Indonesia not yet utilized optimally for ruminant feed. The area of the rice straw is quite

* Corresponding author: srisukaryani@gmail.com

wide and spread, proper management is needed. But rice straw as feed has a limiting factor i.e. low crude protein, high crude fiber, lignin, and cellulose content [1–3]. Rice straw contains crude protein about 3.80 % [4] and fiber around 39.81 %, lignin around 15 % [5] and silica about 10 % to 15 % [6, 7]. The high fiber content will delay the hydrolysis process by fermentatation, i.e. microbes enzyme in rumen resulting lower absorption [8–10]. Previous study Sukaryani [11] showed that 4 d fermentations of rice straw with MA–11 started by submersion in alkaline solution improved the crude protein from 5.22 % to 8.23 % and decreased the crude fiber from 54.24 % to 38.83 %. While the fermentation of rice straw from 2 d to 6 d by MA–11 starting with submersion of rice straw in the alkaline solution, reducing NDF from 69.22 % to 62.20 % and ADF from 50.21 % become 36.93 % [12].

To overcome the limiting factors of rice straw, an alternative effort that can be done are to provide the treatment of immersion in a solution of alkaline (lime), and the process of the rice straw fermentation by using inoculum microbial cellulolytic and lignocellulolytic (*Trichoderma* sp. and *Aspergillus* sp.). *Trichoderma* sp producing the extracellular prolific protein [15] and it has the ability to produce the enzyme to degradation the cellulose [14, 15]. Similarly, *A. niger* produce some enzymes such as enzyme cellulolytic [14, 15], amylase [18], and glucoamylase [19].

This research aimed to assess the content of lignin and cellulose in fermented rice straw with *A. niger* and *T .mutants* AA1.

2 Materials and methods

The study was conducted in Laboratory Agriculture Faculty Veteran Bangun Nusantara University of Sukoharjo, Central Java, Indonesia.

The material used in the form of rice straw obtained from paddy fields in Karanganyar district, Central Java, Indonesia microbial (*A. niger* and *Trichoderma mutant* AA1), molasses, urea, lime, and water.

The tool used in this study was a bucket and place of manufacture fermentation jar, chopper, blender, and laboratory devices for content analysis lignin and cellulose.

This research uses a design completely randomized design with the pattern of four kinds of treatments and five replications. The treatments are:

T0: The rice straw without fermentation
T1: The rice straw fermented with *A. niger*
T2: The rice straw fermented with *T. mutant* AA1
T3: The rice straw fermented with a combination *T. mutans* AA1 dan *A. niger*
Before fermentation, rice straw was submersed in a lime solution for 2 d.

The variables observed in this study werelignin and cellulose content. The data were analyzed statistically using Analysis of Variance (ANOVA) and followed by Duncan's Multiple Range Test (DMRT).

3 Result and discussions

3.1 Lignin content (%)

The results showed that the lignin content of fermented rice straw with *A. niger, T. mutant* AA1, and the combination were presented in Table 1.

Table 1. Lignin content of rice straw fermented with AA1 *mutan Trichoderma* and *A. niger*

The replication	**Lignin content (%)**			
	T0	**T1**	**T2**	**T3**
1	37.422	25.629	26.915	14.401
2	37.779	27.184	25.780	15.989
3	38.638	26.086	27.601	14.425
4	37.234	26.899	24.161	13.451
5	39.745	26.308	25.860	15.701
Average	38.164 ^{a}	26.421 ^{b}	26.063 ^{b}	14.793 ^{c}

Note: Different superscripts on the same line show highly significant ($p < 0.01$).

Based on Table 1, lignin content in fermented straw with *A. niger* or *T. mutant* AA1 individually or in combination degrade lignin highly significant. After further tested by DMRT showed, the lowest lignin content was reached by T3 (14.793 %), followed by T2 (26.063 %), T1 (26.421 %) and T0 (38.164 %). Treatment T1 and T2 showed lignin content was not significantly different. Visually, the lignin in T2 tends to be lower than T1. This indicates that the *Trichoderma* has higher ability to degrade lignocellulose than the *A. niger*. The content of lignin in T3 is lower significantly than T0, T1, and T2, this phenomenon indicates that both types of fungus can grow and thrive in an atmosphere of the same media and mutual synergy to meet the needs. During the fermentation, *A. niger* and *T. mutant* AA1 produced lignolytic enzyme that degrade lignocellulose, wrapping the cell wall binding micro fibril cellulose and hemicellulose apart, then the lignin in the cell wall that prevents the cellulose down. This result in accordance to Agustini and Efiyanti [20] that the biological treatment of the substrates *Albiziachinensis*, sawdust and palm leaf midrib with fungi FB1 MD–14 isolate lower the lignin content. Likewise, some studies stated fermentation with microbes can degrade lignin to improve the quality and palatability of feed [21–24].

3.2 Cellulose content

The research data from cellulose content listed in Table 2.

Table 2. Cellulose content of rice straw fermented with *T. mutant* AA1 and *A. niger*

The replication	**Cellulose content (%)**			
	T0	T1	T2	T3
1	16.709	17.408	17.860	20705
2	17.812	17.821	18.030	19.142
3	16.012	17.728	18.523	20.225
4	16.725	17.649	18.170	21.046
5	16.479	17.957	18.373	20.369
Average	16.747 ^{a}	17.712 ^{b}	18.191 ^{b}	20.297 ^{c}

Note: Different superscripts on the same line show very significantly different ($p < 0.01$)

Result showed that the treatment of fermentation using *A. niger* and *T. mutant* AA1 very significantly improved the content of cellulose. This is might due to the activity of enzyme produce by cellulolytic has degraded lignocellulose. Cellulose become available for degrading and absorption. According to Prihartini, et al. [25], cellulose content in fermented rice straw tends to improve. In addition by Agustini and Efiyanti [20] that biological treatment that applied to the substrate *A. chinensis* wood, sawdust, and palm leaf midrib with fungi MD–14 FB1 isolate tend to improve cellulose. After being further tested by DMRT, the highest cellulose content was by T3 (20.297 %) and the lowest in T0 (16.747 %).

4 Conclusions

The results of the research can be concluded that:

i. Fermented of the rice straw using *A. niger* and *T. mutant* AA1 with monoculture and a combination of both decreased the lignin and increased cellulose content.
ii. The lowest lignin and highest cellulose content in fermented rice straw occurred in a combination of *A. niger* and *T. mutant* AA1 inoculum.
iii. Fermentation rice straw with *A. niger* and *T. mutant* AA1 in monoculture produced not different lignin and cellulose content.

5 Suggestion

Further research on fermentation can be done by using a combination of fungus *A. niger* and *T. mutant* AA1 to be applied on others agricultural waste.

References

1. M. Bata. Agripet. **8**,2:15–20(2008). [in Bahasa Indonesia]. https://doi.org/10.17969/agripet.v8i2.610
2. D. Aquino, A. Del Barrio, N. X. Trach, N. T. Hai, D. N. Khang, N. T. Toan, et al, in *Sustainable Rice Straw Management*, Springer, Cham, (2020).p.111–129. https://doi.org/10.1007/978-3-030-32373-8
3. N. Ali, Suhartina, A. Muktiani, E. Pangestu, IOP Conference Series: Earth and Environmental Science, **492:**012028(2020). https://doi.org/10.1088/1755-1315/492/1/012028 or http://dx.doi.org/10.1088/1755-1315/492/1/012028
4. S. Ahmed, M. J. Khan, M. Shahjalal, K. M. S. Islam, Asian-Australasian J Anim. Sci, **15**,4:522–527(2002). https://doi.org/10.5713/ajas.2002.522
5. G. Gou, W. Wei, M. Jiang, S. Zhang, T. Lu, X. Xie, et al, Pulp Paper Processing, **6:**133–154(2018). http://dx.doi.org/10.5772/intechopen.79014 or https://www.researchgate.net/publication/328083802_Environmentally_Friendly_Method_for_the_Separation_of_Cellulose_from_Steam-Exploded_Rice_Straw_and_Its_High-Value_Applications/link/5bb6c03aa6fdcc9552d3d8e9/download
6. N. Reddy, Y. Yang, J. agric. food chem. **54**,21:8077–8081(2006). https://doi.org/10.1021/jf0617723
7. M. Bhattacharya, M. K. Mandal, J. Clean. Prod. **186:**241–252(2018). https://doi.org/10.1016/j.jclepro.2018.03.099.
8. X. Tang, G.O. Tayo, Z.L. Tan, Z.H. Sun, L.X. Shen, C.S. Zhou, et al. J. Anim. Sci. **86**,5:1164–1172(2008). https://doi.org/10.2527/jas.2007-0438
9. A. Wahyudi, D. Pamungkas, R. H. Setyobudi, L. Hendraningsih, Z. Vincēviča-Gaile. Proc. Pakistan Acad. Sci. part B,**54**,1:41–45(2017). https://www.paspk.org/wp-content/uploads/2017/03/Organic-Acid-and-Nutrient-Composition-of-Lactic-Acid.pdf

10. A. Wahyudi, L. Hendraningsih, Sutawi, R.H. Setyobudi, M. Mel. IOP Conf. Series: Earth and Environmental Science. **293**:01200(2019). doi:10.1088/1755-1315/293/1/012007
11. S. Sukaryani, Agrisaintifika Jurnal Ilmu-Ilmu Pertanian. **2**,2:160–164(2018).[in Bahasa Indonesia] https://agrisaintifika-fpunivet.ac.id/ojs/index.php/AGRISAINTIFIKA/article/view/108
12. S. Sukaryani, Jurnal Ilmiah Teknosains, **2**,2:91–94(2016). [in Bahasa Indoensia] https://doi.org/10.26877/jitek.v2i2/Nov.1198 or http://journal.upgris.ac.id/index.php/JITEK/article/view/1198/1060
13. A. Rodriguez-Iglesias, M. Schmoll, Scientific Reports, **9**,1:10995(2019). https://doi.org/10.1038/s41598-019-47421-z
14. G. E. Harman, C. R. Howell, A. Viterbo, I. Chet, M. Lorito, Nature reviews microbiology, **2**:43–56(2004). https://doi.org/10.1038/nrmicro797
15. S. Chakraborty, R. Gupta, K. K. Jain, R. C. Kuhad, Bioprocess Biosyst. Eng., **39**:1659–1670(2016). https://doi.org/10.1007/s00449-016-1641-6
16. T. C. D. Santos, G. Abreu Filho, A. R. D. Brito, A. J. V. Pires, R. C. F. Bonomo, M. Franco, Revista Caatinga, **29**,1:223–233(2016). https://doi.org/10.1590/1983-21252016v29n126rc
17. S. Vidmar, V. Turk, I. Kregar, Appl. Microbiol. Biotechnol. **20**:326–330 (1984). https://doi.org/10.1007/BF00270594
18. K. Mihajlovski, M. Rajilić-Stojanović and S. Dimitrijević-Branković, Renew. Energ. **152**:627–633(2020). https://doi.org/10.1016/j.renene.2020.01.101
19. K. Ratanaphadit, K. Kaewjan and S. Palakas, Asia-Pacific J. Sci. Technol. **15**,9:833-842(2010). https://so01.tci-thaijo.org/index.php/APST/article/view/84065
20. L. Agustini, L. Efiyanti, Jurnal Penelitian Hasil Hutan, **33**,1:69–80(2015). [in Bahasa Indonesia]. https://doi.org/10.20886/jphh.2015.33.1.69-80 or http://ejournal.forda-mof.org/ejournal-litbang/index.php/JPHH/article/view/640/622
21. S. Rulianah, H. Profiyanty, Y. Maryanty, B. Irawan, Adv. Sci. Lett. **23**,6:5666–5668(2017). https://doi.org/10.1166/asl.2017.8796
22. B.-S. Xing, Y. Han, S. Cao, J. Wen, K. Zhang, H. Yuan, X. C. Wang, Chemosphere, **250**:126104(2020). https://doi.org/10.1016/j.chemosphere.2020.126104 or http://www.sciencedirect.com/science/article/pii/S0045653520302976
23. M. U. Khan, B. K. Ahring, Biomass Bioenergy, **128**:105325(2019). https://doi.org/10.1016/j.biombioe.2019.105325 or http://www.sciencedirect.com/science/article/pii/S0961953419302740
24. M. Asgher, A. Wahab, M. Bilal, H. M. Nasir Iqbal, Biocatal. Agric. Biotechnol. **6**:195–201(2016). https://doi.org/10.1016/j.bcab.2016.04.003 or http://www.sciencedirect.com/science/article/pii/S1878818116300597
25. I. Prihartini, S. Soebarinoto, S. Chuzaemi, M. Winugroho, Anim. Prod. **11**,1:1–7(2011). [in Bahasa Indonesia]. http://animalproduction.net/index.php/JAP/article/view/215/203

Multiplication Arbuscular Mycorrhizal Fungi in Corn (*Zea mays* L.) with Pots Culture at Greenhouse

Sukmawati Sukmawati[1,*], *Adnyana* Adnyana[2], *Dewa* Nengah Suprapta[2], *Meitini* Proborini[3], *Peeyush* Soni[4], and *Praptiningsih* Gamawati Adinurani[5]

[1]Faculty of Agriculture, Nahdlatul Wathan Mataram University, Jl. Kaktus No.1-3, Mataram 83126, Indonesia
[2]Faculty of Agriculture, Udayana University, Jl. PB Sudirman, Denpasar 80113, Indonesia
[3]Faculty of Mathematic and Science, Udayana University. Jl. PB Sudirman, Denpasar 80113, Indonesia
[4]Department of Agricultural and Food Engineering, Indian Institute of Technology Kharagpur, 21302 Kharagpur, West Bengal, India
[5]Department of Agrotechnology, Merdeka University of Madiun, Jl. Serayu No.79, Madiun 63133, East Java, Indonesia

Abstract. This study was multiplicated by Arbuscular Mycorrhizal Fungi (AMF) indigenous in corn with pots culture at the greenhouse. The research will be conducted from August 2019 to October 2019 in Greenhouse, Laboratory of Microbiology, Mataram University, Indonesia. This research aims to determine the influence of AMF in dry land and the application of fertilizer concentration. This research was conducted isolate exploration in four villages at the Pujut Central district, Lombok, Indonesia i.e. Mertak, Sukadana, Kuta, and Sengkol Village. This research is an experimental study with a completely randomized factorial design with two factors i.e the AMF isolate type and the concentration of Johnson's nutrient solution. The first factor with the level without AMF Isolates, Isolate 1, Isolate 2, and Isolate 3. While the second factor is the Johnson nutrient concentration i.e 50 % and 75 % solution. The results showed that were differences in growth such as crop height and the number of leaves where Isolate 1, gave the highest growth and number of leaves. The identification was obtained the Isolate 1 showed highest spore's density and root infections is Isolate 1 with a spherical shape.

Keyword: Environmentally friendly, increase productivity in dry land, microbial fertilizer, phosphorus uptake

1 Introduction

Arbuscular mycorrhizal fungi (AMF), phylum Glomeromycota, are obligate root symbionts that are present in most terrestrial ecosystems and establish a mutualistic symbiosis with several plant species around the world [1]. They produce structures inside plant roots (e.g.,

* Corresponding author: sukmawatiNW69@gmail.com

arbuscules), thus having an important role in plant mineral nutrition, (e.g., P-uptake, N-uptake and micronutrients- uptake) and water absorption [2, 3], resulting in increased plant growth, resistance, and tolerance to abiotic and biotic stresses, such as soil-borne pathogens and drought [4].

Plants continuously interact with other microorganisms present in their environment [5, 6]. Moreover, they are able to establish mutually beneficial associations with some of these microorganisms present in the rhizosphere. One of the most well studied benefificial plant microorganism associations is that established with certain soil fungi known as AMF [2]. Interestingly, the vast majority of land plants, including most agricultural crop species, are able to establish AM symbiosis [2]. It positively affects plant growth and provides tolerance against biotic and abiotic stresses [7]. AMF is known as soil fungi in the rhizosphere area. AMF was a mutualism fungus with plants root i.e. Ectomycorrhizal and Endomikoriza [2]. Besides that, AMF work in poor soils to uptake and transport nutrients (macro and micro), although mycorrhizal association role related to carbon flux is less defined [8, 9]. AMF may live symbiotically on the roots of approximately 80 % of different plant types [10, 11].

AMF provided many benefits, especially on marginal land, i.e., benefits for ecosystems, crops, and humans. That application for the ecosystem increased to the nutrient cycle, improved the soil structure, and channeling carbohydrates from plant roots to other soil organisms. Furthermore, the advantages for plants increased the absorption of nutrients, mainly Myxedema may emit phosphatase and acid-organic acids. [11], crop resistance and tolerance to the abiotics and biotics such as dryness and soil pathogens. Based on the role, AMF is the potential to develop as a biological fertilizer for grew and productivity. The AMF multiplication developed as a biological fertilizer [12, 13]. AMF multiplication by pots culture and a single culture in laboratory [14, 15]. The reproduction of spores by pot culture used as a source of spores, then sterilized and used as a sterile incarnated source for the development and reproduction of the AMF of axiom in vitro cultures [16, 17].

The AMF multiplication factor, i.e., as compatible hosts, places to grow, and the environment. The no less important is the high power of infection and effectiveness, colonization of host roots rapidly and produce many spores [2, 16]. AMF did not choose a specific host, all plants are potentially infected, but the degree of infectivity and effectiveness differs from each host association with AMF. The AMF added to infected and colonized in root plant, type of host plant were *Sorghum bicolor* L., *Zea Mays* L.,*Panicum maximum* Jacq., *Paspalum notatum* L., *Arachis hypogaea* L., and *Pueraria javanica* Roxb. Benth. [17, 18] with various planting media included soil, sand, peat, clay and zeolite [13]. The selection of Inoculum and host plants is the key to success in the cooperation system of both organisms.

The factors of AMF multiplication was a plant variety, AMF type, and fertilizer application [2, 18]. Dosage Phosphate influenced by the form of Phosphate used, inorganic, as well as organic or insoluble and insoluble in soil [2, 15]. Generally, fertilizers used in AMF spores production was nutrient solutions with low Phosphate levels.

This study aims to determine the influence of the isolates and concentrations of Johnson's nutrients against i) plant growth, ii) soil chemical properties, and iii) spores' density.

2 Methods

This research conducted by a Completely Randomized Factorial Design (CRFD) with three repeated. The first factor with the level without AMF Isolates, Isolate 1, Isolate 2, and Isolate 3. While the second factor is the Johnson nutrient concentration i.e 50 % and 75 % solution. This research was conducted in Faculty of Agriculture, University of Nahdlatul

Wathan, Laboratory of West Nusa Tenggara Assessment Institute for Agricultural Technology and Microbiology Laboratory, Faculty of Agriculture, University of Mataram, Indonesia. That research was from August 2019 to October 2019.

The characteristics three types of isolat based by [19]:

2.2.1 *Glomus* sp.

Spores reddish brown, spores slightly rounded form. The spore wall consists of two layers namely layer A and layer B. Layer A is the first layer of thin membrane. Layer B is the second layer. *Glomus* spp. morphology of microscopy. Shown in Figure 1b

2.2.2 *Gigaspora* sp.

The sprouts produced from the spore wall, have spores such as thorns and have a single spore, a globus-shaped, ovoid or irregular, and a yellowish wall. Its microscopic morphology can be seen in Figure 1a

2.2.3 *Acaulospora* sp.

Spores colored brownish-yellow. The spores' developmental process seems to be from Hyfa but it is not so. First there is a hypha whose tip is enlarged like a spore called hyphae terminus. Show in Figure 1c

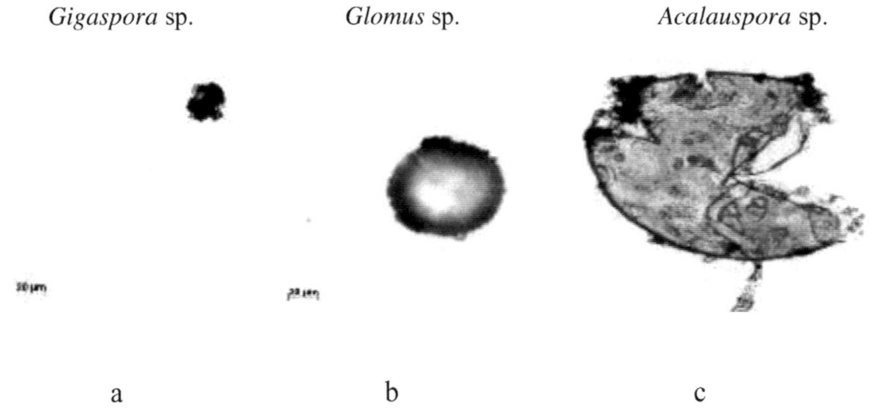

Fig. 1. Type of mycorrhiza

Data Analysis used Analysis of Variance (ANOVA). The conclusion decides based on the F test shown the significance of data and application, then the continued analysis with Tukey's HSD (honestly significant difference) test. The observational variables. The observational variables consist of two parts, i.e., plant growth included high crop, a broad index of leaves, leaf characterization, spore density and mycorrhizal diversity.

3 Results and discussions

3.1 Plant height (cm)

This study observed two times measurements to plants height followed the stages of corn growth. The plant height observed at 14 days after planting (DAP) and 28 DAP. At 14 DAP Corn has entered the V3 to V5 (the number of leaves opened was three strands to five strands), This phase occurs at 10 d to 18 d after germination, and the seminal root grow stopped, furthermore the primary and lateral of the root to be active on the grows below ground level. Ground temperature affected the growing point, while the low temperature would slow out the leaves, increased the number of leaves, and delayed the formation of male flowers [20].

While at 28 DAP of the V6 to V10 leaves was opened in six strands to ten strands. This phase takes place at the time of the plant between 18 d to 35 d after germination. The growth point is already above ground, the development of roots and quick spread on the ground, and the stem enlargement increased rapidly. In this phase will be male flowers (tassel) and the development of cob begins. Plants begin to absorb the nutrients in more quantities. Therefore, fertilization in this phase is necessary to provide nutrient needs for plants.

The results of diversity analysis show there is a noticeable difference in the height of the plant at the age of 28 DAP that is seen in isolates factors, dosing factors and Interaction both.

3.2 Leaf area index

Leaf area index (LAI) indicated crop potentials in photosynthesis. It can directly affect growth and development of plants. The leaves wider in a plant thus maximum absorption of light. According to Gardner and Pearce [20], The wide index range of leaves that are optimal for cultivation plants is from a value of three to a value of five. Based on Table 2, the treatment of phosphate rocks at a dose of 50 % from the recommendation with the application of biological fertilizers as much as two times tends to give better results to the broad index of leaves compared with control and other treatments. Nevertheless, all treatments do not give a noticeable effect on the broad index of leaves.

Plants require large quantities of P, which they obtain as phosphate (Pi) from the solution phase of the soil. While many soils have a high P content, it occurs mostly as complex organic or inorganic forms that are not directly accessible to the plant, and Pi levels in the soil solution are below 10 μM (21). The addition of phosphate nutrients in the corn plant did not encourage an increased increase in the area of the leaf because the phosphate nutrient was not a broad index boundary of the leave. That's factors affected the magnitude of the broad index of leaves include the spacing and supply of nitrogen nutrients. There is no difference in spacing and fertilizer doses of nitrogen nutrient providers used by all the treatments in this experiment.

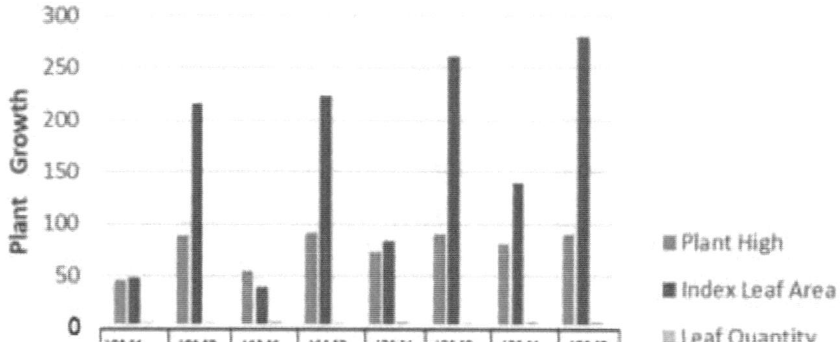

	I0M1	I0M2	I1M1	I1M2	I2M1	I2M2	I3M1	I3M2
Plant High	45.4	88.67	55	91	73	90.67	81.67	90.3
Index Leaf Area	47.86	216.1	39.94	222.7	84	260.8	140	280
Leaf Quantity	5	5	7	5	7	5	6	6

Fig 2. Influence type of isolate and concentration of Johnson nutrie solution on plant growth

3.3 Soil analysis

The distribution of Mycorrhiza is influenced by many factors, such as soil type and structure, nutrient elements Phosphate and Nitrogen in soil, water, pH, and soil temperature. the results was content analysis and soil conditions at all three sampling locations that can be seen in Table 1.

Table 1. Soil analysis

Soil analysis	Nutrient value
The water	8.39
pH	6.83
N (%)	0.30
P (mg kg^{-1})	15.30

The physical properties of other soils are pH. That showed pH is approaching neutral and the analysis of physical and chemical properties the soil aimed to determine the existence of mycosis. The state of the soil is very high in population, colonization and Mycorrhiza.

Based on the results of chemical analysis of soil is known that soil pH found in research area classified as neutral. The content of elements N and P availability of very low conditions.

Mycorrhiza is influenced by several environmental factors such as light, temperature, groundwater content, soil pH, organic matter, and heavy metals and other elements [21]. From the results of soil analysis, have a soil pH content of H_2O (1:1.5) 6.83 (neutral), P Bray-1 15.3 mg L^{-1} (very low), N total 0.30 % (low).

The number of spores and Mycorrhiza type is closely related to soil chemical conditions. In the range of pH 6.85, The number and type of Mycorrhiza is increasing [22]. Generally mycorrhizal is more resistant to soil pH changes. However, the pH changes in the soil rhizosphere have a direct impact on the solubility of Al in the soil. The more acidic soil pH, the rate Al in the soil is increasing, and this has an impact on the decrease in the amount and type of Myoriza. This is because Al is able to inhibit root development. As a result, plants easily experience water, stunted growth, and biomass and low productivity [23].

At soil pH 4.5 to 8.0, P, and C-organic increases, then the number and type of Mycorrhiza will increase. This is because pH determines the ease of nutrient absorbed plants including the element P, where P serves for cell division, assists in the transfer of energy in metabolic activities, resulting in good plant growth, and ultimately assists the development of Mycorrhizal. C-Organic can also guarantee the occurrence of mineralization which results can provide nutrients for symbiotic mybiosis with plants, besides organic matter can induce the growth of hypha mycorrhizal [22].

Based on the results of analysis in the Laboratory (Table 1) on the soilchemical properties indicates the level of soil (pH) of the average land reaches 6.06. Soil chemical properties are known to affect the ability of mycolic fungi associated with plants strongly, this is in accordance with the statement Prihastuti [23], Mycorrhizal can live well in sour soil pH and able to produce organic acids that Freed P fixation.

Phosphor content in soil is known to affect the variety of colonization of microchemical fungi in plant roots. The results of soil analysis against P are available in the soil, indicating very low criteria of 0.30 mg L^{-1}. Soil that contains a high element P is often associated with the decline of Mycorrhizal colonization. The formation of a symbiosis of arbuscular mycosis fungi reaches a maximum if the level P in the soil is not greater than 50 mg L^{-1} [15].

Soils containing high nitrogen and phosphate then the amount of his mycorrhiza is also high. This is because according to Setiadi [24] wherein soil nitrogen comes from soil organic matter, binding by microorganisms from air nitrogen, fertilizer, and rainwater, so that the nitrogen content in the soil is strongly correlated with the amount of Mycorrhizal found. Similarly, the content of phosphate, in which the research was obtained an average value of 0.26. This is because phosphate is a nutrient that is very influential in the presence of Mycophoriza, where phosphate is a macronutrient for plants so that the existence of Mycorrhiza is very helpful in plant growth where hypha Mycorrhizal assists in the absorption of phosphate in the soil. Phosphate that has been absorbed by the external Hypha will soon be converted into a polyphosphate compound and transferred into the internal Hyfa and Arbuskul. In the Arbuskula, polyphosphate compounds are broken down into organic phosphate which are then released into host plant cells.

High or very low moisture and groundwater content are also less good for Mycorrhiza development. Mikoriza develops in humidity and moisture content that is stable, not too high and not too low. When the moisture content and humidity is high or excessive can cause anaerobic conditions that inhibit the development of mycodone because all microform fungi are aerated [25]. Meanwhile, low groundwater content causes dryland conditions. Dryland is very supportive for the development of Mycorrhizal, where the availability of low nutrient on the condition of dry land will optimize the development of Hypha mycorrhizal [26].

Further, Setiadi [24] reported that Mycorrhiza was instrumental in increasing crop tolerance to critical land conditions, drought and heavy metals. The colonization of plant roots with Mycorrhizae may affect communities associated with direct and indirect roots. Direct interactions include the provision of energy-rich carbon compounds, changes in mycorrhizosphere pH, nutritional competitions, and fungal exudation from inhibition or stimulation of compounds. Direct interaction can also occur in the form of mycorrhizal effect on the growth of host crops, the results of root exudation and improvement of soil structure.

3.4 Density and diversity of 'Arbuscular Mycorrhiza Fungi' spore

Diversity of AM fungi with host plant which impart significant impression on plant nutritional status, morphology, genetic expression and symbiotic efficiency in symbiotic relation [27–29].

Mycorrhizal spores' populations i.e., Glomus was high population also thought to be due to environmental conditions that are more suitable, optimal, and compatible in supporting the growth and development of spores. Supported again by the possibility of the absence of antagonistic mushrooms that inhibit mycorrhiza sporulation. Other factors, environmental differences from soil, nutrients, altitude, rainfall, light on both soil examples, allow the difference of spore's density.

In areas that have higher nutrient elements and more loose soil structure found a diversity of type and the number is relatively lower. This is because when the nutrients are insufficient condition, the root of the plant can act as a nutrient absorbent organ so that the plant accumulates nutrient elements in high quantities. The condition will cause a negative response to Mycorriza colonization. This information was emphasized by Subiksa [30], that the fertile media and the increasing P elements in the soil can decrease the activity and infection of mycosis, and even the population will be reduced due to some death.

The Genus Glomus, Acaulospora, and Gigaspora found in the rooting (the rhizosphere) of corn plants allegedly have the suitability of environmental conditions for spores development. This suggests that the genus Glomus and the genus Acaulospora can thrive. According to Brundrett et al. [31], this type of soil and host plant type affects the genus Mycorrhizal found and its effectiveness towards host crops. The dispersion pattern of each MVA mushroom genus is different. Certain genera have a very widespread and a genus of limited distribution. The genus known to have the most widespread dispersion is the genus Glomus.

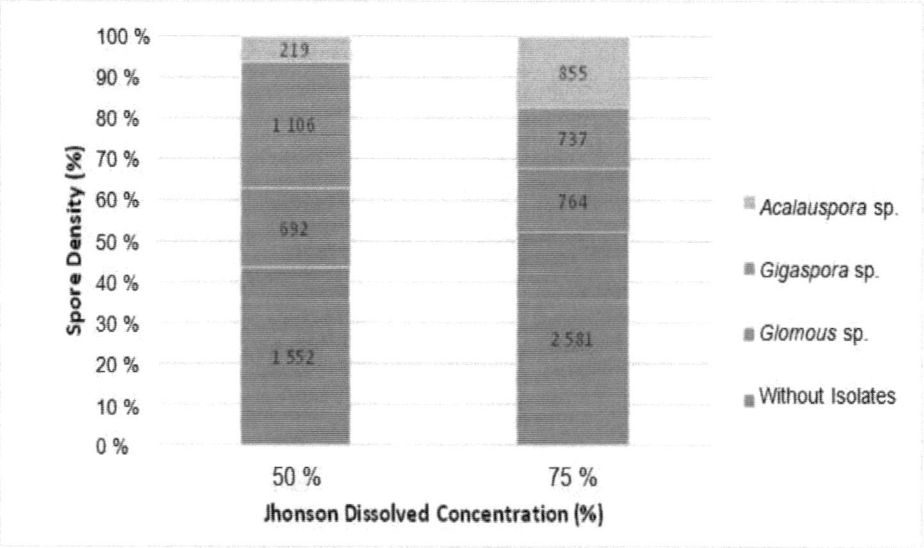

Fig.3. Density of spore inoculation treatment of AMF and Johnson nutrien solution

The Figure 3 above, shows that changes in concentration of the Johnson nutrien solution affect the number of spores. A further 5 % BNJ test was carried out which showed that the concentration of the Johnson nutrient solution 50 % produced the highest number of Gigaspora spores of the three types of spore inoculated. The treatment with high nutrient concentration found a lower number of spores. This is presumably because the nutrients are in sufficient condition so that the roots of the plant can act as a nutrient absorbing organ. Plant will accumulate nutrien elemen in high amounts. This condition will cause negative response to colonization. This information is emphasized In the arbuscular mycorrhizal (AM) symbiosis the reciprocal exchange of nutrients results in a nutritional benefit for both symbionts. The fungus acquires carbon from plant and the plant obtains mineral nutrients from the fungus. While there is evidence for the transfer of phosphorus (P), nitrogen, zinc and copper, current data suggest that P is transferred in the highest quantities and that symbiotic P transfer occurs in the vast majority of AM symbioses [2]. To strongly corelated at change P with colocalized in AM Fungi [32]. Expression of Pt4 and transfer of phosphate are preconditions to establish successful symbiotic relationship and arbuscule formation [33].

4 Conclusion

From the results of the research and the discussion can be withdrawn several conclusions as follows

i. Isolate and dosage factors affect the growth of corn crops such as high crop age 14 DAP and 28 DAP, and Broadleaf index
ii. pH soil, P nutrien content affects the number of spores in the soil, so the higher the concentration of P nutrient, the number of spores in the soil will be low
iii. There was an increse in the number of spores in the soil with the most types *Glomus*, *Gigaspora, Acalauspora*

References

1. Y. Lekberg, S.M. Gibbons, S. Rosendahl, P.W. Rawsey. ISME J. **7**:1423–1433(2013). https://pubmed.ncbi.nlm.nih.gov/23486251/
2. S.E. Smith, D.J. Read. *Mycorrhizal symbiosis*. Third ed. Academic Press. USA(2008).p.769. https://books.google.co.id/books?hl=id&lr=&id=qLciOJaG0C4C&oi=fnd&pg=PP1&dq=6.%09Smith+SE+and+DJ+Read.+2008.+Mycorrhizal+symbiosis.+Third+ed.+Academic+Press.+USA&ots=zrrQlSREIP&sig=basveWgW_gcvBxGEzvFp2nm519w&redir_esc=y#v=onepage&q&f=false
3. A. Hodge, K. Storer. Plant Soil. **386**:1–19(2014). https://link.springer.com/article/10.1007/s11104-014-2162-1
4. L.A. Harrier, C.A. Watson. Pest Manag. Sci. **60**:149–157(2004). https://nph.onlinelibrary.wiley.com/doi/pdf/10.1002/ps.820
5. J.M. Raaijmakers, T.C. Paulitz, C. Steinberg, C. Alabouvette, Y. Moenne-Loccoz. Plant Soil. **321**:341–361(2009). https://link.springer.com/content/pdf/10.1007/s11104-008-9568-6.pdf
6. J.A.López-Ráez M.J., Ozo, J.M. García-Garrido. Botany. **89**:513–22(2011). https://www.nrcresearchpress.com/doi/abs/10.1139/b11-046
7. M.J. Pozo, C. Azcón-Aguilar. Curr. Opin. Plant Biol. **10**:393–398 (2007). https://pubmed.ncbi.nlm.nih.gov/17658291/

8. M-A. Selosse, M. Roy. Trends Plant Sci. **14**:64–70 (2009). https://doi.org/10.1016/j.tplants.2008.11.004 or https://www.sciencedirect.com/science/article/pii/S136013850900017X
9. M. Kayama, T. Yamanaka. Forests. **28**,2:569–583(2016). https://www.mdpi.com/1999-4907/7/11/266
10. H. Koltai, Y. Kapulnik. *Arbuscular Mychorrizas: Physiology and Function.* 2th Edition. Springer.(2010).p.323. https://link.springer.com/book/10.1007/978-90-481-9489-6#about
11. T. Souza. *Handbook of Arbuscular Mycorrhizal Fungi.* Springer(2015).p.153. https://www.springer.com/gp/book/9783319248486
12. S.E. Smith, I. Jakobsen, M. Grolund, F.A. Smith. 2008. Mycorrhizal symbiosis. Third ed. Academic Press USA. https://www.elseiver.com.
13. A.S.d. Santana, U.M.T. Cavalcante, E.V.d.S Sampaio, L.C. Maia. Brazilia J. Botanical. **37**:159–165(2014). http://ojs.uho.ac.id/index.php/green/article/download/3880/2960
14. Husna, D. Mey, T. Yulistiati. Jurnal Agriplus XIII. 3:193–198(2004). [in Bahasa Indonesia] http://ojs.uho.ac.id/index.php/green/article/download/3880/2960
15. A.D. Nusantara, Y.H. Bertham, I. Mansur. *Bekerja dengan Fungi Mikoriza Arbuskula*. SEAMEO BIOTROP: Bogor. (2012).p.83. [in Bahasa Indonesia]. https://biotrop.org/publication/show/bekerja-dengan-fungi-mikoriza-arbuskula
16. P.G. Adinurani, S. Rahayu, L. S. Budi, A. Nindita, P. Soni, M. Mel,. MATEC Web Conference. **164**,01035:1–5(2018). https://doi.org/10.1051/matecconf/201816401035Prapti
17. P.G. Adinurani, S. Rahayu, L. S. Budi, S. Pambudi, P. Soni, IOP Conference Series: Earth and Environmental Science. **293**,012032:1–7(2019). doi:10.1088/1755-1315/293/1/012032
18. N.A. Subekti, Syafruddin, R. Efendi, S. Sunarti. Universitas Lampung. (2011).p. 16–28. [in Bahasa Indonesia]. https://www.coursehero.com/file/18800973/Subekti/
19. INVAM, *International Culture Collection of Vesikular Arbuscular Mycorrhizal Fungi.*West Virginia University. (2005).p.77. https://invam.wvu.edu.
20. F.P. Gardner, R.B. Pierce, R.L. Mitchell. *Physiology of Cultivated Plants.*(1985). hcs.osu.edu.hcs 5621
21. M. Correia, S. Castro, S.R. Echeverria. Aust. J. Bot. **63**,5:387–391(2015). https://www.publish.csiro.au/bt/bt14318
22. I.R. Sastrahidayat, *Rekayasa Pupuk Hayati Mikoriza dalam Meningkatkan Produksi Pertanian* [Mycorrhizal Biofertilizer Engineering in increasing Agricultural Production]. Malang: Universitas Brawijaya Press. (2011).p.238. [in Bahasa Indonesia] https://books.google.co.id/books?hl=id&lr=&id=98KZDwAAQBAJ&oi=fnd&pg=PR5&dq=23.%09Sastrahidayat,+I.+R.+(2011).+Rekayasa+Pupuk+Hayati+Mikoriza+dalam+meningkatkan+Produksi+Pertanian.+Malang:+Universitas+Brawijaya+Press.+&ots=GpPzW9WC_l&sig=i74CXF5m4AKa-3F843o6TneICDI&redir_esc=y#v=onepage&q&f=false
23. Prihastuti, J. Biol. Res. **12**:99–106(2007). [in Bahasa Indonesia]. https://www.researchgate.net/publication/265569928_Isolasi_dan_karakterisa

24. Setia N. Pangaribuan. Jurnal Agro. **1,**1:50–60(2014). [in Bahasa Indonesia] https://www.google.com/url?sa=t&rct=j&q=&esrc=s&source=web&cd=&ved=2ahUKEwig37jLx4DsAhUV7nMBHdAnBgAQFjAAegQIBhAB&url=https%3A%2F%2Fjournal.uinsgd.ac.id%2Findex.php%2Fja%2Farticle%2Fdownload%2F81%2F125&usg=AOvVaw0jqIixNZ9jpaKkVkgiWjBj
25. E. Handayanto, K. Hairiyah, *Biologi Tanah Landasan Pengelolaan Tanah Sehat.* [Soil Biology Foundation for Healthy Soil Management]. Yogyakarta: Pustaka Adipura. (2007). [in Bahasa Indonesia] https://onesearch.id/Record/IOS2847.INLIS000000000006708
26. L. Nurhayati, B. Waryanto, R. Widaningsih,. Outlook Komoditas Pertanian Sub Sektor Tanaman Pangan. Pusat data dan Sistem Informasi Pertanian [Outlook for Agricultural Commodities in the Food Crops Sub-Sector. Data Center and Agricultural Information System] Kementerian Pertanian 2016. (2016).p.103. [in Bahasa Indonesia]. http://perpustakaan.bappenas.go.id/lontar/file?file=digital/192025
27. N. Lu, X. Zhou, M. Cui, M. Yu, J. Zhou, Y Qin, et al. Forest. **6:**734–747(2015). .https://www.mdpi.com/1999-4907/6/3/734/pdf
28. M.B. Oruru, E.M. Njeru. BioMed Res. Int. (2016).p.12. https://www.hindawi.com/journals/bmri/2016/4376240/
29. N. Begum, C. Qin, M.A. Ahanger, S. Raza, M.I. Khan, M. Ashraf, et al. Front Plant Sci. **10**:1068(2019). https://www.ncbi.nlm.nih.gov/pmc/articles/PMC6761482
30. Subiksa, *Penanggulangan Lahan Kritis* [Management of Critical Land] [Thesis Phd]. Institut Pertanian Bogor. Bogor. (2002) [in Bahasa Indonesia] https://www.rudyct.com/PPS702-ipb/04212/igm_subiksa.htm.
31. M. Brundrett, N. Bougher, B. Dell, T. Grove, N. Malayczuk. *Working with Microrhizas in Forestry and Agriculture*. ACIAR Monograph 32. Australian Centre for International Agriculture Research Canberra. (1995).p.374. https://aciar.gov.au/publication/books-and-manuals/working-mycorrhizas-forestry-and-agriculture
32. N. Helber, K. Wippel, N. Sauer, S. Schaarschmidt, B. Hause, N. Requena. Plant Cell.**23,**10:3812-3823(2011). https://www.ncbi.nlm.nih.gov/pmc/articles/PMC3229151
33. F.B. Sessoms, D.S. Floss, S.K. Gomez, N. Pumplin, Y. Ding, V.L. Tremblay, et al. Plant Cell. **27**:1352–1366(2015). http://www.plantcell.org/content/27/4/1352

Modified Off–Season Technology to the Flowering Time and Fruit Yield of *Arumanis* Mango (*Mangifera indica* L.)

Syarif Husen[1,*], *Erny* Ishartati[1], *Muhidin* Muhidin[1], *Devi* Dwi Siskawardani[2], *Anjar* Rizky[1], *Akhmad* Syaifudin[1], *Jumpen* Onthong[3]

[1]Department of Agrotechnology, Faculty of Agriculture and Animal Science, University of Muhammadiyah Malang, Jl. Raya Tlogomas No.246, Malang 65144, Indonesia
[2]Department of Food Technology, Faculty of Agriculture and Animal Science, University of Muhammadiyah Malang,
[3]Department of Earth Science, Faculty of Natural Resources, Prince of Songkla University, 15 Karnjanavanich Rd., Hat Yai, Songkhla 90110, Thailand

Abstract. The purpose of this study was to observe the technology of mango fruit production at off-season used growth regulators. The research was conducted with three treatments namely: i) Paclobutrazol dosage (control, 5 mL L^{-1} per tree, 10 mL L^{-1} per tree and 15 L^{-1} per tree). ii) Ethephon dosages (0 mL L^{-1} per tree, 400 mL L^{-1} per tree, 600 mL L^{-1} per tree and 800 mL L^{-1} per tree) that applied 1 mo after paclobutrazol addition. iii) Paclobutrazol (P) and organic fertilizer (F) application in combination as follows: (control, 5 mL + 10 kg, 10 mL + 15 kg, 15 mL + 20 kg) Randomized Completely Block Designed (RCBD) with five replications was used. The results indicated that i) Paclobutrazol application accelerated the flowering age, shortens the panicle length, and increased fruit yield of the tree. ii) Giving Ethephon accelerated the flowering period, enhanced both the panicles number and yields. The treatment of ethephones 600 mL L^{-1} produced the highest fruit weight, 182.60 kg, respectively iii) The Paclobutrazol and organic fertilizer application indicated significantly ($p < 0.05$) to the panicles number and the fruit weight yield is higher than the control.

Keywords: Ethephon, forcing, organic fertilizer, paclobutrazol, ripener

1 Introduction

Indonesian mango production was increase 2 203 793 t in 2017 to 2 624 791 t in 2018 and the rank is 6th in the world after India, China, Thailand, Pakistan and Mexico [1]. While, in 2011 mango production reached 2 129 608 t, while the exports were 1 485 t, and imports were 989 t. This gap in mango production and export volume indicated that Indonesia's mango is less competitive in the global market [2]. This condition is caused due to i) incompatibility of Indonesian mango quality specifications to world market demand,

* Corresponding author: syarifhusen.hasan@gmail.com

ii) there isn't available comprehensive seed testing system, which able to guarantee the product uniformity, iii) there isn't sustainable breeding program, iv) the farmers didn't apply technological innovation yet, and v) there isn't an institutional system which integrates mango agribusiness component.

Arumanis-143 is one of Indonesia's mango varieties that able to meet market needs for both domestic and export demand. This variety was released in 1985 and in 1990 began to expand widely with the plantation scale business establishment. It was followed in the next 5 yr, *Arumanis*-143 dominated fruit business. However, the productivity and quality of *Arumanis* variety are still low. Technological innovation is needed in regulation out-of-season fruit and fruit wrapping which is still ignored by farmers in Indonesia.

Paclobutrazol is a plant growth regulator and retardant member of triazole fungicide which has antagonist reaction with gibberellin biosynthesis by blocking oxidation from kauren to acid. Paclobutrazol is mainly located in the xylem and accumulated in the leaf [3]. Paclobutrazol can be used to regulate the flowering and fruiting season of *Arumanis* mangoes one week earlier. It is able to enhance flowering branches 8 % to 10 % but has not been able to increase fruiting branches [4]. The Paclobutrazol + organic fertilizer and Ethephon have not been applied yet on *Arumanis* mangoes. Moreover, the addition of fertilizer also important for improving plant growth by repairing soil fertility [5]. This study aimed to determine the technology of mango fruit production at off-season used growth regulators namely Paclobutrazol, Ethephon, and Paclobutrazol + organic fertilizer application on the Arumanis mango plant.

2 Material and methods

The study was conducted in the mango plantation owned by PT. Friga in Pasuruan Regency, Indonesia. The agro-ecological conditions of the location were ± 50 m.a.s.l, the soil type was complex Lathosol soil, climate type D, rainfall 1 332 mm yr^{-1} with 99 rainy days, average temperature 27 °C (21 °C to 34 °C) and 65 % relative humidity. The study was conducted in March to October 2018. The maturity of mango plants that used was 25 yr old. The main materials for plant nutrition were used manure, urea, SP-36, KCL, ZK, NPK and for plant protection used active ingredients azoxitrobil 200 g L^{-1}, diphenoconazole 125 g L^{-1}, moncozeb 80 % with concentration 2 cc L^{-1} to control fungi. The pest control used sipertrin 50 g L^{-1} with concentration 2 cc L^{-1}. The equipment of the pest control used hand sprayer, power sprayer, bucket, stirrer, scale, measuring cup, brush, knife, pruning scissors, and chainsaw), hose, caliper, ruler, scales, bags, plastic, and plant labels.

This research applied Randomized Completely Block Designed on three stages namely: i) Paclobutrazol doses application, (control, 5 mL L^{-1} per tree, 10 mL L^{-1} per tree and 15 mL L^{-1} per tree. ii) Ethephon doses treatment (control, 400 mL L^{-1} per tree, 600 mL L^{-1} per tree, and 800 mL L^{-1} per tree) that applied at 1 mo after paclobutrazol application, and iii) Combination of paclobutrazol (P) and organic fertilizer (F) application (control, 5 mL P + 10 kg F, 10 mL P + 15 kg F, 15 mL P + 20 kg F). Paclobutrazol was applied in dosage according to the treatment and splashed around the plants. Before application, the surrounding soil was cleaned from weeds and other impurities, and implemented in the morning. The parameters observed consisted: i) Flowering (flowering time, flower panicles number, panicle length, panicle width and a number of immature panicles), and ii) crop yields (fruit number, fruit total weight, fruit length and diameter). Furthermore, the data were analyzed by Analysis of Variance (ANOVA) and followed by Least Significant Difference test (LSD).

3 Results and discussions

3.1 Effect of paclobutrazol on flowering and mango yield

The results showed that paclobutrazol accelerated flowering and increased mango yield compared to control (Table 1 and Table 2). The data indicated that faster flowering time after paclobutrazol application, higher more panicles, but crammed panicle widths and a smaller number of immature panicles than without paclobutrazol (control). This result, in accordance to the Totapuri cultivar [6]. Paclobutrazol application gave a significant effect on the fruit number and total weight. But it was no significant difference in the fruit length and diameter. The results were consistent with the study that the application of paclobutrazol 5 mL L^{-1} on a 5 yr old hybrid mango could accelerate the flowering time and increased the number of fruits [7]. Moreover, it also increased Alphonso cultivar [4], Miska cultivar, Mohmoudi cultivar, Totocombo culviar [8] Lechi yield [9]. In Uba mango, paclobutrazol branch tip pruning also enhanched the flowering up to 30 % [10].

Table 1. The effect of paclobutrazol to the flowering parameters

Paclobutrazol (mL L^{-1})	Flowering Parameter				
	Flowering time (d)	Number of panicles	Panicles length	Panicles width	Immature panicles
0	51.27 b	55.6 a	51.27 b	25.4 b	21.6 c
5	49.40 ab	454.6 b	49.40 ab	20.1 a	17.3 ab
10	47.13 a	438.2 b	47.13 a	20.0 a	18.8 b
15	50.13 b	499.6 b	50.13 b	19.4 a	16.1 a
LSD 5 %	2.75	125.95	2.75	3.74	2.24

Noted: The value followed by the same letter in the same column wasn't significantly different according to Duncan's Test ≤ 5 %

Table 2. The effect of paclobutrazol to the fruit yields

Paclobutrazol (mL L^{-1})	Fruit number	Total weight (g)	Length (cm)	Diameter (cm)
0	4.8 a	2.3 a	14.17 a	7.91 a
5	166.6 b	62.7 b	13.26 a	7.48 a
10	176.4 b	65.5 b	13.28 a	7.63 a
15	145.6 b	56.0 b	13.75 a	7.73 a
LSD 5 %	57.12	19.88	1.58	0.36

Noted: The value followed by the same letter in the same column wasn't significantly different according to Duncan's Test ≤ 5 %

3.2 Effect of ethephon on flowering and mango yield

The results indicated that ethephon application increased the number of flower panicles compared without ethephon (control). It also was significantly enhanced the fruit number and yields per tree. Ethephon was reported as one of the buds dormancy breakers that could accelerate flowering in mango [4]. Giving ethephon 400 mL L^{-1} and 600 mL L^{-1} showed a higher yield of mangoes (Tables 3 and Table 4). Blaikie et al. explained that dormant plants could not initiate flowers even though flower buds were induced [3]. The solution is plants have been induced should be given a dormant-breaking agent, therefore it can accelerate the emergence of flower buds. Tandel et al. concluded that spraying ethephon one month after giving paclobutrazol to mangoes, produced the most flowers than control [4]. Giving 400 mL L^{-1} ethephon, a month after being treated with paclobutrazol produces the highest number of flower buds and the fastest flowering time.

Table 3. The effect of ethephon to the flowering parameters

Ethephon (mL L^{-1})	Flowering parameter				
	Number of panicles	Panicles length	Panicles width	Panicles diameter	Immature panicles
0	49.20 a	50.00 c	24.67 a	4.56 a	20.80 b
400	384.40 b	44.27 a	24.53 a	4.48 a	17.27 a
600	315.40 b	45.27b	22.93 a	4.36 a	18.13 a
800	286.80 b	47.67 bc	24.11 a	4.51 a	19.13 a
LSD 5 %	122.80	3.18	2.63	0.15	2.48

Noted: The value followed by the same letter in the same column wasn't significantly different according to Duncan's Test ≤ 5 %

Table 4. The effect of ethephon to the fruit yields

Ethephon (mL L^{-1})	Fruit number	Fruit weight (g)	Weight/tree (kg)
0	59.40 a	397.26 a	23.54 a
400	173.60 b	414.80 a	70.50 bc
600	182.60 c	420.60 a	88.00 c
800	107.60 b	409.60 a	45.00 ab
LSD 5 %	66.84	25	33

Noted: The value followed by the same letter in the same column wasn't significantly different according to Duncan's Test ≤ 5 %

3.3 Effect of paclobutrazol + organic fertilizer on flowering and mango yield

The application of Paclobutrazol followed by the addition of organic fertilizer into the soil markedly accelerated the flowering and mango yield (Tables 5 and Table 6). The results of this study were in accordance with the previous study. Husen et al. stated that the application of paclobutrazol 5 mL L^{-1} on a 5 yr old hybrid mango could accelerate during flowering, and increased the number of fruit yield [5]. Furthermore, Sarker et al. stated that paclobutrazol could improve fruit yield, quality, and accelerate flowering in Amaprali varieties [2]. This result, was similar to the mangosten [11] and Tommy Atkins mango in Ethiopia [12]. Moreover this is due to Paclobutrazol able to increase the total phenolic of terminal buds, and alter the phloem to xylem ratio. This alteration is important in vegetative growth resticting and increasing flowering by assimilate partitioning and nutrient supply [12]

Table 5. The effect of paclobutrazol (p) and organic fertilizer to the flowering

Treatment	**Flowering Parameter**				
	Flowering time (d)	**Number of panicles**	**Panicles length**	**Panicles width**	**Immature panicles**
Control	71.80 b	19.4 a	51.2 a	28.6 b	22.6 b
P 5ml + F10kg	58.20 a	160.8 ab	36.2 b	25.8 b	29.6 c
P 10ml + F 15kg	56.40 a	238.0 b	33.4 a	19.6 a	23.2 b
P 15ml + F 20 kg	57.60 a	210.6 b	36.4 b	18.6 a	18.4 a
LSD 5 %	2.91	150.24	2.39	3.74	4.05

Noted: The value followed by the same letter in the same column wasn't significantly different according to Duncan's Test ≤ 5 %

Table 6. The effect of paclobutrazol (p) and organic fertilizer (f) to the fruit yields

Treatment	**Diameter (cm)**	**Length (cm)**	**Total weight (g)**	**Fruit number**	**Total weight (kg)**
Control	7.44 a	11.90 a	347.8 a	26.4 a	11.6 a
P 5ml + F10kg	7.42 a	11.98 a	388.2 ab	96.4 b	38.1 b
P 10ml + F 15kg	7.51 a	12.34 a	435.6 c	86.2 b	36.1 b
P 15ml + F 20	7.62 a	12.24 a	421.2 bc	142.4 b	52.2 b
LSD 5 %	0.34	0.76	45.03	60.82	21.03

Noted: The value followed by the same letter in the same column wasn't significantly different according to Duncan's Test ≤ 5 %

4 Conclusion

Based on the result, it can be concluded that:

i. Paclobutrazol application accelerated flowering parameters (flowering time, panicle number, panicle length, panicle width, and immature panicle), and increased fruit yield (fruit number and fruit weight) significantly, but insignificant on the length and diameter of fruit.

ii. Giving ethephon accelerated the flowering period, a number of panicles and fruit yields per tree compared to controls. Ethephon concentrations of 400 mL L^{-1} and 600 mL L^{-1} produce the highest weight of the fruit per tree, respectively 70.50 kg and 88.00 kg.

iii. The combination of paclobutrazol and organic fertilizer gave significant effect, and produce higher number of panicles and the fruit yield than control, but insignificant on the fruit diameter and length.

The authors are grateful to the Ministry of Research, Technology, and Higher Education who gave funding Nomor: 015/SP2H/P/K7/KM/2015 in 2015. This study also expresses a big gratitude feeling to the Director of PT. Friga, Pasuruan Regency, Indonesia who facilitated the mango plantation for research.

References

1. Badan Pusat Statistik. *Statistik Tanaman Buah-buahan dan Sayuran Tahunan* [Statistics of Annual Fruit and Vegetable Plants] [Online] from http://www.bps.go.id (2018). [Accessed on 24 July 2020]. [in Bahasa Indonesia]
2. B.C. Sarker, M.A. Rakhim, Bangladesh J. Agric. Res. **43,**1:1–12(2018). DOI: 10.3329/bjar.v43i1.36154
3. S.J. Blaikie, V.J. Kulkarni, W.J. Muller, Scientia Horticulturae. **101,**1–2:51–68(2004). DOI: 10.1016/j.scienta.2003.09.009
4. Y.N. Tandel, N.L. Patel, Karnataka J. Agric. Sci. **24,**3:362–365(2011). https://www.cabdirect.org/cabdirect/abstract/20113240736
5. D.S. Siskawardani, J. Onthong, K. Khawmee, C. Poonpakdee, Agric. Nat. Resour. **50,**4:321–325(2016) DOI: 10.1016/j.anres.2016.01.005
6. K.K. Upreti, Y.T.N. Reddy, S.R. Shivu Prasad, G.V. Bindu, H.L. Jayaram, S. Rajan, Scientia Horticulturae. **150**:414–418(2013). DOI: 10.1016/j.scienta.2012.11.030
7. S. Husen, Kuswanto, S. Ashari, N. Basuki, J. Agric. Food Technol. **2,**9:153–158(2012). https://ei-ado.aciar.gov.au/sites/default/files/HusenEtAl%282012%29InductionFloweringYieldMangoHybridPaclobutrazol.pdf
8. A.O.S.A. Rahim, O.M. Elamin, F.K. Bangerth, ARPN J. Agric. Biol. Sci. **6,**2:55–67(2011). http://arpnjournals.com/jabs/research_papers/rp_2011/jabs_0211_245.pdf
9. A.K. Pandey, P. Singh, S.K. Singh, K. Gupta, Int. J. Current Microbiology Appl. Sci. **6,**8:3280–3288(2017). DOI: 10.20546/ijcmas.2017.608.391
10. G.P. Oliveira, D.L. de Siqueira, L.C.C. Salomão, P.R. Cecon, D.L.M. Machado. Pesquisa Agropecuaria Tropical, **47**:7–14(2017). https://doi.org/10.1590/1983-40632016v4743861
11. K.H. Then, K. Norshafiqah, O.M. Faiz, Int. J. Agric. Innov. Res. **8,**3:245–251. (2019). https://ijair.org/index.php/issues?view=publication&task=show&id=1306
12. T. Yeshitela, P.J. Robbertse, P.J.C. New Zealand J. Crop Hort. Sci. **32,**3:281–293(2004). DOI: 10.1080/01140671.2004.9514307

Entrepreneurship Orientation, Eco-innovation, Information and Communication Technology (ICT) Learning Adoption Capability: A case Study of Food SME's in Central Java, Indonesia.

Vincent Didiek Wiet Aryanto[1,*], *Kunio* Kondo[2], *Yohan* Wismantoro[3], and *Pulung* Nurtantyo Andono[4]

[1]Graduate School of Business Management, Universitas Dian Nuswantoro, Jl. Imam Bonjol No. 207, Semarang 50131, Indonesia
[2]School of Media Science, Tokyo University of Technology Hachioji Campus Tokyo 192-0982 Japan
[3]Department of Management, Faculty of Economics and Business, Universitas Dian Nuswantoro, Jl. Nakula 1 No.5-11 Semarang 50131, Indonesia.
[4]Faculty of Computer Science, Universitas Dian Nuswantoro, Jl. Nakula I No. 5–11, Semarang 50131, Indonesia

Abstract. This study aimed to investigate the entrepreneurship orientation among the Central Java's food small and medium enterprises (SMEs) and its impact on their marketing performance, the business method through eco-innovation. This study also employed eco-innovation and Information and Communication Technology (ICT) learning adoption capability. The respondents of this study were withdrawn purposely from food industries located in Central Java. The location of the study was in Surakarta, Salatiga, Semarang and Kudus. The time frame of the study was from January 2019 to July 2019. The software of SEM version 23 (Structural Equation Modelling) was utilized to test the hypotheses. The findings revealed that all the hypotheses (entrepreneurship orientation, eco-innovation and ICT learning adoption capability) have significant impact on marketing performance.

Keywords: Food industries, green company, marketing performance

1 Introduction

Previous studies revealed that there are relationships between entrepreneurship orientation and marketing performance, those connections are moderated by green company or eco-innovation performance [1, 2]. In contrast, some studies reveal that entrepreneurship orientation does not have a relationship with the company performance and marketing performance [3, 4]. Therefore this study proposed eco-innovation to mediate between entrepreneurship orientation and marketing performance. According to Bagaria et al. the implementation of eco-innovation in food industries can improve food storage technology; introduce energy efficient, low carbon waste and food chain supply; refine processing method to lessen the loss of raw materials; and improve organizational eco-innovation to save food waste cost [5].

The indicators of entrepreneurship orientation both of internal and external locus of control can be elaborated as risk taker, proactive action, aggressive and competitive action

* Corresponding author: vincentaryanto@dsn.dinus.ac.id

[6–8]. Entrepreneurship orientation also incorporates cultural development, new product and service development, technological advancement and green business issues [1, 9]. Entrepreneurship orientation does not only deal with the individual entrepreneur per se, but also deals with the business environment influencing the entrepreneurs.

The eco-innovation is a method based on life-cycle and value added innovation activities incorporating technology-based products as well as non-technology products. The production processes, management, services, and new way of business are manifested in the use of life cycle assessment to result in minimizing risk of pollution, including negative impact in using resources and energy [1, 10]. Furthermore, the researchers stated that eco-innovation consists of organization novelty that has less negative impact on the environment such as change in values, beliefs, norms, knowledge and management activities including change in administration, institutions, regulation and governance systems.

Marketing performance is a part of company performance to aim at augmenting efficiency of the organization. Marketing performance includes sales growth, market share growth, revenue growth, profitability and stock returns. Marketing performance has the dimension of culture, capabilities, process, market performance and financial performance [11].

Information and communication technology (ICT) has enabled to boost firm organization performance. Yunis et al. consider the corporate entrepreneurship influence on firm and marketing performance [12]. Furthermore, the ICT utilization has improved the company's competitiveness. The utilization and adoption of ICT has enhanced the innovative use of company resources. This study was supported by Adamides and Karacapilidis that revealed the utilization of ICT has two dimensions either strategic and operational [13]. Thus, organizations must prepare for the human resources to integrate ICT aspects on their organizational design and planning.

2 Materials and methods

This research utilized a purposive survey method to withdraw from the population of food industries in Central Java. The respondents were those who apply some sort of eco-innovation activities on their business and marketing performance particularly in the level of Small and Medium Enterprises (SMEs). Food related industries were considered to be the respondents considering their rapid growth and self-sustained existence during the financial crisis.

2.1 Entrepreneurship orientation

Entrepreneurship orientation is a process of organizational learning capabilities. Previous studies reveal that there are two approaches for conceptualizing entrepreneurship orientation such as composite dimension and unidimensional approach portrayed by risk taking behavior, innovative behavior and proactive behavior [14, 15]. Another study undertaken by Zhang et al. confirms that those three aspects of entrepreneurship's unidimensional approach have significant and positive impact on marketing performance [16]. In addition, the researchers suggest having better direct influence between entrepreneurship orientations and marketing performance. Fernando & Wah stated that the implementation of eco-innovation towards marketing and firm performance in Malaysia, the firm marketing performance on harmonizing environmental, economic and social performance [17]. The application of eco-innovation has boosted the firm's product quality and production efficiency.

There were two hypothesis applied in this study,

H1: Entrepreneurship Orientation has positive and significant impact on eco-innovation.

H2: Entrepreneurship Orientation has a positive and significant impact on marketing performance.

2.2 ICT learning capability

Previous studies reveal that the use of ICT improves eco-innovation and firm performance significantly; furthermore the use of ICT enhances the organization performance [1, 18, 19]. The adoption of ICT plays an imperative role in promoting a firm's marketing performance. However in the case of Indonesia, Fernando et al. emphasize the use of ICT as the online green network to improve the application of eco-innovation and firm performance since the online green network is considered as the weakness implementation of eco-innovation in Indonesia [17]. ICT learning capability goes hand in hand with the organizational learning capability. It deals as the basis of a managerial approach to bolster learning within an organization. The organizational learning has useful impacts on firm performance [20].

H3: ICT learning capability has positive and significant impact on eco-innovation.

H4: ICT learning capability has a positive and significant impact on marketing performance.

2.3 Eco-Innovation

The innovation capability is the key point of long-period success in business, especially in markets and companies which have capability to innovate and be able to react to challenges of market changes faster and are capable to exploit products and market opportunities [21]. Most of the eco-innovation studies emphasize on describing company performance rather than firm's marketing performance. Fernando et al. explore the application of eco-innovation in Indonesia's furniture industry, in addition their findings the importance of online green network, the implementation of eco-innovation has significant impact on firm performance and sustainable business practices despite the shortage of government initiatives and support [22]. This research is supported by Jogaratnam et al. that enhancing business performance in Taiwan requires a successful application of eco-innovation [23]. Other studies also reveal that integrating eco-innovation in firm strategy is able to enhance the firm's marketing performance [1, 24].

H5: Eco-innovation has a positive and significant impact on marketing performance.

2.4 Sample size

This research was done by collecting 190 respondents in central Java particularly in the city of Surakarta, Salatiga, Semarang and Kudus to test the proposed model of hypotheses. Technique of sampling used purposive to address the food industries that has applied some sort of elements of eco-innovation. The operationalized variables were depicted as in table 1 as follows:

Table 1. Construct measurement

Variable	Concept Definition	Authors
Entrepreneurship Orientation Risk taking behavior Proactive behavior Innovative behavior	Getting to know the direct impact to the firm marketing strategy	Shan et al. [7] Zhang et al. [16] Linton & Kask [8]

(Continued on next page)

Table 1. Continued

Variable	Concept definition	Authors
ICT learning capability IT utilization ICT-based knowledge. Managerial commitment	The utilization of ICT to enable the organization learning capability and improve the firm marketing performance.	Buttol et al. [18] Alegre et al. [20] Green et al. [19] Aryanto et al. [2]
Eco-innovation Green innovation Green food storage technology. Energy efficiency Food waste saving Green food network	The definition of eco-innovation is a method of life-cycle basis and value added basis on innovation activities to incorporate whether technology-based product as well as non-technology product. The production process, management, services, and new way of business are manifested in the use of life cycle assessment to result in minimizing risk of pollution including risk of impact negatively in using resources and energy resources.	Bagaria et al. [5] Aryanto et al. [2] Jogaratnam et al. [23] Hojnik [24]
Marketing performance Sales growth Market share Customer retention Marketing profit	Marketing performance is a part of company performance to aim at augmenting efficiency of the organization. Marketing performance includes sales growth, market share growth, revenue growth, profitability and stock returns. Marketing performance has the dimension of culture, capabilities, process, market performance and financial performance	Aryanto et al. [2] Pimenta de Gama [11]

2.5 Measure instrument

Entrepreneurship orientation variable was measured by three items such as risk taking behavior, proactive behavior and innovative behavior. ICT learning capability was measured by three items such as IT utilization, ICT-based knowledge, and managerial commitment. In addition, Eco-innovation was measured by five items such as green innovation, green food storage technology, energy efficiency, food waste saving and green food network. Finally, the marketing performance variable was measured by four items that were composed of sales growth, market share, customer retention and marketing profit. Research instrument was used by a 7-point scale.

3 Results and discussion

Analysis of the full structural model was undertaken by using SEM-AMOS version 23, H1 mentioned the relationship between entrepreneurship orientation and eco-innovation, the result showed that cr = 2.182 and probability is 0.46. H2 mentioned the relationship between entrepreneurship orientation to marketing performance, the result showed that cr = 2.024 and probability is 0.42. H3 mentioned the relationship between ICT learning adoption capability and eco-innovation, the result showed that cr = 3.094 and probability is 0.02. H4 mentioned the relationship between ICT learning adoption capability and marketing performance, the result showed that cr = 2.505 and probability is 0.11. H5 mentioned the relationship between eco-innovation and marketing performance, the result showed that cr = 2.251 and probability = 0.023. The conclusion was all hypotheses were accepted considering that all cr values > 2 (Table 2).

Table 2. Hypotheses

Hypotheses	Estimate	S.E	CR	P	Conclusion
H1:Entrepreneurship orientation to eco-innovation	0.195	0.98	2.182	0.46	Accepted
H2:Entrepreneurship orientation to marketing performance.	0.167	0.82	2.024	0.42	Accepted
H3:ICT learning adoption capability to eco-innovation	0.348	0.112	3.094	0.02	Accepted
H4:ICT learning adoption capability to marketing performance	0.242	0.096	2.505	0.11	Accepted
H5:Eco-innovation to marketing performance	0.155	0.068	2.251	.023	Accepted

Significant at $p \leq 0.05$

The study found out that entrepreneurship orientation has significant and positive impact on eco-innovation, the entrepreneurs awareness towards the eco-innovation wittingly or not, improve their firm marketing performance, this finding is supported the previous research carried out by Jogaratnam entrepreneurship orientation relies on the paradigm of entrepreneurs vis a vis the eco-innovation. In addition, entrepreneurship orientation also enhances their marketing performance [23]. This is in line with the previous study done by, since entrepreneurship orientation deals with the viable business practices in food industries in Central Java, Indonesia [20].

ICT learning adoption capability has significant impact on eco-innovation, this finding is supported by [1, 18, 19]. The ICT inevitably supports the dissemination of best practices of eco-innovation in food industries in Central Java since the actors are able to share knowledge, information and market access. Furthermore, the adoption of ICT enables food SMEs better access to wider markets and consumers. The ICT adoption in SMEs level needs adjustment and technical assistance from external actors such as universities, consultants etc.

The implementation of eco-innovation in food industries as a matter of fact has improved the marketing performance of the firms. However, some food innovation indicators such as energy efficiency, green food storage technology and green food network are considered weaker than other indicators such as green innovation and waste saving. They experience that the implementation of eco-innovation improves their business performance. This finding is supported by previous study done [1, 11].

4 Conclusion

All the hypotheses (entrepreneurship orientation, eco-innovation and ICT learning adoption capability) have significant impact on marketing performance.

The authors would like to thank *Direktorat Penelitian dan Pengabdian Masyarakat Kementerian Riset, Teknologi Dan Pendidikan Tinggi*, Number of contract 026/L6/AK/PENELITIAN/2019 for granting the sponsorship of this study.

References

1. K. Lumbanbatu, V.D.W. Aryanto. Int. J. Soc. Ecol. Sust. Dev. **6**,4:35–64(2015). https://www.igi-global.com/article/green-practices-implementation-as-prerequisite-to-sustain-firm-competitive-advantages/142146
2. V.D.W. Aryanto, Y. Wismantoro, K. Widyatmoko. Int. J. E-Business Res. **14**,1:21–36(2018). https://www.igi-global.com/chapter/implementing-eco-innovation-by-utilizing-the-internet-to-enhance-firms-marketing-performance/231242
3. A. Rauch, J. Wiklund, G.T. Lumpkin, M. ETP. **33**,3:761–787(2009). https://journals.sagepub.com/doi/abs/10.1111/j.1540-6520.2009.00308.x
4. F. Zahra. Int. J. Mech. Eng. Technol. 8,6:822–830(2017). https://www.google.com/url?sa=t&rct=j&q=&esrc=s&source=web&cd=&cad=rja&uact=8&ved=2ahUKEwjzqdzFitLqAhWEbSsKHU47CYoQFjAAegQIAhAB&url=https%3A%2F%2Fwww.researchgate.net%2Fpublication%2F318339323_The_impact_of_participative_budgeting_on_budgetary_slack_A_meta-analysis&usg=AOvVaw3OlKW2YkxRxcTezRfFex6d
5. I.R.E. Bagaria, E. Croft, A. Dentith, S. Feret, A. Nowack, M. Pancaldi, S. Valkama. Ecowaste 4 Food Interreg Europe. 3–11(2017). https://www.google.com/url?sa=t&rct=j&q=&esrc=s&source=web&cd=&cad=rja&uact=8&ved=2ahUKEwjTqcLUitLqAhUEQH0KHT3eBKQQFjABegQIARAB&url=https%3A%2F%2Fwww.interregeurope.eu%2Ffileadmin%2Fuser_upload%2Ftx_tevprojects%2Flibrary%2Ffile_1505139758.pdf&usg=AOvVaw0D5P0Z161gdZKF6iQIMpav
6. J. Wiklund, D. Shepherd. J. Bus. Ventur. **20**,1:71–91(2005). https://www.google.com/url?sa=t&rct=j&q=&esrc=s&source=web&cd=&cad=rja&uact=8&ved=2ahUKEwi659zfitLqAhXFfH0KHXvbCjoQFjAAegQIBhAB&url=https%3A%2F%2Fwww.researchgate.net%2Fpublication%2F222269148_Entrepreneurial_Orientation_and_Small_Business_Performance_A_Configurational_Approach&usg=AOvVaw3ibmV_gxb9fjL_OIZcb7yS
7. P. Shan, M. Song, X. Ju. J. Bus. Res. **69**,2:683–690(2016) https://www.google.com/url?sa=t&rct=j&q=&esrc=s&source=web&cd=&cad=rja&uact=8&ved=2ahUKEwiV-L_pitLqAhVLWysKHU80D3cQFjAAegQIBRAB&url=https%3A%2F%2Fideas.repec.org%2Fa%2Feee%2Fjbrese%2Fv69y2016i2p683-690.html&usg=AOvVaw2p-1jEq2fOP_Ul5I039wtF
8. G. Linton, K. Johan. J. Bus. Res. **70**:168–176(2017). https://www.google.com/url?sa=t&rct=j&q=&esrc=s&source=web&cd=&cad=rja&uact=8&ved=2ahUKEwjCrofyitLqAhXYSH0KHZQ1CtkQFjAAegQIBBAB&url=https%3A%2F%2Fideas.repec.org%2Fa%2Feee%2Fjbrese%2Fv70y2017icp168-176.html&usg=AOvVaw3M-cyR-J9DcXdXWDew8JTs
9. M. Dayan, R. Zacca, Z. Husain, A. Benedito, J.C. Ryan. J. Bus. Ind. Mark. **31**,5:668–683(2016). https://www.emerald.com/insight/content/doi/10.1108/JBIM-02-2015-0023/full/html
10. R. Kemp, P. Pearson. Policy Brief about Measuring Eco-Innovation. Sage Publication. 98–118(2008). https://www.osti.gov/etdeweb/biblio/21124989

11. A.P. Gama, A.A. Hines. Mark. Intell. Plan. **29**,7:643–661(2011). https://www.emerald.com/insight/content/doi/10.1108/02634501111178677/full/html
12. M. Yunis, A. Tarhini, A. Kassar. J. Bus. Res. **88**:344–356(2018). https://www.sciencedirect.com/science/article/pii/S0148296317305246
13. E. Adamides, N. Karacapilides. J. Innov. Knowl. **90**:2444–5691. (2018). https://www.google.com/url?sa=t&rct=j&q=&esrc=s&source=web&cd=&ved=2ahUKEwjLyaiymd3qAhWX63MBHdHtAWwQFjAAegQIAxAB&url=https%3A%2F%2Ffardapaper.ir%2Fmohavaha%2Fuploads%2F2019%2F11%2FFardapaper-Information-technology-for-supporting-the-development-and-maintenance-of-open-innovation-capabilities.pdf&usg=AOvVaw29IJ7Dm_wb-iY-vqeKHU2S
14. F. Fatimah, D. Rosadi, R. Hakim. Neural Comput. Appl. **31,**397–407(2019). https://link.springer.com/article/10.1007/s00521-017-3011-y
15. C. Lechner, S.V. Gudmundsson. ISBJ. **32,**1:36–60(2012). https://journals.sagepub.com/doi/abs/10.1177/0266242612455034
16. J.A. Zhang, F. Edgar, A. Geare, O'Kane. Ind. Mark. Manag. **59**,131–143(2016). https://www.sciencedirect.com/science/article/pii/S0019850116300220
17. Y. Fernando, W.X. Wah. Sust. Prod. Consumption. **12**:27–43(2017). https://www.sciencedirect.com/science/article/pii/S2352550917300155
18. P. Buttol, R. Bounamici, I. Naldesi, C. Rinaldi, A. Zamagni, P. Masoni. Clean Technol. Environ. Policy. **14**,2:211–221(2012). https://link.springer.com/content/pdf/10.1007/s10098-011-0388-7.pdf
19. K.W.J. Green, P.J. Zlebst, V. Bhadauria, J. Meacham. Ind. Manag. Data Syst. **112**,2:186–205(2012). https://www.emerald.com/insight/content/doi/10.1108/02635571211204254/full/html?queryID=22%2F5407253
20. J. Alegre, K. Sengupta, R. Lapiedra. Int. Small Bus. J. **31**,4:454–470(2013). https://journals.sagepub.com/doi/abs/10.1177/0266242611417472
21. P. Ritala, H. Olander, S. Michailova, K. Husted. Technovation. **35**:22–31(2015). https://www.sciencedirect.com/science/article/pii/S0166497214001060
22. Y. Fernando, M.S. Shaharu, N.A. Wahid. Int. J. Serv. Oper. Manag. **23**,1:43–58 (2016). https://www.inderscienceonline.com/doi/abs/10.1504/IJSOM.2016.073289
23. G. Jogaratnam. Int. J. Hospitality Manag. **60**,104–113(2017). https://fardapaper.ir/mohavaha/uploads/2018/05/Fardapaper-The-effect-of-market-orientation-entrepreneurial-orientation-and-human-capital-on-positional-advantage-Evidence-from-the-restaurant-industry.pdf
24. J. Hojnik, M. Ruzzier, T.S. Manolova. J. Cleaner Prod. **171**, 1312–1323(2018). https://www.sciencedirect.com/science/article/pii/S0959652617324101

Potentials of Gas Emission Reduction (GHG) by the Glass Sheet Industry through Energy Conservation

Washington Purba[1,*], *Erkata* Yandri[1,3], *Roy* Hendroko Setyobudi[1,4], *Hery* Susanto[1], *Satriyo* Krido Wahono[5], *Kiman* Siregar[6,7], *Yogo* Adhi Nugroho[8], *Abubakar* Yaro[9,10], *Kamaruddin* Abdullah[1,2], *Yahya* Jani[11], and *Danny* Faturahman[12]

[1]Graduated School of Renewable Energy, Darma Persada University, Jl. Taman Malaka Selatan No.22, Pd. Kelapa, Duren Sawit, Jakarta 13450, Indonesia
[2]JAP-Energia, Graha Mampang, Lt. 3, Suite 305, Jl. Mampang Prapatan Raya Kav. 100, Jakarta 12760, Indonesia
[3]Center of Renewable Energy, Darma Persada University
[4]Department of Agriculture Science, Postgraduate Program, University of Muhammadiyah Malang, Jl. Raya Tlogomas No. 246, Malang, 65145, East Java, Indonesia
[5]Research Division for Natural Product Technology – Indonesian Institute of Sciences, Jl. Jogja - Wonosari, km 31.5, Gunung Kidul, Special Region Yogyakarta 55861, Indonesia
[6]Department of Agricultural Engineering of Syah Kuala University, Jl.Tgk. Hasan Krueng Kalee No.3, Kopelma Darussalam, Banda Aceh 23111, Indonesia
[7]Indonesian Life Cycle Assessment Network (ILCAN), Kawasan Puspitek Serpong, Tangerang, Banten 15314, Indonesia
[8]Data Processing, *Rumah Paper* – Editage Services, Jl. Tokala No.1, Malang 65146, East Java, Indonesia
[9]Institute of Health Sciences & Research, AHRO Scientific Publishing, 272 Bath Street, Glasgow G2 4JR, Scotland, UK
[10]Strathclyde Institute of Pharmacy & Biomedical Sciences (SIPBS), University of Strathclyde, SPBBS, 161 Cathedral St, Glasgow G4 0RE, Scotland, UK
[11]Department of Built Environment and Energy Technology, Faculty of Technology, Linnaeus University, Lückligs plats 1 (Hus M), 351 95 Växjö, Sweden
[12]Departement Marine Engineering, Faculty of Ocean Technology, Darma Persada University

Abstract. Sheet Glass Industry is one industry that uses 75 % natural gas energy and 25 % electricity. Using the Intergovernmental Panel on Climate Change, IPCC-2006 emission calculation method, the average greenhouses gas (GHG) emissions obtained from the calcination process obtained 112 211 t CO_2 yr^{-1} per plant and an average emission factor (EFkl) of 0.18 CO_2 t^{-1} yr^{-1} of pull. With the technology of converting heat into electrical energy, residual combustion as flue gases has the potential to be used to produce electrical energy. Referring to the analysis and calculation; one of factories has potential to generate 0.8 MW to 3 MW electric energy. It's efficiency of 10 % to 40 % so that it can be calculated as a component of GHG emission reductions whose value is 4.6 t CO_2 yr^{-1} to 18.7 t CO_2 yr^{-1} per plant. With this reduction, each of the GHG emission and emission factors per plant dropped to 93 442 t CO_2 yr^{-1} and 0.16 CO_2 t-pull^{-1}

Keywords: Emission factor, energy efficiency, flue gas, greenhouse gas, global

*Corresponding author: wp.sidadolok@gmail.com

Introduction

Global warming is a term used to describe the gradual increase in the average temperature of the earth's atmosphere and oceans, the value of 0.74 °C ± 0.18 °C (1.33 °F ± 0.32 °F) over the last 100 yr. Increasing the amount of carbon dioxide and other greenhouse gases released by burning fossil fuels, land clearing, agriculture, and other human activities, is believed to be the main source of global warming that has occurred over the past 50 yr. The Intergovernmental Panel on Climate Change (IPCC) concluded that, most of the increase in global average temperatures since the mid-20th century was most likely due to increased concentrations of greenhouse gases (GHGs) due to human activities [1].

The industrial sector provides energy consumption which varies by region, according to differences in industrial output, energy intensity (measured as energy consumed per unit of gross output), and industry composition [2]. Companies can reduce energy consumption in all ways, including improving industrial sector processes to optimize energy use and minimize energy losses, increase the use of cogeneration, and recycle materials and types of fuel to reduce costs and increase efficiency

Energy consumption in our country Indonesia until 2016 is 1 018 MBoe. In contrast to the distribution of global consumption by sector, of the total consumption of the industry is ranked second energy users, namely 30.88 %. The biggest energy consumption is the transportation sector by 43.89 % followed by residential / household 16.62 %, commercial 5.8 % and others. Based on the type of energy used, petroleum is still the largest portion of 43.7 % followed by electricity 19.9 %, natural gas 15.3 %, coal 9.6 %, LPG 8.5 % and biofuels 2.97 % [3].

Concerns from the world community on GHG emissions led to the birth of the Kyoto Protocol in December 1998, followed by the birth of the Paris Agreement at the 21st COP in Paris, 30 November to 12 December 2015. which contained agreements from member countries, which in essence agreed on the temperature threshold globally is below 2 °C, overcoming the maximum temperature change of 1.5 °C from pre-industrial times

Indonesia signed the Paris Agreement on 22 April 2016 and was later ratified through Law Number 16 Year 2016. Indonesia's commitment at COP-21 in Paris to reduce GHG emissions by 29 % in 2030 by its own efforts (CM1) or by 41 % with international assistance (CM2) as outlined in the Nationally Determined Contribution (NDC) document [4]. This momentum is the basis for changing the target for GHG emission reductions in Indonesia by 26 % in 2020. From the 29 % figure, the energy sector received a portion of emission reduction of 314×10^6 t of CO_2 on its own efforts and 398×10^6 t of CO_2 with international support [4].

This paper is intended to analyze and calculate the electrical energy potential of sheet glass flue gas conversion and its contribution to the GHG emission of the industrial sector in Indonesia

2 Literature review

Until 2016, GHG emissions in Indonesia were $1\,515 \times 10^6$ t CO_2 and this figure is down 39 % compared to 2015, and the biggest contribution was from the peat/forest fire sector 88 %, energy by 5 %, land use and forestry sector (Land use, land-use change, and forestry /LULUCF by 27 % while the Industrial sector (Industrial Processes and Production Use/ IPPU) increased by 2 % [4].

Calculation of GHG emissions using the Intergovernmental Panel on Climate Change Guidelines in the 2006 IPCC Guidelines [5, 6] and the application of this methodology has been stipulated in Minister of LHK Regulation Number P.73 / MenLHK / Setjen / Kum.

1/12/2017 dated 29 December 2017 concerning Guidelines for the Implementation and Reporting of Greenhouse Gas Inventories [7].

The sheet glass production process begins with the preparation of raw materials for silica sand, dolomite, cullet and other supporting raw materials which are mixed into one raw material (mixed batch), then put into stages in a furnace with a capacity of 500 t yr^{-1} to 700 t yr^{-1} with a temperature of 1 500 °C , and operates for 24 h, 365 d, 15 yr continuously. In the production process, the biggest use of energy is natural gas (NG) in the smelting of raw materials and electrical energy in the process of forming, cooling and cutting. Figure 1 shows a diagram of the process of producing sheet glass, the use of energy and the resulting flue gas [8].

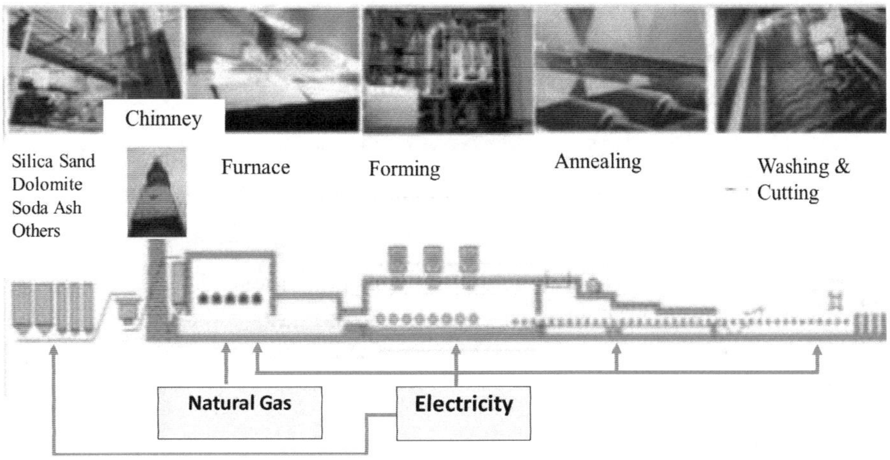

Fig. 1. Glass sheet manufacturing flow process

Raw material that has been melted (molten glass) is flowed on top of tin at a temperature of 600 °C to 800 °C for the formation of thickness and dimensions and is gradually cooled (annealing) with a blower.

The final process is washing, cutting (washing & cutting). In the production process, GHG emissions will occur during the melting process as direct emissions and the use of electricity in the process of forming, cooling and cutting as indirect emissions.

The Equation (1) shows the heat energy from the furnace flue gas.

$$Qhr = m \times Cp \times \Delta T \tag{1}$$

Qhr = Heat (J), m = mass (kg), Cp = Heat Specific (J kg^{-1} . K), dan ΔT = delta Temperatur In/Out (K or C)

Flue gas specific heat with NG fuel as in Table 1.

Table 1. Specific heat from the flue gas with NG fuel [9, 10]

Subtance	Formula	$C_p = a + bt + bt^2 + br^3$ (t in K, C_p in kJ $kmol^{-1}$, K)				RANGE	%ERROR	
		a	B	c	d	(K)	Max	Min
Nitrogen	N_2	28.9	-1.57E-03	8.08E-06	-2.87E-09	273 to 1 800	0.59	0.34
Oxigen	O_2	25.48	1.52E-02	-7.16E-06	1.31E-09	273 to 1 800	1.19	0.28

Table 1. Continue to the next page

Subtance	Formula	$C_p = a + bt + bt^2 + br^3$ (t in K, C_p in kJ $kmol^{-1}$, K)				RANGE	%ERROR	
		a	B	c	d	(K)	Max	Min
Carbon Dioxide	CO_2	22.26	5.98E-02	-3.50E-05	7.47E-09	273 to 1 800	0.67	0.22
Water Vapor	H_2O	32.24	1.92E-03	1.06E-05	-3.60E-09	273 to 1 800	0.53	0.24
Sulfur Dioxide	SO_2	25.78	5.80E-02	-3.81E-05	8.61E-09	273 to 1 800	0.45	0.24

For some types of glass industry, the % of fuel that becomes heat losses in flue gas is 29 %, 30 %, 57 %, 53 %, and 56 % for sheet glass, container, press, insulation fiber, and textile fiber, as in Table 2.

Table 2. The capacity of furnaces and flue gases of several types of glass industry [9, 10]

Glass industry segment	Furnace capacity range	Typical furnace capacity	Natural gas consumption	Waste heat	% natural gas input lost to heat
	(t)	(t)	(Tbtu y^{-1})	(Tbtu y^{-1})	(%)
Flat glass	300 to 1 000	550+	41.10	11.82	29
Container glass	50 to 330	250	45.49	13.65	30
Pressed/brown glass	1 to 300	75	16.82	9.63	57
Insulation fiber glass	20 to 300	100	3.24	1.73	53
Textile fiber glass	100 to 150	100 to 150	11.05	6.14	56

The combustion reaction of a fuel into flue gases is describe [11] in Equation (2):

$$CH_4 + (1+EA/100) \times [2O_2 + 7.52N_2] \rightarrow$$
$$CO_2 + 2H_2O + (1+EA/100) \times 7.52N_2 + (9.52 \times EA/100 \times 21/100)O_2 \quad (2)$$

The percentage of each flue gas as the result of the reaction in Equation 2 with natural gas fuels can be calculated in a number of excess air as shown in Table 3.

Table 3. Composition of combustion flue gases with some % excess air

Flue Gas (%)	Excess Air (EA) in %									
	10	20	30	40	50	60	70	80	90	100
CO_2	8.72	8.05	7.48	6.98	6.54	6.16	5.82	5.51	5.24	4.99
H_2O	17.43	16.1	14.95	13.96	13.09	12.32	11.64	11.03	10.48	9.98
N_2	72.11	72.63	73.09	73.48	73.82	74.13	74.40	74.64	74.86	75.05
O_2	1.74	3.22	4.48	5.58	6.54	7.39	8.14	8.82	9.43	9.98

GHG emission equation from energy use (GHG_{ENG}) [12] is describe in Equation (3).

$$GHG_{ENG} = EC \times EF \quad (3)$$

EC = Energy Consumption, *EF* = Emission Factor (56.2 CO_2 TJ^{-1} for Natural Gas, 77.4 CO_2 TJ^{-1} for Residual Oil).
GHG emission from process (GHG_{IPPU}) [6] is describe in Equation (4).

$$GHG_{IPPU} = Mg \times EF \times (1 - CR) \quad (4)$$

Mg = pull (ton), $EF = 0.21$ kg CO_2 kg^{-1} glass at cullet ratio 10 % to 25 %, CR = cullet ratio.

Assessment of several industrial sector flue gases, conversion technology and investment costs per kW electricity, potential ORC applications for the conversion of flue gas heat to electricity from four industrial intensive sectors in several countries in the EU 27 and GHG emission reduction using electricity converted from gas The flue is shown in Table 4 and Table 5.

Table 4. Conversion of flue gas into electricity in several industrial sectors [13].

Industry	Temperature (°C)	Process	Technology	Efficiency (%)
Glass	230 to 480	Melting	SRC	20 to 43
			ORC	8 to 12
			TEG	5
			KC	15
Aluminium	790 to 1 080	2nd Melting	SRC	20 to 30
Steel	580 to 680	Roll Mill	SRC-CO2-PC	13 to 17
Paper	82 to 140	Dryer	ORC	8 to 12
Food	150 to 350	Boiler	ORC	8 to 12
Chemical	150 to 260	Preheater	ORC	8 to 12
SRC (Steam Rankine Cycle), ORC (Organic Rankine Cycle), KC (Kalina Cycle). SC-CO2-C (Super Critical CO2 Cycle), TEG (Thermo Electric Generator).				

Table 5. ORC application in four sectors in several EU27 countries [14].

Item	Industry	Glass	Cement	Steel	Oil & gas
	Process	Melting	Clinker	Eaf+roll mill	Gcs & gsf
Numbers		58	241	399	
Capacity (Mt yr^{-1})		12	247	217	
ORC Power (MW)		79	574	748	
E Recovery (GWh yr^{-1})	5 000 h	383	2 870	3 740	6 520
	8 000 h	628	4 592	5 984	10 432
Emission (k t CO_2 yr^{-1}	5 000 h	140	1 212	1 351	2 328
	8 000 h	225	1 940	2 162	3 724
Cost Saving (ey y^{-1})	5 000 h	35	260	344	583
	8 000 h	56	416	550	933
Power Specific (r_p)		2.7 (kW t^{-1}) d^{-1}	1.01 (kW t^{-1}) d^{-1}	6.87 (kW t^{-1}) d^{-1}	

3 Methodology

The research methodology uses qualitative and quantitative methods. Quantitative using data for 2014 to 2018 from five different factories both for data on energy consumption, raw materials, and production volume. The data for the 5 yr are averaged, then GHG emissions are calculated using the 2006 IPPC method for the energy sector and IPPU as Figure 2.

GHG uses a combination of quantitative and qualitative, while the physical data of flue gas for 60 d from one of the factories is used as a reference to determine the physical quality of flue gas heat. In the analysis, in addition to using physical / technical research data will also use qualitative methods, namely the chemical stoichiometry calculation to obtain the percentage of oxygen in the flue gas as a basis for the calculation of the resulting flue gas mass. Quantitative data was also carried out with literature studies of several previous studies on the technology of converting hot gas to electrical energy, but not in detail explained electricity from gas conversion is used in the production process and is calculated as a reduction of GHG emissions

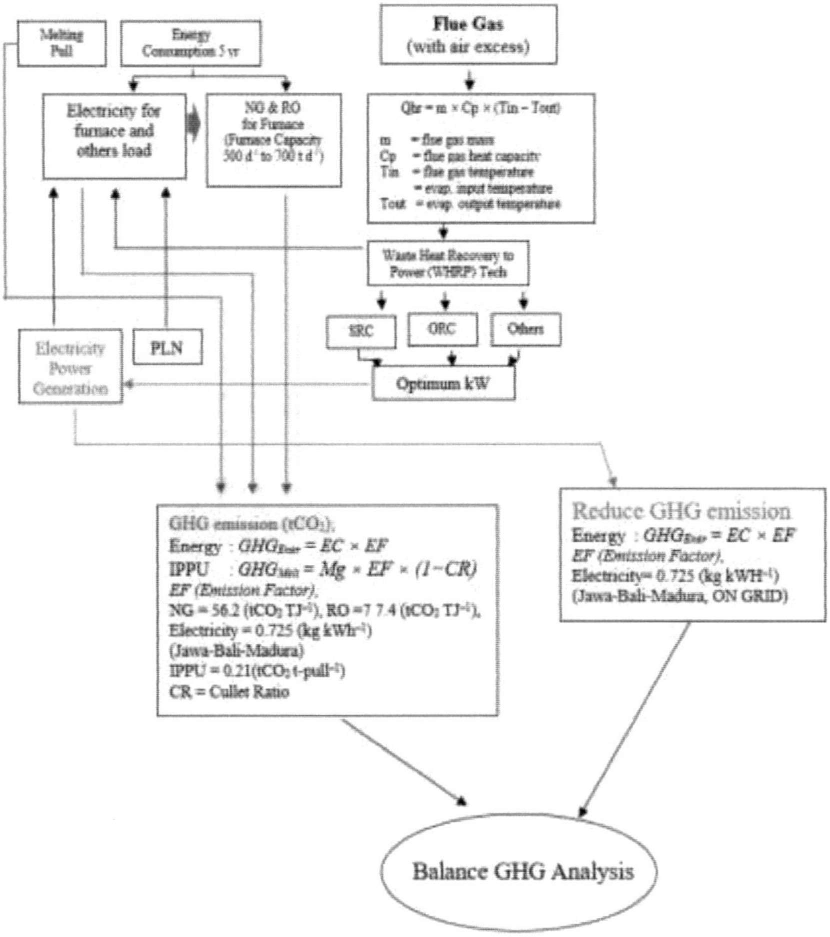

Fig. 2. Research flow

3.1 Data collection

Secondary data sources obtained from average data in 2014 to 2018, both for natural gas (NG) consumption data, residual oil, electricity and total glass volume produced (= pull) from five different factories as Table 6.

Table 6. Energy consumption & glass pull

Description	Unit	2014	2015	2016	2017	2018
Electricity (PLN)	MWh	22 224	21 188	20 632	23 565	24 142
Natural Gas	TJ	1 217	1 211	1 126	1 118	1 175
Residual Oil	TJ	-	8	114	55	2
Pull	t	153 020	152 543	146 259	173 388	171 476
Cullet Ratio	%	35	36	40	35	36

Flue gas temperature data using a source from the Distributed Control System (DCS) in 60 d and PI for one fuel cycle (20 min) such as Figure 3 and Figure 4.

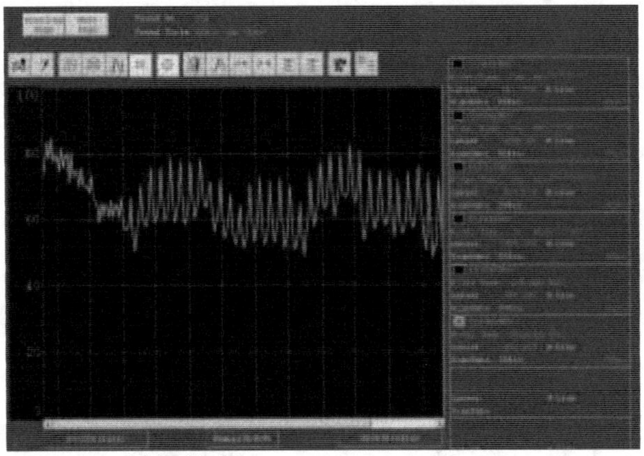

Fig. 3. Flue Gas Temperature within 60 d

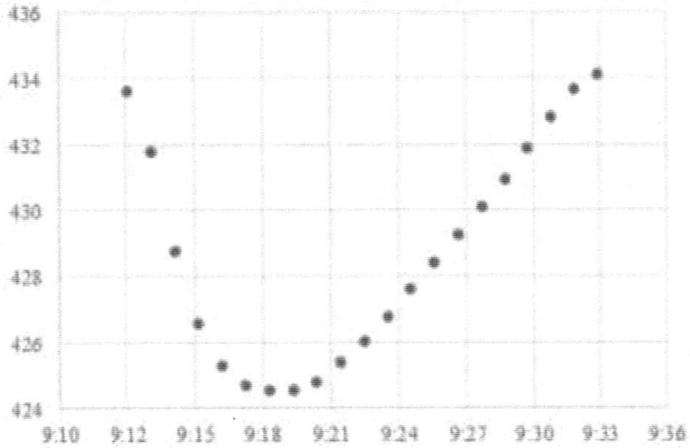

Fig. 4. Flue Gas Temperature within one firing cycle

From the DCS and PI data the maximum temperature is 434 °C, minimum 398 °C and an average of 416 °C.

3.2 Flue gas heat to electricity energy

If the density of NG = 0.74016, 1 NM3 NG = 1.0989 M3, air excess EA% from Table 3, stoichiometry A/F = 9.52, the obtained mass of 1 NM3 NG is 0.8134 kG so that the mass can be calculated each flue gas molecule such as CO_2, H_2O, N_2, O_2 and others with Equations 2 to Equation 9. With this data and the rate of natural gas consumption of 3 880 $NM^3\ h^{-1}$ (one furnace), the calculation of the molecular mass of the flue gas is shown in Table 7.

Table 7. The molecular mass of the flue gas in 1 NM^3 natural gas

Composition	Mass (kg)	%	Cp (kcal kg^{-1})	Unit
CO_2	2.219	10.8	0.200 7	kcal kg^{-1}
H_2O	1.762	8.57	0.480 0	kcal kg^{-1}
N_2	15	72.99	0.248 2	kcal kg^{-1}
O_2	1.399	6.81	0.219 7	kcal kg^{-1}
Others	0.169	0.82	0.230 0	kcal kg^{-1}
Total	**20.55**	**100**	**0.261 9**	**kcal kg^{-1}**

With a molecular mass of 20.55 kg, a specific heat of 0.2619 from Table 6, natural gas consumption of 3 880 $NM^3 h^{-1}$, input temperature of 434 °C, output of 120 °C and putting them into equation 2-1, a flue gas heat energy of is obtained 6.732 kcal h^{-1} and equivalent with 7 830 kW h^{-1}.

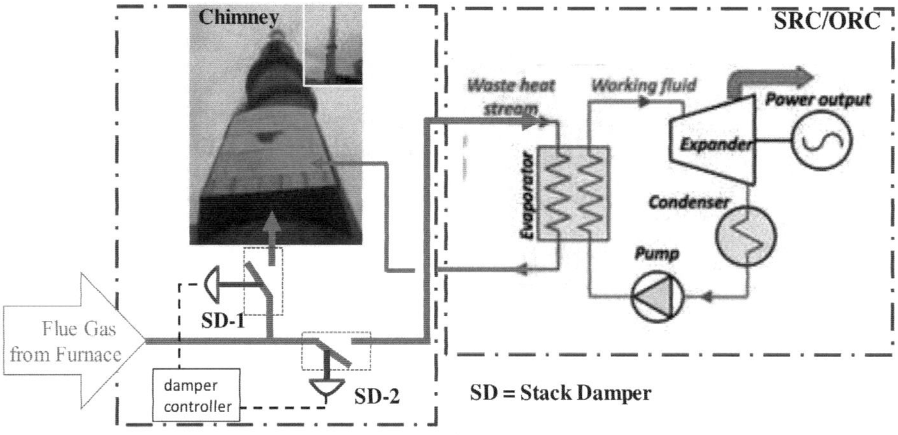

Fig. 5. Flue gas heat flow mechanism into the SRC/ORC

Through the process of converting heat into mechanical energy, electrical energy can be produced at 7 830 kW × the efficiency of converters such as SRC/ORC. Figure 5 shows a simple schematic mechanism of the flue gas heat flow into the SRC/ORC. The flue gas heat that is fed into the SRC evaporator is taken from the furnace output before being discharged into the chimney, through a flow regulating mechanism called a Stack Damper

4 Result and discussion

Using secondary data, a theoretical description of the calculation of GHG emissions and the potential heat of flue gases into electrical energy in the previous discussion, the results are presented in the following graphs.

Average emissions in 2014 to 2018 from five factories were 112 211 t CO_2, and were the biggest contributors to emissions from natural gas energy and 60 % residual oil, followed by the smelting process of raw materials 26 % and electrical energy by 14 %, as in Figure 6. When compared to emissions from electricity with natural gas and oil residuals to total emissions from energy, the percentage is 20 %:80 %. The amount of GHG emissions calculated is only half of the previous research data in Table 5 which is 225 Kt CO2 yr^{-1}

because the table is based on all processes including transportation of raw materials, yields and others within the plant but for this study only uses data in the smelting process

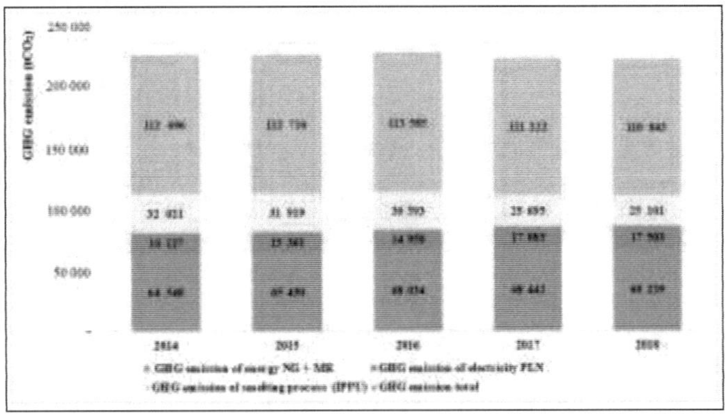

Fig. 6. GHG emission energy & smelting process

Emissions to ton pull, hereinafter referred to as FEkl (FEkl = Sheet Glass Emission Factor) are obtained by dividing the total GHG emissions in Figure 6 against pull in Table 6, as shown in Figure 7. The highest FEkl occurred in 2016 because because one of the factories only operates until July 2017 and used 500 TJ residual oil, almost the same as natural gas of 518 TJ while residual oil emission factors were 32 % higher than natural gas. FEkl decreased from 0.19 to 0.16 starting in 2017 and 2018 because one other new factory starts production in early 2017 has a new furnace and a greater capacity, means that the efficiency of burning a new furnace is better.

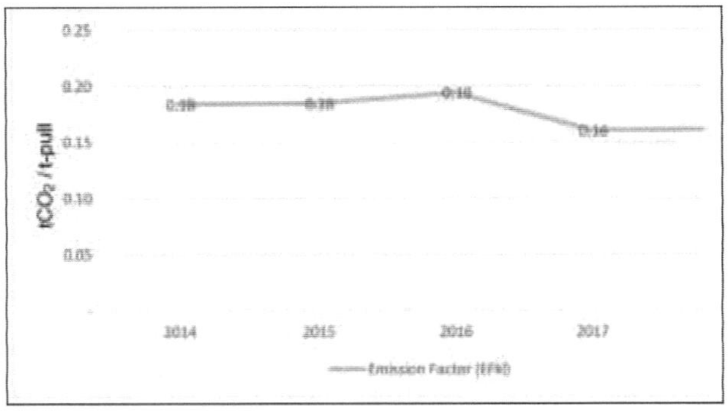

Fig. 7. GHG emissions (tCO_2) / t-pull.

The potential of electricity per year per furnace is 6 472 MWH up to 26 889 MWH from the flue gas heat of 7 830 kWh with an efficiency of SRC / ORC 10 % to 40 % as Figure 8. The higher the efficiency, the greater the electrical energy produced.

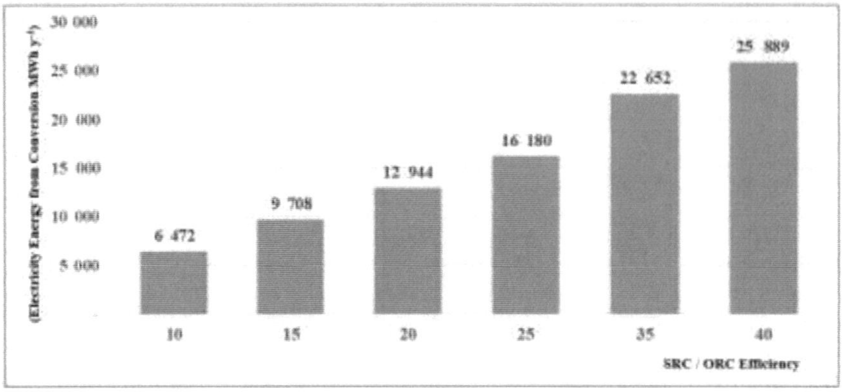

Fig. 8. Potential Electricity Energy (1 yr) from Flue Gas Conversion 7 830 kW-h per $NM^3 h^{-1}$ Natural Gas at SRC / ORC efficiency of 10 % to 40 %.

Calculation of the potential for reducing GHG emissions from the use of converted electricity, using the equation and the emission factor of the power plant from a PLN source of 0.725 t CO_2 MWh^{-1} (grid Ja-Ma-Li) as shown in Figure 9.

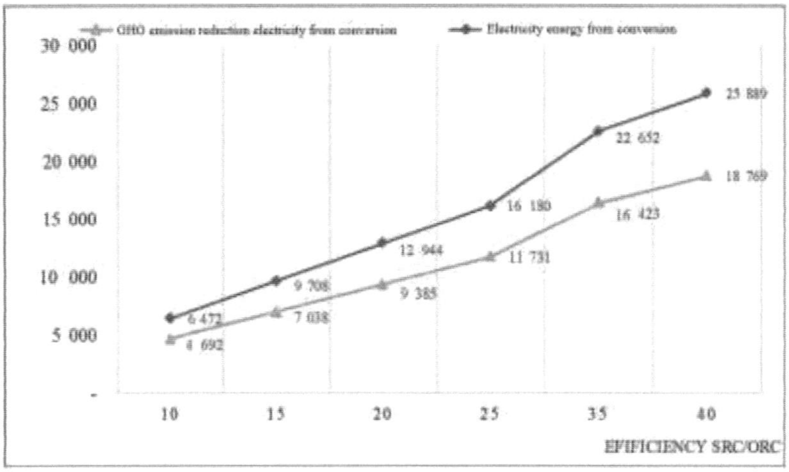

Fig. 9. Potential annual GHG Emission reductions from converting electricity

The converted electricity used in Figure 9 is a reduction in emissions, and will indirectly reduce the average emission factor from 0.18 t CO_2/ t pull on Figure 7 to 0.17 t CO_2 to 0.15 t CO_2 / t-pull, it means that the average FEkl is down by 11 %, shown in Figure 10.

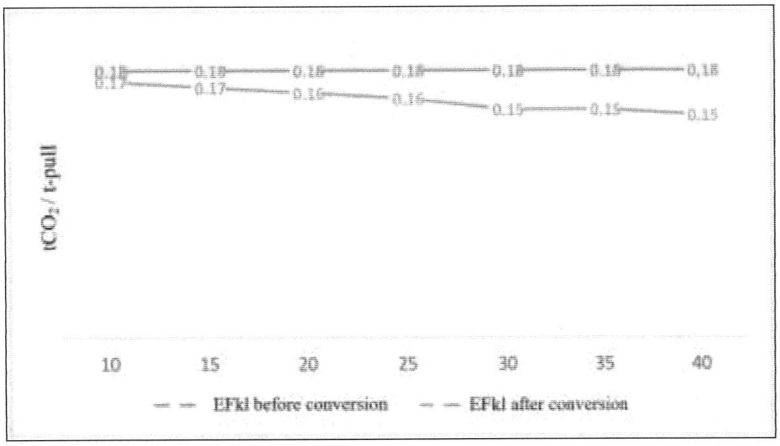

Fig. 10. GHG emissions per ton-pull (FEkl) before and after the use of converted electricity

5 Conclusion

Utilization of flue gas that is converted into electrical energy in the sheet glass industry with a pull capacity of 500 t to 700 t has the potential to generate electricity from 6 472 MWH yr^{-1} to 25 889 MWH yr^{-1}, depending on the efficiency of the SRC/ORC used. The converted electricity will indirectly reduce GHG emissions by a factor of 0.18 t CO_2/ t pull to 0.17 t CO_2 to 0.15 t CO_2/ t pull, or an average reduction of 11 %, meaning that every single furnace in the glass industry can contribute to reducing emissions 11 % of emissions by utilizing flue gas into electrical energy that is used alone, and in emission total per year will reduce 112 211 t CO_2 yr^{-1} to 93 442 t CO_2 yr^{-1} per plant.

References

1. Encyclopedia Britannica. *Global Warming - Climatic Variation Since The Last Glaciation*. [Online] from https://www.britannica.com/science/global-warming/Climatic-variation-since-the-last-glaciation (2020). [Accessed 23 July 2020].
2. L. Miró, S. Brückner, L.F. Cabeza, Renewable and Sustainable Energy Reviews, **51**:847–855(2015). https://www.doi.org/10.1016/j.rser.2015.06.035
3. Ministry of Energy and Mineral Resources Republic of Indonesia, *2017 Handbook of Energy & Economic Statistics of Indonesia*, Jakarta: ESDM (2017), pp. 72. https://www.esdm.go.id/assets/media/content/content-handbook-of-energy-economic-statistics-of-indonesia-2017-.pdf
4. KEMEN-LHK, *Laporan Inventarisasi GRK dan Monitoring, Pelaporan dan Verifikasi 2017* [GHG Inventory and Monitoring, Reporting and Verification Report 2017], Jakarta: Kemen-LHK (2017), pp. 117. [in Bahasa Indonesia]. http://ditjenppi.menlhk.go.id/reddplus/images/adminppi/dokumen/lap_igrk_2018.pdf
5. D.R. Gómez, J. Wattersson, B.B. Americano, C. Ha, G. Marland, E. Matsika et al., *Stationary combustion*. Energy, 2006 IPCC Guidelines for National Greenhouse Gas Emissions Inventories, Geneva : Intergovernmental Panel on Climate Change (2006). https://www.ipcc-nggip.iges.or.jp/public/2006gl/pdf/2_Volume2/V2_2_Ch2_Stationary_Combustion.pdf
6. L. Hanle, P. Maldonado, E. Onuma, M. Tichy, H.G. van Oss, *Mineral Industry Emisison*, 2006 IPCC Guidelines for National Greenhouse Gas Inventories, IPCC

(2006). pp. 1–40. https://www.ipcc-nggip.iges.or.jp/public/2006gl/pdf/3_Volume3/V3_2_Ch2_Mineral_Industry.pdf

7. KEMEN-LHK, *Peraturan Menteri Lingkungan Hidup dan Kehutanan Republik Indonesia nomor P.73/Menlhk/Setjen/Kum.1/12/2017 tentang Pedoman Penyelenggaraan dan Pelaporan Inventarisasi Gas Rumah Kaca Nasional* [Regulation of the Minister of Environment and Forestry of the Republic of Indonesia Number P.73 / Menlhk / Setjen / Kum.1 / 12/2017 Regarding Guidelines for Implementing and Reporting the National Greenhouse Gas Inventory]. Jakarta: Kemen-LHK (2018). No. 163, 2018, pp. 249. [in Bahasa Indonesia]. http://ditjenpp.kemenkumham.go.id/arsip/bn/2018/bn163-2018.pdf
8. MTPV-Designing Energy for the Future. *Glass Manufacturers - Convert Waste Heat Into Clean Energy* [Online] https://www.mtpv.com/markets/glass/ (2020). [Accessed 23 July 2020].
9. I.B. Page, *Waste heat recovery: technology and opportunities in US Industry*. US: Industrial technologies program US Department of Energy - Industrial Technologies Program, pp. 1–112(2008). https://www1.eere.energy.gov/manufacturing/intensiveprocesses/pdfs/waste_heat_recovery.pdf
10. B. Inc, *Waste Heat Recovery: Technology Opportunities in the US Industry*. US: US Department of Energy, (2008). pp. 1–112. https://www1.eere.energy.gov/manufacturing/intensiveprocesses/pdfs/waste_heat_recovery.pdf
11. Engineering ToolBox, *Stoichiometric Combustion*. [Online] from: https://www.engineeringtoolbox.com/stoichiometric-combustion-d_399.html (2003). [Accessed 23 July 2020].
12. BAPPENAS. *Pedoman Teknis Perhitungan Baseline Emisi Gas Rumah Kaca Sektor Berbasis Energi* [Technical Guidelines for Baseline Calculation of Energy-Based Greenhouse Gas Emissions], Jakarta: Bappenas), (2014), pp. 60. http://ranradgrk.bappenas.go.id/rangrk/admincms/downloads/publications/Pedoman_teknis_penghitungan_baseline_emisi_GRK_sektor_berbasis_energi.pdf
13. A. Amiri, M.R. Vaseghi, IEEE Transactions on Industry Applications, **51**,1:13–19(2015). https://ieeexplore.ieee.org/abstract/document/6877735/
14. F. Campana, M. Bianchi, L. Branchini, A. De Pascale, A. Peretto, M. Baresi, et al., Energy Conversion and Management. **76**:244–252(2013). https://www.doi.org/10.1016/j.enconman.2013.07.041

The Use of Rumen Contents as Bio–Activators for Fermentation in Goat Manure Fertilizer Production

Wehandaka Pancapalaga[1,*], *Suyatno* Suyatno[1] and *David* Sedlacek[2]

[1]Department of Animal Science, Faculty of Agriculture and Animal Science, University of Muhammadiyah Malang, Jl Raya Tlogomas no 246, Malang 65145, East Java, Indonesia
[2]Department of Life Siences, Faculty of Science, Masaryk University, Zerotinovo namesti 9, Brno 601 77, Czech Republic

Abstract. Slaughterhouse waste cause environmental problems. Rumen content comprise a lot of enzymes of potential bacteria as bio–activators in the process of making goat manure fertilizer. The study aimed to evaluate the quality of goat manure fertilizer added with rumen content. The materials used were goat manure and cattle rumen contents. The research method uses a Completely Randomized Designed (CRD). The percentages of cattle rumen contents treatments i.e. 0 %, 5 %, 10 % and 15 %. The variables measured were Nitrogen (N), Phosphor (P), Potassium (K), Carbon organic, C/N ratio and moisture content. The results showed, the addition of rumen contents had a very significant effect ($p < 0.01$) on C organic, P, N, and C/N, but had no significant ($p > 0.05$) on K and moisture content. The results showed that the best treatment was addition of 15 % rumen, that had N 1.59 %, P 0.15 %, K 0.72 %, C organic 30.60 %, C/N 19.24 %, and moisture content 45.86 %. This content had fulfilled SNI 19–7030–2004 for goat manure fertilizer. It could be concluded that giving rumen contents could increase N, P, K, C organic and decrease C/N and moisture content.

Key words: Bio–activator, environment friendly, organic fertilizer, prevent river pollution, utilization slaughterhouse waste

1 Introduction

One of the slaughterhouse wastes is the contents of the rumen, in the form of grass that has not been fermented and fully digested by the cattle. Cattle rumen also contains nutrients that are used by microbes as an energy source. The rumen also functions as a bio–activator to accelerate the process of maturation of organic compost [1], because the rumen contents comprised many microorganisms, including bacteria (10^9 mL^{-1} to 10^{12} mL^{-1} rumen fluid), protozoa (10^5 mL^{-1} to 10^6 mL^{-1} liquid rumen), and a number of fungi [2].

* Corresponding author: pancapalaga1966@gmail.com

Abundant population of microorganisms living in the rumen is very valuable to organic farms. The rumen contents of ruminants slaughterhouse such as cattle and goats were usually dispose, even though it is rich in nutrients. The waste is actually very potential when used for the process of making liquid and solid organic fertilizer, including as a bioactivator in the process of making goat manure fertilizer.

Organic fertilizer needs ingredient that has optimal initial C/N ratio of 10 to 20, a moisture content of 30 % to 60.5 % and good porosity of the material. Whereas, the rumen contents of cattle only have a C/N ratio between 6.44 to 13.71 [3] and the value of the C/N ratio of goat manure is generally still above 30. According to [4], if the C/N ratio is too low, the excess nitrogen will be not used by microorganisms, so it cannot be assimilated. Therefore, Nitrogen will be lost through volatility as ammonia or denitrified. On the contrary, the C/N ratio too high, the biological activity of microorganisms will be reduced. It takes several cycles of microorganisms to degrade compost, thus it takes a long time for composting and resulting lower quality fertilizer.

Based on the description above, it was necessary to study the quality of goat manure fertilizer with rumen contents addition as a bio–activator for fermentation.

2 Material and methods

The fermentation process was conducted for 21 d in the University of Muhammadiyah Malang (UMM), East Java, Indonesia experimental farm. The material used was manure taken from Etawa Buck (*Capra aegagrus hircus*). The fresh rumen content were taken from 2 yr old bulls slaughtered in Gadang Malang Slaughterhouse (*Rumah Potong Hewan*).

The research method used was an experimental method. The experiment was carried out by adding rumen contents to goat manure for fermentation. The design of the experiment used a Completely Randomized Designed (CRD). The treatments were addition of rumen content (0 %, 5 %, 10 % and 15 %) of percentage of goat manure used.

The research variables measured were the content of N, P, K, C organic, C/N ratio, and moisture content. K content was measured by the Spectrophotometry, nitrogen content was measured using the Kjeldahl, and phosphorus content was measured using the Atomic Absorption Spectrophotometric (AAS). The C/N ratio was obtained using the calculation of C/N × 100 %.

The collected data was analyzed by one–way variance analysis. If there are differences between treatments, further tests are performed using Duncan's Multiple Range Test (DMRT). SPSS version 19 software was used to facilitate analysis.

Each experiment unit consist of 2 kg Buck wet manure, 10 % commercial starter solution ($v\ v^{-1}$) and 5 % molasses. Then, rumen contents were added according to the treatment (0 %, 5 %, 10 % and 15 %) ($v\ v^{-1}$) for composting. The anaerobic composting process was done for 20 d, temperature and organoleptic were checked every 3 d.

3 Results and discussions

The analysis result of the quality of goat manure fertilizer combined with rumen contents at various levels are presented in Table 1.

Table 1. Laboratory analysis of the quality of goat manure fertilizer

No	Parameter	Percentage of rumen contents addition				SNI 19–7030–2004
		0 %	5 %	10 %	15 %	
1.	Total of Nitrogen (%)	1.49	1.49	1.49	1.59	≥ 0.40
2.	Phosphor (%)	0.13	0.14	0.13	0.15	≥ 0.10
3.	Potassium (%)	0.62	0.63	0.69	0.72	≥ 0.20
4.	C organic (%)	31.65	31.30	31.20	30.60	9.8 to 32
5.	C/N	21.2	21.0	20.9	19.24	10 to 20
6.	Moisture content (%)	47.744 5	47.63	47.41	45.86	50

***SNI 19–7030–2004**: Indonesian standard for compost and organic manure

Goat manure fertilizer quality parameters with the addition of rumen contents in various levels were illustrated from the N, P, K, C organic content, C/N ratio and moisture content are explained as follows:

3.1 Total nitrogen content

Table 1 showed that the N content of goat feces fertilizer added by 15 % rumen content resulting in highest content than the others. This was presumably because the addition of 15 % rumen contents, the microorganisms in the rumen were able to work more optimally to decompose goat manure. Nitrogen in the rumen or in goat manure used by microorganisms as a food source to form new cells. Microorganisms decomposed proteins and other nitrogenous organic matter, and ammonia release occurred. After the composting process, the microbes died and become a source of N in goat manure fertilizer. According to [5], the higher the N content of the composting base material, the easier it is to undergo a decomposition process, and the higher the total N compost level will be. The increase in nitrogen content in the composting process is also influenced by the activity of microorganisms in the composting.

3.2 Phosphorus content

Table 1 stated that the phosphorus content from the goat manure increase along with the addition of rumen contents, with the highest P content occurred in 15 % rumen content addition. It was expected that by adding rumen contents more, the organic material would also be higher. Therefore, microorganisms would remodel the goat manure and increase of phosphorus availability. According to [6], the phosphorus content was also influenced by the high nitrogen content. The higher the nitrogen, the the higher the P content.

3.3 Potassium content

Table 1 showed that the potassium content in goat manure which was added 15 % rumen content was highest than the others. This increased in potassium was due to the addition of 15 % rumen content which had more forage residue. The more forage residu, the higher K content. Potassium was used by microorganisms for its activity during the fermentation process. According [7] an increase in potassium due to the composting process will release K^+ ions from cation exchange and decomposition of organic matter dissolved in rumen contents.

https://doi.org/10.1051/e3sconf/202122600048

3.4 C organic content

The content of C organic fertilizer can improve soil properties. C organic functions as a means of photosynthesis in plants and also as a buffer for the supply of nutrients for plants.

Table 1 stated that the C organic content of goat manure fertilizer which was given a higher level of rumen content contained higher C organic compared to the others. This could be due to the addition of rumen contents in goat manure that also added the number of decomposing bacteria (bio–activator) to the material so that it could accelerate the C organic decomposition process. The higher rumen content addition, the higher C organic in goat manure fertilizer. Moreover [8] stated that the quality of organic matter greatly determines by the speed of decomposition and mineralization of organic matter.

3.5 C/N ratio

The composting process is a process of decomposition of organic matter from a high C/N ratio to a low C/N ratio (15 to 25) with an effort to activate the microbial activity of a remodel or decomposer (bacteria, fungi, and actinomycetes). A C/N ratio level is used to assess the maturity of compost produced during the composting process. C/N ratio can be used to predict the mineral mineralization rate of organic matter [9].

Table 1 showed that the C/N ratio of goat manure fertilizer added by the rumen content 15 % was higher than the others. It was due to differences in the percentage of rumen biogas contents in each treatment affecting the work of bacteria in the degradation process. The less rumen content was added, the less worked the bacteria. The small number of bacteria in the rumen contents made the goat manure fertilizer was not ready to use, while nitrogen was used to form organism cells in the degradation process. According to [8] too high ratio of C/N compost resulting N immobilization, because all inorganic nitrogen available in the soil will be converted into organisms' bodies in organic form. Then, there is competition for nitrogen between host plants and bacteria. When the C/N ratio is less than 20, it means that there has been a release of nitrogen from organic matter due to decomposition into the soil [10].

3.6 Moisture content

The resulting moisture content was the lowest by the addition of 15 % rumen content. In composting process, the moisture content would decrease along with the level of compost maturity. According to [6], the efficiency of the composting process will be reduced if the moisture content was too low. Optimal moisture content in composting was 45 % to 55 %. If the moisture content exceeds 60 %, the volume of air was reduced, odors will be generated due to aerobic conditions and decomposition becomes slow.

4 Conclusions

N, P, K, C organic, C/N ratio and moisture content of goat manure fertilizer in this study fulfilled SNI 197030–2004 requirements. The best quality of goat manure fertilizer was occurred from 15 % rumen content addition.

References

1. M.P. Bernal, J.A. Alburquerque, R. Moral. Bioresour. Technol.. **100**,22: 5444–5453(2009). https://doi.org/10.1016/j.biortech.2008.11.027
2. Suhardjadinata, D. Pangesti, T. Tedjaningsih. Jurnal Agro. **5**,1:29–47(2018). [in Bahasa Indonesia]. https://doi.org/10.15575/1675
3. R.A. Wulandari, Y. Trihadiningrum. *Proses komposting limbah padat rumah potong hewan dengan metode Aerobik dan AAO (Anaerobik–Anoksik–Oksik)*. [Composting process of slaughterhouse solid waste by Aerobic and AAO (Anaerobic–Anoxic–Oxy) Methods]. Seminar nasional 2014–Waste management II, (Surabaya, Indonesia 2014). Seminar nasional 2014–Waste management II. [in Bahasa Indonesia]. http://digilib.its.ac.id/ITS-Master-33103150001233/36601
4. N. Kamolmanit, A. Reungsang. J. Water Environ. Technol. **4**,1:33–50(2006). https://doi.org/10.2965/jwet.2006.33
5. R. Guo, G. Li, T. Jiang, F. Schuchardt, T. Chen, Y. Zhao, et al. Bioresour. Technol **112**:171–178(2012). http://dx.doi.org/10.1016/j.biortech.2012.02.099
6. S. Uçaroğlu, U. Alkan. J Air Waste Manag Assoc. **66**,3:288–295(2016). http://dx.doi.org/10.1080/10962247.2015.1131205
7. X.Z. Yu, F.F. Zhang. Appl. Environ. Biotechnol. **2**,1:29–36(2017). http://doi.org/10.26789/AEB.2017.01.004
8. S. Trivedi, O. Chahar, K. Mehta. Journal of Ecology and Environmental Sciences. **6**,1:147–153(2015). http://www.bioinfopublication.org/viewhtml.php?artid=BIA0002501
9. N. Zhu. Bioresour Technol. **98**,1:9–13(2007). https://doi.org/10.1016/j.biortech.2005.12.003
10. M. Clarholm, U. Skyllberg, A. Rosling. Soil Biol. Biochem. **84**: 168–176(2015). https://doi.org/10.1016/j.soilbio.2015.02.019

E3S Web of Conferences **226**, 00049 (2021) https://doi.org/10.1051/e3sconf/202122600049
ICoN BEAT 2019

Exploration and Characterization of "Uwi" Plant (*Dioscorea* sp.) in East Java Uplands, Indonesia

Wuryantoro Wuryantoro[1,*], *Ratna* Mustika Wardhani[1], *Indah* Rekyani Puspitawati[1], *Praptiningsih* Gamawati Adinurani[1], and *Bohari* Mohammad Yamin[2]

[1]Department of Agrotechnology, Faculty of Agriculture, Merdeka University of Madiun, Jl. Serayu no.79, Madiun City 63133, East Java, Indonesia
[2]Department of Chemistry, Faculty of Science and Technology, Kebangsaan University of Malaysia, 43600 UKM, 43600 Bangi, Selangor, Malaysia

Abstract. "Uwi" *(Dioscorea* sp.*)* is a tuber plant species with potentials to support the national food diversity program because it has high tolerance when planted in upland areas, which are wide in Indonesia and beyond. The study aims to rescue germplasm of "Uwi" plants through exploration, identification, and collection for the sake of national long–term goal of providing alternative food sources to support food security. The research began with exploration of the west upland area of East Java, and continued with specimen planting for characterization testing in the following year. Exploration resulted in 127 samples of variants observed throughout uplands in Madiun, Ponorogo, Ngawi, Magetan and Pacitan, East Java, Indonesia with each result of 28, 26, 9, 20 and 44 variants. Five big groups of *Dioscorea* family namely *Dioscorea alata, Dioscorea aculeata, Dioscorea hispida, Dioscorea pentaphylla* and *Dioscorea bulbifera* were mapped out as a result. Using SPSS cluster analysis, *D. alata* as the most variant were specified into 20 groups and further gathered into seven clusters based on its main botanical characteristics of leaf shape, leaf color, stem color, tuber color, tuber shape, and skin color of the tuber. Meanwhile, *D. aculeata* and *D. pentaphylla* were each detailed into two sub–clusters and other types were each detailed into one.

Key words: Characterization, cluster analysis, food securiy, germ plasm, tubers plant potentials

* Corresponding author: wuryantoro@unmer-madiun.ac.id

1 Introduction

Indonesia's food provision is deteriorating due to the vast functional shift of agricultural land nationwide, and it has gone so low that even the international Food and Agriculture Organization (FAO) considers it serious in the global hunger index. It is said that there will be food scarcity in near future as a result of various events, such as environmental damage, land conversion, high fossil fuel prices, and climate change. In fact, the nation's critical land area has reached 24 303 294 ha [1] and continues to grow. Food is an essential need, and therefore requires intense implementation to secure it.

One critical problem currently exists in Indonesia regarding major food commodity is the high reliance on rice – it is so high that scarcity often occurs. This is not the case in ancient times. Back then, not all regions in Indonesia consumed rice since many of them were affixed to their local staples. That Indonesia is naturally an archipelagic country with a variety of ecosystems actually makes it perfect for staple diversification. Most staple foods are carbohydrate–based due to its fulfilling and energizing effect, thus the reason w–hy rice is popular. Its domination, however, diminish the chance of other commodities to grow; the low development on other carbohydrate–based sources is evident. Considering that food security relations and sustainable agriculture development using local planning will result in multidimensional roles and impact food security in terms of availability, access, and utilization [2], attempts made to increase food self–sufficiency should not only be focused on rice and wheat, but also be supported by other types of strategic commodities, e.g. tubers, food–producing trees – like sago (*Metroxylon sagu* Rottb), bread fruit [Artocarpus altilis (Parkinson) Fosberg], "aren" [Arenga pinnata (Wurmb) Merr.] – and other multipurpose plant species. When consumption is generated from various sources and, at the same time, specific food source commodities are maintained, food security shall be steady. There is no single food ingredient in the world that contains all the nutrients human body needs ideally. In order to meet the nutritional needs, people should also consume carbohydrate sources other than rice. There are many types and variety of potential bulbs spread all nationwide over as alternative food for rice substitute. Researches on food, therefore, should emphasize more on its availability, accessibility, and security. Study [5] shows that food provision in accordance with each region's potential based on what is available in each respective region will greatly facilitate the community in meeting their needs. Compared to rice cultivation, which tends to exploit irrigated land leading to quick land deterioration, upland (non irrigation) cultivation should be capable of not only supporting non–rice diversification food products, but also protecting the environment, with proper management and adequate cultivation system. What occurs in Beneue State of Nigeria [9], where yam is the main staple food consumption as well as source of income and employment generation, can occur in Indonesia, too.

"Uwi" plant (later referred as "uwi"), due to its tolerant nature to grow upland and its high productivity, is a potential source of carbohydrate in food diversification. A member of the genus Dioscorea, it is generally a climbing shrub with heart–shaped leaves. Producing tubers in the soil, some may also have hanging (aerial) ones. To obtain high production, planting at the beginning of the season is highly recommended [11, 12]. Result of Boutau Yam (*D. alata*) study indicates that the optimum harvesting time is the ninth month after sowing, since it gives the highest value of energy, carbohydrates, and dry matters [14]. The research shows that in Nigeria, the production and consumption of "uwi" continued to increase from 1996 to 2006 with a relative increase of more than 400 %, indicating the potential of "uwi" as food for the future is getting more important [10]. Production was majorly constrained by lack of capital, scarcity/high cost of seed yams, high cost of labor, and pests and diseases infestation [13]. Research in Madiun residence, East Java, Indonesia also shows that *D. alata* is the type with the most diversity and a broad distribution [15].

The high diversity in "uwi" should allow production of various types of processed foods. These non–rice plants, after processed into flour, can take part as an analog rice material. Their nutritional values, combined with other materials, should be adjustable to suit various preferences. For illustration, "uwi" bulb contains low glycemic index value the way analog rice of potatoes, cassava flour, and carrots do [16], while ones in purple sweet potatoes and white bread is 67 [17]. Referring to [6] studies, the starch extracted from purple *Dioscorea alata* and yellow ones grown in Jambi Province, Indonesia have different physical properties – starch of purple bulb has lower swelling power but higher solubility compared to the other. The protein content of U–type strains of *D. alata* is relatively higher compared to other tuber plants, which is a useful dietary nutrient [7]. It is even mentioned that diosgenin and aglycone, chemicals found in Dioscorea, are used commercially in pharmaceutical industry [8]. There are other types of *D. alata* that need more in–depth study.

Despite its benefits and potentials for agricultural development, "uwi" is getting rare. Fearing that its God–gifted genetic source (germplasm) should extinct in near future, efforts to conserve it – as well as other carbohydrate source plants – should be engaged. Their strategic values for long–term food security must be priceless, considering how the population continues to increase contradicting the decreasing agricultural land both quantitatively and qualitatively. Therefore, it is necessary to research "uwi" for the purpose of exploring, identifying, and preserving its germplasm to make future food security real.

2 Materials and methods

2.1 Place and time

The research was conducted in December 2016 to July 2017, located in the experimental station which is the property of the Faculty of Agriculture of Merdeka University of Madiun, East Java, Indonesia.

2.2 Materials and tools

The materials used in this research are tuber results of the previous year's observation, media of soil planting and fertilizer, and pesticide. The equipment used includes various observation tools consisting of a color chart, roll meter, scales, and data recording equipment.

2.3 Experimental design

The previous year's observation has 127 result sample collection, taken from five districts of the ex–residential area of Madiun, East Java, Indonesia, consisting of five groups of Dioscorea species in descending quantity order *D. alata, D. aculeata, D. hispida, D. pentaphylla*, and *D. bulbifera*. Each of these samples was planted in a planter bag using the same media. The research employed Complete Randomized Design with three replications. Observations were conducted towards the main botanical characteristics including leaf shape, leaf color, leaf base color, thorn stem, aerial tubers, shoot color, tuber shape, branch tuber, tuber skin color, and tuber color.

2.4 Data analysis

Samples were distinguished by *alata* species (*D. alata*) and non–*alata* species (*D. pentaphylla, D. aculeata, D. hispida,* and *D. bulbifera*), and data obtained were analyzed using SPSS cluster analysis to see the proximity of one sample to another. The analyses were based on the main distinguishing characters mentioned above.

3 Result and discussion

From observing "uwi" crops in five regencies of ex–residency of Madiun, 127 variants were noted. The most variants were found in Pacitan district, followed by Madiun, Ponorogo, Magetan, and Ngawi. This indicates that the higher the district is, the more types of "uwi" grow. It was also found that "uwi" was rarely cultivated by farmers, and some

 https://doi.org/10.1051/e3sconf/202122600049

variants were even difficult to obtain in some places. Apparently, farmers in some areas of Java and Sumatra prefer cultivating cassava, sweet potato, and taro [14].

After planting under controlled land conditions, the results of the identification showed that of the five species found, *D. alata* species has the most variants. It is presumably due to its good taste, a trait of white tubers, that this species is fairly likeable.

The data were then clustered into two groups – *alata* and non–*alata* – since the first group varied more than the latter. Generally, they are quite easy to distinguish from their winding directions, the shape of their leaves, the shape of their tubers, and the existence of stem thorns. This study has discovered that 20 variants in *alata* group produced much larger bulb yields in sizes, indicating that they contain rich germplasm potential to food provision.

The classification using SPSS software resulted in plant grouping as in Figures 1–4. In Figure 1, the dendrogram shows how 20 variants of *D. alata* taken from five different areas are organized into seven clusters, mainly characterized by their bulb shapes and bulb colors, which is estimated to be due to different nutrient contents in each type. Further information regarding the seven clusters is detailed in Figure 2.

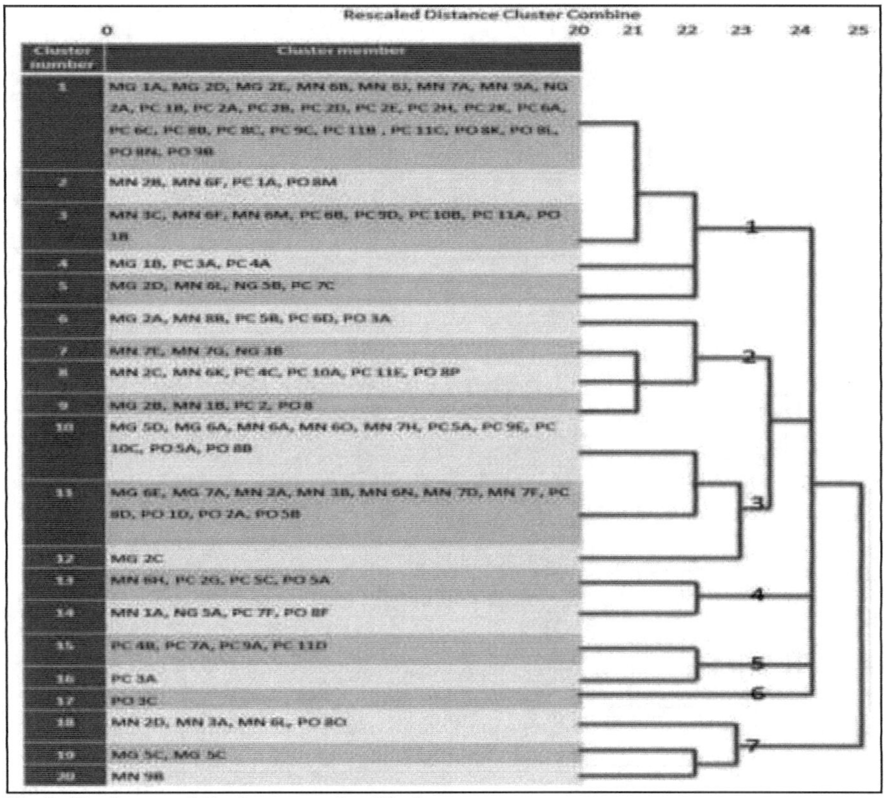

Fig. 1. Dendogram of *D. alata* group (modified)

Notes. MN: Accession from Madiun, PO: Accession from Ponorogo, PC: Accession from Pacitan, MG: Accession from Magetan, NG: Accession from Ngawi.

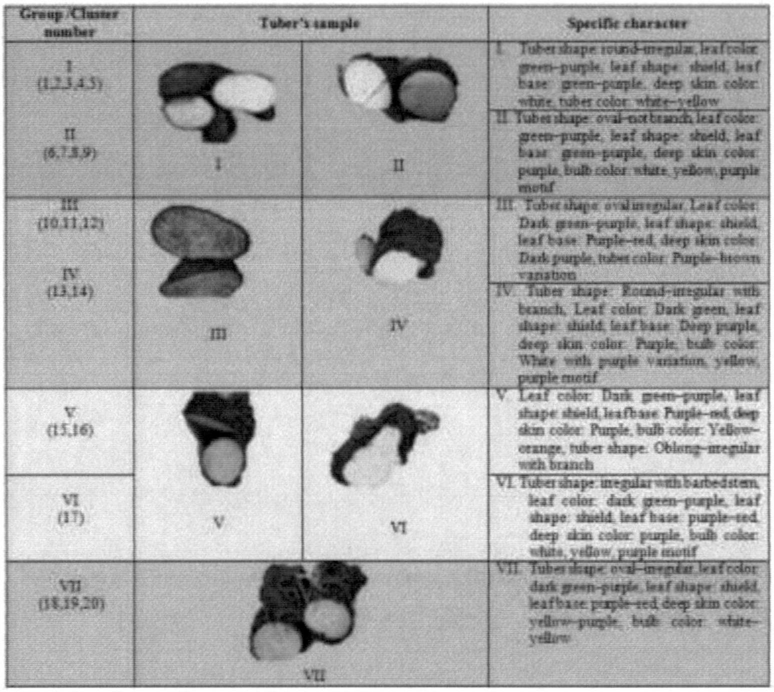

Fig. 2. Characteristics of seven clusters of *D. alata*

A similar case occurred in clusters II, III, IV and VII, which is the difference is due to different nutrients in each type. Characteristics of cluster V are only found only in Pacitan, and ones of group VI are only found in Ponorogo. Above evidences indicate that variants obtained from aforementioned regions are actually related to ones researched by Trimanto regarding "uwi" in Nganjuk, another region in East Java, Indonesia [9].

As for non–*alata* group, 22 variants are organized into *D. aculeata*, *D. hispida*, *D. Pentaphylla*, and *D. Bulbifera* – Figure 3 shows its dendogram in clusters. Out of all clusters, *D. aculeata* (cluster I) is the largest one, characterized by shape, number, and size of tubers. A bunch of fewer than five larger, round–shaped bulbs belong to G*embolo*, while smaller five–or–more bulb bunches go with *Gembili* (thus the two sub–clusters). *Gembili* are then further set apart to be *Gembili Brul* for the smaller tubers and *Gembili Teropong* for the bigger ones. Specifically for *Gembolo*, it is also characterized by the number of thorns both on the stem and roots. Another indication of this cluster is that the wider the leaf is, the larger the tuber will be. *D. hispida* (*Gadung*) was not found in the assessed collection and put into one separated cluster. *D. pentaphylla (Uwi Katak)* has two sub–clusters characterized by not only size and shape of tubers but also origins. *Bulbifera (Uwi Gandul)*, despite difference in origins, is in one same cluster due to their similarities. Points above ground the detailing in Figure 4.

 https://doi.org/10.1051/e3sconf/202122600049

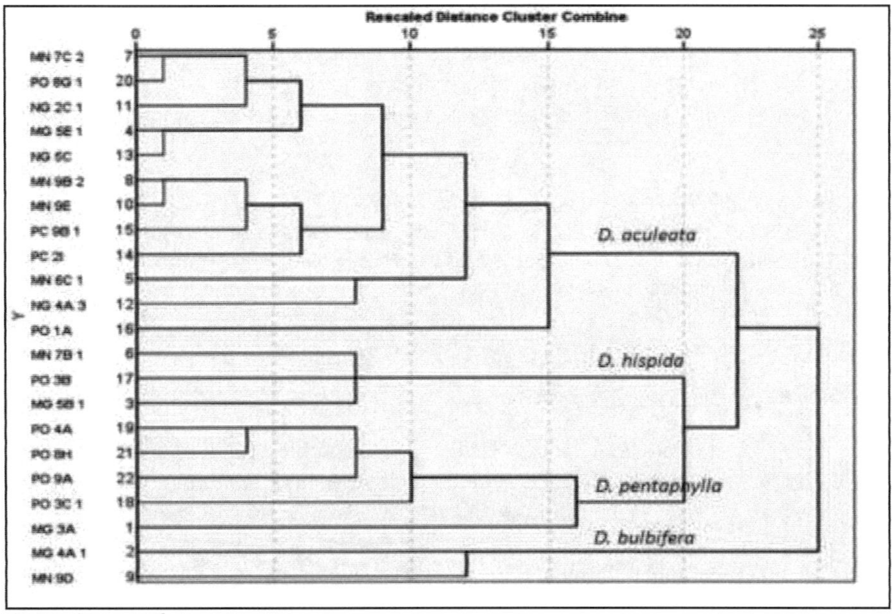

Fig. 3. Dendrogram of non–*alata* group

Notes: MN : Accession from Madiun, PO : Accession from Ponorogo, PC : Accession from Pacitan, MG : Accession from Magetan, NG : Accession from Ngawi

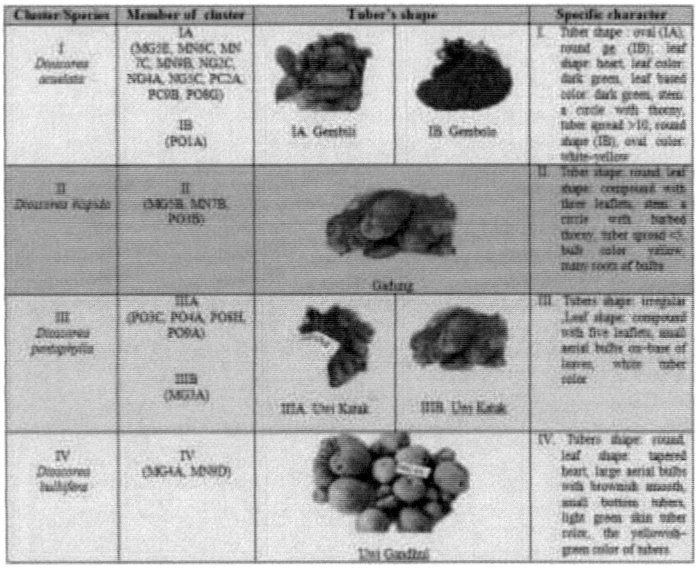

Cluster/Species	Member of cluster	Tuber's shape		Specific character
I *Dioscorea aculeata*	IA (MG5E, MN6C, MN 7C, MN9B, NG2C, NG4A, NG5C, PC2A, PC9B, PO8G) IB (PO1A)	IA. Gembili	IB. Gembolo	I. Tuber shape : oval (IA), round ge (IB); leaf shape: heart, leaf color: dark green, leaf based color: dark green, stem: a circle with thorny, tuber spread >10, round shape (IB), oval color: white-yellow
II *Dioscorea hispida*	II (MG5B, MN7B, PO3B)	Gadung		II. Tuber shape: round leaf shape: compound with three leaflets, stem: a circle with barbed thorny, tuber spread <5, bulb color: yellow, many roots of bulbs
III *Dioscorea pentaphylla*	IIIA (PO3C, PO4A, PO8H, PO9A) IIIB (MG3A)	IIIA. Uwi Katak	IIIB. Uwi Katak	III. Tubers shape: irregular ,Leaf shape: compound with five leaflets, small aerial bulbs on-base of leaves, white tuber color
IV *Dioscorea bulbifera*	IV (MG4A, MN9D)	Uwi Gandhul		IV. Tubers shape: round, leaf shape: tapered heart, large aerial bulbs with brownish smooth, small bottom tubers, light green skin tuber color, the yellowish-green color of tubers

Fig. 4. Characteristic of *D. aculeata*, *D. hispida*, *D. pentaphylla* and *D. bulbifera*.

4 Conclusions

From the analyses presented, it is conclusive that "uwi", with a vast variants available locally and rich in carbohydrates, is a strong candidate to support national food diversification. When strategically developed, it should contribute well in sustaining food provision along with other commodities. Since the government's policy is proven to be essential in increasing rice productivity, one to expand access to credit markets for "uwi" farmers is also key to implementing food security programs. Efforts to manage "uwi" effectively and efficiently are also key to sustainable food security, and it calls for further role of education in nutrition and disease studies more effectively and more efficiently.

The authors would like to express their gratitude to DRPM Kemenristekdikti for funding this study under Contract Number 034/SP2H/K2/KM/2017. The authors would also like to thank the Rector of Merdeka University of Madiun, East Java, Indonesia as well as the Dean of the Faculty of Agriculture and all team members and students for all the supports.

References

[1] Kementerian Lingkungan Hidup dan Kehutanan Republik Indonesia. *Statistik kementrian lingkungan hidup dan kehutanan* [Statistics of the ministry of environment and forestry]. Pusat data infomasi Jakarta. p. 1–20(2015). [in Bahasa Indonesia]. [Online] from https://scholar.google.com/scholar?q=Kementerian+Lingkungan+Hidup+dan+Kehutanan+Republik+Indonesia+2015+Statistik+kementrian+lingkungan+hidup+dan+kehutanan+%28Jakarta%3A+Pusat+data+infomasi+Jakarta%29+1-20 (2015). [Accessed on October 2018].

[2] M. Noer. Int. J. Adv. Sci. Eng. Inf. Technol. **6**,3:277–280(2016). https://www.researchgate.net/profile/Melinda_Noer/publication/304455451_Bridging_Food_Security_and_Agriculture_Development_through_Regional_Planning/links/5d82cae0458515cbd1984204/Bridging-Food-Security-and-Agriculture-Development-through-Regional-Planning.pdf

[3] E. Wicaksono. Int. J. Adv. Sci. Eng. Inf. Technol. **4**,5:322–325(2014). https://doi.org/10.18517/ijaseit.4.5.427

[4] M. Aslam, S. Rasool. Int. J. Adv. Sci. Eng. Inf. Technol. **4**,3:19–23(2014). https://doi.org/10.18517/ijaseit.4.3.388

[5] A.D. Santoso, Warji, D.D. Novita, Tamrin. J. Tek. Pertan. Lampung. **2**,1: 27–34(2013). [in Bahasa Indonesia]. http://jurnal.fp.unila.ac.id/index.php/JTP/article/viewFile/143/151

[6] Ulyarti, Lavlinesia, D. Fortuna, Surhaini. Int. J. Adv. Sci. Eng. Inf. Technol. **6**,4: 456–459(2016). https://www.researchgate.net/publication/307850296_The_Study_of_Physico-Chemical_Properties_of_Dioscorea_alata's_Starch_From_Jambi_Province

[7] F.D. Baah, B. Maziya–Dixon, R. Asiedu, I. Oduro, W.O. Ellis. J. Food Agric. Environ. **7**,2:373–378(2009). http://dspace.knust.edu.gh/bitstream/123456789/9766/1/Nutritional%20and%20Biochemical%20composition%20of%20D.%20alata.pdf

[8] B. Dutta. J. Pharmacogn. Phytochem. **3**,4:15–18(2015). http://www.phytojournal.com/archives/2015/vol3issue5/PartA/7.1.pdf

[9] N. Verter, V. Becvarova. Eur. Sci. J. **10**,31:27–42(2014). https://www.researchgate.net/publication/269992864_Yam_Production_as_Pillar_of_Food_Security_in_Logo_Local_Government_Area_of_Benue_State_Nigeria

[10] A.G. Ironkwe, R. Asiedu. African J. Root Tuber Crop. **11**,1:56–64(2014). https://scholar.google.co.id/scholar?hl=id&as_sdt=0%2C5&q=Women+Farmers+in+Seed+Yam+Production%E2%80%AF%3A+Implication+for+Increased+Productivity+and+Sustainable+Yam+Improvement+in+Southeastern+Nigeria&btnG=

[11] F.O. Tobih, L.U. Okonmah, A.A. Omoloye. Int. J. AgriScience. **1**,3:178–184(2011). https://www.cabdirect.org/cabdirect/abstract/20123239776

[12] A.M.N. Benjamin. Eur. J. Res. Soc. Sci. **2**,3:1–9(2014). https://www.idpublications.org/wp-content/uploads/2014/08/AN-EVALUATION-OF-FARMERS%E2%80%99-ADOPTION-OF-YAM-MINI-SETT-TECHNIQUE-IN-CROSS-RIVER-STATE-NIGERIA.pdf

[13] J.T. Omojola. Adv. J. Agric. Res. **2**,6:99–103(2014). https://fliphtml5.com/ocso/wdjl/basic

[14] E.N. Ellong, C. Billard, D. Pétro, S. Adenet, K. Rochefort. J. Exp. Biol. Agric. Sci. **3**,2:139–150(2015). https://www.researchgate.net/publication/275723386_Physicochemical_nutritional_and_sensorial_qualities_of_Boutou_Yam_Dioscorea_alata_varieties

[15] Wuryantoro, R.M. Wardhani, I.R. Pupitawati. *Plasma nutfah tanaman Uwi, karunia Tuhan yang harus diselamatkan* [Germplasm uwi plant, God's gift that must be saved]. Prosiding Seminar Hasil Penelitian Tahun 2017. p. 1–20(2017). [in Bahasa Indonesia]. https://scholar.google.com/scholar?q=Wuryantoro%2C+Wardhani+R.+M.+and+Rekyani+I.+2017+Plasma+Nutfah+Tanaman+Uwi%2C+Karunia+Tuhan+Yang+Harus+Diselamatkan+Prosiding+Seminar+Hasil+Penelitian+Tahun+2017+1-20

[16] T. Anggraini, V.J. Putri, Neswati, Yuliani. Int. J. Adv. Sci. Eng. Inf. Technol. **6**,5:723(2016). http://insightsociety.org/ojaseit/index.php/ijaseit/article/view/762/pdf_257

[17] E. Aritonang, A. Siagian, F. Izzati. Int. J. Adv. Sci. Eng. Inf. Technol. **7**,2: 580–586(2017). https://www.researchgate.net/publication/316986520_Mixed_cooked_rice_with_purple_sweet_potato_is_potential_to_be_the_low_glycemic_index_food_and_staple_food_alternative

Sweet Potatoes: Development and Potential as Alternative Food Ingredients in Karanganyar Regency, Indonesia

Yoesti Silvana Arianti*, and *Yos* Wahyu Harinta

Faculty of Agriculture, Department of Agribusiness, Universitas Veteran Bangun Nusantara, Jl. Letjend Sujono Humardani No.1, Sukoharjo 57521, Indonesia

Abstract. Sweet potato is a local food that can reduce dependence on rice and flour consumption. It is the main source of carbohydrates, after rice, cassava, flour and corn. The benefits and potency of sweet potatoes as alternative food needs to be developed, especially in rural areas. However, sweet potato has not been considered as an important and high economic value commodity in Indonesia. Sweet potatoes have been used as food and non-food raw materials in developed countries i.e. noodles, fried sweet potatoes, desserts, confectionery, soy sauce, flour, wine, vinegar, nata de coco, bioethanol and others. Around 89 % of sweet potatoes in Indonesia are cultivated for providing food to rural communities, the rest are used for industrial raw materials and animal feed. The content of sweet potatoes includes carbohydrates, vitamins, and minerals. Yellow/orange sweet potato is rich in beta-carotene and purple sweet potato contains anthocyanin (antioxidants). Sweet potato production is still limited to traditional food that is less attractive compared to flour products. Meanwhile, intermediate products have been developed including flour, instant flour, and starch that can be used as a substitute for flour in pastry products, wet cakes, breads, and noodles.

Keyword: Consumption, flour, food diversification, intermediate product

1 Introduction

Diversification program has been planned by the Indonesian government since 1974. This is due to the high consumption of rice by the community. Apart from rice, Indonesian people also consume a lot of wheat and various processed wheat. Therefore, the import figures for each year are quite high [1]. Food diversification will provide high benefits if it is able to explore, develop and optimize the utilization of local food sources and local wisdom. Indonesia has various types of food plants that can be grown under any conditions, but their existence is not really considered because they are not a staple food ingredient [2].

Amount one of the local food sources that can be developed is sweet potato. The ideal consumption of sweet potatoes in the Indonesian *PPH–Pola Pangan Harapan* [Food Hope Pattern] for 2012 is 100 g per capita d^{-1}. Meanwhile, the real consumption of Indonesian about 40 g per capita d^{-1} or around 6 % of ideal consumption. Local food ingredients as

* Corresponding author: silvanayusti@gmail.com

Indonesia's wealth need to be preserved for their cultivation and its benefits. At present, the grains group still dominates, so root consumption needs to be encouraged and increased thus the new PPH score reaching 75.5 in 2012 can meet the target score of 93.9 in 2014 and 100 in 2020 [5].

Sweet potato (*Ipomoea batatas* L. Lam.) is widely cultivated in various countries, including Indonesia, which ranks sixth as an important food ingredient after rice, wheat, potatoes, corn and cassava. Apart from being used as a source of carbohydrate food, sweet potatoes can also be used as industrial raw materials, raw materials for animal feed and alcohol production. This annual plant has a low glycemic index (IG) that is good for diabetics [6]. Food with a low glycemic index can be beneficial in reducing the incidences of chronic diseases [7]. Sweet potatoes may be beneficial to persons with type two diabetes, resulting from the high fiber and manganese content, which could aid in stabilizing blood sugar levels and reduce insulin resistance [8]. In addition, sweet potatoes is one of the Indonesian foodstuff rich in dietary fiber and olygosaccharides [9].

Besides having a role in the supply of food, sweet potatoes also can be used as industrial raw materials and animal feed. But the use of sweet potatoes is limited, that are considered as low products (inferior) such as raw materials for sauces and traditional foods such as boiled/fried sweet potato, *kolak*, *getuk*, *timus* and chips. Therefore, the utilization and consumption of sweet potatoes need to be improved through innovative diversification products that are attractive, nutritious, and have added value with simple technology [10]. The development of sweet potato products requires the support of the continuity and availability of these commodities. This paper aimed to reveal the growth of harvested area, production and productivity of sweet potato. It is expected to provide implications for sweet potato development opportunities in the future to support the success of accelerated food consumption diversification program in Karanganyar Regency.

2 Methods

The data and information used in this paper were derived from secondary data from various sources as well as a review of research in the field of food diversification. The main data used was the growth of harvested area, production and productivity of sweet potatoes in Karanganyar Regency, Indonesia.

Data collected by the documentation method that is studying, recording archives or data relating to research problems [11]. This method is carried out by taking data from relevant agencies, including Central Java Province BPS–*Badan Pusat Statistik* [Indonesia Statistics Agency], Karanganyar Regency Statistics, Karanganyar Regency Agriculture Office, Karanganyar Regency Food Security Agency and other secondary data derived from a review of research. Data were analyzed using a descriptive analysis method which is used to explain the facts and findings of the survey [11].

3 Results and discussions

3.1 Growth of harvested area, production and productivity of sweet potatoes in Karanganyar regency

3.1.1 Harvest of sweet potatoes

Sweet potatoes harvested areas in Karanganyar regency ranks fourth out of 25 districts and cities in Central Java Province. The average of sweet potato harvested areas in Karanganyar

Regency in the period of 2012 to 2017 was 815 ha. The average share or contribution of the sweet potato harvest area in Karanganyar Regency compared to Central Java Province, Indonesia was 10.67 %.

Table 1. Harvested area (ha) of sweet potatoes in Karanganyar Regency in 2012 to 2017

	Year					
	2012	**2013**	**2014**	**2015**	**2016**	**2017**
Harvested area (ha)	758	642	860	868	868	894
Share (%)	9.48	6.41	9.50	12.27	12.27	14.08
Average growth per yr^{-1} (%)						4.516

Source: Statistics Indonesia, 2013 to 2018

The average growth of sweet potatoes harvested area per year in Karanganyar Regency is 4.52 %. Sweet potato harvested area in Karanganyar Regency ranks fourth after Semarang Regency, Magelang Regency and Wonosobo Regency [4]. Sweet potatoes have the advantage that is capable of adapting in less fertile and dry areas.

3.1.2 Sweet potato production

In 2012 to 2017 the average sweet potato production in Karanganyar Regency was or 30.26 t. These conditions make Karanganyar Regency the largest contributor of sweet potato production in Central Java Province, Indonesia. The contribution of sweet potato production in Central Java Province was 10.67 %. The average annual growth rate for sweet potatoes in Karanganyar Regency is 3.81 %.

Table 2. Production (t) of sweet potatoes in Karanganyar

	Year					
	2012	2013	2014	2015	2016	2017
Production (t)	32.2130	27.0590	23.8760	31.0760	31.0760	36.2500
Share (%)	19.29	14.73	13.31	20.54	20.54	24.99
Average growth yr^{-1} (%)						3.806

Source: Statistics Indonesia, 2013 to 2018

Sweet potato varieties cultivated by farmers in Karanganyar Regency are cv Manohara and cv Japanese Sweet Potato. Sweet potato yields in Karanganyar Regency were mostly sold entirely by farmers (79.66 %), amount 16.96 % were sold and those used for self-consumption (not sold) amounted to 3.38 % [11]. Based on an interview with one of the sweet potato farmers in Karanganyar Regency, the current yield of sweet potatoes can reach 40 t ha^{-1} to 45 t ha^{-1} with an average selling price of IDR 3 000 kg^{-1}. The production cost is around IDR 40 000 000 thus the income is approximately IDR 12 000 000 mo^{-1} and the profit is IDR 80 000 000 for a period of 6 mo [12].

3.1.3 Sweet potato productivity

Sweet potato productivity in Karanganyar Regency tended to increase with an average of 37.254 t ha^{-1} during 2012 to 2017. The average productivity growth per year is 1.55 %.

Table 3. Productivity (t ha^{-1}) sweet potatoes in Karanganyar Regency 2012 to 2017

	Year					
	2012	**2013**	**2014**	**2015**	**2016**	**2017**
Production (t)	32.2130	27.0590	23.8760	31.0760	31.0760	36.2500
Share (%)	19.29	14.73	13.31	20.54	20.54	24.99
Average growth yr^{-1} (%)						3.806

Source: Statistics Indonesia, 2013 to 2018

Karanganyar Regency's productivity is in the second position after Brebes Regency. Based on the results of the Balai Penelitian Tanaman Aneka Kacang dan Umbi (Balitkabi) Malang showed that the productivity of sweet potatoes can reach 25 t ha^{-1} to 35 t ha^{-1} and the productivity of sweet potatoes in Karanganyar Regency can already exceed that number.

3.1.4 Opportunities for the development of sweet potatoes in supporting the local food based diversification program

Based on the data, the development of production, productivity and size of sweet potato in Karanganyar Regency is on a positive trend. Continuity of production supports the demand for sweet potatoes for food, feed ingredients, and raw materials for the processing industry (agroindustry). Sweet potato agroindustry needs to be developed because its current product is less varied so it is considered an inferior product. Based on [13], Indonesian and Sub-Saharan Africa are a like in sweet potato consumption that are mostly consumed by rural communities for breakfast menus or fried, steamed or baked snacks.

The development and existence of local commodity-based agro-industry as an effort to diversify food is very dependent on the availability of main raw materials. Therefore, increasing the production of local commodities needs to be considered along with the development of the agroindustries. Agroindustry, which is usually small and medium scale enterprises, plays an important role in rural social and economic development, plays a role in GDP, and is able to survive times of economic crisis compared to large companies [14]. Challenges for diversification of food based on local food include:

i. Rice is a superior food commodity

The paradigm that develops in society in general is rice as a staple food which has a higher social status because it is a superior commodity. For Indonesians, foods such as potatoes and sweet potatoes are less popular than rice. Moreover, the government provides rice subsidies for the community, allowing an increase in the amount of rice consumption compared to other foodstuffs. [15]

ii. Concept of eating

For the people of Indonesia, rice consumption can be done at least three times to four times a day. Especially for people living in urban areas, consuming rice indicates that their income is higher than others. Therefore, the government can campaign for more profit if people consume staple foods other than rice. For example, sweet potatoes have lots of vitamins and minerals, lower sugar content and higher carbohydrates than rice [16].

iii. Rice is more nutritious and easy to process

Rice has advantages over other carbohydrate sources such as higher energy and protein compared to corn and cassava. In addition, the taste tastes better, easier and more practical and takes less time to process.

iv. Non-rice food processing technology and its promotion is still limited

Innovation in processing technology, attractive packaging and intensive promotion will produce sweet potatoes that have higher quality, attractive, liked and affordable by the public.

Sweet potatoes is very suitable to be cultivated in the tropics. The opportunity for developing sweet potatoes in Karanganyar Regency is very large due to the relatively stable production and productivity of sweet potatoes. This advantage can be used for complementing feed stuff and supplying raw material of industrial enterprise [17]. But some improvement are needed such as government policies for supporting the growth of the sweet potato industry, to set forth related technical specifications, to carry out research and improvement on new post harvest technology of sweet potato, to improve the operational system of enterprises and explore the market volume products, to do research and development better quality equipments as well as to increase the extraction rate of starch and improve its quality and better food criteria, solve the problem of environmental pollution [18]

The development of agro-industries, especially those that process sweet potatoes, is feasible to be developed by the government. Rural-based agro-industry can reduce the rate of urbanization, can easily be established with minimal skill. They also contribute substantially to the country's gross domestic product, export earnings and development of employment opportunies [19].

Diversification of sweet potato products needs to be done to elevate the degree to superior food, for example processed into flour, noodles, milk, and others [20]. Attractive packaging also needs so the people will be interested in buying sweet potato products. The existence of these steps is expected to help the government program on diversification of food made from local food, thus it can reduce the consumption of rice and flour in Indonesia.

4 Conclusion

Based on the results of the study, it concluded that the development of sweet potatoes in Karanganyar Regency is still wide open. Sweet potatoes can be used as an alternative diversification to substitute rice which contains many nutrients that are good for the body. Intensification efforts can be made to increase sweet potato production. In addition, counseling and training on processed sweet potato product preparations can be informed to the public. It aims to make people get a greater profit than just selling fresh sweet potatoes.

References

1. S. Budijanto, N.D. Yuliana. J. Dev. Sust. Agric. **10**:7–14(2015). https://www.jstage.jst.go.jp/article/jdsa/10/1/10_7/_pdf/-char/ja
2. M.A. Trisia, K. Osozawa, H. Bai. *How to Feed 311 Million of Indonesian People by 2050? Advancing Local Food Adaptation and Food Security Policy*. KnE Life Sciences. ICoA Conferences Proceedings. **3**:49–54(2016). https://www.knepublishing.com/index.php/KnE-Life/article/view/417
3. J. Kumara. J. Adv. Manag. Sci. **2,**1:65–71(2014). https://www.researchgate.net/publication/275595148_Staple_Food_Balance_Sheet_Coefficient_of_Variation_and_Price_Disparity_in_Indonesia

4. Badan Pusat Statistik. *Provinsi Jawa Tengah dalam Angka 2018* [Central Java Province in Figures 2018] [online] from https://jateng.bps.go.id/publication/2018/08/16/73e4da7b5902cb91ff591bc7/provinsi-jawa-tengah-dalam-angka-2018.html (2018). [Accessed on 29 March 2019]. [in Bahasa Indoensia]
5. Pusat Data dan Sistem Informasi Pertanian. Buletin Konsumsi Pangan. **10**,2:1–78(2019). [in Bahasa Indonesia]. http://epublikasi.setjen.pertanian.go.id/arsip-buletin/53-buletin-konsumsi/677-buletin-konsumsi-vol-10-no-2-2019
6. A.M. Tangapo, D.I. Astuti, P. Aditiawati. Agric. Nat. Resour. **52,**4:309–316(2018). https://www.sciencedirect.com/science/article/pii/S2452316X18300681
7. P.S.B-Singh, C.K. Riley, A.O. Wheatley, H.I.C. Lowe. J. Nut. Metab. **2011**:1–6 (2011). https://www.ncbi.nlm.nih.gov/pmc/articles/PMC3205609/
8. E. Aritonang, A. Siagian, F. Izzati. Int. J. Adv. Sci. Eng. Information Technol. **7,**2:580–586 (2017). https://core.ac.uk/download/pdf/194818326.pdf
9. M. Astawan, S. Widowati. Indonesian J. Agric. Sci. **12,**1:40–46(2011). http://ejurnal.litbang.pertanian.go.id/index.php/ijas/article/view/583
10. G. Erlina, R. Yulifianti, M. Jusuf. Jurnal Pangan. **23,**2:194–207(2014). [in Bahasa Indonesia]. http://www.jurnalpangan.com/index.php/pangan/article/download/63/57
11. N. Nilamsari. Wacana, **8,**2:177-181(2014). [in Bahasa Indonesia] https://pdfs.semanticscholar.org/1d5a/7144eeabb9b9c3d60ccb5461df09d3a02868.pdf
12. J. Hamdani, C. Wirawan. Procedia Economics and Finance. **4**:223–233 (2012). https://www.sciencedirect.com/science/article/pii/S2212567112003371
13. J. Low, A. Ball, S. Magezi, J. Njoku, R. Mwanga, M Andrade,K. Tomlins, T.V. Mourik. African J. Food Agric. Nut. Dev. **17**,2:11955–11972(2017). https://www.ajol.info/index.php/ajfand/article/view/155126
14. D.R. Panuju, K. Mizuno, B.H. Trisasongko. J. Saudi Soc. Agric. Sci. **12**:27–37(2013). https://www.sciencedirect.com/science/article/pii/S1658077X12000136
15. A. Widyanti, I. Sunaryo, A.D. Kumalasari. JISSAAS. **20,**1:93–103(2014). https://agris.fao.org/agris-search/search.do?recordID=PH2014000946
16. Y. Widodo, S. Wahyuningsih, A. Ueda. Procedia Chemistry. **14**:493–500(2015). https://www.sciencedirect.com/science/article/pii/S1876619615000674
17. M. Jusuf, E. Ginting. Energy Procedia. **47**:173–179(2014). https://www.sciencedirect.com/science/article/pii/S1876610214002276
18. M.O. Agwu. European J. Sust. Dev. **3,**1:101–114(2014). https://www.researchgate.net/publication/304417683_Issues_Challengesand_Prospectsof_Small_and_Medium_Scale_Enterprises_SMEs_in_Port-Harcourt_City
19. H. Kusumayanti, N.A. Handayani, H. Santosa. Procedia Environ. Sci. **23**:164–167(2015). https://www.researchgate.net/publication/271772250_Swelling_Power_and_Water_Solubility_of_Cassava_and_Sweet_Potatoes_Flour

Study of Agronomic Characteristics of Robusta Coffee at Coffee Plantations in Temanggung, Indonesia

Yohana Theresia Maria Astuti[1,*], *Enny* Rahayu[1], *Tri* Nugraha Budi Santosa[1], *Dian* Pratama Putra[1], *Agus* Solifudin[2], *Yureana* Wijayanti[3], and *Marcus* Fittkow[4]

[1]Agriculture Faculty, Institut Pertanian Stiper, Jl. Nangka II, Sleman, Special Region Yogyakarta 55283, Indonesia
[2]Coffee Enterpreuner, *Dusun* Mandang, *Desa* Sucen, *Kecamatan* Gemawang, Temanggung 56283, Indonesia
[3]Faculty of Engineering, Bina Nusantara University, Jl. K.H.Syahdan No.9, Kemanggisan, Palmerah, Jakarta 11480, Indonesia
[4]Hochschule Ruhr West, University of Applied Sciences, Duisburger Str. 100, Mulheim 45479, Germany

Abstract. This research aims to study the condition of the land, its relation to the character of the coffee plant in the farmers' coffee plantation in the Sucen Village, Gemawang District, Temanggung, Indonesia. The research was carried out at a community coffee plantation in Sucen Village, Temanggung, Central Java, Indonesia. The research was conducted using descriptive and inferential statistics. Observation of performance with 30 samples was conducted by random sampling technique in three clones. Land suitability analysis was carried out at three observation points. The result showed that the vegetative character of BP 409 clones is better than BP 288 and BP 358. However, the highest production was obtained at BP 288. Land suitability in Sucen Village remains in the inappropriate criteria, which can be improved through land conservation and balanced fertilization.

Keywords: Character of the coffee plant, *Coffea canephora*, increase coffee productivity, land feasibility, sustainable coffee agriculture

1 Introduction

Robusta coffee (*Coffea canephora* Pierre ex A.Froehner) grows optimally on land with an average annual temperature of 21 °C to 25 °C. Rainfall is 2 000 mm yr^{-1} to 3 000 mm yr^{-1} with two to three dry months. Good soil drainage with a texture of loam, sandy loam, silt loam. The effective depth of more than 100 cm with moderate or more soil KPK, soil pH 5.5 to pH 6.0, salinity less than 1. Total N, P and K > low, slope < 8 %, surface rock < 5 %, moderate / low erosion hazard level [1].

A review of the effect of shade on the growth and production of arabica coffee in Ethiopia showed that the shade stimulated changes in the physiology of coffee plants, including increasing photosynthesis and adding leaf area indexes. Thus the coffee

* Corresponding author: astuti_maria2000@yahoo.com

plantations under shade have a larger and heavier coffee production compared to those without shade. Shade also increases the dry matter of coffee fields, which means it plays a role in storing carbon stocks [2]. The response of arabica coffee plants to the shade of fruit trees in Kenya showed that the shade increased the level of photosynthesis of coffee leaves as well as the opening of stomata [3]. Research on the shade of coffee microclimate conditions showed that the shade decreases light intensity by 38 % [4]. Research on responses to the drought that an increase in arabica coffee plantation yields in Brazil is related to higher plant tissue water potential as a result of opening smaller stomata.
Thus, the biochemical reaction of plants can take place well [5]. Research on the content of micronutrients in the soil in relation to the content in robusta coffee plants in Uganda results obtained that the content of micronutrients in the soil is greater than the content of the same elements in the body of the plant, namely the leaves and the lowest in the coffee fruit [6]. Research on nutrient requirements in arabica coffee in Ethiopia that there was a decline in the quality of coffee on the condition of decreasing soil fertility on community plantations due to increased population [7]. Research on the effect of temperature and water on coffee cultivation showed that temperature affects photosynthesis and vegetative growth. It also affects the ripening of fruit [8].

Coffee is one of the leading commodities in Indonesia, with positive impacts in socio-economic [9], health science [10, 11], and others. Coffee performance is not only in beans. But it is also shown by the side product, among others by solid waste, namely husks and pulps [12–14]. The area of coffee in Indonesia reaches 1 227 787 ha which includes smallholder plantations covering 1 179 769 ha (96 %) with production of 599 902 t (94 %), state plantations covering 22 525 ha (2.0%) with production of 19 922 t (3. 2 %), and private plantations covering an area of 25 493 ha (2.0 %) with a production of 17 715 t (2.8 %), bringing the total production to 637 539 t, which is spread throughout the provinces in Indonesia [15].

Temanggung is one of the farmers' coffee plantation centers in Central Java, Indonesia with an area of 9 262.02 owned by around 36 222 farmers. The problem faced by farmers in Temanggung is the low productivity of farmer's coffee in Temanggung, which is around 0.331 t ha^{-1} yr^{-1} [16]. This productivity is still low compared to the average coffee productivity in Indonesia, which is 0.792 t ha^{-1} yr^{-1} [15].

In research related to productivity, initial data is needed on the vegetative and generative characteristics of coffee and land suitability. As far as the literature study has been done, so far research has not been carried out on vegetative and generative characteristics of Temanggung robusta coffee plants and land suitability in Temanggung. This study aims to evaluate the characteristic of a coffee plant on land in Sucen Village, Gemawang District, Temanggung, Indonesia. The urgency of this research is useful as preliminary data for coffee research in Temanggung, in an effort to develop community coffee plantation. Robusta coffee plant characteristic data can be used to recommend action that supports sustainable agriculture, soil health and crops, so that plants can produce in the long term, according to the productive age of the coffee plant.

2 Materials and methods

2.1 Time and place

The study was conducted in November 2018 to December 2018 in Sucen Village, Gemawang District, Temanggung, Central Java, Indonesia. Clones of BP 409, BP 288 and

BP 358 at the people's coffee plantation in Mandang, Sucen Village, Gemawang District, Temanggung Regency, Central Java, Indonesia.

2.2 Research method

Research using descriptive and inferential statistics. Plant samples used 30 samples taken by random sampling technique. Distinguished based on three coffee plant clones on three location points in the sub village of Mandang. The clones which observed are BP 409, BP 288, BP 358. The coordinates of location points are: Location 1 is 7° 10' 48" S 110° 10' 20" E; Location 2 is 7° 10' 43" S 110° 10' 10" E; Location 3 is 7° 10' 43" S 110° 10' 44" E. Inferential statistics using Analysis of Variance which be continued by Duncan Multiple Range Test to find out the significant difference between data.

2.3 Research parameters

Vegetative characters were plant height, number of branches, number of leaves per branch, leaf area index, the content of N, P, K leaves. Generative characters were estimated production and estimated productivity is obtained. Land suitability analysis of Mandang sub-village, Sucen village, Gemawang District, Temanggung: rainfall, C-organic, the texture of the soil, soil pH, land and EC of soil solution, N, P, K, and WV of soil.

3 Results and discussions

3.1 Vegetative characters

The results of research analysis on the vegetative characteristics of robusta coffee in people's coffee plantations namely plant height, stem diameter, canopy diameter, number of branches, number of leaves, chlorophyll level, and leaf area are presented in the following table.

Table 1. Effect of different locations on robusta coffee plant height (cm) in various clones in 2018

Clones	Location points			Average
	Location 1	Location 2	Location 3	
BP 409	160.10	162.40	139.80	154.10a
BP 288	158.20	140.40	142.8	147.13a
BP 358	152.10	145.90	136.90	144.96a
Average	156.80A	149.56AB	139.83B	-

Note: The average number followed by the same letter in the same column or row shows that there is no real difference according to the DMRT at the 5 % real level.; (-): There is no significant interaction.

Table 2. Effect of different locations on robusta coffee stem diameter (cm) in various clones in 2018

Clones	Location points			Average
	Location 1	Location 2	Location 3	
BP 409	9.74a	6.29bc	5.06cd	7.03
BP 288	7.51b	5.35cd	5.25cd	6.03
BP 358	5.92bcd	5.15cd	4.33d	5.13
Average	7.72	5.59	4.88	+

Note: The average number followed by the same letter in the same column and row shows that there is no real difference according to the DMRT at the 5 % real level.; (+): There are significant interactions.

Table 3. Effect of different locations on the diameter (cm) of the robusta coffee canopy in various clones in 2018

Clones	Location points			Average
	Location 1	Location 2	Location 3	
BP 409	291.00a	230.30bcd	198.00cde	258.03
BP 288	245.60abc	190.90de	253.60ab	305.85
BP 358	218.00bcde	178.20e	173.40e	393.62
Rerata	251.53	199.80	208.33	+

Note : The average number followed by the same letter in the same column and row shows that there is no real difference according to the DMRT at the 5 % real level. (-): There is no significant interaction.

Table 4. Effect of different locations on the number of branches of Robusta coffee plants in various clones in 2018.

Clones	Location points			Average
	Location 1	Location 2	Location 3	
BP 409	3.20	4.40	6.50	4.70b
BP 288	4.90	6.30	5.00	5.40b
BP 358	6.90	7.80	6.50	7.06a
Average	5.00A	6.16A	6.00A	-

Note: The average number followed by the same letter in the same column or row shows that there is no real difference according to the DMRT at the 5 % real level. (-): There is no real interaction.

Table 5. Effect of different locations on the number of robusta coffee plant leaves in various clones in 2018

Clones	Location points			Average
	Location 1	Location 2	Location 3	
BP 409	231.63	269.33	273.13	258.03b
BP 288	392.63	322.66	202.26	305.85b
BP 358	391.23	464.86	320.93	406.82a
Average	338.50A	352.28A	254.34B	-

Note: The average number followed by the same letter in the same column or row shows that there is no real difference according to the DMRT at the 5 % real level. (-): There is no significant interaction.

Table 6. Effect of different locations on chlorophyll content (units) of robusta coffee plant leaves in various clones in 2018.

Clones	Location points			Average
	Location 1	Location 2	Location 3	
BP 409	59.10	55.73	59.17	58.00a
BP 288	54.60	55.34	54.15	54.69b
BP 358	51.89	50.09	53.04	51.67b
Average	55.19A	53.72A	55.45A	

Note: The average number followed by the same letter in the same column or row shows that there is no real difference according to the DMRT at the 5 % real level. (-): There is no significant interaction.

Table 7. Effect of different locations on the area of robusta coffee plant leaves (cm^2) in various clones in 2018

Clones	Location points			Average
	Location 1	Location 2	Location 3	
BP 409	276.76cd	339.95ab	364.15a	326.96
BP 288	250.43cd	300.93abc	298.08abc	283.15
BP 358	229.61cde	184.97de	173.42e	196.00
Average	252.25	275.28	278.55	

Note: The average number followed by the same letter in the same column and row shows that there is no real difference according to the DMRT at the 5 % real level. (-): There is no significant interaction.

Table 1 and Table 5 show that difference location affect on plant height and the number of leaves. Locations 1 and locations 2 are better than location 3. Similarly, difference clone significantly affect the number of leaves. BP 358 clones have the highest number of leaves compared to BP 409 and BP 288 clones. Whereas all clones show the same plant height.

Table 4 and Table 6 show that different location do not affect on the number of branches and chlorophyll content. While the difference clone significantly affected the number of branches and the chlorophyll content. BP 358 clone shows the best number of branches compared to BP 409 and BP 288 clones, whereas BP 409 clones have the best chlorophyll content compared to BP 288 and BP 358 clones.

Table 2 shows that BP 409 clone in Location 1 had the best stem diameter compared to the stem diameter of coffee plants in the same clone with different location and other clones in all locations. Table 3 shows that BP 409 clones at Location 1 and BP 288 clones at Location 1 and Location 3 have better canopy diameter compared to the same varieties at different location and BP 358 clones at all locations. Table 7 shows that BP 409 clones and BP 288 clones in Location 2 and Location 3 have better leaf area compared to BP 358 clones in all locations. The results show that different location affect vegetative characters. This is because of the difference although location 1 has a greater slope than location 2, but because location 1 is closer to the settlement, it receives more attention. Location 3 has a slope and sand content similar to location 1, but the type of andisol andisol has silica content from amorphous clay so that the water retention capacity is higher, storage capacity is reduced, so that erosion is faster, and soils are eroded more. As a result, the coffee plant in location 1 has a better vegetative performance than the vegetative performance of coffee plant in location 3. Likewise, location 2 has a latosol soil type, similar to location 1. But has a lower sand content. As a result, this soil is also experiencing greater erosion than location 2, as a result the vegetative performance of location 1 coffee plants is better than the vegetative performance of coffee plants at location 2.

3.2 Generative characters

The results of analysis of variance showed that there was no real interaction between differences in location and clones in their effect on estimated production of one period of harvest. The results of the analysis are presented in Table 8.

Table 8. Effect of different locations on estimated production (kg) per robusta coffee plant in various clones in 2018

Clones	Location points			Average
	Location 1	Location 2	Location 3	
BP 409	1.76	0.65	0.52	0.98b
BP 288	9.59	4.30	2.46	5.45a
BP 358	2.23	2.75	2.58	2.52b
Average	4.53A	2.57A	4.53A	-

Note: The average number followed by the same letter in the same column or row shows that there is no real difference according to the DMRT at the 5% real level. (-): There is no significantl interaction

Table 9. Effect of location differences on estimated robusta coffee productivity (kg ha^{-1} yr^{-1}) in various clones in 2018.

Clones	Location points			Average
	Location 1	Location 2	Location 3	
BP 409	880.85	328.79	261.17	490.30b
BP 288	4 799.40	2 153.06	1 232.08	2 728.20a
BP 358	1 115.71	1 376.41	1 294.73	1 262.32b
Average	2 265.32A	1 286.12A	2 265.32A	-

Note: The average number followed by the same letter in the same column or row shows that there is no real difference according to the DMRT at the 5 % real level. (-): There is no significant interaction

Table 8 and Table 9 show that differences in location do not significantly affect coffee production estimates. While the differences in clones have a significant effect on estimated production, best at BP 288. From vegetative and generative performance it can be seen that although vegetative performance in various locations shows different growth, but has the same productivity. While differences in clones still indicate differences in production. This shows that production is much influenced by clone differences.

Table 10. Results of analysis of N, P and K content of robusta coffee plant leaves in Sucen Village, Gemawang, Temanggung in 2018

Parameter	Clones		
	BP 409	BP 288	BP 358
Total N	1.48	2.18	2.24
Total P	0.11	0.14	0.15
Total K	1.31	1.12	1.57

Table 10 shows that the highest nitrogen content in leaves of BP 358 coffee clone compared to the nitrogen content in BP 409 and BP 288 coffee clones. The highest potassium and phosphorus content in leaves of BP 358 coffee clone compared to the potassium and phosphor content in BP 409 and BP 288 coffee clones.

3.3 Soil characters

Table 11. Results of soil analysis in the people's coffee plantation in Mandang, Sucen Village, Gemawang, Temanggung in 2018

Parameters	Unit	Location points		
		Location 1	Location 2	Location 3
Coordinate		7° 10' 48" S 110° 10' 20" E	7° 10' 43" S 110° 10' 10" E	7° 10' 43" S 10° 10' 44" E
Slope		28	19	25
Height	m (above sea level)	740	700	640
Temperature	°C	29	29	26
WV	$g\ cm^{-3}$	1.20	1.22	0.80
Struktur		Rounded lumps	Rounded lumps	Cornered lumps
Type of soil		Prediction: Latosol Rodik (*Typic Dystrudepts)*	Prediction: Latosol Rodik (*Typic Dystrudepts)*	Prediction: Andisol Okrik
Texture				
Sand	%	52	27	49
Dust	%	31	32	31
Clay	%	17	41	20
pH (H2O)		6.20	5.97	6.11
pH (KCl)		3.69	3.92	3.92
Soil Solution EC	$\mu s\ cm^{-1}$	12	52	37
C-organic	%	1.13	1.79	1.53
Total N	%	0.04	0.09	0.10
Available K	$mg\ kg^{-1}$	19	19	18
Available P	$mg\ kg^{-1}$	18	11	11

Average annual rainfall in 2018 was 3118.7 mm. It was indicated high rainfall in Gemawang district, Temanggung. However, based on the analysis of climate types according to Schmidt & Ferguson was slightly wet. Therefore Sucen village has a sloping topography, accompanied by high rainfall (high humidity), with type C climate (slightly wet). The land in Sucen village has various types of soil, identified two types of soil namely: Latosol (Location 1 and Location 2) and Andisol (Location 3). Latosol soil in Sucen is developed from volcanic material, Clay content of ≥ 40 %, crumbs, loose and homogeneous in color. This is consistent with the results of the analysis at Location 2, which is 41 % Clay fraction and 27 % Sand fraction. However, at Location 1 with the same type of soil has 17 % Clay fraction and 52 % Sand fraction. This is because the Clay faction is lighter than Sand so that the Clay faction is carried away by rain downwards. Based on the height of the land, Location 1 with a height of 740 m above sea level and Location 2 with a height of 700 m above sea level with a slope of ± 20°, so that the loss of surface soil is very high. High erosion changes the dynamics of the soil texture. The results of the C-Organic analysis at Location 1 have the lowest C-Organic at 1.13 %. This is consistent with the results of texture analysis, the lack of bond between soil particles which is dominated by the Sand faction. Andisol Land in Mandang sub-village also develops from volcanic material. The texture found in the soil with 49 % sand, 31 % dust and 20 % clay. It is thought to be caused by high erosion in the area and the characteristic nature of Andisol soil crumbs. The results of soil pH analysis in this study show that there is a difference between the actual and potential soil pH. pH at each location that is classified as an acid. The difference between the actual pH and the potential pH is due to the high rainfall so that the ions that retain the pH of the soil is immediately leached (calcium and phosphate). In other cases, soils found in Locations 2 and location 3 have lower P-availability compared to Locations 1. It is possible that Phosphate ions present in Locations 2 (Latosol soil) are lost due to nutrient leaching or absorbed on the clay surface with Kaolinite minerals, whereas at Location 3 the Andisol soil is absorbed by the surface of the allophane mineral. This condition affect the growth of roots because coffee roots are effective at a depth of 0.3 m with a range 1.5 m distance from the stem [17].

On the availability of Nitrogen, each location has different N-available content. whereas the results of the C-Organic analysis of Location 2 have low C-Organic content. C/N of land Location 1 = 28.25, at Location 2 = 19.88 and at Location 3 = 15.3. Based on information from the local community, coffee farmers fertilize their land with Urea and Ponska each 250 g per tree plus ash. Therefore, the results of the analysis of N, P and K of very low soils supported by the results of an analysis of pH and slope of land that reaches 10°, shows low nutrient content due to leaching by water flow, which causes nutrient-poor soils. Such land conditions cause less than optimal production results. The results of the production analysis show that BP 288 has the highest results compared to the others. The people's coffee production in Temanggung is quite good although soil analysis shows that the soil is nutrient-poor. The average of productivity of coffee in Sucen, Temanggung is 1.493 t ha^{-1} yr^{-1} which higher than the average of coffee productivity in Indonesia which is 0.792 t ha^{-1} yr^{-1} [16]. Increased production can be done by improving soil through land conservation and fertilization.

4 Conclusion

The conclusions that can be drawn from the analysis of the result are:

i. The vegetative character of BP 409 clones is better than that of BP 288 and BP 358. Even so, the highest production is obtained at BP 288.
ii. Land suitability in Mandang Subvillage is still included in the inappropriate criteria, which can be improved through land conservation and balanced fertilization.

Thank you for Institut Pertanian Stiper; Thank you for students of Agriculture Faculty of Instiper students namely Bobby Rifkia Putra, Regan Ophelio, Surya Nugroho, Muhammad Ferly, Dani Syahputra, Deni Wahyu Anggara, Lambang Tri Wibowo, Gusfian, Arifin, Akhyar, for contributions in this research as the technical team.

References

1. S. T. Dermawan, I.M. Mega, T.B. Kusmiyarti, E-Jurnal Agroekoteknologi Tropika. **7**, 2: 230–241(2018). [in Bahasa Indonesia] https://ojs.unud.ac.id/index.php/JAT/article/view/39365/23845

2. A. D. Bote, P. C. Struik, J. Hort. Foresty. **3**,11:336–341 (2011). https://edepot.wur.nl/192807
3. M. W. Mithamo, *Effect of Intercropping Coffee With Fruit Trees on Coffee Eco-Physiological and Soil Factors at Coffee Research Foundation in Ruiru*. [Thesis], Kenyatta University, Kenya, (2013).p.92. https://pdfs.semanticscholar.org/a500/3ad722685e2b6e559325fdef25504d32b9dc.pdf?_ga=2.138594481.36078382.1595587080-1249195893.1568265303
4. P. Siles, J. M. Harmand, P. Vaast, Agroforest Syst. **78**:269–286 (2010). http://orton.catie.ac.cr/repdoc/A11602i/A11602i.pdf
5. F.M. DaMatta, R.L. Cunha, W.C. Antunes, S.C.V.Martins, W.L. Araujo, A.R. Fernie, *et al*., New Phytologist, **178**:348–357 (2008). https://www.locus.ufv.br/bitstream/handle/123456789/12111/DaMatta_et_al-2008-New_Phytologist.pdf?sequence=1&isAllowed=y
6. K. Ngugi, P. Aluka, D. Maina, Annu. Res. Rev. Biol. **9**,2:1–13 (2016). https://www.journalarrb.com/index.php/ARRB/article/view/25890/48455
7. A. Melke, F. Ittana, AJEA. **5**,5:400–418 (2015). https://www.journaljeai.com/index.php/JEAI/article/view/365/299
8. E.A. daSilva, P. Mazafera, Am. J. Plant Sci. Biotechnol. **2**,2:32–41(2008). http://www.globalsciencebooks.info/Online/GSBOnline/images/0812/AmJPSB_2(1&2)/AmJPSB_2(2)32-41o.pdf
9. N. Viartasiwi, A. Trihartono. Coffee Sci. **15**,e151687:1–5(2020). https://doi.org/10.25186/.v15i.1687 http://www.coffeescience.ufla.br/index.php/Coffeescience/article/view/1687
10. I.D.A.R. Dewanti, P. E. Lestari, R. Budirahardjo, D. Setyorini, R.W.E. Yani, S. Wibisono, et al., Coffee Sci. **14**,4: 477–483(2019). https://doi.org/10.25186/cs.v14i4.1619 http://www.coffeescience.ufla.br/index.php/Coffeescience/article/view/1619
11. I.D.A. Susilawati, S. Suryono, P. Purwanto, J. Burlakovs , A. Yaro, Coffee Sci. **15**:e151637(2020). https://doi.org/10.25186/.v15i.1637 http://www.coffeescience.ufla.br/index.php/Coffeescience/article/view/1637
12. R.H. Setyobudi, S. K. Wahono, P.G. Adinurani, A. Wahyudi, W. Widodo, M. Mel, et al. MATEC Web of Conference. **164**,01039:1–13(2018). https://doi.org/10.1051/matecconf/201816401039
13. R.H. Setyobudi, L. Zalizar, S.K. Wahono, W. Widodo, A. Wahyudi, M. Mel, et al. IOP Conference Series: Earth and Environmental Science. **293**,012035:1–25(2019). doi:10.1088/1755-1315/293/1/012035 or https://iopscience.iop.org/article/10.1088/1755-1315/293/1/012035/pdf
14. D. Damat, R. Anggriani, R.H. Setyobudi, P. Soni, Coffee Sci. **14**,4:493–500 (2019). https://doi.org/10.25186/cs.v14i4.1625 http://www.coffeescience.ufla.br/index.php/Coffeescience/article/view/1625
15. Direktorat Jenderal Perkebunan Indonesia, *Statistik Perkebunan Indonesia 2017–2019* [Tree Crop Estate Statistics of Indonesia 2017–2019]; Kopi [coffee], Kementerian Pertanian, Jakarta (2018).p.79.[in Bahasa Indonesia] https://drive.google.com/u/1/uc?id=1S8UHnTgSVwKgqEJB5WXFSp6903sRiiWH&export=download
16. T. Risandewi, Jurnal Litbang Provinsi Jawa Tengah, **11**,1:87–102(2013).[in Bahasa Indonesia] https://www.academia.edu/28775907/ANALISIS_EFISIENSI_PRODUKSI_KOPI_ROBUSTA_DI_KABUPATEN_TEMANGGUNG_Studi_Kasus_di_Kecamatan_Candiroto

17. F.M. DaMatta, C.P. Ronchi, M.Maestri, R.S. Barros, J. Plant Physiol. **19**,4:485–510(2007) https://www.scielo.br/pdf/bjpp/v19n4/a14v19n4.pdf

AUTHOR INDEX

Abdullah Kamaruddin 121, 190, 361
Abimanyu Haznan 326
Adhi Nugroho Yogo 121, 190, 361
Adhywirawan Sutaro Ganjar 164
Adnyana Adnyana 338
Ainul Yakin Engkus 333
Akmar Zakaria Zainul 245
Alam Parvez 130
Alifsmara Joelyna Falvocha 245
Alimin Ilham 27
Alun Mustafa Yoga 62
Anam Choirul 83
Andriani Yosie 326
Andriawan Soni 164
Anne Olga 83, 95, 158
Arfian Anditya Sudjarwo Wisnu 151
Asmawati Asmawati 69
Astri Muliasari Ade 18
Azus Shony Azar Moh 83
Budi Handayani Catur 43
Budiono Rohmad 301
Burlakovs Juris 190
Chan Yefri 190
Chang Zhong–wen 164
Christela Melviana Agustine 53, 278
Damat Damat 43, 83
Dani Septia Erfan 174, 181
Destryana Amilia 278
Devi Anggraini Apriliana 319
Dewi Oryza Sativa Rima 287
Dian Kurniasari Yossy 113
Diananing Rika 278
Didiek Wiet Aryanto Vincent 354
Dwi Purbajanti Endang 301, 333
Dwi Ratnani Rita 296
Dwi Siskawardani Devi 348
Eshpulatov Nodir 10
Faizi Polontalo Nadiyah 245
Fatchur Rochmah Hidayati 18
Faturachman Danny 95
Faturahman Danny 361
Fittkow Marcus 392
Fridolini Fridolini 121
Gamawati Adinurani Praptiningsih 18, 237, 301, 338, 378
Hadiyanto Hadiyanto 77, 151, 245, 296
Handajani Hany 164
Harini Noor 83
Hartati Sri 43
Hartawati Hartawati 113
Haryadi Imam 170
Hasanah Nurul 267
Hendroko Setyabudi Roy 121
Hendroko Setyobudi Roy 53, 69, 77, 95, 190, 361
Heri Purnomo Agus 27
Heryadi Rudy 308
Horas Manahan Sinaga Rizal 138
Husen Syarif 348
Ibragimov Matkarim 10
Ikhwan Ali 113
Illi Mohamad Puad Noor 278
Indratmi Dian 113
Intan Niken Tari Agustina 43, 253
Ishartati Erny 348
Ismoyowati Ismoyowati 202
Isnatin Umi 237
Izzati Munifatul 77
Jani Yahya 361
Kala Mahaswa Rangga 267
Kemala Dewi Ratih 18
Khalikov Abdulhak 1
Khewkhom Netnapis 174, 181
Kondo Kunio 354
Krido Wahono Satriyo 95, 121, 190, 361
Kurnia Setiawan Chandra 253
Kurniati Novitri 258
Kusumawati Asmarani 209
Kusumawati Rinta 27
Kuswardhani Nita 83
Lesmana Rahayu Indra 209
Malik Islam Filardli Abdullah 245
Mangihut Simatupang Togar 138
Marianah Marianah 69
Mel Maizirwan 53, 62, 296
Mohammad Yamin Bohari 378
Motalova Martina 319
Muchtasjar Bunyamin 77
Muhammad Muhammad 237
Muhammad Nur Syukri 308
Muhidin Muhidin 348
Mukti Wibowo Yari 151
Muryanto Muryanto 326
Mustika Wardhani Ratna 378
Nengah Suprapta Dewa 338
Novianto Bangun 121
Nugraha Budi Santosa Tri 392
Nur Aziz Fuad 301
Nur Wulan Mulatmi Septi 319
Nurdiansyah Yanuar 221
Nurdin Yusuf Muhamad 38, 231
Nurtantyo Andono Pulung 354
Onthong Jumpen 348
Oryza Sativa Mutyarsih 258
Pancapalaga Wehandaka 373
Parlindungan Siallagan Manahan 138
Paul Nick 27
Pratama Putra Dian 392
Pratitis Asri 27
Prima Indra Budianto Gregorius 151

AUTHOR INDEX

Proborini Meitini 338
Pujiasmanto Bambang 105
Purba Washington 95, 121, 361
Purmini Purmini 258
Purnomo Djoko 105
Rachmawati Diana 83, 158
Rachmi Esyanti Rizkita 53
Radjabov Abdurahman 10
Rahayu Enny 392
Rahman Hakim Riza 164
Rahmawati Khusnul 226
Rakoto Malala Andoniana 18
Rekyani Puspitawati Indah 378
Retnoaji Bambang 130
Rimmer Mike 27
Rizky Anjar 348
Rohsari Indah Utami Amaliyah 62
Ronald Mavuso Boy 164
Rosana Dewi Tria 213
Rosyidi Hamdi 158
Rukminastiti Masrifah Atika 170
Saka Rani Devitra 221
Santosa Ribut 278
Sedlacek David 373
Setyaningrum Desy 105
Setyaningrum Haris 170
Shaywont Pratami Febrianti 319
Silvana Arianti Yoesti 386
Siregar Kiman 221, 361
Solifudin Agus 392
Soni Peeyush 237, 338
Stankevièa Karina 43
Suaidah Suaidah 326
Sudradjat Sudradjat 231
Sudrajat Sudrajat 38
Sugiwati Sri 326
Sugiyanto Didik 190
Sukaryani Sri 333
Sukiyono Ketut 258
Sukmawati Sukmawati 338
Sukorini Henik 174, 181
Supriyanto Supriyanto 221
Supriyono Supriyono 105
Susanto Herry 95, 190
Susanto Hery 361
Susilo Adib 170
Suswoyo Imam 202
Sutawi Sutawi 158, 287
Suwandi Suwandi 62
Suyatno Suyatno 373
Syaifudin Akhmad 348
Tarik Ibrahim Jabal 158, 287
Theresia Maria Astuti Yohana 392
Theresia Sri Budiastuti Maria 105
Tri Pujiastuti Endang 95
Triwahyuni Eka 326
Tupa Ronauli Silalahi Fitriani 138
Turkadze Tsitsino 301
Umi Abida Farahsani 130
Vincevièa–gaile Zane 77, 202, 333
Wahyu Harinta Yos 333, 386
Wahyudi Ahmad 83
Wardani Irma 213
Wely Asmoro Novian 253
Wibowo Haryo 121
Wibowo Singgih 27
Widayat Widayat 151
Widhianto Agung 267
Widiyanto Widiyanto 296
Widodo Wahyu 190, 202
Widyastuti Retno 253
Wijayanti Yureana 392
Wijayanto Feri 221
Wirasdenty Aigahayunindy Feby 174
Wismantoro Yohan 354
Witono Yuli 83
Wuryantoro Wuryantoro 378
Yandri Erkata 95, 121, 190, 361
Yani Yahya 95
Yaro Abubakar 69, 190, 226, 361
Yuniawan Isyanto Agus 38, 231
Zalizar Lili 226